Global Culture and Sport Series

Series Editors: Stephen Wagg, International Centre for Sport and Culture, De Montfort University, UK, and David Andrews, University of Maryland, USA.

The Global Culture and Sport series aims to contribute to and advance the debate about sport and globalization by engaging with various aspects of sport culture as a vehicle for critically excavating the tensions between the global and the local, transformation and tradition and sameness and difference. With studies ranging from snowboarding bodies, the globalization of rugby and the Olympics, to sport and migration, issues of racism and gender, and sport in the Arab world, this series showcases the range of exciting, pioneering research being developed in the field of sport sociology.

Toby Miller • Joan Pedro Carañana

Global Sports Go Green—Or Do They?

palgrave
macmillan

Toby Miller
Profesor Distinguido
Instituto Tecnológico de Monterrey
Guadalajara, Mexico

Joan Pedro Carañana
Permanent Professor, Department of
Journalism and New Media
Universidad Complutense de Madrid
Madrid, Spain

ISSN 2662-3404 ISSN 2662-3412 (electronic)
Global Culture and Sport Series
ISBN 978-3-032-14206-1 ISBN 978-3-032-14207-8 (eBook)
https://doi.org/10.1007/978-3-032-14207-8

Cover Art: Ana Rodríguez

This Palgrave Macmillan imprint is published by the registered company Springer Nature Switzerland AG.
The registered company address is: Gewerbestrasse 11, 6330 Cham, Switzerland

Acknowledgments

Toby thanks Jenine Abboushi, Jessamyn Abel, Akuavi Adonon, Pal Ahluwalia, Eva Aladro Vico, the late Manuel Alvarado, the late Peter Anderson, Dave Andrews, Paulina Aroch, Jesús Arroyave, Sarah Berry, Prudence Black, Jules Boykoff, Edward Buscombe, Mike Butterworth, Belinda Carpenter, Victoria Carpenter, Ben Carrington, Jo Clifford, Cole, Richard Collins, Joselyne Contreras, Stuart Cunningham, the late Johnny Darling, André Dorcé, Claire Duffy, Natalie Fenton, Des Freedman, Chris Gaffney, Néstor García Canclini, Gabriela Gómez, Michael Gordon-Smith, Bill Grantham, Maggie Gray, Ric Gruneau, Richard Higgott, Brett Hutchins, Nathalie Hyde-Clarke, Steve Jackson, Helen Jefferson Lenskyj, Alisa Jones, Ketan Joshi, Noel King, Geoff Lawrence, Salvador Leetoy, Marie Leger, Libby Lester, Justin Lewis, Emilienne Limón, Jo Littler, the late Ana López, Javier López-Ferrández, Joe Maguire, the late Alan Mansfield, the late Randy Martin, Daniel Mato, Rick Maxwell, Tony May, Vicki Mayer, Alec McHoul, Jim McKay, Gynna Farith Millan Franco, Caitlin Miller, Lainey Paloma Miller, Tani/Tilly Miller, Txingurri/Naranja/Orange Miller, Daniela Inés Monje, Stephen Muecke, Rob Nixon, Jed Novick, the late Tom O'Regan, Rune Ottosen, Esra Özcan, Josep Ramon Pedro Carañana, Mauro Porto, Isabel Cristina Ramírez, Paula Requeijo Rey, Garry Rodan, Rob Roggema, Ana Rosas Mantecon, Andrew Ross, David Rowe, Jorge Saavedra Utman, Alfredo Sabbagh Fajardo, Víctor Sampedro Blanco, Tom Scanlon, Paola Seminara, Federico Guillermo Serrano-López, Rita Shanahan, Paul Smith, Jennifer Smith Maguire, Olga Lucía Sorzano, Rob Steen, Damion Sturm, Gordon Tait,

Anamaria Tamayo Duque, Tor Törnqvist, Enrique Uribe Jongbloed, Aimée Vega Montiel, Larry Wenner, Garry Whannell, Gary Wickham, Sven Daniel Wolfe, Ferruh Yilmaz, George Yúdice, anonymous reviewers, and everyone at the press involved with this book.

Joan thanks Begoña Gutiérrez-Martínez María Marqués, José R. Pedro, Neus Carañana, Els d'allà raere, Juan González, Carlos Salido, Dimonis de Campanar, Ana Sànchez Andrés, ULEPICC-ESPAÑA, Los Olivos, Javi Cone, José María Tortosa, Paola Seminara, Xaxo, Sandra Nieto, Gerard Salom, Luis Nitrihual, Pilar Martínez, and José Durán Rodríguez.

Contents

List of Figures

List of Tables

CHAPTER 1

Introduction: The Slow Violence of Carbon Sports

Sports' seemingly untrammeled competition and unpredictable drama can divide, unite, distract, and excite. They lay powerful claims to incarnate nature through the apparent purity of scoring, sprinting, or swimming on a metaphorically 'level playing field' that can even 'overturn (for a moment) historical hierarchies' through an event's 'unpredictability, its uncontrollability, its sublime indifference to all extraneous narratives' (Marqusee 2012). For instance, in each match of association football (football), 'we know beforehand the place, the numbers and personalities of the players, the rules of the game, and the fact that everything will take place within the area of the great pitch—except that in a game of football we do not know until the very end who will win' (Eco 1981: 160–61). And other things are going on. Sports matter 'in business, in sex, in politics, in the general style of performance. All these domains are affected by sport's criteria of "excellence", effort and record-breaking' (Baudrillard 1993: 8).

In Terry Eagleton's words,

> Sport … stands in for all those noble causes—religious faith, national sovereignty, personal honour, ethnic identity—for which, over the centuries, people have been prepared to go to their deaths. Sport involves tribal loyalties and rivalries, symbolic rituals, fabulous legends, iconic heroes, epic battles, aesthetic beauty, physical fulfilment, intellectual satisfaction, sublime spectaculars, and a profound sense of belonging. (Eagleton 2007: 26)

T. Miller, J. P. Carañana, *Global Sports Go Green—Or Do They?*, Global Culture and Sport Series,
https://doi.org/10.1007/978-3-032-14207-8_1

A mixture of order and surprise, of control and guaranteed drama, is especially true of the romantic propaganda that cradles mega-events—huge, lengthy, environmentally destructive contests such as the men's World Cup of football and the summer Olympics. They occur in a wide variety of places and attract significant investment from febrile states, corporations, and the *bourgeois* media. Sports are *the* global culture industry.

Psychologically, such occasions are heralded as expressions of a purportedly universal brother- and sister-hood of humanity ("Sports Mega-Events" 2006). Nationally, they supposedly stimulate pride in the face of world competition, opportunities for hosts to 'showcase our spirit, our character, our history, our culture' ("Meares" 2024). Commercially, they are said to boost economic development through infrastructural projects and tourism (McCartney et al. 2010; Preuß 2015). Environmentally, they are promoted as responsible, even exemplary actors (FIFA 2019; International Olympic Committee 2018; World Sailing, n.d.; Formula 1 2022; Fédération Internationale de l'Automobile 2023). All those assertions combine to sell sporting events to politicians, functionaries, taxpayers, and journalists around the globe.

Sports have obvious social correlatives of class, gender, ideology, obedience, opportunity, race, spectatorship, and so on. But they also depend, variously, on animal enslavement, betting, communications, contracts, equipment, kinesiology, media coverage, medicine, merchandizing, sponsorship, stadia, tax "arrangements," technique, technology, training, travel, and "vitamin supplements." Far from being straightforward instantiations of innate ability, sports exemplify the will to channel and discipline nature; to make it competitive and financial. Yet by contrast with extraction and manufacturing, they have generally been regarded as ecologically sustainable, with residues of blood, digits, grief, injury, joy, memory, mud, sweat, tape, and tears rather than pollution. Sports are assiduously promoted as environmental pastimes (Warren 2020)—baseball parks in the US are known as "Cathedrals of Green" (Grant, Grant 2014). But these are carbon events: each year, pro sports are thought to emit 350 million tons of CO_2e, about 1% of the world's total (Carbon Literacy Project n.d.) and it's estimated that since 2000, 'carbon emissions were between 2.0 and 4.5 million tonnes of CO_2e for the Olympics and 1.65 to 3.63 million tonnes of CO_2e for the Men's Football World Cup' (Gogishvili and Müller 2025).

Consider the environmental impact of building stadia. After the US and China, concrete is the world's largest emitter of carbon. Construction is the source of a third of energy use and CO_2 emissions and responsible for massive debris (United Nations Environment Programme and Global Alliance for Buildings and Construction 2025: 13). And big venues, per mega-sports, often house coffee shops, bars, restaurants, playgrounds, and museums in addition to sporting fields. All require ongoing maintenance, repair, sanitation, power, water, and publicity (Lehne and Preston 2018; Watts 2019). Little priority—if any—is given to environmental considerations (Francis et al. 2023) in these monuments to craven capitalistic excess and state servitude.

Spectators to mega-events largely travel in unsustainable ways and generate astounding levels of waste: each person at the Olympics and the World Cup averages between 0.25 and 7 kilograms a day via packaging, plastics, paper, and fast food (Zafari et al. 2025). Meanwhile, over two hundred sponsorship contracts ensure carbon sports receive US $5.6 billion a year from extractivist polluters seeking to burnish their reputations by association with allegedly green activities (New Weather Institute 2024).

Carbon mega-events are therefore both directly and indirectly environmentally destructive, via the occasions themselves and their promotion of unsustainable industries. In Wray Vamplew's words, '[n]othing sport does has positive consequences for the climate or the environment' (Vamplew 2021: 400).

This book interrogates the four most-prominent such competitions—the World Cup of men's football, the summer Olympics, America's Cup sailing, and Formula One motorsport—not as universal, nationalistic, or bountiful, but as carbon commodity signs, with environmental destruction their clandestine signatures. We don't address cricket's Indian Premier League or Gringolandia's National "Football" League (NFL). Those instances of capitalism in one country make 96% and 98% of their money, respectively, in host nations ("A Global" 2024).

This is a book of scholarly tendency, drawing on both scientific inquiry and progressive commitments. Our critique of carbon competitions emphasizes their ecological impact on air, animals, athletes, land, residents, and water, and the relationship of sporting rhetoric to reality and pro-business policies to popular resistance. We examine the history, economy, and meaning of sports to provide context, with extensive case studies and discussion of the environmental destructiveness and greenwashing practices of carbon mega-events and their sponsors.

This Introduction looks at sports in general—their past and present, their current political economy, and their role in greenwashing—followed by a proposal to understand these phenomena.

Sporting Society

Sports are the globe's ninth-biggest industry: valued at $2.65 trillion, they are growing faster than the overall economy and above the Standard & Poor's 500 norm (Anani 2024). In 2021, private-equity corporations invested $51 billion in sports ("Saudi Arabia" 2023). The three biggest US pro leagues each had 2022 revenues in excess of $10 billion ("Saudi Arabia" 2023). In 2024, *Forbes* magazine's list of the "The World's Most Valuable Sports" saw a collective $289 billion bestowed on franchises, almost four times the figure of a decade earlier and PwC's global survey of executive leeches saw predictions of 7.3% growth over the next three to five years (2024: 6). Long-term broadcasting deals guarantee cash flow (over $56 billion in 2023) even when sponsorship contracts suffer because of the greed, corruption, and incompetence of highly geared entities (The Business 2024; SportBusiness 2023; "Is it" 2009; Földesi 2014; Transparency International 2016).

The last decade has witnessed dozens of new professional competitions in the US; the United Arab Emirates announcing plans for sports to account for 0.5% of their economies by 2031; and the Saudis pouring hundreds of millions into golf, football, and boxing. Three-quarters of the world's consumers enjoy sports, with live television the most popular format (YouGov 2023). Eight-five percent of Internet users routinely watch online or via TV (Global Web Index 2019). Internationally, the genre is second to politics among factual programs and stations preferred by men, and equal-first for men under thirty-five. All large international languages have dedicated sports media (Horky and Nieland 2013) and the topic occupied four of Wikipedia's twenty most-read English pages in 2024. Fifty-nine percent of US newspaper readers focus on sports, which dominate TV and streaming ratings in the Anglo US (Schenk n.d.; Gough 2023). The *Guardian*'s most-read stories online from 2010 to 2014 saw football rank behind only world news, well ahead of British politics (Sedghi et al. 2014). UK teenagers followed sports above all other topics in 2024.

Those figures generate a basic query: what is all the fuss about?

Sports began in two ways: on the one hand, as state and popular pleasure derived from more ordered, if similarly violent, activities (in the Colosseum); on the other, as unregulated, unruly peasant and proletarian joy, relaxation, and violence (in the commons). They performed contradictory roles, 'reinforcing, for example, social hierarchies, performing shared values, and playing out deep cultural tensions' (Scanlon 2021: 4).

In the ancient Mediterranean, the body was the *locus* of a combat with pleasure and pain that supposedly disclosed the inner truths of ruling-class men and schooled them in the mastery needed to control their own drives—and others'. Hedonism, austerity, and a readiness to govern could be achieved and combined through training:

> The metaphor of the match, of athletic competition and battle, did not serve merely to designate the nature of the relationship one had with desires and pleasures, with their force that was always liable to turn seditious or rebellious; it also related to the preparation that enabled one to withstand such a confrontation. (Foucault 1986: 72)

Managing male lust and decadence was homologous to achieving sporting and political success, signifying provisional suitability to rule. In each case, pleasure/hegemony could lead to failure unless accompanied by regular physical training and examination of one's conscience (Foucault 1986: 72–77, 104, 120, 197–98, 212). Consider how Euryalus taunted Odysseus in Homer's investigations of class tensions (Scanlon 2018):

> No, stranger, I would not say you are like a man skilled in any contests of the many sorts that exist among men, but are like someone who is used to a ship with many oarlocks, a leader of sailors who are also merchantmen with his mind on a load, an overseer of cargoes and of gain got by greed. You do not resemble an athlete. (Homer. 1967: 8.159–64)

Socrates declared that 'care of the body is a single craft' in 'two parts: gymnastics and medicine,' with physical self-discipline akin to shepherding legislation in politics (Plato 1997: 1507). Aristotle favored regular, ongoing flirtations with sex and sports—tests as well as pleasures (1998: 232). Isocrates noted that collective exercise was thought to help the young 'abstain from many vices' (1929: 133). In his "Erotic Essay," Demosthenes regarded it as a key means of distinguishing oneself from 'slaves and aliens' (2015: n.p.).

Roman ethics regarded 'gymnastics and athletic and military training' as 'an integral part of the education of a free man' (Foucault 1988: 56–57). Marcus Aurelius admired athletic control of the body (Aurelius 1862: CLXIV) and early Christianity offered guides to physical education, a means of governing fleshly desires by distinguishing control from excess and good versus evil to ensure one's ultimate ascension to heaven. Athletic discipline became a metaphor, a description, and a preparation for monastic or pre-marital life. Virginity was deemed a troublesome gift of grace, akin to the ascetic denial required of gymnasts (Foucault 2021: 27, 307, 341–43).

In Isocrates' view, Olympic triumph was

> admired by the whole world … the Greeks made display of their wealth, strength of body, and training … not only the athletes were the objects of envy … the cities of the victors became renowned. … Expenditures in the Olympian Festival … enhance the city's reputation throughout all Greece. (Isocrates. 1945: 195)

Sports were a means of avoiding reproductive necessities and excluding women and slaves from civic leadership (Foxhall 1998; Engels 2010: 351). They became central to an ethics of the self designed to inculcate and evaluate free men's physical and political fitness to rule others. Some of those functions remain, in addition to organizing, disciplining, and entertaining the popular classes, representing the nation—and making money.

The phenomenology of sports is very much about dominion over space and time (Barthes 1997: 76), albeit operating under the promise and premise of 'egalitarian and apolitical agency' (Stoddart 1988: 651). A long, dialectical history connects sporting violence to order and pleasure to passivity, per Norbert Elias investigating the expansion of rule-governed contests between individuals and teams from the sixteenth century via European travel, conquest, colonization, governmentality, and exemplification. Sentiment and behavior came to be codified, supplanting excess and self-laceration with temperate auto-critique. High tension and low risk blended popular appeal with public safety—a utilitarian calculus of time and joy that helped train and control gentry, workers, and colonists, articulating them to organized sports in a way akin to diverting pets' predatory instincts onto toys (Elias and Dunning 1986; Krauze 1994; Nussbaum 2006: 371).

Elias discerned a largely agreeable sublimation and civilization of dangerous drives in such developments—a sanctioned release from the travails of life, with sentiment and behavior governed by sports' magical attractions of tension and catharsis:

> Battle lust and aggressiveness … find socially permitted expression … in competitive sports. And they are manifest above all in 'spectating' … the daydream-like identification with some few people who, in a moderate and precisely regulated way, are allowed to act out such affects. … active, pleasurable aggression is transformed into a more passive and restrained pleasure. (Elias 1978: 240)

Montaigne argued that '[g]ood governments take care to assemble the citizens and bring them together for sports and amusements as well as for the serious functions of piety; sociability and friendliness are thereby increased' (Montaigne 1966: 131). In 1618, Britain's James I promulgated a "Declaration" listing "Lawful Sports," later revised by Charles I. Designed to be read in churches, it insisted that 'the meaner sort who labour hard all the week' needed rest and recreation free from Puritanical prohibitions on joy, recommending 'archery for men, leaping, vaulting, or any other such harmless recreation' (James I and Charles I 1914). For Herbert Spencer, rough games for youngsters were agreeably displaced by 'masculine dignity' and 'feminine modesty,' channeling adults into rule-governed activities (Spencer 2005: 468).

Encounters in English villages had provided opportunities for young men to settle scores in ways that would be considered quite brutal today; participants were often disabled and sometimes killed (Elias and Dunning 1986). This was the era of a sport called shin-kicking (Underdown 1985: 64); perhaps not so 'harmless,' as Toby discovered courtesy of a reactionary politico's son in a rugby scrum (Leighton 2005 explains shin-kicking's comeback). A 'genuinely anarchic affair, a foretaste of a classless society' threatened market and state power (Eagleton 2010) until destruction of the commons, the consolidation of private property, and industrialization divided sports from the environment. With that enclosure of what had been the commons, the passage from a public to a private good made sports subject to capitalism (Galeano 2016). For example, football 'used to be played on village greens and anyone who wanted to watch could; football was a public good. Then walls were built around the playing field and the spectacle was privatized' (Varoufakis 2002: 230).

Per Stuart Hall, sports emerged 'in conjunction with, and as a consequence of, the destruction of an earlier culture, partly by imposition, partly unconsciously, within the larger process of the adaptation to industrial work and urban life' (Hall 2016: 43), developing into 'one of the essential elements of industrial life,' with 'the human being … a kind of machine … an efficient piece of apparatus which is henceforth unacquainted with anything but the harsh joy of exploiting his body and winning' (Ellul 1964: 383). This is the worker's world of Taylorism and Fordism parodied in *Modern Times* (Charlie Chaplin 1936), a life of factory-like repetition, boredom, and control.

In Max Weber's words:

> where violent conflict changes to "competition," whether for Olympic wreaths … or for social honor or gain … regulations serve as "rules of the game" determining the forms of conflict. … gradually increasing "pacification," in the sense of the reduction in the use of physical force, only reduces but does not ever wholly eliminate the appeal to the use of force … [which] has been increasingly monopolized by the coercive apparatus of a certain kind of association. (Weber 1981: 173)

Modernity, masculinity, sociality, imperialism, and pastoral nostalgia came to characterize games as places where capitalist and colonial divisions of labor and new technology met and science, myth, and religion pronounced upon the body and nature—part of a long heritage emphasizing the paradox that productive workers must be entertained as well as disciplined. Governed and ungoverned, pacific and riotous forms blended, mixing volunteer with waged labor and amateur with professional management in a search for controlled, policed aggression (Weber 1992: 112; Eichberg 1986). Jon Stratton has explained the dual, paradoxical views of nature that governed this change. One aspect amounted to:

> a celebration of the environment, of the natural world as being natural and indeed for being natural. In more general, cultural terms the English nineteenth-century intellectual middle-class had a certain revulsion against the growing industrialised urbanism and found in nature a nostalgia for a lost pastoral past … sport, in the form of institutionalised and delimited physical activity, provided a training for the 'natural' body and a discipline which was a part of 'civilising' the mind. The body became the site of a complex resolution. On the one hand there was the attempt to eradicate nature in the form of the 'natural' child through the inculcation and

> internalisation of bourgeois values. On the other hand, however, nature as a general concept was valorised over the 'unnatural' depredations of city life. (Stratton 1986: 88–89)

Teachers in élite British schools saw codified sports as a means of indoctrinating future leaders of the nation and Empire in discipline and morality. This standardization emphasized homosocial heterosexuality, gentlemanliness, modesty, loyalty, fair play, bravery, self-control, a balance between mind and physique, sports for their own sake, patriotism, imperialism, and Anglocentrism. Matthew Arnold noted the emergence into British life of 'games and sports which occupy the passing generation of boys and young men' to help generate 'a better and sounder physical type for the future to work with.' This 'muscular Christianity' was as important to the nation's future as coal and wealth (Arnold 2003). A similar logic propelled Emerson to declare sports 'lessons in the art of power' (Emerson 2000: 1698).

Capitalism took hold of sports not only by enclosing grounds and charging for attendance but transforming play into spectacle, enabling media coverage of action, advertising, and promotional messages and generating national, international, and transnational characters. These twin cultural-commercial domains simultaneously, symbiotically, supplied each other with money, audiences, promotion, and content (Bourdieu 1999: 16). Sports were increasingly shaped by the media, spectacularized by commerce, employed to deliver audiences to sponsors, and linked to technological change.

Joseph Maguire typifies the resultant sporting body as an object of discipline, a mirror, a site of domination, and a form of communication: remodeled through diet and training; encouraging mimetic conduct via the purchase of commodities; exercising power through physical force on the field and—potentially—off it; and incarnating an expressive totality, balletic and beautiful, awkward and ugly (Maguire 1993). For Umberto Eco, successful athletes have 'hypertrophized one organ,' turning the 'body into the seat and exclusive source of a continuous play ... a monster ... dedicated to total instrumentalization' (Eco 1987: 161). An endless production line of sporting bodies—at their "best"—is generated for states, corporations, colleges, and associations to exploit. Athletes 'make their bodies through labour, sport and play, but they do not make them in circumstances of their own choosing' (Shilling 1991: 665).

Small percentages of young people from proletarian and peasant backgrounds may become fleetingly wealthy and famous, flickering as incendiary but ephemeral signs of class, gender, and racial mobility. Their 'peculiar amalgam of ritual, repetition, ingrained twitches, drilled skill, known strengths and feared limits,' striving for an unreachable eternity, is all too human. For in 'the cruel judgement' so typical of sport, athletes end up 'old. And done … ebbed beyond usefulness' (Mitchell 2024). Shriveled bodies break on wheels of lies, distortions, and fantasies constructed and manipulated by craven colleges, corporate owners, hypocritical rather than Hippocratic medical staff, mindless media, and tragic delusions of invulnerability and bourgeois individualism. Intense bodily achievement and vulnerability leave athletes dirty, disconsolate, furious, injured, joyous, sweaty, tearful, undressed, unguarded, and unconscious. Inevitably beaten, bloodied, and bowed, they are cashiered in favor of tall, testosteronal, triumphant replacements.

Tolstoy's *Confession* detailed, rather guiltily, the self-confidence generated by his young body, followed by its somatic demoralization:

> My muscles were growing and strengthening, my memory was richer. My capacity to think and comprehend was increasing. I was growing and developing, and, feeling this power within myself, it was natural for me to believe that there was a law governing the world, in which I could find the answers to the questions of my life. But the time came when I stopped growing; I felt that I was no longer developing but was drying up, my muscles were growing weaker, my teeth falling out. (1987: 90)

AE Housman's lamentations for the fragility of life read similarly. Consider "To an Athlete Dying Young," his meditation on youth's wartime finality:

> The time you won your town the race
> We chaired you through the market-place;
> Man and boy stood cheering by,
> And home we brought you shoulder-high.
> Today, the road all runners come,
> Shoulder-high we bring you home,
> And set you at your threshold down,
> Townsman of a stiller town.
> (Housman 1919: XIX)

Dylan Thomas might have been imploring athletes to keep going when he wrote:

> Do not go gentle into that good night,
> Old age should burn and rave at close of day;
> Rage, rage against the dying of the light. (Thomas 2014)

It can be no surprise that anxieties over masculinity and the domination of territory have underpinned governmental investment in sports, with numerous institutions dedicated to discharging wayward male impulses or channeling them into military preparedness through state-sanctioned exercise: physical culture, 'strenuous living,' social Darwinism, rational recreation, and neoclassical French romanticism among them (Smith Maguire 2006; Mangan 1981).

Arnold penned the famous expression, 'the battle of Waterloo was won in the playing fields of Eton'; but a profound concern with military preparedness saw him warn that 'disasters have been prepared in those playing-fields as well as victories' (Arnold 1910: 255). Pierre de Coubertin revived the ancient Olympics as an international festival of male athletes and diplomats that could cultivate 'la musculature morale de l'homme' [man's moral musculature] (de Coubertin 1913: 12), redeeming French masculinity after the shocks of the Franco-Prussian conflict a quarter of a century earlier. The South African War saw British imperialists clamoring for physical education of the male working class to improve fitness for military service (Birley 1995: 1–2).

毛泽东 [Mao] worried that the Chinese 'nation is wanting in strength. The military spirit has not been encouraged; [t]he physical condition of the population deteriorates daily. … If our bodies are not strong we will be afraid as soon as we see enemy soldiers' (1917). 蔣介石[Chiang Kai-Shek] called on 'sports to save the nation' (quoted in Xu 2008: 12). Rosa Luxemburg drew on them to describe class vulnerability in the face of state and military domination (Luxemburg 2004: 218) and Hitler and Pétain introduced physical-fitness tests for the young in Nazi Germany and Vichy France (Houlihan 1997: 61–64).

US studies of the mid-1950s found flaws in young people's 'muscular strength and flexibility' due to a reliance on cars, elevators, and school buses by contrast with Europe's more pedestrian life. A national schools exercise program was proposed (Kraus and Hirschland 1953) and Dwight

D Eisenhower created the President's Council on Youth Fitness in 1956. Four years later, President-elect John F Kennedy alerted readers of *Sports Illustrated* to a 'growing softness, our increasing lack of physical fitness.' These tendencies supposedly constituted 'a threat to our security' that must be addressed, per Ancient Greece's Olympic quest to forge and sustain 'a vigorous state.' After all, 'struggles against aggressors throughout our history have been won on the playgrounds and corner lots and fields of America' (1960–61). Decades later, Donald J Trump's *National Youth Sports Strategy* worried that 'a majority of youth are not moving enough.' It detailed 'surveillance systems' for monitoring children playing sports (Office of the Assistant Secretary for Health 2019: 2, 37–44). His 2025 "Presidential Fitness Test" for school pupils aimed to improve 'our economy, military readiness, academic performance, and national morale' and 'emphasize the importance' of 'military readiness' (Trump 2025). Such initiatives coincided with increased 'child mortality; chronic physical, developmental, and mental health conditions; obesity; sleep health; early puberty; limitations in activity; and physical and emotional symptoms' (Forrest et al. 2025).

Other benefits are routinely imagined as part of these fantasies, per Durkheim's belief that sports could 'balance and relieve … serious life' (Durkheim 1961: 361). Functionalist theorists herald sports' allegedly integrative capacities, their supposed capacity to harmonize populations, incarnating and strengthening the social order (Simmel 1949). A 'merging of the one in the whole that may be learned from sport' (Cooley 1910: 34) via 'microcosmic images of the social macrocosm' (Merton 1968: 190). Society is said to be a self-regulating system that can restore equilibrium when faced with internal or external shocks, thanks to shared values, epitomized by sports, which help things stay together rather than fall apart (Orsini 2024: 281–360; Loy and Booth 2000; Delaney 2015). Team games form a healthy civil society: a fit, productive, domestically pacific population, capable of expressing powerful feelings in law-abiding ways while standing ready to serve the nation at war (Parsons et al. 1955: 129). For TH Marshall, a 'common right' to sports (Marshall 1950: 82) provided physiological, psychological, and sociological outlets for tensions, especially for young men, while Gandhi's search for the "Conquest of Sex" recommended 'brisk walking exercise in the open air early in the morning and at night before going to bed' (Gandhi 1943: 83–84).

The first *Middletown* study declared that high-school loyalties developed and class differences slackened in 1920s Muncie, Indiana, with the arrival of basketball (Lynd and Lynd 1956: 212–13, 485). A century later, Michael Sandel divined that when 'capitalism and community collide, … professional baseball, football, basketball, and hockey provide a source of social glue and civic pride' (Sandel 2005: 81). Neoclassical economists argue for universal broadcasting of major events on the grounds that folks talked to one another about the shared experience of viewing. This so-called social capital allegedly diminished when only a privileged few could watch (Boardman and Hargreaves-Heap 1999: 168, 178).

Such fantastical desires have real political currency. In the rather despairing words of corrupt US Vice President Spiro Agnew, '[S]port, is one of the few bits of glue that holds society together, one of the few activities where young people can proceed along traditional avenues' (quoted in Lipsyte 1971). Today, the US government advises that 'organized sport participation can reduce youth violence and crime' and develop 'internal personal qualities, characteristics, and skills such as goal setting, emotional control, self-esteem, and work ethic' (Office of the Assistant Secretary for Health 2019: 49–50).

UK politician Jeremy Hunt puts sports in the cultural vanguard: '[W]e want to nurture the values that are common in sport, but perhaps not as common outside sport as they should be' (Hunt 2010). A House of Lords Select Committee is a true believer: '[P]articipation in sport and recreation can foster greater social and community cohesion and benefits both the national and local economies' (2021: 4).

The European Commission avows that sports are a

> growing social and economic phenomenon which makes an important contribution to the European Union's [EU] strategic objectives of solidarity and prosperity and generates important values such as team spirit, solidarity, tolerance and fair play, contributing to personal development and fulfilment. It promotes the active contribution of EU citizens to society and thereby helps to foster active citizenship. (2007)

The Association of Southeast Asian Nations (ASEAN) says: 'More than as a means to a healthy body and mind, sports promotes the values of respect, inclusion, fairness and unity' (n.d.). The Organization of American States sought 'to promote greater social cohesion through sport' (2023) and the African Union argues that sports enable:

> African integration based on the ideals of Pan-Africanism and African Renaissance good governance, democracy, shared values, common heritage, respect for human rights, justice, rule of law gender equality, inclusion of persons with disabilities and safeguarding of women, girls and children from non-accidental harm. (African Union Sports Council n.d.)

The United Nations claims sport 'brings individuals, communities and countries together and, in doing so, often bridges cultural, ethnic and national divides' (2020: 2); its Office on Drugs and Crime maintaining sport 'contributes to peace, enables sustainable development, creates jobs and plays a vital role in promoting healthy lifestyles. It inspires, teaches and brings people, young and old, together'. The UN supports the International Olympic Committee's (IOC) "Olympic Truce" to 'harness the power of sport to promote peace, dialogue and reconciliation' (2024).

These assumptions have always been problematic. They fail to engage 'power, class and their manifestations' via such questions as:

> why women do not participate to the same extent as men, why governments underwrite sport, why corporations use sport for advertising, why athletes take drugs, why nationalism and sexism appear to be fostered by sports, and so forth. Nor can we glimpse a way, or even a reason, to alter the existing system. (Lawrence and Rowe 1986: 21)

Sports' very existence has long been criticized as much as it has been embraced. Aristotle feared over-zealous training could exhaust young people (1998: 232), while Isocrates problematized track and field being held in higher esteem than knowledge:

> I marvel that so many cities judge those who excel in the athletic contests to be worthy of greater rewards than those who, by painstaking thought and endeavour, discover some useful thing, and that they do not see at a glance that while the faculties of strength and speed naturally perish with the body, yet the arts and sciences abide for eternity, giving benefit to those who cultivate them. (Isocrates 1945: 463)

Marcus Aurelius warned against talking about sports too much, lest one appear trivial (Aurelius 1862: CLXIV). A millennium later, John of Salisbury worried that juggling had a negative impact on 'unoccupied minds ... pampered by the solace of some pleasure' (quoted in Zyvatkauskas

2007). In the sixteenth century, UK god-botherers outlawed what were seen as plebeian pastimes liable to curate "Papist" and superstitious beliefs; sports should be the almost exclusive sphere of property-owners. After the English Civil War, the Puritan victors regarded football matches as dangerous opportunities for Cavaliers to organize royalist opposition (Underdown 1985: 47, 261, 270). In eighteenth-century Britain, sporting pursuits were not deemed to be in the national interest, as a consequence of their unruly use of space and wasteful idleness (Neeson 1996: 51, 178).

Thorstein Veblen problematized sports because their complex trajectory traversed 'hostile combat, through skill, to cunning and chicanery, without its being possible to draw a line at any point.' This tendency was clear from the conduct of star athletes, their 'rant and swagger and ostensible mystification' (Veblen 2007: 285–86). Between Russia's 1905 rebellion and the 1917 revolution, factory owners embraced football as a distraction for potentially restive workers. Maxim Gorky criticized attempts 'to gain the sympathies of the workers' by 'organising football teams' (Gorky 1952: 258–59). Trotsky noted that working-class revolutionary possibilities were 'skilfully restrained and suppressed' through cathexis onto sports (Trotsky 1925). Early Soviet policies were dedicated to health, fitness, art, and equality, but the notion of athlete-*qua*-machine triumphed under Stalin (Harte 2020: 20–24).

The same year as the *Middletown* adventure, a Carnegie Foundation report into US and Canadian universities found that 'college sports have been developed from games played by boys for pleasure into systematic professionalized athletic contests for the glory and, too often, for the financial profit of the college' (Pritchett 1929: x). And when the *Middletown* authors returned to Indiana a decade after their first visit, debts incurred for the school stadium had made it an unpopular liability (Lynd and Lynd 1965: 291–92).

Richard Hoggart emphasized the weakness of functionalist fantasies that sports produced a fit population, ready and able to work and express powerful feelings in law-abiding ways:

> When we say that adolescence must often be a time of opposition and rebellion we should realise that this will often mean *real* rebellion, not something that can be fairly easily piped-off, by providing physical exercise or some kind of strenuous sport or initiative-test. (Hoggart 1965: 35)

Consider carbon sporting events' allegedly positive impact on mental and physical health, reducing health costs, via much-vaunted 'demonstration effects,' whereby people watching high-level athletes miraculously wish to emulate them, or at least enhance their own physical fitness. Study after study shows such claims to be as specious as they are oft-repeated. The evidence for better health and social relations thanks to hosting mega-events just isn't there (Cummins et al. 2018; Dowling and Harris 2022; Bauman et al. 2021; National Audit Office 2022; Dowling 2025).

Agnes Heller criticized the automatic association of aggression and sports, where a contest can either be deemed socially useful to manage a natural tendency or the opposite—but always as something elemental, essential (Heller 1979: 26) rather than historically and geographically situated. As if to make her point, during Agnew's time, pitched battles were fought on US campuses between college athletes and anti-Vietnam War demonstrators: the right used sports to rally reactionaries while the left taunted "football" players, some of whom favored assassinating black progressives (Shepherd 2023: 85, 147, 177, 188). In 1970, the California State Superintendent of Public Instruction notoriously advised that 'the love of clean, competitive sports is too deeply imbedded in the American matrix, too much a part of the warp and woof of our free people, ever to surrender to the burning-eyed, bearded draft-card burners who hate and envy the athlete because he is something they can never be—a man' (Rafferty 1970). Consider Erving Goffman's description of the 'complete unblushing male in America: a young, married, white, urban, northern, heterosexual Protestant father of college education, fully employed, of good complexion, weight, and height, and a recent record in sports' (Goffman 1986: 491–92).

The "other" side to this was chillingly outlined by WEB Du Bois:

> The concept of the European gentleman was evolved: a man well-bred and of meticulous grooming, of knightly sportsmanship and invincible courage even in the face of death; but one who did not hesitate to use machine guns against assegais and to cheat "niggers"; an ideal of sportsmanship which reflected the Golden Rule and yet in the face of this, not only in business and in industry within white countries, but all over Asia and Africa, arose lying, murder, theft, rape, deception and degradation of the same sort and kind which leaves a world aghast when it reads today of what the Nazis did in Europe. (Du Bois 2022: 208)

The reality is that a gendered, racialized nationalism suffuses sporting desires, fears, and plans (Brentin and Cooley 2015). Schumpeter argued that it 'satisfies the need for surrender to a concrete and familiar super-personal cause, the need for self-glorification and violent self-assertion' (Schumpeter 1966: 12). Less prosaically, George Orwell derided nationalism as 'the lunatic modern habit of identifying oneself with large power units and seeing everything in terms of competitive prestige' (Orwell 1945). For May Joseph, '[t]here has never been a pure space within the nation-state' (Joseph 1995: 3), even though 'nation-ness is the most universally legitimate value in the political life of our time' (Anderson 2006: 12).

Reactionary ethno-nationalists have long maintained that nations exist because of supposedly indelible ties: origin stories, languages, customs, races, and religions (Herder 2002; Smith 2000). Some of those traditions indeed developed organically, but most were invented and governed, in hybrid form. The materiality and ideology of the nation generally derived from the Industrial Revolution and colonialism, which brought places together that had not previously been linked through the interdependence of capitalist and governmental organizations; colonial occupation, kidnapping, sexual violence, and slavery; the commodification of everyday life; and the unity, division, power, and domination generated by nation- and empire-binding technologies and institutions, most notably policing, the media, and public education. They enabled those who may never have met to regard themselves as alike (Nandy 1998; Hobsbawm and Ranger 2002; Gellner 1988; Anderson 2006). Cultural components of belonging, commitment, difference, hegemony, and identification were incarnated via the ideological state apparatus of school, propaganda, patriotism, imperialism, religion, xenophobia, and sports (Althusser 1971), a conceptual buttress and a preparation for service to state and capital. Stefano Harney makes the point with reference to Trinidad and Tobago:

> [D]espite the cultivation of nationalism, … critics and writers alike still face a nation whose citizens by and large do not imagine themselves chiefly as members of a modern nation-state … [a] World Cup football qualifying match against the United States illustrates too well the fragile state of national commonality, and the breathless efforts of the intelligentsia to find that commonality. (Harney 2006: 73)

Dallas Smythe described sports as 'a representation of the human condition,' an 'image of skillful use of trained bodies ... or gross sex aggression ... competing for survival without the benefit of accepted law ... a form of folk-drama ... a sardonic morality story, or perhaps a means of reassuring cynical viewers that life is fixed from the start' (Smythe 1954). For Herbert Marcuse, they represented a '[c]onspicuous social mobilization of aggressiveness, the militarization of the affluent society. This mobilization goes far beyond the actual draft of manpower ... no longer the "classical" heroizing of killing in the national interest, but rather its reduction to the level of natural events and contingencies of daily life' (Marcuse 2009: 195). Orwell protested that 'international sporting contests lead to orgies of hatred ... the most savage combative instincts are aroused' (Orwell 1945). In Robert Merton's eyes:

> [W]hen the aim of victory is shorn of its institutional trappings and success becomes construed as "winning the game" rather than "winning under the rules of the game," a premium is implicitly set upon the use of illegitimate but technically efficient means. The star of the opposing football team is surreptitiously slugged; the wrestler incapacitates his opponent through ingenious but illicit techniques; university alumni covertly subsidize "students" whose talents are confined to the athletic field. The emphasis on the goal has so attenuated the satisfactions deriving from sheer participation in the competitive activity that only a successful outcome provides gratification. (Merton 1968: 189)

Violence is always at risk when institutions encourage 'public emotions' during sporting contests (Nussbaum 2013: 203). That happens on and off the pitch, due to the vicarious aggressiveness of supporters and touchline parents abusing players, referees, and others (Salinas Arango Arango and Andrea 2018; Sandel 2007: 53). Iris Marion Young found the notion that sports could 'enhance democracy or contribute to a solidarity of strangers' absurd (Young 2000: 162). In

> most major [team] sports we find an emphasis on retaliation for injury ... players are thought wimpy and unmanly to the extent that they do not strike back so far as the rules permit (and a little beyond that). Even though ... injuring one player does not take away the injury to another, it is a different story if one focuses not on injury but on ranking and humiliation: the retaliatory hit is plausibly seen as taking away the humiliation of the first hit. (Nussbaum 2016: 25)

Ernst Bloch maintained that 'in backward bourgeois conditions sport often stultifies the mind, and for this very reason is promoted from above' (Bloch 1995: 452). Theodor Adorno thought watching games promoted 'a retrogressive and sometimes even infantile type of person' (Adorno 1945: 213). In *Nineteen Eighty-Four*, Orwell's proles were distracted from their conditions of existence because they read 'rubbishy newspapers containing almost nothing except sport, crime, and astrology' (Orwell 1949: 24). Fifty years ago, the *New York Times* thundered that sports were an "Opiate of the Masses" that must be brought to an end because of their unhappy role 'in the seizure of men's souls, in the fanning of fanaticism, in the proliferation of tedium' that 'stimulates bellicosity' (Shepard 1974).

Pierre Bourdieu argued that sports' blend of violence and discipline was a classic working-class pastime—fatally so in terms of political efficacy (Bourdieu 1978). Eco saw them as an 'escape machine geared for the entertainment of the masses' (Eco 1981: 167–68). Guy Debord railed against the 'struggles between illusory qualities designed to generate fervent allegiance to quantitative trivialities' epitomized by sports' 'farcical competitions' (Debord 2004: 82) and Jean Baudrillard derided their 'childish notion of self-transcendence' (Baudrillard 1993: 8). Félix Guattari thought the fleeting moments of *jouissance* in spectatorship, when the self was lost in orgasmic joy through identification with collective triumph, were forever scarred by violent capitalism, because they relied on conquering others (Guattari 1996: 156). Christopher Lasch decreed that sports could 'obliterate awareness of everyday reality … not by dimming awareness but by raising it to a new intensity of concentration' (Lasch 1991: 100). From Perry Anderson's Olympian heights, '[w]here there is sport, infantilism is rarely far away' (Anderson 2011: 276), while Bruno Latour and disciples regarded fans as mere atoms, with banal collective responses (Latour et al. 2012) blinded by a 'collective narcissism' (Burke 2004: 69). Edward Said worried that

> consciousness of sports, with its scores and history and technique and all the rest of it, is at the level of sophistication that is almost terrifying, especially if you compare it with the lack of awareness of what's going on in the world. That's where you get the sense that the investment is being made in those things that distract you from realities that are too complicated. (Said 1993: 23)

Irigaray saw sports as 'an extraordinary opportunity which encourages the stereotypes of the male world: a competitive relationship between the same—one organized around the appropriation of an object belonging to one and all, yet none' (Irigaray 2004: 93). For Kristeva, the media created a quest for male sports stars to 'test their phallic endurance' (Kristeva 2024: 91).

The point of quoting and citing so many renowned theorists and critics, on either side of the good/bad division, is to demonstrate that sports are serious business. They rarely, if ever, simply stand for themselves as autotelic, organic forms and pleasures. Rather, they do ideological work, from numerous epistemological and political perspectives. And it's noteworthy that political economy and the environment are largely missing from those endorsements and critiques.

Carbon Sports Money

Time-series analysis discloses that mega-events have led to a worldwide 'increase in the wealth shares of the top 1% and 10%, often at the expense of the bottom 50%' (Ivanov and Ashyrov 2024). And whereas the proportion of carbon mega-events 'hosted by autocracies fell from 36% in 1945–88 to 15% in 1989–2012,' it now sits at 37%, courtesy of China, Russia, and the wealthy Arab world ("International Sporting" 2022; Matras 2024).

Claims that publicly financed professional sports have positive multiplier effects on economic growth are arrant nonsense. Studies of governments underwriting relocated sports teams, subsidizing stadia, attracting international events, and the like have shown nothing but fiscal spuriousness (Nunn and Rosentraub 1997). Thirty billion dollars of such subvention over thirty years have produced few if any economic benefits for the wider population, as tax burdens mount to cover stadium upkeep (Coulibaly et al. 2022; Marire 2022). The employment connected with carbon mega-events, from building stadia to attracting visitors, may simply displace people from other work and add little to extant tourist destinations (Bogaards 2022: 18).

Carbon sports mega-events are 'economic monsters' (Gruneau and Horne 2016: 2) in their capacity to dupe states and seduce sponsors: the expense of staging World Cup and Olympic events rose sixty-fold between 1964 and 2018—thirteen times more than the global economy—with a fifty-billion dollar shortfall for public investment (Müller et al. 2022b, 2023). The \$220–300 million to stage the Qatar World Cup Finals make those cumulative figures look almost rational (Flynn 2022; Foxman and

Nair 2022). For its part, the 2021 America's Cup regatta returned just 72 cents for every dollar spent (Fresh Info 2021). Formula One threatens cities with legal action if they disclose their costs, but in 2017, best estimates put the price of hosting at a billion dollars a decade, albeit not all from public sources and with some cost variation, depending on the need to construct new race circuits (Sylt 2017). Most recently, five "football" stadia built in the US between 2020 and 2023 cost taxpayers $750 million (Bradbury et al. 2023; Moore 2024). Such realities have rarely stood in the way of city, state, and national governments enthusiastically providing welfare to private interests (McBride et al. 2024). Buoyed by boosterish arguments, governments of all complexions rush to compete for the privilege of hosting athletic jamborees, often ignoring cultural critiques, protests, environmental-impact statements, and cost-benefit analyses.

Meanwhile, private governing bodies thrive. The Fédération Internationale de Football Association (FIFA) generated perhaps $6.4 billion from 2019 to 2022 and anticipated $11 billion in receipts over its next cycle (Thomas 2024; FIFA n.d.); the IOC attracted $7.6 billion between 2017 and 2021 (2022a); and Formula One's 2023 revenues reached $3.2 billion, up 25% in a year (Brittle 2024). Growth is their watchword. FIFA drones on and on about it in the modestly titled planning document, *Football Unites the World* (FIFA n.d.); the IOC undertakes to be a world leader in 'urban growth,' insisting that any bid to host the Games include guarantees of 'long-term socio-economic development' ("Olympic Highlights" 2024; "Electing Olympic" 2023); America's Cup Event Ltd. emphasizes growth (n.d.); World Sailing's strategy for 2025–29 deploys the word twenty-five times (2024); and growth is incarnate in new Formula One racing teams ("FIA Officially" 2023). This grisly automaton-like litany is reminiscent of Edward G Robinson's gangster in *Key Largo* (John Huston 1948). When asked by Humphrey Bogart and Lionel Barrymore's characters what he wants, the *buchón* seems uncertain until prompted, then comes out with 'That's it. I want more.' Mindless acquisitiveness matches his mendacity and violence.

Numbers matter not only as indices of sports' "value," but as signs of broader issues of communications and intellectual property. The last twenty-five years have seen significant struggles between different fractions of capital to obtain consumer subscriptions for broadband, cable, satellite, telephony, and Wi-Fi. Sports became part of this conflict as coverage slipped away from broadcast, free-to-air television. Rupert Murdoch notoriously, indecorously said his companies 'use sports as a "battering ram" to obtain subscribers' (quoted in Milliken 1996). In the late 1990s, France's Canal+ estimated that 40% of its audience signed on just to watch football (Williams 1998: M3; Williams 1999: 104). By 1999, this enticement saw the rights to cover European matches on TV cost over $2 billion (Croci and Ammirante 1999: 500).

The terrain shifted when newer online formations sought to attract young viewers with few affiliations to broadcasting or cable. In the embarrassingly mindless metaphors deployed by the industry, a 'full funnel approach' claimed the 'ability to drive right through to commerce' (quoted in Weprin 2024). Amazon was taking advantage of commercial TV's loss of advertising revenue to buy sports rights at depressed prices. It planned to draw subscribers to Prime to watch and to buy goods while distracting attention from its labor practices and ecological record and gaining general goodwill (Hutchins et al. 2019; Poindexter 2021). Streamers even became partners with cable sports networks to dominate how programs were watched online (Szalai 2024). From the other direction, Warner Bros. Discovery, Disney, and Fox announced plans to stream sports together (Steinberg 2024). In place of Murdoch's 'battering ram,' the new grotesque cliché is that sports provide the media with a 'core competitive differentiation' (Vernon et al. 2024).

Enabled by the Summer Olympics and football's European Championships (the Euros), 2024 saw over $60 billion expended on rights, up 12% on the previous year (SportBusiness 2024). In the US, 16.5% went to streamers (Genovese 2024), which spent ten billion worldwide on sports (Ampere 2025: 6). Apple TV purchased global coverage of Major League Soccer for a decade from 2023 at $250 million per year, matching its deals for Major League Baseball. Amazon Prime captured certain Yankees home games, NFL fixtures (for ten billion), and European football's Champions League, and announced emerging interests in motorsport and women's basketball. Google bought Sunday NFL rights for fourteen billion from 2024. Netflix started League coverage that year, "wrestling" in 2025, and the women's World Cup of football from 2027, investing five billion over a decade (Silverman and Ourand 2022; Battaglio 2022; Kafka 2021; Weprin 2021, 2024; Germano et al. 2022; R Porter 2024). DAZN handed FIFA a billion euros to broadcast the inaugural Club World Cup in 2025 ("DAZN" 2024).

Formula One's income from media rights grew 7.9% in 2024 to $1.18 billion as it sought to switch from US cable to streaming (Agini 2025). The commodification of viewers as objects sold to advertisers was vital. In celebrating a record expansion of sports media in 2024, Ampere hailed the 'ample opportunity for broadcasters and brands to acquire … audiences' (2025: 3); this became ever more resonant with sponsors, as a means of ampliying their product placement (12).

The implications for our environment are dire. For instance, 2021's Super Bowl TV commercials produced two million tons of CO_2 from 6.3 billion spectators—26 million online—and 64 billion references on "social" media. A million such impressions, as they are known, equate to a metric ton of CO_2 (L 2024). Each minute a person uses TikTok, 2.63 grams of carbon are emitted (Musto 2024). An hour of streaming to mobile gadgets requires more electricity than two new refrigerators (Mills 2013). Even on standby, they exhaust hundreds of terawatt hours a year, akin to coal-fired power plants (International Energy Agency 2014). People watching football on these wee devices increase their footprint ten-fold in comparison with television (Carbon Trust 2013, n.d., 2016; TePoel 2017).

By 2023, the world's data centers consumed as much electricity as Britain ("Can Computing" 2023) and were responsible for 4.4% of overall US energy use (Shehabi et al. 2024). There are over ten thousand of these "clouds" across the globe, about half in the Americas and a third in Asia. Clouds require around 55 gigawatts of energy and 300 m square feet of land ("The Data-Centre" 2025). Artificial intelligence's environmental impact is intense—each ChatGPT request takes ten times the power of a Google query (EPRI 2024). Data centers also need a lot of water—official disclosures indicate that the Google cloud used 21.198 million liters in 2022, an increase of 20% on 2021, and Microsoft's 6.435 million, up 34% (Google 2023; Microsoft 2022). Google, Microsoft, and Apple are building dozens of data centers in parched areas of the world, denying local people much-needed reserves (Source Material 2025). And because they employ very few people once construction is complete, the ongoing economic benefits to host cities are minimal (Commonwealth of Virginia, Joint Legislative Audit and Review Commission 2024). *En passant*, Apple, Meta, Google, and Microsoft understated their 2023 greenhouse gas emissions by 662% (O'Brien 2024).

Greenwashing

Since the advent of the non-governmental organizations (NGOs) Friends of the Earth and Greenpeace in 1970, companies and governments have felt the need to respond to public concerns over pollution by producing economic, ethical, legal, and philanthropic discourses designed to fend off democratic intervention and market products and services as green (Hemphill 2004; Logsdon 2004; Delmas and Cuerel Burbano Delmas and Burbano 2011). Many corporations even brand themselves as stakeholders in the green economy, promoting sustainability and capitalism as natural partners while relying on an obedient media for cover (Lewandowsky et al. 2015).

They crave a 'social license to operate,' an invidious expression coined by the United Nations Commission on Sustainable Development (1998) to recommend that businesses and governments treat local communities as "stakeholders" in the exploitation of territory (Wilburn and Wilburn 2011), part of that multiple oxymoron 'corporate social responsibility' (CSR), a concept created in the 1950s and shepherded into mainstream capitalism by the Committee for Economic Development (CED) (Bowen 2013). CSR seeks to generate new markets, control labor, deliver positive public relations, and heighten brand recognition (Fifka and Jaeger 2020; Thomson and Boutilier 2011). By following the CSR pathway, capital plans not only to obtain free rides from carbon-sports host nations but to cultivate the image of a responsible international subject whose exemplary liberal self-governance means it does not require democratic regulation. One of the discourse's tactics is to favor 'environmental, social and governance' (ESG) investing, a concept invented by banks under the rubric of a "Global Compact" supported by the UN (Swiss Federal Department of Foreign Affairs and United Nations 2004).

CSR and its awkward progeny ESG obscure a multitude of harms, the environment notable among them. For example, the CED insists on confronting climate change via 'market principles, the acceleration of innovation, and preservation of US competitiveness' (n.d.). Corporations counter environmental activism by deploying the language of sustainability, implementing amply affordable parts of any critique, then moving on (Marshall et al. 2012). As a consequence, ESG norms do nothing for the climate—and magnify greenwashing (Treepongkaruna et al. 2024; Todaro and Torelli 2024). Even the *Economist* has felt forced to say: the 'human face that CSR applies to capitalism goes on each morning, gets increasingly smeared by day and washes off at night' (Crook 2005: 4), asking '[c]an profit-seeking companies really help save the planet?' ("The Fundamental" 2022).

States are coevally complicit. While most issue propaganda announcing that global heating must fall below 1.5 °C, they 'produce more than double the amount of fossil fuels' compatible with that necessity (Stockholm Environment Institute et al. 2023). Eighty-five percent of nations failed to produce biodiversity plans in 2024. Needless to say, the US hasn't even signed the relevant convention (Dunne and Greenfield 2024). The reality is that fossil fuels' 80% share of global energy in 2023 was down just 2% on a decade earlier (International Energy Agency 2024: 24).

A particular story describes these tendencies. In 1983, Jay Westerveld snuck into the Beachcomber resort on a Fijian island to towel down after surfing. He found a card that read:

> Save Our Planet: Every day, millions of gallons of water are used to wash towels that have only been used once. You make the choice: A towel on the rack means, 'I will use again.' A towel on the floor means, 'Please replace.' Thank you for helping us conserve the Earth's vital resources. (quoted in Sullivan 2009)

Clearly, the place was skimping on the electrical, materials, and labor costs of cleaning. Westerveld gave this self-promotion a name—greenwashing (Motavalli 2011; Watson 2016).

This figure is from a Nordic hotel chain that prides itself on having invented the practice (Fig. 1.1).

Westerveld sees the problem he first noted forty years ago developing rather than receding:

> [C]hemical polluters have evolved novel ways to cloak polluting activities past an increasingly uncritical, unconcerned and pathologically distracted public, oftentimes garnering popular support for toxic activities through carefully crafted rebranding. In the 21st century, we're being groomed to see massive pesticide use and selective slaughter of wildlife as ecological steps forward. (Westerveld 2024)

Today, the Beachcomber company he infiltrated has an *Environmental & Social Charter: Our 52 Commitments* (2024)—no doubt a model ESG initiative.

Fig. 1.1 (Source: Toby Miller)

The *New York Times* defines greenwashing as 'when a company tries to portray itself as more environmentally minded than it actually is … in the rush to be seen as green, companies often exaggerate claims, or simply make things up' (Gelles 2015). For the *Oxford English Dictionary*, the term can be simplified as the 'creation or propagation of an unfounded or misleading environmentalist image' (2002). Nielsen defined it this way: '[W]hen a company purposely makes misleading sustainable claims or overstates progress against stated goals' ("The Changing Climate" 2022), while the Federal Trade Commission has categorized greenwashing as 'deceptive[ly] eco-friendly' (Fair 2022).

Among peak inter-governmental bodies, the UN points out that 'greenwashing promotes false solutions to the climate crisis that distract from and delay concrete and credible action' (n.d.). The Intergovernmental Panel on Climate Change notes the tendency's ubiquity when capitalist methods of combating our crisis are pursued (2022: 545, 1550, 1586, 1599). ASEAN identifies 'conduct that deceives or misleads consumers into believing that the supplier's practices or goods and services, are more environmentally positive or have greater environmental benefits, than is indeed the case' ("Study on Greenwashing" 2023). The EU refers to vague, selective, and exaggerated 'sustainability-related claims relating to all aspects of the ESG spectrum' about products and services (European Banking Authority et al. 2022: 3, 5–6). It argues that across commodity chains, 'misleading environmental communication, presents a significant impediment to achieving decarbonisation and sustainability goals and undermines consumers' ability to make informed choices, diminishing their empowerment as rights holders' (FRA 2024). The Organisation for Economic Co-operation and Development (OECD) nominates moments in consumer perception when 'claims are believed to be unsubstantiated and possibly inauthentic' (2011: 5).

Greenpeace specifies 'nature-inspired imagery or green buzzwords that rarely hold up when looked at in more depth' (Das 2022), a

> deliberate attempt to cover up or distract from the damage a company is doing to the environment … delay or avoid the action we so badly need to avoid catastrophic climate change. … the false perception that planet-critical problems are being tackled, even solved, when they are not. … Greenwashing is our Pied Piper to climate doom. (Gelmini 2021)

Friends of the Earth warns that corporations deploy 'the language of sustainability, or co-opt terms from grassroots movements, twisting their use to serve their own ends' (2024: 1). The Natural Resources Defense Council defines greenwashing as 'a way for companies to continue or expand their polluting as well as related harmful behaviors, all while gaming the system or profiting off well-intentioned, sustainably minded consumers' (Lindwall 2023). For ClientEarth, it allows them 'to appear socially responsible while continuing to operate as they wish'

("Greenwashing" 2023). WWF offers a *Guide to Greenwashing* to help people spot and analyze the problem (n.d.).

Latitudinal and longitudinal studies indicate that many companies engage in 'selective disclosure' about their environmental records, only sharing positive information (Marquis et al. 2016). RepRisk (the clue is in the name) issues annual reports on greenwashing, which it identifies as 'risky' for firms. In the decade from 2012, it found many spectacular incidents, most notably in the Global North. And major climate disasters tied to corporate greenwashing increased 25% in 2023 on the year before. By 2024, exceedingly risky incidents had 'surged by over 30%' (RepRisk 2024).

A third of the globe's largest two thousand corporations say they are "committed" to net-zero emissions. But per their current policies and programs, 93% of those greenwashers will not achieve that goal (Accenture 2022). In their quest to avoid democratic intervention, peak capitalist organizations promote self-regulation of greenwashing (International Organization of Securities Commissions 2023; International Financial Reporting Standards 2024). Meanwhile, 2023 found three-quarters of US corporations admitting to greenwashing; the global figure was 58%. The biggest growth area is financial firms that fund extraction (Flynn 2023). The success of far-right anti-science in US politics, coupled with the timorous policies of social democrats, appeared to herald a new era of proud opposition to environmentalism, but many firms continued to set ambitious targets to counter climate change ("The Remarkable" 2025; PwC 2025).

Claims to ecological virtue emanate from numerous sporting institutions, governments, and companies, through published internal audits, marketing, responses to criticisms, lobbying against democratic regulation, and cultivation of journalistic chorines. Two hundred and fifty-five public, private, and mixed projects of international development utilized sports greenwashing in 2008, a 93% increase over five years. A significant proportion involved corporations, frequently via *faux* grassroots organizations (Levermore 2010; Silk et al. 2005; Trendafilova et al. 2013). Regrettably, determinedly, carbon sports lustfully seek, and gleefully accept, sponsorship from polluting industries, imbuing corporations with leisure's pleasurable connotations. The sports involved adopt an unwarranted clean and green image to bestow on their sponsors, associating both sides with environmentalism (Beck 2009: 103). We'll see how that occurs in the chapters to come.

Methods

Massively contradictory, non-linear changes have created and affected sports over centuries, particularly the earth-shattering concatenation of political-economic events since the eighteenth century: shifts from absolute monarchy to democracy, imperialism, colonialism, slavery, industrialization, war, nationalism, postcolonialism, globalization, and financialization. This is especially true of the last hundred and fifty years, as a consequence of monopoly capitalism, trade unions, anti-trust, parliamentary socialism, Leninist, Stalinist, Maoist, and Eurocommunist tendencies, Keynesian demand-and-monetary management, state socialism, monetarism, neoliberalism, urbanization, human rights, social movements, famine, the insertion of China and India into the world system, the increased importance of trade in services, AIDS, climate change, Keynesianism's return, de-industrialization, COVID-19, the disappearance and reappearance of tariff and non-tariff barriers, and so on. Those contexts form a backdrop to the chapters that follow.

These big-ticket items are often excluded from the social sciences. In a quest for minuscule precision, 'abstracted empiricism,' a 'fetishism of the Concept,' adheres to methods that overdetermine any serious engagement with broader and deeper matters (Mills 2000: 50). Reducing topics to tiny aspects serves the interests of capital. Such 'administrative research' implicitly or explicitly supports technological innovation, demand, marketing, and property relations—capitalist efficiency and governmental normativity (Lazarsfeld 1941). There is a profound problem with such instrumental reason: 'to find means for the goals one adopts at any given time' amounts to a 'socially conditioned tendency' (Horkheimer 2012: 4–5) that is 'methodologically and theoretically weak and primarily concerned with vocational and/or industry objectives' (McKay 2015: 547–48). For example, dutiful followers claim that environmental problems occur absent the operation of pure market forces: once prices are placed on such negative externalities as pollution, everything can be put to rights. This has become a key argument for commodifying communal space (Hardin 1968).

Under prevailing social relations, capitalism rules the economy, with the third sector or civil society representing and empowering community interests (Srinivas 2022; Shivji 2007). Using that cover, a corporate takeover of life in wealthy countries since the 1970s has been 'licensing capital, leashing labor, demonizing the social state and the political, attacking equality, promulgating freedom' (Brown 2019: 12). It has led to privatized—but still publicly subsidized—services (Miller 2007). Again and again, democratically unaccountable entities, from reactionary corporations to religious zealots, are magically entrusted to undertake self-regulation and ameliorate inequality in a conjuncture laden with powerful structural determinations. The Global North runs international finance but the South 'produce[s] the majority (about sixty percent) of world GDP' (Ali 2022) and a fifth of the planet consumes more than three-quarters of its resources via seemingly unassailable, unquestionable, unquenchable sovereign desires (McMichael 2011).

The dominant Anglo scholarly literature on mega sports events shows minimal engagement with their negative impacts (Thomson et al. 2019). A cosmically, comically limited array of epistemological and political-economic concerns (Müller et al. 2022a) routinely disavows civil-society angles in favor of capitalistic ones. Much research on corporate-sports relationships seeks greater "efficiency" (Carlini et al. 2025), amounting to little more than an 'advertorial' (Paché 2020: 92) for capitalist interests in the name of the general good. There are alternatives.

Conventional neoclassical economics and political economy differ in their nature and method: hegemonic economic formalism versus substantive materialism (Wilk and Cliggett 2007: 3–14). The formalist view allows for three purportedly legitimate actors: consumers to desire, companies to supply, and the state to police property. Its lodestone is the imagined individual rationality of firms and their customers. Formalists claim supply and demand determine price, which in turn equals value. They deride or ignore labor and democratic action calculated to increase equality.

But as Hall explained, markets

> do not work mysteriously by themselves or "clear" at their optimum point. Only by bracketing out of the calculation the yawning differences between the relative wealth which buyer and seller bring into the exchange can they be called "fair." No "hidden hand" guarantees the common good. They require the external power of state and law to establish and regulate them. But the discourse provides its subjects with a "lived" "imaginary relation" to their real conditions of existence. This does not mean that markets are simply manufactured fictions. Indeed, they are only too real! They are "false" because they offer partial explanations as an account of whole processes. (Hall 2011: 716)

By contrast, critical political economy pays heed to power, notably as exercised by capital, labor, and the state, and to history, in terms of money, technology, social relations, and the law. It starts from everyday production, stressing that value is generated by workers who rarely gain due benefit from it—unlike the owners of capital. This approach is grounded in a labor-and-use theory of value, not an exchange one.

Political economy identifies and criticizes the privilege that derives from inherited, speculative, and productive capital, each of which limits life chances for others, versus work, which it regards as the crucial component of business. A commitment to egalitarianism is fundamental, while a focus on solidarity and cooperation relies on a relatively optimistic view of the ability to play nicely with one another. Political economists believe human agency can change the world—that people are *subjects* of history, not merely its objects. Horkheimer identified three core elements for developing alternatives to capitalism: explanations that account for phenomena without necessarily accepting them as given; normativity, suggesting how matters might be otherwise; and applications derived from counter-hegemonic thinking and social movements (Horkheimer 2012).

But this has not always amounted to a progressive project. A key point of such analysis should be struggle, given the differing interests of the *bourgeois* and popular classes. But due to the typically overweening power of state and capital, conflict has frequently been subordinated by analysts to a focus on dominant social forces. An ironically functionalist Marxism finds the world replicating established authority in a seemingly inexorable manner, with structure appearing to override agency.

That emphasis on ownership and control can disable critical political economy, ironically diminishing its putatively key starting point. Organization charts of the intersection of state and capital élites—corporations snaking across the industrial, financial, governmental, and cartographic world—are valuable, but may miss the dynamics of particular circumstances.

In addition to that tendency, focusing on production and growth can ignore nature, other than as something to be transformed. And as a consequence of critical political economy's concentration on the industrial proletariat, it is traditionally weak, sometimes hopeless, on gender, race, colonialism, culture, region, religion, and social movements. That emphasis all-too-often ignores or devalues women, peasants, slaves, children, domestic labor, the disabled, informal work, and our fellow-animals.

By contrast, this book blends political economy, textual analysis, and ethnography with cultural studies and environmental materialism. The idea is to highlight social relations, organizational propaganda, lived experience, and ecological impact. Such work combines social, cultural, and ecological theories and methods to focus on established lines of force and authority as well as the margins of power. It queries the who, what, when, where, why, and how of material politics and privilege—most notably, *cui bono*? [who benefits?], a question rarely posed in the *bourgeois* press or policy debates (Atkinson 2015: 5).

We operate from the assumption that the rubric of social justice includes—but may be subsumed by—environmental justice. The search for equality and fairness must include protection of the planet as much as public health or guaranteed income, and draw on an international vision of social relations (Newman 2005; Nash and Fernandez 1983). This is per Edgar Morin and Anne Brigitte Kern's recognition thirty years ago that we live in a 'polycrisis,' where climatic, financial, and political disasters interweave and overlap (1993: 73). Without citing their work or the original definition, even the World Bank now uses the term (2024).

Reading also matters: academic research, activist manifestos, social-movement publications, political fiction, company disclosures, regulatory findings, state inquiries, stock-market prices, financial and industry journalism, union policy briefs, textual and content analysis, legal cases, memoir, fiction, poetry, drama, and governmental, inter-governmental, and commercial statistics. Critical interpretation and varied sources are necessary in order to evaluate arguments and empirical material. That may involve 'reading between the lines and against the grain' in ways that risk missing the value of denotative understanding (Frow 2019: 8–9) but can add to our comprehension of duplicitous corporate and state conduct. Per Benjamin's original concept:

> There is no document of civilization which is not at the same time a document of barbarism. And just as such a document is not free of barbarism, barbarism taints also the manner in which it was transmitted from one owner to another. A historical materialist therefore dissociates himself [*sic.*] from it as far as possible. He regards it as his task to brush history against the grain. (Benjamin 2003: 392)

Roger Chartier's notion of the 'world of text … [as] a world of objects and performances' (Chartier 2005: 38–39) is helpful. He proposes a tripartite approach to meaning, *viz.* a reconstruction of 'the diversity of older readings from their sparse and multiple traces'; a focus on 'the text itself, the object that conveys it, and the act that grasps it'; and an identification of 'the strategies by which authors and publishers tried to impose an orthodoxy or a prescribed reading on the text' (Chartier 1989: 157, 161–63, 166).

Discerning where, when, and how conjunctures emerge, function, and change requires more than armchair theorizing or mathematical quixotry: shifts can be made visible through ethnographic research, participating in society up, down, and sideways to disclose the often-invisible complexities of political economy that color the everyday. Consider Laura Nader's advocacy of 'studying up' to comprehend 'how power and responsibility are exercised' as well as 'activities whereby the victims of the system made their victimization visible' (Nader 1972: 284–85); the weight of history and geography, per Eric Wolf (1982); and George Marcus' multi-sited ethnography, which draws equally on the world system and quotidian life (Marcus 1995) to analyze the 'transnational capitalist class' (Sklair 2016).

This ushers in research that decenters—without forgetting—Enlightenment forms, instantiating the Global South as a *center* of knowledge, rather than its object (Lins Ribeiro and Escobar 2006); a zone for intellectuals and activists 'to learn from, not just about' (Connell 2020: viii). Examples include action research, which Orlando Fals-Borda pioneered sixty years ago in Colombia (Fals-Borda 1988). It blends ethnography, adult education, political economy, and activism through alliances between scholars and citizens. And there is a longstanding history of anticolonial, African-American, and gendered political economy and ethnography (Meghji 2021; Bohrer 2019). From early Marxism to feminist economics and decolonial critiques, we can discern such tendencies running counter to conventional academia in the Global North (Newman 2005; Berik and Kongar 2021; Cupples and Grosfoguel 2018; Bulbeck 1998).

One way to think of political economy from an everyday perspective is to recognize socialism, for example, as an ordinary, albeit often invisible or unacknowledged, part of existence. The clue is in the word itself: socializing the risks and costs of life, rather than individualizing them. Hence taxation funding education, environmental protection, health, mass transit, the military, postal services, power, prisons, rail, research, roads, telecommunications, water—and sports. These services are socialist when they have been removed from the market or subdued democratically, per the regulation of food and pharmaceuticals.

Socialism also inflects daily work. Individual labor is negated by capitalist private property, which expropriates the work of the many by the few. But despite the time- and work-discipline of capitalism, employees may develop individuality, relate to nature as subjects of labor, and construct a sense of collectivity in their work that transcends its direction by the *bourgeoisie* (Thompson 1967). For capitalist production partially negates itself because its own ideology—self-determination—potentially brings about the expropriation of the few by the many, thereby negating its own negation of work (Otani 2018).

Close inspection of different states, industries, and labor experiences reveals an array of different capitalisms, riven with struggle and possibilities for radical politics, but without the traditional binaries of acclaim or denunciation from Schumpeterian or Marxist angles. Luis Reygadas shows that apologists and critics for capitalism alike have difficulty understanding transformations from rigidly hierarchical factory labor to supposedly more horizontal cultural work (Reygadas 2021).

Almost a century ago, before his rightward turn, the noted novelist John Dos Passos wrote what can be construed as a brief multi-sited ethnography, using textual analysis and political economy with ecological tones. Here are his prefatory remarks to the *U.S.A.* trilogy:

> U.S.A. is the slice of a continent. U.S.A. is a group of holding companies, some aggregations of trade unions, a set of laws bound in calf, a radio network, a chain of moving picture theaters, a column of stock quotations rubbed out and written in by a Western Union boy on a blackboard, a public library full of old newspapers and dogeared history books with protests scrawled on the margins in pencil. U.S.A. is the world's greatest river valley fringed with mountains and hills. U.S.A. is a set of big-mouthed officials with too many bank accounts. U.S.A. is a lot of men buried in their uniforms in Arlington Cemetery. (Dos Passos 1937: vii)

Dos Passos pithily engaged crucial elements of political-economic analysis, such as administration, banking, corruption, law, the media, topography, and equally materialist everyday observations (readers contesting history and stock prices erased and substituted).

In that spirit, Marshall Berman avowed that '[l]ife studies is one of the big things Marxism is for' (Berman 2017) and Marvin Harris introduced 'cultural materialism' as 'the simple premise that human life is a response to the practical problems of earthly existence' (Harris 1979: ix), later fleshed out by Raymond Williams' focus on:

> the production (rather than only the reproduction) of meanings and values by specific social formations, on the centrality of language and communication as formative social forces, and on the complex interaction both of institutions and forms and of social relationships and formal conventions. (Williams 2005: 500)

Can political economy, textual analysis, and ethnography open the sealed set of capitalism to disclose its morphology and inner workings, engaging but also transcending Marxism's traditional subjects of history—the state, corporations, and unions—to examine daily interaction, personal life, small-scale structures, and climate change (Reygadas 2021)?. Can it encompass non-capitalistic economic activity, such as altruism, barter, and communalism, via action research that regard all the people involved as political economists, writers/readers, and ethnographers? (Gibson-Graham 2006a, 2006b). And center the environment? That's my wager, per Williams insisting forty years ago that 'no political development is now more necessary than a convergence of the labour movement and the ecology movement' (2020).

To do so necessitates understanding today's key environmental political theories, which vary between anthropocentric, ecocentric, ecofeminist, and ecosocialist forms (Swanton 2010; Groves 1995; Pepper 2000). The four schools differ over values (which entities qualify for moral consideration and matter most), rights (protecting individuals and groups), and consequences (responsibility for diminished wellbeing).

Anthropocentrism privileges human interests. It has a powerful heritage: Kant regarding humans as 'an entirely different being from things, such as irrational animals, with which one can do as one likes' (Kant 2006: 15), Hegel disavowing nature's 'tedious chronicle,' where there is 'nothing new under the sun' (Hegel and Friedrich 1988: 50, 154, 61), and William James sharing with us that 'nature is but a name for excess' (1909: 63). Its acme came when Radovan Richta and collaborators famously announced that 'man-made environments are replacing the natural' (1969: 23), part of what Isaac Asimov called the drive 'to withdraw from nature' (Asimov 1964).

That position dominates state, capitalist, scholarly, and popular discourse, largely excluding theoretical and empirical work done in, by, and for the Global South. Sometimes it is subcutaneously dominant—symbolically tempered to form part of an ideology of conservation that 'recognizes fully the right of the present generation to use what it needs and all it needs of the natural resources now available' but 'recognizes equally our obligation so to use what we need that our descendants shall not be deprived of what they need' (Pinchot 1910: 98). The assumption is that non-human nature has no moral standing (and hence no rights) in the event of environmental change, other than when people are also affected (Curry 2006). This "Human Exceptionalism Paradigm" holds that alphabetic people are the sole living creatures capable of making culture (Catton and Dunlap 1978).

Thanks to that perspective, and the spread of capitalism and state socialism, our interaction with the environment has become one of domination, even when a special relationship to the natural world is asserted. Contemporary anthropocentrism features such "reasonable" perspectives as ecomodernism, which cheerily declares that 'knowledge and technology, applied with wisdom, might allow for a good, or even great, Anthropocene' (*An Ecomodernist* n.d.). The state must simply create conditions for decentralized green governance: small-scale institutions can design and monitor sustainable practices when governmental oversight and management are neither feasible nor efficient. Political-economic arrangements allegedly encourage ecologically sound relationships between enterprises and communities that share ecosystems. The two groups monitor what works and what doesn't, modifying methods of acquiring resources and sharing information (Ostrom 2000: 47).

Such anthropocentrism formed the framework for sustainable development established by the 1987 World Commission on Environment and Development and enshrined in the 1992 Rio Declaration on Environment and Development as a human right 'to a healthy and productive life.' Advocates may urge people to buy responsibly and recycle, but their gendered notion of virtue favors a hegemonic masculine mode of self-reliance, embodies a neoliberal focus on individual responsibility rather than collective and state-based action, and rejects participatory, deliberative democracy in favor of a moralistic, plutocratic republicanism (Arias-Maldonado 2007; Barry 2006; Latta 2007; MacGregor 2006). This 'transactional ecology' (Westerveld 2024) accords equivalent value to the avowedly compatible ends of economic growth, social progress, ecological survival, and, in more recent interpretations, cultural sustainability (Curry 2006): 'environmental degradation and resource exhaustion are being diagnosed as management problems rather than as a crisis or breakdown; this management exercise then becomes a new source of dynamism for capitalism' (Goldman and Schurman 2000: 567).

There are radical alternatives. Ecocentrism argues that 'some or all natural beings, in the broadest sense, have independent moral status' (Curry 2006: 64). So the 2008 Ecuadorian Constitution guarantees the rights of nature/*Pachamama* (Constitución Ecuador 2008) and New Zealand/Aotearoa grants rights to natural topographic features as ancestors (Corlett 2025). Ecofeminism identifies gender as crucial to environmental destruction, due to masculinist commitments to "development" and "growth" and stereotypes of nature and women as "irrational" forces that must be controlled (d'Eaubonne 1978). Ecosocialism highlights the incompatibility of capitalism with the preservation of nature, the earth reshaped and its lifespan shortened by industrial capital (Luxemburg 1970: 335).

We've thrown in our lot with environmental materialism, a tendentious syntagm that brings these alternative views together with science and Marxism. Such a combination of theoretical and empirical rigor with socialist thought was far from unusual in the first fifty years of the twentieth century. It fell victim both to Cold War ideology, which saw many US scientists suffer Red-baiting, and the horror of state-socialist human rights violations, which forced progressives to re-evaluate the Second World. Today it attracts opprobrium from essentialist ideologues who associate environmental hazards with a putative human nature and mandatory economic activity rather than actually existing, changeable social forces.

Nevertheless, renewed interest in environmental materialism is emerging across the human sciences. Both a scholarly method and an activist logic, it focuses on work and the environment (Barber et al. 2018; Bertrand 2019; Davis 2017), drawing on ecofeminism and ecosocialism to help explain the ecological consequences of corporate and governmental policies and projects. Environmental materialism measures survival separately from monetary exchange, prioritizing sustainability and the democratic control of business over growth and profit. Over the last five decades, the theory has connected nature to capital and labor as constitutive variables of analysis, favoring the regulation of business via ecological principles (Fisk 1980; Foster 1999; Benton 1996; Brevini and Lewis 2018; Goldman and Schurman 2000; Martinez-Alier 2012; Maxwell et al. 2015; O'Connor 1998).

An adequate materialist approach must engage with ecological theories and research from the social sciences, the humanities, law, and science (Carmichael 2022; Gibbs et al. 2010). While political economy offers a conceptual backdrop to the current conjuncture, environmental materialism can pose specific '*cui bono*?' questions about public and private investments in carbon sports, examining the operation and malfeasance of state and capital in terms of their ecological impact (Miller 2019). And it can push us to think about the environment across generations, frontiers, and species in ways that go beyond the productivist growth fetish shared by neoclassical and Marxist, formalist and material, models alike.

Conclusion

Carbon sports are ecologically '*weak, hollow, and half-hearted*.' Favoring 'economic concerns over environmental concerns' (Wilson and Millington 2020: 2), they are complicit, directly and indirectly, with what Robert Nixon calls the 'slow violence' of ecological destruction, a

> violence that occurs gradually and out of sight, a violence of delayed destruction that is dispersed across time and space, an attritional violence that is typically not viewed as violence at all ... neither spectacular nor instantaneous, but rather incremental and accretive, its calamitous repercussions playing out across a range of temporal scales. (Nixon 2011: 2)

In Veblen's words, '[T]he most noticeable effect of the sportsman's activity [is] to keep nature in a state of chronic desolation by killing off all living things whose destruction he can compass' (Veblen 2007: 287).

Bourdieu criticized the '*escapism of Wertfreiheit* (value-free-ness),' of merely 'relaying the discourse which, in sport as elsewhere, the object produces about itself' (Bourdieu 1999: 15). Discerning and resisting slow sporting violence, while seeking ways to overturn its disastrous tendencies, animates the work of environmental materialism you hold in your hands. It is the slightest of counters to the carbon sports that cloud our world and play with our future—a dirty business 'increasingly run on extractive lines for the benefit of predatory rentiers, autocrat-backed sovereign wealth funds and private-equity ghouls' (Timms 2025).

Bibliography

NB—Links from US Science Have Been Removed By Its Government

"A Global Gold Rush is Changing Sport." (2024, July 25). *Economist* https://www.economist.com/leaders/2024/07/25/a-global-gold-rush-is-changing-sport.

"Can Computing Clean Up its Act?" (2023, August 16). *Economist* https://www.economist.com/science-and-technology/2023/08/16/can-computing-clean-up-its-act.

"DAZN Secures Exclusive Fifa Club World Cup Rights." (2024, December 4). *BBC* https://www.bbc.com/sport/football/articles/crl387gr18ro

"Electing Olympic and Paralympic Hosts: Targeted Dialogue Explained." (2023, November 27). International Olympic Committee https://olympics.com/ioc/news/electing-olympic-and-paralympic-hosts-targeted-dialogue-explained.

"FIA Officially Launches an Application Process for Prospective Formula 1 Teams." (2023, February 2). Fédération Internationale de l'Automobile https://www.fia.com/news/fia-officially-launches-application-process-prospective-formula-1-teams.

"International Sporting Events are Increasingly Held in Autocracies." (2022, November 17). *Economist* https://www.economist.com/graphic-detail/2022/11/17/international-sporting-events-are-increasingly-held-in-autocracies.

"Is it Recession-Proof?" (2009, February 12). *Economist* https://www.economist.com/international/2009/02/12/is-it-recession-proof.

"Meares: Stop Bickering Over Brisbane's Olympics." (2024, August 10). *ESPN* https://global.espn.com/olympics/story//id/40806298/australian-anna-meares-stop-bickering-brisbane-olympics.

"Olympic Highlights 15/11/2024." (2024, November 15). International Olympic Committee https://olympics.com/ioc/news/olympic-highlights-15-11-2024.

"Saudi Arabia is Spending a Fortune on Sport." (2023, August 10). *Economist* https://www.economist.com/briefing/2023/08/10/saudi-arabia-is-spending-a-fortune-on-sport.

"Sports Mega-Events: Social Scientific Analyses of a Global Phenomenon." (2006). *Sociological Review* 54, no. 2: 1–187.

"Study on Greenwashing in Online Marketing Funded by CCCS Finds Use of Vague Environmental Claims and Confusing Technical Jargon." (2023, November 16). ASEAN https://asean-competition.org/read-cases-study-on-greenwashing-in-online-marketing-funded-by-cccs-finds-use-of-vague-environmental-claims-and-confusing-technical-jargon.

"The Changing Climate of Sustainability Has Reached a Critical Moment." (2022, November 30). Nielsen https://nielseniq.com/global/en/insights/analysis/2022/the-changing-climate-of-sustainability-has-reached-a-critical-moment/.

"The Data-Centre Investment Spree." (2025, February 5). *Economist* https://www.economist.com/business/2025/02/05/the-data-centre-investment-spree-shows-no-signs-of-stopping.

"The Fundamental Contradiction of ESG is Being Laid Bare." (2022, September 29). *Economist* https://www.economist.com/leaders/2022/09/29/the-fundamental-contradiction-of-esg-is-being-laid-bare.

"The Remarkable Rise of "Greenhushing"." (2025, July 29). *Economist* https://www.economist.com/business/2025/07/29/the-remarkable-rise-of-greenhushing.

Accenture. (2022). *Accelerating Global Companies Toward Net Zero by 2050* https://www.accenture.com/content/dam/accenture/final/capabilities/strategy-and-consulting/strategy/document/Accenture-Net-Zero-By-2050-Global-Report-2022.pdf#zoom=40.

Adorno, Theodor W. (1945). "A Social Critique of Radio Music." *Kenyon Review* 7, no. 2: 208–17.

African Union Sports Council. (n.d.). "Introduction to Sport for Development." https://sportscouncil.au.int/en/introduction-sport-development.

Agini, Samuel. (2025, March 15). "Formula 1 Owner Bets on Brad Pitt's Pulling Power to Shift into US Mainstream." *Financial Times* https://www.ft.com/content/61b0156f-0025-4f2b-8037-e7477c2e92c1?utm_source=semafor.

Ali, Mona. (2022, April 27). "Regime Change? The Evolution and Weaponization of the World Dollar." *Phenomenal World* https://www.phenomenalworld.org/analysis/regime-change/.

Althusser, Louis. (1971). *Lenin and Philosophy and Other Essays.* Trans. Ben Brewster. New York: Monthly Review Press.

America's Cup Event Ltd. (n.d.). *36th America's Cup Presented by Prada Auckland 2021: America's Cup Event Limited Final Event Report.* https://www.americascup.com/files/m5301_Americas-Cup-Event-Limited-Final-Event-Report_FA_2.pdf.

Ampere. (2025). *A Year of Sport 2024: Full Stream Ahead in 2025.* https://19957038.fs1.hubspotusercontent-na1.net/hubfs/19957038/2024%20in%20Sports-1.pdf?utm_medium=email&_hsenc=p2ANqtz-_dXEmIwhUZXNR8DZTyCGUGWDN8MjPoO5EcDUzk62EM2P1PCObUjAH9Y8GOaZ2ESRWnL-vWSERAUBeKBETg6joUvYAj0w&_hsmi=307835695&utm_content=307835695&utm_source=hs_automation.

Anani, Zak. (2024, December 9). "The True Size of the Global Sports Industry." GlobalSportsInsightshttps://gis.sport/news/the-true-size-of-the-global-sports-industry/.
Anderson, Benedict. (2006). *Imagined Communities: Reflections on the Origin and Spread of Nationalism*, rev. ed. London: Verso.
Anderson, Perry. (2011). *The New Old World*. London: Verso.
An Ecomodernist Manifesto. (n.d.) https://www.ecomodernism.org/manifesto-english.
Arias-Maldonado, Manuel. (2007). "An Imaginary Solution? The Green Defence of Deliberative Democracy." *Environmental Values* 16, no. 2: 233–52.
Aristotle. (1998). *Politics*. Trans. C. D. C. Reeve. Indianapolis: Hackett Publishing.
Arnold, Matthew. (1910). *Essays in Criticism: Third Series*. Boston: The Ball Publishing Co.
Arnold, Matthew. (2003). *Culture and Anarchy*. Project Gutenberg http://www.gutenberg.org/dirs/etext03/7cltn10.txt.
Asimov, Isaac. (1964, August 16). "Visit to the World's Fair of 2014." *New York Times* http://www.nytimes.com/books/97/03/23/lifetimes/asi-v-fair.html.
Association of Southeast Asian Nations. (n.d.). "Overview." https://asean.org/our-communities/asean-socio-cultural-community/sports/.
Atkinson, Anthony B. (2015). *Inequality: What Can Be Done?* Cambridge, Mass: Harvard University Press.
Aurelius, Marcus. (1862). *Meditations*. Trans. George Long. https://lexundria.com/m_aur_med/1/lg.
Barber, Daniel A., Lee Stickells, Daniel J. Ryan, Maren Koehler, Andrew Leach, Philip Goad, Deborah van der Plaat, Cathy Keys, Farhan Karim, and William M. Taylor. (2018). "Architecture, Environment, History: Questions and Consequences." *Architectural History Review* 22, no. 2: 249–86.
Barry, John. (2006). "Resistance is Fertile: From Environmental to Sustainability Citizenship." *Environmental Citizenship*. Ed. Andre Dobson and Derek Bell. Cambridge, Mass.: MIT Press. 21–48.
Barthes, Roland. (1997). "Of Sport and Men." Trans. Scott MacKenzie. *Canadian Journal of Film Studies/Revue canadienne d'études cinématographiques* 6, no. 2: 75–83.
Battaglio, Stephen. (2022, June 14). "Why Apple TV Has Become the Exclusive Streaming Home for Major League Soccer." *Los Angeles Times* https://www.latimes.com/entertainment-arts/business/story/2022-06-14/apple-tv-will-become-the-exclusive-streaming-home-for-major-league-soccer.
Baudrillard, Jean. (1993). *The Transparency of Evil: Essays on Extreme Phenomena*. Trans. James Benedict. London: Verso.
Bauman, Adrian E., Masamitsu Kamada, Rodrigo S. Reis, Ricardo P. Troiano, Ding Ding, Karen Milton, Niamh Murphy, and Pedro C. Hallal. (2021). "An

Evidence-Based Assessment of the Impact of the Olympic Games on Population Levels of Physical Activity." *The Lancet* 393, no. 10298: pp. 456–64.

Beachcomber Resorts & Hotels. (2024, September). *Environmental & Social Charter: Our 52 Commitments* https://www.beachcomber-hotels.com/en/environmental-responsibility.

Beck, Ulrich. (2009). *World at Risk*. Trans. Ciaran Cronin. Cambridge: Polity.

Benjamin, Walter. (2003). "On the Concept of History." *Walter Benjamin: Selected Writings Volume 4, 1938–1940*. Ed. Howard Eiland and Michael W. Jennings. Cambridge, Mass.: Harvard University Press. 389–400.

Benton, Ted, ed. (1996). *The Greening of Marxism*. New York: Guilford.

Berik, Günseli and Ebru Kongar, eds. (2021). *The Routledge Handbook of Feminist Economics*. Abingdon: Routledge.

Berman, Marshall. (2017). *Modernism in the Streets: A Life and Times in Essays*. Ed. David Marcus and Shellie Sclan. London: Verso.

Bertrand, Aliénor. (2019). "A Rupture Between Human Beings and Earth: A Philosophical Critical Approach to Coviability." *Coviability of Social and Ecological Systems: Reconnecting Mankind to the Biosphere in an Era of Global Change. Vol. 1: The Foundations of a New Paradigm*. Ed. Olivier Barrière, Mohammed Behnassi, Gilbert David, Vincent Douzal, Mireille Fargette, Thérèse Libourel, Maud Loireau, Laurence Pascal, Catherine Prost, Voyner Ravena-Cañete, Frédérique Seyler, and Serge Morand. Cham: Springer. 269–84.

Birley, Derek. (1995). *Playing the Game: Sport and British Society, 1910–45*. Manchester: Manchester University Press.

Bloch, Ernst. (1995). *The Principle of Hope Volume Two*. Ed. Neville Plaice, Stephen Plaice, and Paul Knight. Cambridge, Mass.: MIT Press.

Boardman, Anthony E. and Shaun P. Hargreaves-Heap. (1999). "Network Externalities and Government Restrictions on Satellite Broadcasting of Key Sporting Events." *Journal of Cultural Economics* 23, no. 3: 167–81.

Bogaards, Rodney. (2022, March 29). *Brisbane Olympic Games 2032: Will the Olympics Deliver a Net Benefit for the Australian Community?* Parliamentary Library Research Paper https://www.aph.gov.au/-/media/05_About_Parliament/54_Parliamentary_Depts/544_Parliamentary_Library/Research_Papers/2021-22/BrisbaneOlympicGames2032NetBenefit-epub.epub?la=en&hash=CA50DAC37173B3329EB354307454202BBEE4907A.

Bohrer, Ashley J. (2019). *Marxism and Intersectionality: Race, Gender, Class and Sexuality Under Contemporary Capitalism*. Bielefeld: [transcript].

Bourdieu, Pierre. (1978). "Sport and Social Class." *Social Science Information* 17, no. 6: 819–40.

Bourdieu, Pierre. (1999). "The State, Economics and Sport." Trans. Hugh Dauncey and Geoff Hare. *France and the 1998 World Cup: The National Impact of a World Sporting Event*. Ed. Hugh Dauncey and Geoff Hare. London: Frank Cass. 15–21.

Bowen, Howard R. (2013). *Social Responsibilities of the Businessman.* Iowa City: University of Iowa Press.

Bradbury, John Charles, Dennis Coates, and Brad R. Humphreys. (2023). "Public Policy Toward Professional Sports Stadiums: A Review." *Journal of Policy Analysis and Management* 43, no. 3: 899–937.

Brentin, Dario and Laurence Cooley. (2015). "Sport and Nationalism: A Review of the Literature." *Studies on National Movements* 3 http://snm.nise.eu/index.php/studies/article/view/0306s.

Brevini, Benedetta and Justin Lewis, eds. (2018). *Climate Change and the Media.* New York: Peter Lang.

Brittle, Cian. (2024, February 29). "F1 Revenue Soars 25% YoY to US$3.2bn." *SportsPro* https://www.sportspro.com/finance-investment/finances/f1-2023-revenue-accounts-financial-year-liberty-media/.

Brown, Wendy. (2019). *In the Ruins of Neoliberalism: The Rise of Anti-Democratic Politics in the West.* New York: Columbia University Press.

Bulbeck, Chilla. (1998). *Re-Orienting Western Feminisms: Women's Diversity in a Postcolonial World.* Cambridge: Cambridge University Press.

Burke, Peter. (2004). *Languages and Communities in Early Modern Europe: The 2002 Wiles Lectures Given at Queen's University, Belfast.* Cambridge: Cambridge University Press.

Carbon Literacy Project. (n.d.). *Sport* https://carbonliteracy.com/sport/#:~:text=The%20global%20sports%20industry%20is,the%20'Race%20to%20Zero'.

Carbon Trust. (2013, August 9). "Carbon Trust Unlaces the 'Carbon Bootprint' of Watching Football." https://www.carbontrust.com/news-and-insights/news/carbon-trust-unlaces-the-carbon-bootprint-of-watching-football.

Carbon Trust. (2016). *The "Carbon Bootprint" of Euro 2016: Which Nation's Fans Have the Lowest Carbon Footprint When Watching Games* https://www.carbontrust.com/news/2016/06/the-carbon-bootprint-of-euro-2016-which-nations-fans-have-the-lowest-carbon-footprint-when-watching-games/.

Carbon Trust. (n.d.). *Carbon Bootprint of the FA Community Shield* https://www.carbontrust.com/media/360767/carbon-bootprint-infographic.pdf.

Carlini, Joan, Adele Pavlidis Alana Thomson, and Clare Morrison. (2025). "Delivering on Social Good—Corporate Social Responsibility and Professional Sport: A Systematic Quantitative Literature Review." *Journal of Strategic Marketing* 33, no. 3: 323–36.

Carmichael, Andy. (2022). "Time for Practice; Sport and the Environment." *Managing Sport and Leisure* 27, no. 3: 189–98.

Catton, William R., Jr., and Riley E. Dunlap. (1978). "Environmental Sociology: A New Paradigm." *American Sociologist* 13, no. 1: 41–49.

Chartier, Roger. (1989). "Texts, Printings, Readings." *The New Cultural History.* Ed. Lynn Hunt. Berkeley: University of California Press. 154–75.

Chartier, Roger. (2005). "Crossing Borders in Early Modern Europe: Sociology of Texts and Literature." Trans. Maurice Elton. *Book History* 8: 37–50.

Committee for Economic Development. (n.d.). *Climate, Energy & Nature* https://www.conference-board.org/us/committee-economic-development/climate-energy-environment.

Commonwealth of Virginia, Joint Legislative Audit and Review Commission. (2024). *Data Centers in Virginia: Report to the Governor and the General Assembly of Virginia Commission Draft* https://jlarc.virginia.gov/pdfs/reports/Rpt598-2.pdf.

Connell, Raewyn. (2020). *Southern Theory: The Global Dynamics of Knowledge in Social Science.* Abingdon: Routledge.

Constitución de Ecuador. (2008). http://www.asambleanacional.gob.ec/documentos/constituciondebolsillo.pdf.

Cooley, Charles Horton. (1910). *Social Organization: A Study of the Larger Mind.* New York: Charles Scribner's Sons.

Corlett, Eva. (2025, January 30). "Taranki Mounga: New Zealand Mountain Granted Same Legal Rights as a Person." *Guardian* https://www.theguardian.com/world/2025/jan/30/taranaki-mounga-new-zealands-second-highest-mountain-granted-same-legal-rights-as-a-person.

Coulibaly, Thierry Yerema, Mihoko Tegawa Wakamatsu, and Shunsuke Managi. (2022). "The Use of Geographically Weighted Regression to Improve Information from Satellite Night Light Data in Evaluating the Economic Effects of the 2010 FIFA World Cup." *Area Development and Policy* 7, no. 4: 463–81.

Croci, Osvaldo and Julian Ammirante. (1999). "Soccer in the Age of Globalization." *Peace Review* 11, no. 4: 499–504.

Crook, Clive. (2005, January 22). "The Good Company." *Economist*: Survey 3–4.

Cummins, Steven, Charlotte Clark, Daniel Lewis, Neil Smith, Claire Thompson, Melanie Smuk, Stephen Stansfeld, Stephanie Taylor, Amanda Fahy, Trish Greenhalgh, and Sandra Eldridge. (2018). "The Effects of the London 2012 Olympics and Related Urban Regeneration on Physical and Mental Health: The ORiEL Mixed-Methods Evaluation of a Natural Experiment." *Public Health Research* 6, no. 12 https://www.ncbi.nlm.nih.gov/books/NBK532750/pdf/Bookshelf_NBK532750.pdf.

Cupples, Julie and Ramón Grosfoguel, eds. (2018). *Unsettling Eurocentrism in the Westernized University.* New York: Routledge.

Curry, Patrick. (2006). *Ecological Ethics: An Introduction.* Cambridge: Polity Press.

Das, Leah. (2022, April 12). "Greenwash: What It is and How Not to Fall for It." Greenpeace https://www.greenpeace.org.uk/news/what-is-greenwashing/.

Davis, Whitney. (2017). "Visuality and Vision: Questions for a Post-Culturalist Art History." *Estetika: The Central European Journal of Aesthetics* 54, no. 2: 238–57.

d'Eaubonne, Françoise. (1978). *Écologie/féminisme: Révolution ou mutation?* Paris: Éditions ATP.

Debord, Guy. (2004). *The Society of the Spectacle.* Trans. Ken Knabb. London: Rebel Press.

de Coubertin, Pierre. (1913). *Essais de psychologie sportive* https://fr.wikisource.org/wiki/Essais_de_psychologie_sportive/Texte_entier.

Delaney, Tim. (2015). "The Functionalist Perspective on Sport." *Routledge Handbook of the Sociology of Sport.* Ed. Richard Giulianotti. London: Routledge. 18–28.

Delmas, Magali A. and Vanessa Cuerel Burbano. (2011). "The Drivers of Greenwashing." *California Management Review* 54, no. 1: 64–87.

Demosthenes. (2015). *The Complete Works of Demosthenes.* Hastings: Delphi Classics.

Dos Passos, John. (1937). *U.S.A.: I. The 42nd Parallel. II. Nineteen Nineteen. III. The Big Money.* New York: Modern Library.

Dowling, Matthew and Spencer Harris. (2022, July 26). "London 2012 Olympics: How it Boosted Medal Winning but Failed to Inspire a Generation." *The Conversation* https://theconversation.com/london-2012-olympics-how-it-boosted-medal-winning-but-failed-to-inspire-a-generation-187383.

Dowling, Matthew. (2025). "London 2012: Did it Inspire a Generation?" *Events and Society: Bridging Theory and Practice.* Ed. Mike Duignan. Abingdon: Routledge. 64–72.

Du Bois, W. E. B. (2022). *W. E. B. Du Bois: International Thought.* Ed. Adom Getachew and Jennifer Pitts. Cambridge: Cambridge University Press.

Dunne, Daisy and Patrick Greenfield. (2024, October 15). *COP16: More Than 85% of Countries Miss UN Deadline to Submit Nature Pledges* https://www.carbonbrief.org/cop16-countries-miss-un-deadline-to-submit-nature-pledges/.

Durkheim, Émile. (1961). "The Solidarity of Occupational Groups." *Theories of Society: Foundations of Modern Sociological Theory. Volume I.* Ed. Talcott Parsons, Edward Shils, Kaspar D. Naegle, and Jesse R. Pitts. New York: Free Press of Glencoe. 356–63.

Eagleton, Terry. (2007). *The Meaning of Life: A Very Short Introduction.* Oxford: Oxford University Press.

Eagleton, Terry. (2010, June 15). "Football: A Dear Friend to Capitalism." *Guardian* https://www.theguardian.com/commentisfree/2010/jun/15/football-socialism-crack-cocaine-people.

Eco, Umberto. (1981). *The Role of the Reader: Explorations in the Semiotics of Texts.* London: Hutchinson.

Eco, Umberto. (1987). *Travels in Hyperreality.* Trans. William Weaver. London: Picador.

Eichberg, Henning. (1986). "The Enclosure of the Body—On the Historical Relativity of 'Health', 'Nature' and the Environment of Sport." *Journal of Contemporary History* 21, no. 1: 99–121.

Elias, Norbert and Eric Dunning. (1986). *Quest for Excitement: Sport and Leisure in the Civilizing Process.* Oxford: Basil Blackwell.

Elias, Norbert. (1978). "On Transformations of Aggressiveness." *Theory and Society* 5, no. 2: 229–42.

Ellul, Jacques. (1964). *The Technological Society.* Trans. John Wilkinson. New York: Vintage Books.

Emerson, Ralph Waldo. (2000). *The Essential Writings of Ralph Waldo Emerson.* Ed. Brooks Atkinson. New York: The Modern Library.

Engels, Friedrich. (2010). *The Origin of the Family, Private Property and the State.* London: Penguin.

EPRI. (2024, May 28). *Powering Intelligence: Analyzing Artificial Intelligence and Data Center Energy Consumption* https://www.epri.com/research/products/000000003002028905.

European Banking Authority, European Insurance and Occupational Pensions Authority, and European Securities and Markets Authority. (2022). *ESAs Call for Evidence on Better Understanding Greenwashing* https://www.esma.europa.eu/sites/default/files/library/esas_call_for_evidence_on_greenwashing.pdf

Fair, Leslie. (2022, April 8). "$5.5 Million Total FTC Settlements with Kohl's and Walmart Challenge "Bamboo" and Eco Claims, Shed Light on Penalty Offense Enforcement." Federal Trade Commission https://www.ftc.gov/business-guidance/blog/2022/04/55-million-total-ftc-settlements-kohls-and-walmart-challenge-bamboo-and-eco-claims-shed-light.

Fals-Borda, Orlando. (1988). *Knowledge and People's Power: Lessons with Peasants in Nicaragua, Mexico and Colombia.* New Delhi: Indian Social Institute; New York: New Horizons Press.

Fédération Internationale de l'Automobile. (2023). "FIA Purpose Driven." https://purposedriven.fia.com/page/frontpage.

FIFA. (n.d.). *Football Unites the World: Strategic Objectives for the Global Game: 2023–2027* https://digitalhub.fifa.com/m/38577b5ce4c3cc23/original/FIFA-Strategic-Objectives-for-the-Global-Game_2023_2027_EN.pdf.

FIFA. (2019). *2018 FIFA World Cup Russia™ Sustainability Report* https://img.fifa.com/image/upload/ya7pgcyslxpzlqmjkykg.pdf.

Fifka, Matthias and Johannes Jaeger. (2020). "CSR in Professional European Football: An Integrative Framework." *Soccer & Society* 21, no. 1: 61–78.

Fisk, Milton. (1980). "Materialism and Dialectic." *Critique: Journal of Socialist Theory* 12, no. 1: 97–116.

Flynn, Henry. (2022, November 28). "Qatar Has Spent Well Over $200 Billion On A Flawed World Cup." *Forbes* https://www.forbes.com/sites/henryflynn/2022/11/27/qatar-has-spent-well-over-200-billion-on-a-flawed-world-cup/.

Flynn, Jack. (2023, November 14). "20+ Shocking Greenwashing Statistics [2023]." *Zippia* https://www.zippia.com/advice/greenwashing-statistics/.

Földesi, Gyöngyi Szabó. (2014). "The Impact of the Global Economic Crisis on Sport." *Physical Culture and Sport* 63, no. 1: 22–30.

Formula 1. (2022). "More Efficient, Less Fuel, and Carbon Net Zero—7 Things You Need to Know about the 2026 F1 Engine Regulations." https://www.formula1.com/en/latest/article.more-efficient-less-fuel-and-carbon-net-zero-7-things-you-need-to-know-about.ZhtzvU3cPCv8QO7jtFxQR.html.

Forrest, Christopher B., Lauren J. Koenigsberg, Francis Eddy Harvey, Mitchell G. Maltenfort, and Neal Halfon. (2025). "Trends in US Children's Mortality, Chronic Conditions, Obesity, Functional Status, and Symptoms." *Journal of the American Medical Association* https://doi.org/10.1001/jama.2025.9855.

Foster, John Bellamy. (1999). "Marx's Theory of Metabolic Rift: Classical Foundations for Environmental Sociology." *American Journal of Sociology* 105, no. 2: 366–405.

Foucault, Michel. (1986). *The Use of Pleasure: The History of Sexuality Volume Two*. Trans. Robert Hurley. New York: Vintage.

Foucault, Michel. (1988). *The Care of the Self: Volume 3 of The History of Sexuality*. Trans. Robert Hurley. New York: Vintage.

Foucault, Michel. (2021). *Confessions of the Flesh: The History of Sexuality Volume 4*. Ed. Frédéric Gros. Trans. Robert Hurley. New York: Pantheon.

Foxhall, Lin. (1998). "Pandora Unbound: A Feminist Critique of Foucault's *History of Sexuality*." *Rethinking Sexuality: Foucault and Classical Antiquity*. Ed. David H. Larmour, Paul Allen Miller, and Charles Platter. Princeton: Princeton University Press. 122–38.

Foxman, Simone and Adveith Nair. (2022, November 18). "What Qatar Built for the Most Expensive World Cup Ever." *Bloomberg* https://www.bloomberg.com/graphics/2022-what-qatar-built-for-the-world-cup/?leadSource=uverify%20wall&embedded-checkout=true.

FRA European Union Agency for Fundamental Rights. (2024, March 7). *Enforcing Consumer Rights to Combat Greenwashing* https://fra.europa.eu/en/publication/2024/enforcing-consumer-rights-combat-greenwashing?page=2#read-online.

Francis, Annes Elsa, Matthew Webb, Cheryl Desha, Sharyn Rundle-Thiele, and Savindi Caldera. (2023). "Environmental Sustainability in Stadium Design and Construction: A Systematic Literature Review." *Sustainability* 15, no. 8 https://doi.org/10.3390/su15086896.

Fresh Info. (2021). *36th America's Cup Impact Evaluation: Final Report* https://www.majorevents.govt.nz/dmsdocument/15674-36th-americas-cup-impact-evaluation-report.

Friends of the Earth. (2024, January). *From False Promises to Real Solutions: How to Navigate Greenwashing by Corporations and Governments* https://friendsoftheearth.eu/wp-content/uploads/2024/02/FRI-23-False-Promises_WEB-corrected.pdf.

Frow, John. (2019). *On Interpretive Conflict*. Chicago: University of Chicago Press.

Galeano, Eduardo. (2016, September 16). "Fútbol a sol y a sombra." *Don Patadon* https://www.don-patadon.com/2015/02/futbol-sol-y-sombra-de-eduardo-galeano.html.

Gandhi, M. K. (1943). *Conquest of Self (Being Gleanings from His Writings and Speeches*. Comp. R. K. Prabhu and U. R. Rao. Bombay [Mumbai]: Thacker & Co.

Gelles, David. (2015, October 17). "Social Responsibility That Rubs Right Off." *New York Times* https://www.nytimes.com/2015/10/18/business/energy-environment/social-responsibility-that-rubs-right-off.html.

Gellner, Ernest. (1988). *Plough, Sword and Book: The Structure of Human History*. Chicago: University of Chicago Press.

Gelmini, Stefano. (2021, June 29). "We're Living in a Golden Age of Greenwash." Greenpeace https://www.greenpeace.org.uk/news/golden-age-of-greenwash/.

Genovese, Jack. (2024, October 11). "Why Do US Sports Media Rights Keep Climbing?" *SportsPro* https://www.sportspromedia.com/insights/opinions/sports-tv-rights-values-usa-nba-nfl-cfp-nhl-ampere-analysis-data/.

Germano, Sara, Anna Nicolaou, and Hannah Murphy. (2022, December 22). "Google Agrees NFL Streaming Deal as Big Tech Chases Sports Rights." *Irish Times* https://www.irishtimes.com/business/2022/12/22/google-agrees-nfl-streaming-deal-as-big-tech-chases-sports-rights/.

Gibbs, Carole, Meredith L. Gore, Edmund F. McGarrell, and Louie Rivers III. (2010). "Introducing Conservation Criminology: Towards Interdisciplinary Scholarship on Environmental Crimes and Risks." *British Journal of Criminology* 50, no. 1: 124–44.

Gibson-Graham, J. K. (2006a). *The End of Capitalism (As We Knew It): A Feminist Critique of Political Economy*. Minneapolis: University of Minnesota Press.

Gibson-Graham, J. K. (2006b). *A Postcapitalist Politics*. Minneapolis: University of Minnesota Press.

Global Web Index. (2019). *Sports Around the World* https://iccopr.com/wp-content/uploads/2019/03/Sports-Around-the-World-report.pdf.

Goffman, Erving. (1986). *Stigma: Notes on Management of Spoiled Identity*. New York: Touchstone.

Gogishvili, David and Martin Müller. (2025). "Mega-Events and the Climate: What's the Game?" *Current Issues in Sport Science* 10, no. 2 https://doi.org/10.36950/2025.2ciss074.

Goldman, Michael and Rachel A. Schurman. (2000). "Closing the "Great Divide": New Social Theory on Society and Nature." *Annual Review of Sociology* 26: 563–84.

Google. (2023). *Environmental Report 2023* https://www.gstatic.com/gumdrop/sustainability/google-2023-environmental-report.pdf.
Gorky, Maxim. (1952). *Collected Works in Ten Volumes. Volume VIII: The Artamonovs.* Trans. Helen Altschuler. Moscow: Progress Publishers.
Gough, Christina. (2023, March 10). "Sports on U.S. TV—Statistics & Facts." *Statista* https://www.statista.com/topics/2113/sports-on-tv/#topicOverview.
Grant, Thomas L, Jr. (2014). "Green Monsters: Examining the Environmental Impact of Sports Stadiums." *Environmental Law Journal* 25, no. 1: 149–76.
Groves, Richard H. (1995). *Green Imperialism: Colonial Expansion, Tropical Island Edens and the Origins of Environmentalism 1600–1860.* Cambridge: Cambridge University Press.
Gruneau, Richard and John Horne. (2016). "Mega-Events and Globalization: A Critical Introduction." *Mega-Events and Globalization: Capital and Spectacle in a Changing World Order.* Ed. Richard Gruneau and John Horne. London: Routledge. 1–28.
Guattari, Pierre-Félix. (1996). *The Guattari Reader.* Ed. Gary Genosko. Oxford: Blackwell.
Hall, Stuart. (2011). "The Neo-Liberal Revolution." *Cultural Studies* 25, no. 6: 705–28.
Hall, Stuart. (2016). *Cultural Studies 1983: A Theoretical History.* Ed. Jennifer Daryl Slack and Lawrence Grossberg. Durham: Duke University Press.
Hardin, Garrett. (1968). "The Tragedy of the Commons." *Science* 162, no. 3859: 1243–248.
Harris, Marvin. (1979). *Cultural Materialism: The Struggle for a Science of Culture.* New York: Random House.
Harte, Tim. (2020). *Faster, Higher, Stronger, Comrades! Sport, Art, & Ideology in Late Russian and Early Soviet Culture.* Madison: University of Wisconsin Press.
Harney, Stefano. (2006). *Nationalism and Identity: Culture and the Imagination in a Caribbean Diaspora.* Jamaica: University of the West Indies Press.
Hegel, Georg Wilhelm Friedrich. (1988). *Lectures on the Philosophy of World History. Introduction: Reason in History.* Trans. H. B. Nisbet. Cambridge: Cambridge University Press.
Heller, Agnes. (1979). *On Instincts.* Trans. Mario Fenyö. Assen: Van Gorcum.
Hemphill, Thomas A. (2004). "Corporate Citizenship: The Case for a New Corporate Governance Model." *Business and Society Review* 109, no. 3: 339–61.
Herder, Johann Gottfried von. (2002). *Philosophical Writings.* Trans. and Ed. Michael N. Forster. Cambridge: Cambridge University Press.
Hobsbawm, Eric and Terence Ranger, eds. (2002). *La invención de la tradición.* Trans. Omar Rodríguez. Barcelona: Editorial Crítica.
Hoggart, Richard. (1965). "The World We Offer." *Health Education Journal* 23: 34–40.
Homer. (1967). *The Odyssey.* Trans. Albert Cook. New York: WW Norton.

Horkheimer, Max. (2012). *Critique of Instrumental Reason.* Trans. Matthew J. O'Connell *et al.* London: Verso.

Horky, Thomas and Jörg-Uwe Nieland, eds. (2013). *International Sports Press Survey 2011 (Sport & Communication Volume 5).* Hamburg: Horky Sport & Communication.

Houlihan, Barrie. (1997). *Sport, Policy and Politics: A Comparative Analysis.* London: Routledge.

House of Lords Select Committee on a National Plan for Sport and Recreation. (2021, December). *A National Plan for Sport, Health and Wellbeing* https://committees.parliament.uk/publications/8094/documents/84813/default/.

Housman, A. E. (1919). *A Shropshire Lad* https://www.gutenberg.org/ebooks/5720.

Hunt, Jeremy. (2010, June 28). "Sports Keynote Speech" https://www.gov.uk/government/speeches/sports-keynote-speech.

Hutchins, Brett, Bo Li, and David Rowe. (2019). "Over-the-Top Sport: Live Streaming Services, Changing Coverage Rights Markets, and the Growth of Media Sport Portals." *Media, Culture & Society* 41, no. 7: 975–94.

Intergovernmental Panel on Climate Change. (2022). *Climate Change 2022: Mitigation of Climate Change* https://www.ipcc.ch/report/ar6/wg3/downloads/report/IPCC_AR6_WGIII_FullReport.pdf.

International Energy Agency. (2014). *More Data, Less Energy: Making Network Standby More Efficient in Billions of Connected Devices* https://www.iea.org/publications/freepublications/publication/MoreDataLessEnergy.pdf.

International Energy Agency. (2024). *World Energy Outlook 2024* https://iea.blob.core.windows.net/assets/47a9a222-78e4-4c43-9bab-977b4ad5326b/WorldEnergyOutlook2024.pdf.

International Financial Reporting Standards. (2024, June). *Feedback Statement IFRS® Sustainability Disclosure Standards* https://www.ifrs.org/content/dam/ifrs/project/issb-consultation-on-agenda-priorities/agenda-consultation-feedback-statement-june-2024.pdf.

International Olympic Committee. (2018). *Carbon Footprint Methodology for the Olympic Games*https://library.olympic.org/Default/doc/SYRACUSE/184686/carbon-footprint-methodology-for-the-olympic-games-international-olympic-committee?lg=fr-FR.

International Olympic Committee. (2024, July 22). *Olympic Truce* https://www.olympics.com/ioc/olympic-truce.

International Organization of Securities Commissions. (2023, December). *Supervisory Practices to Address Greenwashing: Final Report* https://www.iosco.org/library/pubdocs/pdf/IOSCOPD750.pdf.

Irigaray, Luce, ed. (2004). *Luce Irigaray: Key Writings.* New York: Continuum.

Isocrates. (1929). *Isocrates in Three Volumes. II.* Trans. George Norlin. London: William Heinemann/New York: GP Putnam's Sons.

Isocrates. (1945). *Isocrates in Three Volumes. III.* Trans. Larue Van Hook. Cambridge, Mass.: Harvard University Press/London: William Heinemann.

Ivanov, Denis and Gaygysyz Ashyrov. (2024). "Wealth Inequality and Mega Events." *European Journal of Political Economy* 85 https://doi.org/10.1016/j.ejpoleco.2024.102615.

James I and Charles I. (1914). "The King's Majesty's Declaration to His Subjects Concerning Lawful Sports to be Used." *Documents Illustrative of English Church History.* Ed. Henry Gee and William John Hardy. New York: Macmillan. 528–32.

Joseph, May. (1995). "Diaspora, New Hybrid Identities, and the Performance of Citizenship." *Women and Performance* 7, no. 2/8, no. 1: 3–13.

Kafka, Peter. (2021, March 18). "Amazon's $10 Billion NFL Deal is Huge—And a Sign That Sports is Staying on TV for a While Longer." *Vox* https://www.vox.com/recode/22338794/amazon-nfl-10-billion-tv-analysis.

Kant, Immanuel. (2006). *Anthropology from a Pragmatic Point of View.* Trans. and Ed. Robert B. Louden. Cambridge: Cambridge University Press.

Kraus, Hans and Ruth P. Hirschland. (1953). "Muscular Fitness and Health." *Journal of the American Association for Health, Physical Education, and Recreation* 24, no. 10: 17–19.

Krauze, Enrique. (1994, June 28). "México en un balón." *El País* https://elpais.com/diario/1994/06/28/opinion/772754409850215.html.

Kristeva, Julia. (2024). *Revolution in Poetic Language.* Trans. Margaret Waller. New York: Columbia University Press.

L, Jennifer. (2024, February 9). "Beyond Touchdowns and Trophies: Unveiling the Carbon Footprint of Superbowl LVIII." *Carbon Credits* https://carboncredits.com/beyond-touchdowns-and-trophies-unveiling-the-carbon-footprint-of-superbowl-lviii/.

Lasch, Christopher. (1991). *The Culture of Narcissism: American Life in an Age of Diminishing Expectations.* New York: WW Norton.

Latour, Bruno, Pablo Jensen, Tommaso Venturini, Sébastian Grauwin, and Dominique Boullier. (2012). "The Whole is Always Smaller Than its Parts: A Digital Test of Gabriel Tarde's Monads." *British Journal of Sociology* 63, no. 4: 590–615.

Latta, P. Alex. (2007). "Locating Democratic Politics in Ecological Citizenship." *Environmental Politics* 16, no. 3: 377–93.

Lawrence, Geoffrey and David Rowe. (1986). "Towards a Sociology of Australian Sport." *Australian Sport. Essays in the Sociology of Australian Sport.* Ed. Geoffrey Lawrence and David Rowe. Sydney: Hale & Iremonger. 13–45.

Lazarsfeld, Paul F. (1941). "Remarks on Administrative and Critical Communications Research." *Studies in Philosophy and Social Science* 9, no. 1: 2–16.

Lehne, Johanna and Felix Preston. (2018). *Making Concrete Change: Innovation in Low-Carbon Cement and Concrete*. Chatham House https://www.chathamhouse.org/sites/default/files/publications/2018-06-13-making-concrete--change-cement-lehne-preston-final.pdf.

Leighton, Neil. (2005, June 3). "Old Shin Sport Alive and Kicking." *BBC* https://news.bbc.co.uk/2/hi/uk_news/england/gloucestershire/4605157.stm.

Levermore, Roger. (2010). "CSR for Development Through Sport: Examining its Potential and Limitations." *Third World Quarterly* 31, no. 2: 223–41.

Lewandowsky, Stephan, Naomi Oreskes, James S. Risbey, Ben R. Newell, and Michael Smithson. (2015). "Seepage: Climate Change Denial and its Effect on the Scientific Community." *Global Environmental Change* 33: 1–13.

Lindwall, Courtney. (2023, February 9). "What is Greenwashing?" NRDC https://www.nrdc.org/stories/what-greenwashing.

Lins Ribeiro, Gustavo and Arturo Escobar, eds. (2006). *World Anthropologies: Disciplinary Transformations Within Systems of Power*. Oxford: Berg.

Logsdon, Jeanne M. (2004). "Global Business Citizenship: Applications to Environmental Issues." *Business and Society Review* 109, no. 1: 67–87.

Loy, John W. and Douglas Booth. (2000). "Functionalism, Sport and Society." *Handbook of Sports Studies*. Ed. Jay Coakley and Eric Dunning. London: Sage Publications. 8–27.

Luxemburg, Rosa. (1970). *Rosa Luxemburg Speaks*. Ed. Mary-Alice Waters. New York: Pathfinder Press.

Luxemburg, Rosa. (2004). *The Rosa Luxemburg Reader*. Ed. Peter Hudis and Kevin B. Anderson. New York: Monthly Review Press.

Lynd, Robert S. and Helen Merrell Lynd. (1956). *Middletown: A Study in Modern American Culture*. New York: Harcourt, Brace and Company.

Lynd, Robert S. and Helen Merrell Lynd. (1965). *Middletown in Transition: A Study in Cultural Conflicts*. New York: Harcourt Brace Jovanovich.

MacGregor, Sherilyn. (2006). *Beyond Mothering Earth: Ecological Citizenship and the Politics of Care*. Vancouver: University of British Columbia Press.

Maguire, Joseph. (1993). "Bodies, Sportscultures and Societies: A Critical Review of Some Theories in the Sociology of the Body." *International Review for the Sociology of Sport* 28, no. 1: 33–52.

Mangan, J. A. (1981). *Athleticism in the Victorian and Edwardian Public School: The Emergence and Consolidation of an Educational Ideology*. Cambridge: Cambridge University Press.

Marcus, George. (1995). "Ethnography in/of the World System: The Emergence of Multi-Sited Ethnography." *Annual Review of Anthropology* 24: 95–117.

Marcuse, Herbert. (2009). *Negations: Essays in Critical Theory*. Trans. Jeremy J. Shapiro. London: MayFlyBooks.

Marire, Juniours. (2022). "Effect of Government Spending on Total Factor Productivity in South Africa." *European Journal of Economics, Law and Social Sciences* 6, no. 1: 435–52.

Marquis, Christopher, Michael W. Toffel, and Yanhua Zhou. (2016). "Scrutiny, Norms, and Selective Disclosure: A Global Study of Greenwashing." *Organization Science* 27, no. 2: 483–504.

Marqusee, Mike. (2012, August 7). "London 2012: Spare Us the Jingoistic Olympic Hype." *Guardian* https://www.theguardian.com/commentisfree/2012/aug/07/london-2012-olympic-hype.

Marshall, Alasdair, Richard Telofski, Udechukwu Ojiako, and Max Chipulu. (2012). "An Examination of "Irregular Competition" Between Corporations and NGOs." *Voluntas* 23, no. 2: 371–91.

Marshall, T. H. (1950). *Citizenship and Social Class and Other Essays.* Cambridge: Cambridge University Press.

Martínez-Alier, Joan. (2012). "Environmental Justice and Economic Degrowth: An Alliance Between Two Movements." *Capitalism Nature Socialism* 23, no. 1: 51–73.

Matras, Tomasz. (2024). "'Less Democracy Is Sometimes Better': Sportswashing and Hosting World Championships by Non-Democratic Countries, 1993–2022." *International Journal of the History of Sport* 41, no. 6: 531–48.

Maxwell, Richard, Jon Raundalen, and Nina Lager Vestberg, eds. (2015). *Media and the Ecological Crisis.* New York: Routledge.

McBride, James, Noah Berman, and Melissa Manno. (2024, July 20). "The Economics of Hosting the Olympic Games." *Council on Foreign Relations* https://www.cfr.org/backgrounder/economics-hosting-olympic-games.

McCartney, Gerry, Sian Thomas, Hilary Thomson, John Scott, Val Hamilton, Phil Hanlon, David S. Morrison, and Lyndal Bond. (2010). "The Health and Socioeconomic Impacts of Major Multi-Sport Events: A Systematic Review." *British Medical Journal* 340 https://doi.org/10.1136/bmj.c2369.

McKay, Jim. (2015). "Assessing the Sociology of Sport: On Revisiting the Sociological Imagination." *International Review for the Sociology of Sport* 50, nos. 4–5: 547–52.

McMichael, Philip. (2011). "Revaluing Low Carbon Cultures: Learning to Live with the Earth." *Development* 54, no. 2: 169–71.

Meghji, Ali. (2021). *Decolonizing Sociology.* Cambridge: Polity Press.

Merton, Robert K. (1968). *Social Theory and Social Structure.* New York: The Free Press.

Microsoft. (2022). *2022 Environmental Sustainability Report* https://query.prod.cms.rt.microsoft.com/cms/api/am/binary/RW15mgm.

Miller, Toby. (2007). *Cultural Citizenship: Cosmopolitanism, Consumerism, and Television in a Neoliberal Age.* Philadelphia: Temple University Press.

Miller, Toby. (2019). "La crisis ambiental: El Marxismo contra el antropocentrismo filosófico." *Das Kapital. Marx, actualidad y crítica*. Ed. Paulina Aroch Fugellie, Enrique Gallegos, Miriam Madureira, and Felipe Victoriano. Mexico City: Siglo Veintiuno/Universidad Autónoma Metropolitana, Unidad Cuajimalpa. 200–14.

Milliken, Robert. (1996, October 15). "Sport is Murdoch's 'Battering Ram' for Pay TV." *Independent* https://www.independent.co.uk/sport/sport-is-murdoch-s-battering-ram-for-pay-tv-1358686.html.

Mills, C. Wright. (2000). *The Sociological Imagination*. New York: Oxford University Press.

Mills, Mark P. (2013). *The Cloud Begins with Coal: Big Data, Big Networks, Big Infrastructure, and Big Power*. National Mining Association and American Coalition for Clean Coal Electricity https://www.tech-pundit.com/wpcontent/uploads/2013/07/CloudBeginsWithCoal.pdf?c761ac.

Mitchell, Kevin. (2024, November 20). "Farewell, Rafael Nadal: It Was My Pleasure to Know One of Sport's Most Gracious Champions." *Guardian* https://www.theguardian.com/sport/blog/2024/nov/20/farewell-rafael-nadal-one-of-sports-most-gracious-champions-tennis.

Montaigne, Michel de. (1966). *The Complete Essays of Montaigne*. Trans. Donald M. Frame. Stanford: Stanford University Press.

Moore, Dan. (2024, May 8). "Taxpayers Are About to Subsidize a Lot More Sports Stadiums." *The Atlantic* https://www.theatlantic.com/ideas/archive/2024/05/sports-stadium-subsidies-taxpayer-funding/678319/.

Motavalli, Jim. (2011, February 12). "A History of Greenwashing: How Dirty Towels Impacted the Green Movement." *AOL News* https://www.aol.com/article/2011/02/12/the-history-of-greenwashing-how-dirty-towels-impacted-the-green/19628686/?gen=1.

Müller, Martin, David Gogishvili, Sven Daniel Wolfe, Christopher Gaffney, Miriam Hug, and Annick Leick. (2023). "Peak Event: The Rise, Crisis and Potential Decline of the Olympic Games and the World Cup." *Tourism Management* 95 https://doi.org/10.1016/j.tourman.2022.104657.

Müller, Martin, Sven Daniel Wolfe, David Gogishvili, Christopher Gaffney, Miriam Hug, and Annick Leick. (2022a). "The Mega-Events Database: Systematising the Evidence on Mega-Event Outcomes." *Leisure Studies* 41, no. 3: 437–45.

Müller, Martin, David Gogishvili, and Sven Daniel Wolfe. (2022b). "The Structural Deficit of the Olympics and the World Cup: Comparing Costs Against Revenues Over Time." *Environment and Planning A: Economy and Space* 54, no. 6: 1200–218.

Musto, Julia. (2024, October 29). "Every Email, TikTok and Text We Send is Killing the Environment. Here's How." *Independent* https://www.independent.co.uk/climate-change/email-tiktok-environment-climate-change-b2637645.html.

Nader, Laura. (1972). "Up the Anthropologist—Perspectives Gained from Studying Up." *Reinventing Anthropology*. Ed. Dell H. Hymes. New York: Pantheon. 284–311.

Nandy, Ashis. (1998). "The Twilight of Certitudes: Secularism, Hindu Nationalism, and Other Masks of Deculturation." *Postcolonial Studies* 1, no. 3: 283–98.

Nash, June and Maria Patricia Fernandez, eds. (1983). *Women, Men and the International Division of Labor*. Albany: SUNY Press.

National Audit Office. (2022). *Grassroots Participation in Sport and Physical Activity* https://www.nao.org.uk/wp-content/uploads/2022/07/Grassroots-participation-in-sport-and-physical-activity.pdf.

Neeson, J. M. (1996). *Commoners: Common Right, Enclosure and Social Change in England, 1700–1829*. Cambridge: Cambridge University Press.

Newman, Michael. (2005). *Socialism: A Very Short Introduction*. Oxford: Oxford University Press.

New Weather Institute. (2024, September 18). *Dirty Money: How Fossil Fuel Sponsors Are Polluting Sport* https://www.newweather.org/wp-content/uploads/2024/09/Dirty-Money-How-Fossil-Fuel-Sponors-Pollute-Sport-18-09-2024.pdf.

Nixon, Robert. (2011). *Slow Violence and the Environmentalism of the Poor*. Cambridge, Mass: Harvard University Press.

Nunn, Samuel and Mark S. Rosentraub. (1997). "Sports Wars: Suburbs and Center Cities in a Zero-Sum Game." *Journal of Sport & Social Issues* 21, no. 1: 65–82.

Nussbaum, Martha C. (2006). *Frontiers of Justice: Disability, Nationality, Species Membership*. Cambridge, Mass: The Belknap Press of Harvard University Press.

Nussbaum, Martha C. (2013). *Political Emotions: Why Love Matters for Justice*. Cambridge, Mass.: The Belknap Press of Harvard University Press.

Nussbaum, Martha C. (2016). *Anger and Forgiveness: Resentment, Generosity, Justice*. New York: Oxford University Press.

O'Brien, Isabel. (2024, September 15). "Data Center Emissions Likely 662% Higher Than Big Tech Claims. Can it Keep Up the Ruse?" *Guardian* https://www.theguardian.com/technology/2024/sep/15/data-center-gas-emissions-tech.

O'Connor, James R. (1998). *Natural Causes: Essays in Ecological Marxism*. New York: Guilford.

Office of the Assistant Secretary for Health. (2019). *National Youth Sports Strategy* https://health.gov/sites/default/files/2019-10/National_Youth_Sports_Strategy.pdf.

Orwell, George. (1945, December). "The Sporting Spirit." *Tribune* https://orwell.ru/library/articles/spirit/english/e_spirit.

Orwell, George. (1949). *Nineteen Eighty-Four* https://files.libcom.org/files/1984.pdf.

Organisation for Economic Co-operation and Development. (2011). *Environmental Claims: Findings and Conclusions of the OECD Committee on Consumer Policy.* OECD Green Growth Papers 2011-01. https://www.oecd-ilibrary.org/docserver/5k9h3633prbq-en.pdf?expires=1733738709&id=id&accname=guest&checksum=B641F414D2BB4A02D244BFF937445B3F.

Organization of American States. (2023, March 6). "OAS and ACES Europa Promote Sport as a Tool for Social Change." https://www.oas.org/en/media_center/photonews.asp?sCodigo=FNE-123173.

Orsini, Alessandro. (2024). *Sociological Theory: From Comte to Postcolonialism.* Cham: Palgrave Macmillan.

Ostrom, Elinor. (2000). "Reformulating the Commons." *Swiss Political Science Review* 6, no. 1: 29–52.

Otani, Teinosuke. (2018). *A Guide to Marxian Political Economy: What Kind of a Social System is Capitalism?* Trans. Michael Schauerte. Cham: Springer.

Paché, Gilles. (2020). "Sustainability Challenges in Professional Football: The Destructive Effects of the Society of the Spectacle." *Journal of Sustainable Development* 13, no. 1: 85–96.

Parsons, Talcott and Robert F. Bales in collaboration with James Olds, Morris Zelditch, Jr., and Philip E. Slater. (1955). *Family, Socialization and Interaction Process.* Glencoe: Free Press.

Pepper, David. (2000). "Environmentalism." *Understanding Contemporary Society: Theories of the Present.* Ed. Gary Browning, Abigail Halcli, and Frank Webster. London: Sage Publications. 445–62.

Pinchot, Gifford. (1910). *The Fight for Conservation.* New York: Doubleday, Page & Company.

Plato. (1997). *Plato: The Complete Works.* Ed. John M. Cooper. Assoc. Ed. D. S. Hutchinson. Indianapolis: Hackett Publishing.

Poindexter, Owen. (2021, June 21). "Amazon's Aggressive Expansion into Sports Continues." *Front Office Sports* https://frontofficesports.com/amazon-wants-all-the-live-sports/.

Porter, Rick. (2024, January 23). "Ted Sarandos: WWE Deal Doesn't Change Netflix's Live Sports Stance." *Hollywood Reporter* https://www.hollywoodreporter.com/business/business-news/netflix-wwe-deal-doesnt-change-live-sports-strategy-1235805365/.

Preuß, Holger. (2015). "A Framework for Identifying the Legacies of a Mega Sport Event." *Leisure Studies* 34, no. 6: 643–64.

Pritchett, Henry S. (1929). "Preface: Athletics, An Element in the Evolution of the American University." *American College Athletics.* Bulletin 23. Howard J. Savage, Harold W. Bentley, John T. McGovern, and Dean F. Smiley.

New York: The Carnegie Foundation for the Advancement of Teaching. v–xxii.

PwC. (2025, March). *PwC's Second Annual State of Decarbonization Report: Essential Ingredients for Achieving Your Climate Ambition* https://www.pwc.com/us/en/services/esg/library/assets/pwc-sustainability-decarbonization-2025.pdf.

Rafferty, Max. (1970). "The Indestructible Sport." *Newton Graphic* https://archive.org/stream/athletethe196970unse/athletethe196970unse_djvu.txt.

RepRisk. (2024, October). *A Turning Tide in Greenwashing? Exploring the First DeclineinSixYears*https://www.reprisk.com/research-insights/reports/a-turning-tide-in-greenwashing-exploring-the-first-decline-in-six-years.

Reygadas, Luis. (2021). *Otros capitalismos son posibles.* Mexico City: Universidad Autónoma Metropolitana, Iztapalapa.

Richta, Radovan and a Research Team. (1969). *Civilization at the Crossroads: Social and Human Implications of the Scientific and Technological Revolution*, 3rd ed. Prague: International Arts and Sciences Press.

Said, Edward. (1993). "An Interview with Edward Said." *boundary 2* 20, no. 1: 1–25.

Salinas, Arango, Natalia Andrea. (2018). "Encrucijada de la violencia asociada al fútbol: Entre el desagrado y la complacencia." *Trabajo Social* 20, no. 1: 49–68.

Sandel, Michael J. (2005). *Public Philosophy: Essays on Morality in Politics.* Cambridge, Mass.: Harvard University Press.

Sandel, Michael J. (2007). *The Case Against Perfection: Ethics in the Age of Genetic Engineering.* Cambridge, Mass.: Harvard University Press.

Saudi Arabia. (2024). *Vision & Legacy* https://saudi2034bid.com/vision-legacy/#tournament-concept.

Scanlon, Thomas F. (2018). "Class Tensions in the Games of Homer: Epeius, Euryalus, Odysseus, and Iros." *Bulletin of the Institute of Classical Studies* 61, no. 1: 5–20.

Scanlon, Thomas F. (2021). "Overview and Approaches." *The Oxford Handbook of Sport and Spectacle in the Ancient World.* Ed. Alison Futrell and Thomas F. Scanlon. Oxford: Oxford University Press. 3–13.

Schenk, James. (n.d.). "What is the Most Popular Section in a Newspaper?" *AuthorsCast* https://authorscast.com/what-is-the-most-popular-section-in-a-newspaper.

Schumpeter, Joseph Alois. (1966). *Imperialism & Social Classes: Two Essays.* Trans. Heinz Norden. Cleveland: Meridian Books.

Sedghi, Ami, George Arnett, and Chris Moran. (2014, April 17). "*The Guardian*'s Top 100: Which Articles Have Been Most Popular?" *Guardian* https://www.theguardian.com/news/datablog/2014/apr/17/the-guardians-top-100-which-articles-have-been-most-popular.

Shehabi, Arman, Sarah Josephine Smith, Alex Hubbard, Alexander Newkirk, Nuoa Lei, Md AbuBakar Siddik, Billie Holecek, Jonathan G. Koomey, Eric R. Masanet, and Dale A. Sartor. (2024, December). *2024 United States Data Center Energy Usage Report* https://eta.lbl.gov/publications/2024-lbnl-data-center-energy-usage-report.

Shepard, Richard F. (1974, November 17). "Time to End Sports, Opiate of the Masses." *New York Times* https://www.nytimes.com/1974/11/17/archives/time-to-end-sports-opiate-of-the-masses-will-it-never-end.html#:~:text=Karl%20Marx%20once%20described%20religion,in%20the%20proliferation%20of%20tedium.

Shepherd, Lauren Lassabe. (2023). *Resistance from the Right: Conservatives and the Campus Wars in Modern America.* Chapel Hill: University of North Carolina Press.

Shilling, Chris. (1991). "Educating the Body: Physical Capital and the Production of Social Inequalities." *Sociology* 25, no. 4: 653–74.

Shivji, Issa G. (2007). *Silence in NGO Discourse: The Role and Future of NGOs in Africa.* Nairobi: Fahamu.

Silk, Michael, David L. Andrews, and Cheryl L. Cole, eds. (2005). *Sport and Corporate Nationalisms.* Oxford: Berg.

Silverman, Alex and John Ourand. (2022, June 14). "MLS Goes with Apple in Landmark 10-Year Global Media Rights Deal." *Sports Business Journal* https://www.sportsbusinessjournal.com/Daily/Issues/2022/06/14/Media/MLS-TV-rights.aspx.

Simmel, Georg. (1949). "The Sociology of Sociability." Trans. Everett C. Hughes. *American Journal of Sociology* 55, no. 3: 254–61.

Sklair, Leslie. (2016). "The Transnational Capitalist Class, Social Movements, and Alternatives to Capitalist Globalization." *International Critical Thought* 6, no. 3: 329–41.

Smith, Anthony D. (2000). *The Nation in History: Historiographical Debates about Ethnicity and Nationalism.* Oxford: Polity Press.

Smith Maguire, Jennifer. (2006). "Exercising Control: Empowerment and the Fitness Discourse." *Sport, Rhetoric, and Gender: Historical Perspectives and Media Representations.* Ed. Linda K. Fuller. New York: Palgrave Macmillan. 119–30.

Smythe, Dallas. (1954). "Reality as Presented by Television." *Public Opinion Quarterly* 18, no. 2: 143–56.

Source Material. (2025, April 9). "Big Tech's Data Centre Push Will Take Water from the World's Driest Areas." https://www.source-material.org/amazon-microsoft-google-trump-data-centres-water-use/.

Spencer, Herbert. (2005). *Essays on Education and Kindred Subjects.* London: Everyman's Library/New York: Dutton.

SportBusiness. (2023). *Global Media Report 2023* https://www.sportbusiness.com/2023/11/global-value-of-sports-media-rights-nears-56bn-sportbusiness-report-reveals/.

SportBusiness. (2024). *Global Media Report 2024* https://www.sportbusiness.com/global-media-report-2024/.

Srinivas, Nidhi. (2022). *Against NGOs: A Critical Perspective on Civil Society, Management and Development*. Cambridge: Cambridge University Press.

Steinberg, Brian. (2024, April 29). "Fox, Disney, Warner Will Tap Streaming Sports Venture in TV Upfront (EXCLUSIVE)." *Variety* https://variety.com/2024/tv/news/fox-disney-warner-tv-streaming-sports-venture-upfront-ad-sales-1235984530/.

Stockholm Environment Institute, Climate Analytics, E3G, International Institute for Sustainable Development, and United Nations Environment Programme. (2023). *The Production Gap: Phasing Down or Phasing Up? Top Fossil Fuel Producers Plan Even More Extraction Despite Climate Promises* https://doi.org/10.51414/sei2023.050.

Stoddart, Brian. (1988). "Sport, Cultural Imperialism, and Colonial Response in the British Empire." *Comparative Studies in Society and History* 30, no. 4: 649–73.

Stratton, Jon. (1986). "Australia—This Sporting Life." *Power Play: The Commercialisation of Australian Sport. Essays in the Sociology of Australian Sport*. Ed. Geoffrey Lawrence and David Rowe. Sydney: Hale & Iremonger. 85–114.

Sullivan, John. (2009, August 1). "'Greenwashing' Gets His Goat." *Times Herald-Record*https://www.recordonline.com/story/news/2009/08/01/greenwashing-gets-his-goat/51926672007/.

Swanton, Christine. (2010). "Heideggerian Environmental Virtue Ethics." *Journal of Agricultural & Environmental Ethics* 23, nos. 1–2: 145–66.

Swiss Federal Department of Foreign Affairs and United Nations. (2004). *The Global Compact. Who Cares Wins: Connecting Financial Markets to a Changing World* https://web.archive.org/web/20220901042632/https://www.unepfi.org/fileadmin/events/2004/stocks/who_cares_wins_global_compact_2004.pdf.

Sylt, Christian. (2017, March 13). "The $1 Billion Cost of Housing an F1 Race." *Forbes* https://www.forbes.com/sites/csylt/2017/03/13/the-1-billion-cost-of-hosting-an-f1-race/.

Szalai, Georg. (2024, January 17). "Amazon to Buy Minority Stake in Diamond Sports, Sinclair to Pay $495M in Settlement." *Hollywood Reporter* https://www.hollywoodreporter.com/business/business-news/amazon-diamond-sports-investment-reorganization-sinclair-settlement-1235790585/.

TePoel, Dain. (2017). "Digital Sport History, with Costs: An Ecocentric Critique." *Journal of Sport History* 44, no. 2: 350–66.

Thomas, Dylan. (2014). *The Poems of Dylan Thomas: Centenary Edition*. Ed. John Goodby. New York: New Editions.

Thomas, Ian. (2024, July 18). "Inside Evolving FIFA 2026 World Cup Plan to Cash in on the Rise of High-End Sports Tourism." *CNBC* https://www.cnbc.com/2024/07/18/fifa-2026-world-cup-plan-to-cash-in-on-high-end-sports-tourism.html.

Thompson, E. P. (1967). "Time, Work-Discipline, and Industrial Capitalism." *Past & Present* 38: 56–97.

Thomson, Alana, Graham Cuskelly, Kristine Toohey, Millicent Kennelly, Paul Burton, and Liz Fredline. (2019). "Sport Event Legacy: A Systematic Quantitative Review of Literature." *Sport Management Review* 22, no. 3: 295–321.

Thomson, Ian and Robert G. Boutilier. (2011). "Social License to Operate." *SME Mining Engineering Handbook*. Ed. Peter Darling. Littleton: Society for Mining, Metallurgy and Exploration. 1779–796.

Timms, Aaron. (2025, July 13). "How the Rightwing Sports Bro Conquered America." *Guardian* https://www.theguardian.com/sport/ng-interactive/2025/jul/13/sports-bro-announcer-pat-mcafee-espn.

Todaro, Dina Lucia and Riccardo Torelli. (2024). "From Greenwashing to ESG-Washing: A Focus on the Circular Economy Field." *Corporate Social Responsibility and Environmental Management* 31, no. 5: 4034–046.

Transparency International, ed. (2016). *Global Corruption Report: Sport*. Abingdon: Routledge.

Treepongkaruna, Sirimon, Hue Hwa Au Yong, Steen Thomsen, and Khine Kyaw. (2024). "Greenwashing, Carbon Emission, and ESG." *Business, Strategy and the Environment* 33, no. 8: 8526–539.

Trendafilova, Sylvia, Kathy Babiak, and Kathryn Heinze. (2013). "Corporate Social Responsibility and Environmental Sustainability: Why Professional Sport is Greening the Playing Field." *Sport Management Review* 16, no. 3: 298–313.

Trotsky, Leon. (1925). *Where is Britain Going?* https://www.marxists.org/archive/trotsky/britain/wibg/ch08.htm.

Trump, Donald J. (2025, July 31). *Fact Sheet: President's Council on Sports, Fitness, and Nutrition, and the Reestablishment of the Presidential Fitness Test* https://www.whitehouse.gov/fact-sheets/2025/07/fact-sheet-presidents-council-on-sports-fitness-and-nutrition-and-the-reestablishment-of-the-presidential-fitness-test/.

Underdown, David. (1985). *Revel, Riot, and Rebellion: Popular Politics and Culture in England 1603–1660*. Oxford: Oxford University Press.

United Nations. (n.d.). *Greenwashing—The Deceptive Tactics Behind Environmental Claims* https://www.un.org/en/climatechange/science/climate-issues/greenwashing.

United Nations. (2020). *Recovering Better: Sport for Development and Peace Reopening, Recovery and Resilience Post-COVID-19* https://social.desa.un.org/sites/default/files/migrated/22/2020/12/Final-SDP-recovering-better.pdf.

United Nations Commission on Sustainable Development. (1998). *Chapeau for Business and Industry*. Background Paper No. 1 https://www.un.org/documents/ecosoc/cn17/1998/background/ecn171998-bp1.htm.

United Nations Environment Programme and Global Alliance for Buildings and Construction. (2025). *Global Status Report for Buildings and Construction 2024/25—Not Just Another Brick in the Wall* https://wedocs.unep.org/20.500.11822/47214.

Vamplew, Wray. (2021). *Games People Played: A Global History of Sport*. London: Reaktion Books.

Varoufakis, Yanis. (2002). *Foundations of Economics: A Beginner's Companion*. New York: Routledge.

Veblen, Thorstein. (2007). *The Theory of the Leisure Class*. Oxford: Oxford University Press.

Vernon, Jezz, Catherine Johnson, Joanne Evans, Rob Collier, Richard Ellis, Abhaya Sumanasena, Graham Mills, Awnit Kumar, Tim Jacks, Bertrand Moullier, and Alex Taylor. (2024). *Future of TV Distribution: A Report Prepared by the University of Exeter with the University of Leeds, MTM, 3 Reasons and Real Wireless*. Department for Culture, Media and Sport https://assets.publishing.service.gov.uk/media/672cafe262831268b0b1a2f4/Future_of_TV_Distribution_FINAL__7_Nov_2024_-accessible.pdf.

Warren, Gina. (2020). "Mega Sports Events Have Mega Environmental and Social Consequences." *Missouri Law Review* 85, no. 2 https://papers.ssrn.com/sol3/papers.cfm?abstractid=3463119.

Watson, Bruce. (2016, August 20). "The Troubling Evolution of Corporate Greenwashing." *Guardian* https://www.theguardian.com/sustainable-business/2016/aug/20/greenwashing-environmentalism-lies-companies.

Watts, Jonathan. (2019, February 25). "Concrete: The Most Destructive Material on Earth." *Guardian* https://www.theguardian.com/cities/2019/feb/25/concrete-the-most-destructive-material-on-earth.

Weber, Max. (1981). "Some Categories of Interpretive Sociology." Trans. E. E. Graber. *Sociological Quarterly* 22, no. 2: 151–80.

Weber, Max. (1992). *The Protestant Ethic and the Spirit of Capitalism*. Trans. Talcott Parsons. London: Routledge.

Weprin, Alex. (2021, August 11). "Sports Rights' Streaming Wave May Finally End Pay-TV Bundle." *Hollywood Reporter* https://www.hollywoodreporter.

com/business/business-news/sports-rights-tv-amazon-disney-viacomcbs-1234995721/.

Weprin, Alex. (2024, January 12). "Amazon Wants to Turbocharge Its Sports Advertising Business, Too." *Hollywood Reporter* https://www.hollywoodreporter.com/business/business-news/amazon-sports-advertising-business-growth-1235786459/.

Westerveld, Jay. (2024, April 16). "Earth Day 2024: Greenwashing is the Order of the Day." *The Chronicle* https://www.chroniclenewspaper.com/opinion/columns/earth-day-2024-greenwashing-is-the-order-of-the-day-FF3304676.

Wilburn, Kathleen M. and Ralph Wilburn. (2011). "Achieving Social License to Operate Using Stakeholder Theory." *Journal of International Business Ethics 4, no. 2:* 3–16.

Wilk, Richard R. and Lisa Cliggett. (2007). *Economies and Cultures: Foundations of Economic Anthropology*, 2nd ed. Cambridge, Mass.: Perseus.

Williams, Michael. (1998, January 19–25). "Ad Deals Strike a Cultural Match." *Variety*: A8.

Williams, Michael. (1999, September 20–26). "Soccer Kicky for Congloms." *Variety*: 1, 104.

Williams, Raymond. (2005). *Culture and Materialism.* London: Verso.

Williams, Raymond. (2020, September 1). "Ecology and the Labour Movement." Raymond Williams Society https://raymondwilliams.co.uk/2020/09/01/part-two-ecology-and-the-labour-movement-by-raymond-williams/.

Wilson, Brian and Brad Millington. (2020). "Introducing a Sociological Approach to Sport, Environmental Politics, and Preferred Futures." *Sport and the Environment: Politics and Preferred Futures.* Ed. Brian Wilson and Brad Millington. Bingley: Emerald. 1–28.

Wolf, Eric R. (1982). *Europe and the People Without History.* Berkeley: University of California Press.

World Bank. (2024). *Poverty, Prosperity, and Planet Report 2024: Pathways Out of the Polycrisis.* Washington: World Bank.

World Sailing. (n.d.). *Sustainability Agenda 2030: A Bold Ambition for Sailing's Contribution to Global Sustainability* https://d7qh6ksdplczd.cloudfront.net/sailing/wp-content/uploads/2022/01/18094553/SustainabilityAgenda2030-23247.pdf.

World Sailing. (2024). *Ready for the Future: World Sailing Strategy 2025–2029* https://d7qh6ksdplczd.cloudfront.net/sailing/wp-content/uploads/2024/11/13112135/Ready-for-the-Future-2024-2029-World-Sailing-Strategy.pdf.

WWF. (n.d.). *WWF Guide to Greenwashing* https://www.wwf.org.uk/learn/guide-to-greenwashing.

Xu, Guoqi. (2008). *Olympic Dreams: China and Sports, 1895–2008.* Cambridge, Mass.: Harvard University Press.

YouGov. (2023). *The Global Sports Media Landscape: Sports Whitepaper 2023* https://commercial.yougov.com/rs/464-VHH-988/images/YouGov-Sports-Whitepaper-v4.pdf.

Young, Iris Marion. (2000). *Inclusion and Democracy.* New York: Oxford University Press.

Zafari, Zahra, Abooali Golzary, Kiana Rouhi, and Omid Mansourihanis. (2025). "From Conventional Approaches to Circular Systems: Evolution of Waste Management in Mega-Sporting Events." *Journal of the Air & Waste Management Association* https://doi.org/10.1080/10962247.2025.2462005.

Zyvatkauskas, Caz. (2007, February 3). "Theatre Critic." *Economist*: 18.

CHAPTER 2

The World Cup of Carbon Football

Football is more popular than Jesus and John Lennon combined: we are told that half the population viewed the 2010 men's World Cup on television; by 2018, the same number watched the sport regularly ("Almost Half" 2011; Global Web Index 2019). In 2022, five out of the globe's 8 billion people followed the tournament through various media; the same proportion watched the 2024 Euros (UEFA 2024). Football is an extraordinarily valuable media property—the rights to screen it grew 11.7% in 2024 alone (SportBusiness 2024). As a consequence, European clubs increased their annual revenue for 2023 by €2.9 billion to 26.8 billion (UEFA 2025). The UN has declared May 25 "World Football Day."

What are people watching—and why?

Gianni Infantino, head of FIFA, makes extraordinary claims for the sport's significance:

> Football unites the world like no other sport, and the FIFA World Cup is the perfect showcase for a message of unity and inclusion, as well as providing an important illustration of how different cultures can be together and can learn and better understand one another.

T. Miller, J. P. Carañana, *Global Sports Go Green—Or Do They?*, Global Culture and Sport Series,
https://doi.org/10.1007/978-3-032-14207-8_2

Mind you, prior to the 2022 Finals, he also said:

> Today I feel Qatari. Today I feel Arabic. Today I feel African. Today I feel gay. Today I feel disabled. Today I feel [like] a migrant worker ... because I know what it means to be discriminated [against], to be bullied, as a foreigner in a foreign country. As a child I was bullied— because I had red hair and freckles ... I feel like a woman too! (quoted in Ingle 2022)

Putting that eccentricity aside, while storing it as a guide to personal and institutional solipsism and phenomenological confusion, football *is* akin to a universal currency, a *lingua franca*—common ground. In Dave Zirin's words, it's 'the closest thing we have to a connective cultural tissue that binds our species across national and cultural borders' (2011). Functionalists admire the sport's integrative capacities (Biel et al. 2024); for Vargas Llosa, it had 'the status of a lay religion' (1997: 520). England's Premier League claims the 'power of football to change lives gives people inspiration and opportunity' (2025: 1). Go anywhere in the world with a ball under your arm, or a ripping sporting yarn close to your glass, and you'll never walk, sit, or stumble alone; for Jacques Attali, football constituted 'le seul vrai sujet de conversation planétaire' [the only true topic of universal conversation] (Attali 2001: 262). Ulrich Beck claimed the sport's capitalist ecumenism supported multicultural peacefulness (2009) and in Gabriel García Márquez's *Noticia de un secuestro* [*News of a Kidnapping*] (1996), a fictionalized report-cum-ethnography of narco-lawlessness, prisoners and guards connect with one another through football, transcending the haze and violence that otherwise characterized their lives together.

This chapter looks at the sport's history and significance and the ecological destructiveness of its principal competitions, notably the men's World Cup. We pay particular attention to the 2022 Finals, future events, and the greenwashing of such sponsors as Coca-Cola and Aramco.

Football, Football

Football can indeed inspire. Lisa Tetzner's 1932 children's book *Der Fußball* saw the sport resisting Nazism (Zipes 2023: 169). In one of his more utopic moments, Brecht joyously noted: 'We have our eye on those huge concrete pans, filled with 15,000 men and women of every variety of class and physiognomy, the fairest and shrewdest audience in the world' (1964). Althusser viewed stadia as sites of oscillation between exuberant arrogance and passive conformity (1971: 162) and Lennon experienced a political pleasure when crowds sang The Beatles' 'All Together Now' (1971).

Weber referred to 'the cool objectivity of the association' that 'promotes the precise placement of the individual in the purposive activity of… a football team' (1985: 11). For Georg Lukács, proletarians embraced the sport as a spectacular extension of their industrial experience (1972). Sergei Eisenstein suggested football was a 'symbol of joint battle and cooperation' (1987: 284). Jean-Paul Sartre used it to explain how people could be beholden to a capitalist club but still exercise individual rights (2004: 450). Amílcar Cabral thought teams modeled the unity of purpose yet diversity of identity needed for revolutionary change (1979: 4) and Frantz Fanon referred to their 'spirit of sporting loyalty and friendly coexistence' (2018). For Agnes Heller:

> A good football team can raise the game of each of its members; the individual grows into the group. Human life is full of such instances of 'growing up into' the group, and they are by no means limited to the coming-of-age period—on the contrary, they are characteristic of human life. (1984: 31)

Hall saw an 'overwhelming mirror of football' in the social world (1998: 191). The Situationists celebrated when *soixante-huitards* took over the offices of the Fédération Française de Football [French Football Federation] (Solidarity 2001) and Le Comité d'Action des Footballeurs [Footballers Action Committee] called for efforts to 'Libérer le football de la tutelle de l'argent des pseudo-mécènes incompétents qui sont à l'origine du pourrissement du football' [Free football from the financial domination of incompetent pseudo-leaders, who are bringing it down] (quoted in Correia 2018: 978).

Per the Introduction's account of sports more generally, football has always had other sides, given its heritage of the British pacifying, adopting, and codifying unruly male working-class pastimes and spreading them across the world in close concert with colonial, maritime, military, and commercial interests (Elias and Dunning 1986; Krauze 1994).

In *Utopia*, Thomas More called for a ban on football, because it encouraged the poor to invest their money unwisely. Shakespeare had the Earl of Kent warn King Lear to cease behaving like a 'base football player' (1994: 1062). The ever-severe Quaker George Fox admonished his 1654 football-playing readers that:

> you are glorying in your strength, ye are boasting, ye are vapouring, ye are disdaming, ye are despising, you are envying, you are in pride, you are in ambition, with the light which Christ: hath enlightned you withal, you are to be condemned from God, and all these practises I see them all, that manhood, and all that fame, and all that renown, it must be laid down in the dust.

H.G. Wells drew a homology between teams and armored divisions (1902: 184), Kipling disparaged 'muddied oafs at the goals' (1902), David Hare's grandmother Euphemia referred to the sport as 'eleven grown men chasing a wee bit of leather' (quoted in Hare 2015: n.p), and Wittgenstein problematized the seemingly synonymous use of the word "good" in the formulations "a good football player" and "a good fellow," because of the complexities that arise when words are redisposed to connote ethical value rather than empirical description (1965).

Orwell famously described football as 'war minus the shooting' (1945), with shared ties to violence and imperialism. That pithy description came from his observation of a 1945 "goodwill" British tour by commando Дина́мо Москва́ [Dynamo Moscow] (Dmowski 2015). Orwell's colonial and militia experience of the sport in Burma, India, and Spain was of uncontrolled and passionate derision expressed by one section of a crowd toward another. He feared that such 'contests lead to orgies of hatred' because their inherent competitiveness symbolized and stimulated the desire to defeat the other (1945). Adorno and Horkheimer articulated a 'left-half at football, a black-shirt, a member of the Hitler Youth, and so on' (1979: 164).

For Gramsci, the sport was 'a model of individualistic society' that 'demands initiative, but an initiative which keeps within the framework of the law,' where 'the capitalist regime's economic individualism has transformed the whole way of life' (1994: 73–74). In *Nineteen Eighty-Four*, 'the proles' were fascinated; matches 'filled up the horizons of their minds' (Orwell 1949: 41). Jorge Luis Borges despaired that people 'remember insipid things like a football match' in place of poetry (2013: 321). Max Horkheimer argued that 'so-called entertainments, which have taken over the heritage of art, are today nothing but popular tonics, like swimming or football' (1973: 289–90). For Baudrillard, '[p]ower is only too happy to make football bear a facile responsibility, even to take upon itself the diabolical responsibility for stupefying the masses' (1983: 14).

The British monarchy started attending the Cup Final to exploit a 'festival of the proletariat,' merging spectacle and ritual under the family's avowedly benevolent gaze (Hobsbawm 1989: 107). Roberto daMatta lamented that Brazil's 'mortar of soccer ... based on performance rather than on substance—builds real ramparts against social dissatisfaction and unrest in a system based on conformism and marked by an alarming lack of coherence between its ideas and its actions' (1991: 136). Alain Badiou expressed concern that:

> serious newspapers compare the celebrations that accompanied France's footballing victory at the World Cup to the demonstrations that followed the liberation of Paris in 1945. Why not compare them to the taking of the Bastille, or the Long March? These days, 'celebration' is the name for something like a 'counter-demonstration.' (2007: 107)

Eagleton called football 'a scheme to distract the populace from political injustice and compensate them for lives of hard labour' (2010). Fred Halliday waxed nostalgic for 'a time before public discourse was dominated by footballers' (2009) and Eduardo Galeano traced the sport's history as 'un triste viaje del placer al deber. A medida que el deporte se ha hecho industria, ha ido desterrando la belleza que nace de la alegría de jugar porque sí ... el fútbol profesional condena lo que es inútil, y es inútil lo que no es rentable' [a sad journey from pleasure to duty. As the sport has industrialized, it has banished the beauty that came with playing for its own sake ... professional football condemns whatever is useless, and defines as useless whatever can't be commodified] (1995: 6). Raymond Chandler said of another code, '[t]he smell of old dust hung in the air as flat and stale as a football interview' (1949).

But Galeano also saw condescension in stereotypes that condemned the sport:

> The scorn of many conservative intellectuals comes from their conviction that soccer worship is precisely the superstition people deserve. Possessed by the ball, working stiffs think with their feet, which is entirely appropriate, and fulfill their dreams in primitive ecstasy. Animal instinct overtakes human reason, ignorance crushes culture, and the riffraff get what they want. In contrast, many leftist intellectuals denigrate soccer because it castrates the masses and derails their revolutionary ardor. Bread and circus, circus without the bread: hypnotized by the ball, which exercises a perverse fascination, workers forget who they are and let themselves be led about like sheep by their class enemies. (2013: 82–83)

Beyond consciousness—false or otherwise—property relations intervene. For much of the twentieth century, Western European football teams were mostly small city-based businesses, drawing on athletes who had grown up close to the ground. Representing and stimulating local cultures, they were run rather like not-for-profit concerns, albeit sometimes dominated by local corporations (in Germany) or *petit-bourgeois machistas* (in Britain). A shift came in the 1990s via media "reform" and the New International Division of Cultural Labor (NICL) (Miller et al. 2001).

Huge amounts of money became available to football as television was deregulated across Europe, permitting the advent of stations dedicated to single genres, such as sports. New communications technologies and consumer electronics proliferated and élite competitions maximized their media coverage. In the decade from 1990, a period of minimal macroeconomic inflation, European prices for football rights grew 800%. As we saw in the Introduction, this was a means of securing subscribers to cable, the internet, satellite, and Wi-Fi (Miller 2010a).

The English Premier League started in 1992, generating £205 million in its first year. Brian Glanville presciently renamed it 'the Greed is Good League' (2018). Three decades on, revenues had increased by 2850% (Maguire and Philippou 2024). Internationally, the Premier League attracted $4.5 billion a season for the right to screen matches live from 2022–25, up almost 20% on 2019–22 (Harraghy 2024). In the USA, NBC Sports paid the League $450 million a season from 2022 to 2028, five times the 2013 figure (Crafton 2024; Vlessing 2024). UK rights from 2025–2029 went to TNT and Sky for £6.7 billion ("Premier League Agrees" 2023). These amounts may seem large, but are less imposing next to the price of producing high-end drama: over £4 million per hour in Britain (British Screen Forum 2021), while US series can cost a hundred million dollars to initiate and more than ten million per episode (FilmLA 2019a, b). As an extreme example, Amazon paid $250 million for the right to adapt *Lord of the Rings* for TV, then $465 million to produce the initial season (Hibberd 2021).

At the same time as television was deregulated, the NICL was stimulated by Jean-Marc Bosman's successful appeal to the European Court of Justice (ECJ) against his suspension in 1990 by the Union Royale Belge des Sociétés de Football/Koninklijke Belgische Voetbalbond [Royal Belgian Football Association] for seeking an overseas transfer. Bosman's case was founded on the right to freedom of movement for EU workers. Five years later, the Court decided in his favor (*Union Royale* 1995). In reaction, the Union des associations européenes de football [Union of European Football Associations] (UEFA) and FIFA had this to say:

> Sport, like culture, or other economic activities such as agriculture, cannot be reduced to a single commercial dimension without severe detriment. … it is a fundamental feature of national identity that FIFA and UEFA are determined to safeguard. ("FIFA and UEFA" 1999)

The two agencies argued that a sport they had worked hard to commodify was not a commodity, and they called for the preservation of national heritage by undemocratic, non-state actors—themselves. Subcomandante Marcos sardonically "prophesied" that 'for real, soccer is going to be a sport and not a business' (2017: 235). Facing the threat of fines from the European Commission, FIFA discontinued rules restricting the number of foreigners who could play in national leagues. Player mobility increased sharply across Europe and the gap in talent between wealthy teams and also-rans widened. As a consequence, managers, agents, players, network executives, and institutional investors grew committed to football's expansion (Duval and Van Rompuy 2016). In 2024, a further ECJ decision on a case brought by Lassana Diarra freed players to leave clubs while their contracts remained in play (*Fédération internationale* 2024). Justice for Players (n.d.) announced a multi-billion-dollar class-action lawsuit on behalf of a hundred thousand players whose livelihoods had been adversely affected by FIFA's illegal restrictions (Hughes 2025).

Taken together, this deregulation of sports, the media, and labor produced extraordinary riches for a privileged few. Many teams fell prey to fictive capital, becoming sources of asset inflation for rentiers to service other debts through cash flow from television money and merchandising (Wartuny 2024). In 2024, *Forbes* valued Real Madrid at $6.6 billion, Manchester United at $6.55 billion. That year saw Madrid become the first team to generate a billion euros in revenue, in part thanks to selling "Personal Seat Licenses" for 76 million that merely guaranteed the right to buy season tickets for three decades. The *Deloitte Football Money League* calculated the revenue of Europe's richest thirty clubs in 2024 at €11.6 billion, 6% more than the previous year (2025). Top-flight English football garnered £12.25 billion in 2024 (Garrahan 2025).

A commodifying tendency continues to intensify, nowhere better exemplified than in betting. Online gambling on sports was slow to become legal and/or ubiquitous, but exploded into global popularity in the 2020s ("How Sports" 2024). The *Economist* hailed it as a welcome burst of 'freedom' ("America's Gambling" 2024). Conversely, *The Lancet* Public Health Commission on Gambling noted that customers were predicted to lose $700 billion a year by 2028:

> Sponsoring and partnering with professional sports organisations provides gambling operators with marketing opportunities with huge new audiences. This far-reaching and interdependent corporate ecosystem collectively wields substantial influence over policy and has multiple points-of-contact through which to leverage the behaviour of consumers. (Wardle et al. 2024: 1)

Opening weekend broadcasts of the 2023–24 football season circulated 10,999 gambling messages on British TV and radio—115 per hour (Rossi et al. 2023). That hyper-commodification rapidly accelerated: the beginning of the 2024–25 competition saw a 165% increase in such promotions (Pirks and Grundy 2024).

More than half of Premier League teams promote gambling outlets on their clothing, for a "consideration" of $135.43 million (Austin 2025). Across the game, hundreds of clubs are sponsored by such firms (Buzzoni et al. 2025). Their principal targets are gamblers in countries that prohibit the addiction or money launderers. The Premier League and others have routinely carried the names of groups involved in offshore betting and sex trafficking (Committee on Women, Children, Family Relations and Gender Equality 2022; Auclair 2021, 2022, 2023; Kennedy and Southern 2024; Auclair et al. 2024b). The implicit signage? 'Welcome, criminals!' and 'The laundry will remain open' (Brown et al. 2025). Few legislative attempts have been made to deal with the implications for public health and organized crime (Ukhova et al. 2024). In its laughable way, the UK's Gambling Commission claims to oppose money laundering, but does nothing about it (2025; Auclair 2025).

Football's almost-cosmic duality also has environmental homologies. On the surface, it appears the least ecologically malevolent of pastimes: players need a ball, some space, and other people who like to run about. But the 'destructive effects of the society of the spectacle' (Paché 2020) have an impact: 895 international matches between the 1990 World Cup Finals in Italy and qualifying stages of the 2024 Euros provided devastating evidence of a surge, just in terms of professional travel—excluding spectators, executive hangers-on, and equipment. Football is now directly responsible for 0.3–0.4% of the world's annual carbon emissions—equivalent to Denmark (Ashoo 2021). The *Wall Street Journal* advises that the sport generated over 30 million metric tons of CO_2 in 2024, mostly through meeting energy "needs" via fossil fuels (Forbes 2025), and equivalent to burning 150 million barrels of oil. Each game at the World Cup Finals emits between 44,000 and 72,000 tCO_2e—per the amount from 30,000–40,000 cars on British roads every year (New Weather Institute with Scientists for Global Responsibility 2025a).

Consider also how football's basic equipment is made: in accordance with the Australian state's murderous policies, generations of kangaroos have been slaughtered as raw material for 'k-leather' boots, while balls are manufactured in Pakistan from synthetic leather derived from oil, rubber and cotton from trees and plants, and leather and glue from animals killed before their natural lifespan (Kang 2020; Preston 2020; Godson 2024). The 60 million footballs sold in 2010 traveled

> from the stitching rooms of Sialkot to the professional football pitches of Europe and America. First you get the sub-subcontractors—the stitching centers, the backroom workshops, the one-man businesses. Add to that the subcontractors, the transport firms, the customs offices, the sports equipment giants, the advertising industry, the sports good retailers and the department stores. The chain converts a 63-rupee ball into a product costing more than €100. (Kazim 2010)

And think of the concrete, water, and chemicals involved in ground construction and maintenance; food consumed at games; electricity required to watch fixtures and bet or report electronically; and the impact of tourism (Malhado and Rothfuss 2013): '[n]o other sport sees fans travel so frequently and in such great numbers' (Konyn 2024).

In turn, football is affected by the climate chaos it has helped to unleash. Unprecedented storms endanger pitches and high temperatures jeopardize players' health (The Climate Coalition 2018). The insurance company Zurich UK's climate model suggests 40% of British football grounds will be imperiled by climate change from 2050, one in five flooded ("Nearly Half" 2023). World Cup venues in Europe and the USA are predicted to lose $800 million through rain and heat ("Climate X" 2024).

The biggest cause of this planetary horror is sponsorship, which is responsible for 75% of the sport's emissions (New Weather Institute with Scientists for Global Responsibility 2025a). These greenwashers use football as a response to ecological crisis and shame. For example, shortly after the Spanish energy and petrochemical firm Repsol spilt thousands of barrels of oil across 1400 hectares of the Peruvian Pacific coast in 2022, killing native life and destroying the livelihoods of thousands of people, it signed on to sponsor the men's national team ("Repsol" 2023). The project's slogan, also associated with supporting junior and women's representatives, was 'Veamos el futuro' [We see the future]. Closer to home, Repsol was about to conclude an agreement to supply Athletic Club de Bilbao with energy that supposedly affirmed the corporation's commitment to decarbonization ("Petronor" 2024). Meanwhile, the UK's Advertising Standards Authority (ASA) found Repsol had distorted its environmental record by highlighting the firm's synthetic and biofuel operations, a tiny fraction of its basic business—fossil fuels (Hodgson with Wilson 2023; "ASA Ruling on Repsol" 2023).

The sport lacks consistent policies ensuring competitions, supporters, teams, and stadia reduce emissions (Daddi et al. 2022). Consider 2004's Football Association Cup Final in Wales. Most attendees drove from England, traversing 43 million kilometers and multiplying the match's carbon footprint by a factor of eight. Large amounts of energy were also required to prepare and distribute the highly processed commodities that visitors drank and ate (Collins and Flynn 2008). Such problems also apply to regular home-and-away fixtures (Dosumu et al. 2017). Travel to Europe's 2019 Champions League and Europa League Finals produced 35,000 tons of CO_2 (Lamborelle and Rios 2019). Although many fans know they add to emissions through transportation, their conduct remains largely unchanged (Sant'Anna School of Advanced Studies, Institute of Management 2020).

A study of the 2016–17 English Premier League season disclosed that local team travel alone generated approximately 1134 tons of CO_2-eq (Tóffano Pereira et al. 2019). In 2022–23, clubs undertook eighty-one short-haul flights over two months. There were at least 116 across the 2024–25 season (Lockwood 2023a; "Premier League Clubs' Flight" 2025). Many flights send planes with no passengers to collect players. In the case of Arsenal, North London's labor aristocracy, travel has included a fourteen-minute executive-jet flight to Norwich; for Manchester United, a ten-minute plane ride to Leicester (de Menezes 2015; Stanton et al. 2021). Needless to say, Arsenal speaks of 'improving its environmental performance and the prevention of pollution' (n.d.) and United announces '[e]veryone at the club is committed to tackling environmental and social issues at regional, national and international levels, using the Manchester United brand to leverage support and create awareness of the issues facing the planet' (n.d.).

These clubs also fly around the globe in the off-season for lucrative paydays against minor opposition in the USA and Asia, accounting for tens of thousands of air miles (Lockwood 2024), while in continental Europe, Barcelona avows that 'volem contribuir a disminuir l'impacte del canvi climàtic' [we want to contribute to diminishing the impact of climate change]. It has deals with numerous environmentally disastrous companies, such as betting, automotive, soft-drink, brewing, courier, apparel, and cryptocurrency firms (n.d.). Paris Saint-Germain drones on about how environmentally friendly it is. Owned by Qatar, the team flies across France on that nation's airline rather than taking sustainable options. When questioned about this in 2022, management "jokingly" said the team had considered a 'sand-yacht' as an alternative ("PSG Coach" 2022). Heaven forbid chaps should travel by train. The club's most famous player of the day, Lionel Messi, tootled around the globe fifty-two times that summer, accounting for thousands of tons of emissions ("PSG" 2022). He is part of the 1% of the population responsible for monstrous emissions from private jets, on average 2074 tons of carbon a year. It would take a person in the poorest 50% 2000 years to match that figure (Murphy and Simon 2021; Oxfam International 2024). These wealthy polluters imperil our planet. The vast majority of the world's population does not: 'carbon footprint inequality' (Nielsen et al. 2024).

Thanks to 177 new fixtures, Europe's three principal continental club tournaments—the Champions, Europa, and Conference Leagues—were responsible for 1.5 billion air miles in 2022–23 and an expected 2 billion for 2024–25. That travel exhaled almost half a million tons of greenhouse gases, equating to 4000 return trips to the moon (Lockwood and Hickson 2023). Professionals playing in Europe and representing far-distant countries elsewhere may travel over 150,000 kilometers internationally each year (Fédération Internationale des Associations de Footballeurs Professionnels 2024: 8).

All this occurs under the design and supervision of UEFA, which claims to favor sustainability, and praises itself for 'climate action and advocacy' (n.d.). That moralism is enabled by saying sustainability is about fair and equal pay—clearly desirable goals, but highlighted to the exclusion of fan, player, and executive-hanger-on travel from carbon calculations. Following the 2024 Euros, Qatar Airlines handed UEFA €500 million to act as the

Union's exclusive aviation advertiser for the next six years (Cronin 2024; Fossil Free Football et al. 2024: 2):

> the commercial side of Uefa's affairs suggests the intent of greening up its flagship tournament, while undoubtedly sincere in some corners of the organisation, is being undermined by the governing body's continuing dependency on the millions coming from oil-hungry industries. (Auclair 2024)

So the Union insists on 'no government interference in the running of football' (quoted in Roan 2024) while permitting states to own teams and sponsor competitions.

The 2016 Euros in France generated the same carbon emissions as a hundred residents create in a year: the stadia produced 1542 tons of waste, just over a third of which was recycled (Lamborelle and Rios 2019). The 2020 Euro Finals could almost have been designed to create maximal ecological chaos—they were scheduled for Germany, Hungary, Scotland, England, Azerbaijan, Holland, Italy, Denmark, Russia, Ireland, and Spain (McKie et al. 2019). Postponement because of COVID-19 did not make for a more environmentally rational itinerary based in one country (UEFA 2020). The 2024 tournament emitted 320,000 tons of CO_2e: 250,000 from spectators' travel, the remainder thanks to UEFAcrats (UEFA 2024: 11). Among the nations competing, Germany (the host) and Switzerland (a neighbor) were the only countries that eschewed flying. Competition sponsors included oil and auto companies, airlines, and fossil-fuel financiers.

The World Cup

But have no fear; FIFA's here. The Association boasts that over the last decade, it 'has continued to strengthen its requirements and programmes related to environmental protection, encompassing issues such as climate change, waste management, recycling, sustainable construction and procurement' and taken 'concrete actions to protect our planet, in particular around FIFA's flagship events' (quoted in Gray 2023). FIFAcrats promote recycling, waste management, and synthetic turf made from recycled trees in place of natural grass, thereby minimizing water usage and avoiding fertilizers and pesticides. But tire-rubber infill unleashes numerous contaminants injurious to people and the wider environment; toxic nano-plastics proliferate (Hua et al. 2024). And these executives do not insist on the capital outlays needed to install vital environmental technology (Walters and Tacon 2011; GA[e]Engineering n.d.). Because their "efforts" exclude travel and sponsorship, principal sources of environmental despoliation, FIFA's climate claims are hollow.

And the men's World Cup Finals? The Mexico City earthquake of 1985 saw at least 10,000 people lose their lives, 30,000 injured, and thousands more rendered homeless (National Institute of Standards and Technology, n.d.). National and international footballing authorities reassured themselves that despite the chaos, the 1986 Finals could and should go ahead, because they would bring joy, pride, and foreign investment. Elena Poniatowska replied, 'Yeah, sure' (1995: 101). That tragic quip encapsulates a wider truth.

Since 1966, 90% of Cup hosts have run deficits, with costs this century averaging around $5 billion. Governmental "investment" in these carbon-sports jamborees is startling:

2022 Qatar $220–300 billion
2018 Russia $14 billion
2014 Brazil $12 billion
2010 South Africa $3.6 billion
2006 Germany $5 billion

Sources: Lyjak (2023) and Fett (2020)

Public subvention of the 2014 Finals saw:

> a series of infrastructure, mobility and stadia construction projects ... carried out at a cost of US$10 billion, of which 84 per cent came from the public sector, despite promises to the contrary. ... [B]usiness groups associated with local governments profited, either through exploiting commercial sport facilities or by expanding new fronts for real estate capital near facilities and infrastructure projects. On the other hand, the low-income excluded populations were the main losers, either by being removed from the areas affected by the projects, or as a result of the misuse of public resources. (Cusce Nobre 2017: vi)

The Finals make money for FIFA, thanks to public subvention and funding from sponsors. Governments receive minimal direct revenue in return (Müller et al. 2022), and reliance on the public purse doesn't ensure democratic participation. In 2025, FIFA clarified its desire to transmogrify the kind of person attending matches. Sponsors and other corporations were permitted to purchase blocks of tickets for clients and executive lizards, whose successes in governing the labor process would be rewarded by package deals with tour organizers (Habtemariam 2024a). A hidden part of ahistorical claims to astronomical economic benefits would be stadia and cities laden with visiting oligarchs.

Perhaps chastened by how much the Cup has cost others, prior to the 2026 tournament, FIFA released its 'first ever pre-tournament FIFA World Cup™ Economic Impact Assessment, which looks at the estimated economic output for Canada.' Miraculously, the Association announced the Finals would 'contribute up to CAD 3.8bn in positive economic output for Canada' ("FIFA World Cup 26™" 2024). No explanation was given for this dramatic break from decades of history—why Canada was so blessed.

One reason was the generosity of its provincial and federal governments. A few months earlier, the Cup's Vancouver hosts had estimated their subsidy at between $483 and $581 million, double the 2022 prediction and the City of Toronto at around $380 million, ten times the original figure ("2025 Budget" 2025). The federal government pledged $220 million to the two cities and the province of Ontario 'up to $97 million' to Toronto. There was major subvention for the 2026 Finals by Canada's co-hosts—Seattle $26 million ("Session is Done" 2024), Missouri $40 million ("Missouri Gov." 2025), and Mexico City $348 million toward its mass-transit budget, an increase of 186%, 12.5% beyond the norm for public works (Ionescu 2024; Vasquez 2024)—this at a time when infrastructural investment was collapsing, other than in fossil fuels, and public-sector interest payments were soaring (Alarcón 2025). The Mexican government announced it would remodel four airports to qualify them as 'Aeropuertos FIFA' (López-Castro 2025). When the Building and Wood Workers' International (BWI) sought to inspect labor conditions during renovation of the Azteca stadium and construction of new amenities, FIFA blocked the union's entry (MacInnes 2025). It's a familiar story to find FIFA/*raison d'état* ignoring or bludgeoning people—500 families had been evicted when the Azteca was constructed (Pistolesi and Pelliccia 2024).

Meanwhile, LA's 2026 Host Committee hired consultants who said what those people say: the event would 'deliver nearly a quarter of a billion dollars in increased wages,' thanks to expenditure of $594 million, which would also bring in millions to public revenue (Weinstein et al. 2024). Those numbers were dwarfed when Infantino advised that the 2025 Club World Cup and 2026 Finals would create 200,000 jobs, adding $40 billion to the US economy (Schaerlaeckens 2025).

Meanwhile, the World Cup's environmental record is striking. In *France 24*'s words, claims to 'la neutralité carbonne' [carbon neutrality] are just rhetoric in what has become a 'Coupe de monde du "greenwashing"' [World Cup of greenwashing] (Cabot 2022). From solar-powered stadia to free public transportation, the 2006 World Cup's "Green Goal" promised a 'climate-neutral' event, saving 100,000 tons of CO_2 via projects in India and South Africa and minimal transportation, energy, water, and refuse (Mitchell 2007). But the data excluded international travel, a feature of environmental audits that focus on one country rather than overall ecological impact (Collins et al. 2009). Because claims about the 2006 tournament rang so hollow, FIFA set up an Environmental Forum to "green" stadia, training grounds, accommodation, amenities, and so on, in accordance with the UN Environmental Programme and its Global Forum for Sports and the Environment (Collins et al. 2009). As ever, neither proliferation nor efficacy have been demonstrated (Death 2011; Cornelissen et al. 2011).

Norwegian and South African scientists studying the probable environmental impact of the 2010 Finals issued stern warnings (McCarthy 2009; Republic of South Africa et al. 2009). But South Africa's *Accelerated and Shared Growth Initiative* promised prosperity from staging the event, with President Thabo Mbeki hailing a 'development World Cup' for the continent that would presage an 'African renaissance'. His government used economistic, nationalistic, and anthropocentric arguments against science: the necessity of development, being the best in the world, and the fantasy that those desires could be alchemically combined with green objectives. Government tender documents invited competition to offset the impending 2010 flight footprint, but no contracts were issued. This was hardly surprising from a state that largely neglected statutes mandating environmental assessments of projects and the use of alternative energy (Ahmed and Pretorius 2010).

Cape Town proposed upgrading its principal stadium, but FIFA preferred new construction (Engar and du Toit 2023). Local communities were excluded from planning (Paul Nkemngu 2012). While mass transit improved, freeways received most investment (Estrada 2010). As a consequence of poor internal transportation infrastructure and dependence on coal-fired power, a significant carbon impact was inevitable (Paul Nkemngu 2012). Relatively few tourists visited, limiting emissions—and any positive multiplier effects (Hyde-Clarke et al. 2014).

The South African Finals sponsors included Anheuser-Busch—a company guilty of air pollution, failed chemical safety, and inadequate transparency—and Visa, which emits close to half a million tons of CO_2 each year. Its cards will take centuries to degrade. Most end up as landfill, contributing to vast contamination because of their plastic, metallic, and electronic componentry. Visa's Match Planner "smartphone" application provided World Cup scores and retail information, a further energy drain. The company uses implausible amounts of energy every second in its digital transactions and has also been accused of raising prices by pressuring banks to punish firms that use other debit companies (Bautista-González 2023).

The 2014 Finals in Brazil were supposed to be played in green stadia—a veritable *Copa Verde* [Green Cup] (Estrada 2010). Needless to say, estimates of the Cup's environmental impact often exclude construction, but the 2014 Finals emitted 2.72 million tons of carbon through construction of stadia that have attracted few spectators since (Crabb 2018; Sturrock 2018). The inevitable clearance of the poor and the unhoused had dramatic impacts on city centers, while offset programs designed to minimize FIFA bootprints were unsuccessful (Crabb 2018; Sturrock 2018). Brazil repeated South Africa's subterfuge of 2010, excluding international and internal transportation from pre-event predictions. In the end, over three and a half million tourists accounted for 84% of emissions associated with the Cup (Spanne and *The Daily Climate* 2014; "Calculan costo" 2014). Yet again, FIFA's CSR smog did little, if anything, for ordinary people (Woods and Stokes 2019; Inteligência Esportiva Research Institute and Play The Game 2023; Silic 2022). But hooray for BP. Its carbon-management division arranged for ticket holders to register their carbon bootprint, with the prospect of two registrants winning seats to the 2014 Final. The idea of establishing one's very own carbon footprint had been invented and popularized by the firm's public-relations people to divert attention from its nefarious activities and make consumers feel culpable for our climate crisis (Kaufman 2020; Solnit 2021). The plan was set in motion via the tagline 'What on earth is a carbon footprint? Everybody in the world has one' (quoted in Doyle 2011: 200; "BP Ad" 2003). Some members of this group called 'Everybody' leave larger residues than others, as we'll see in the next chapter.

Even the Association acknowledged that 2018 in Russia generated well over 2.1 million carbon tons, with building a core problem (FIFA 2019: 68). And 2022 was astonishing. Qatar emits the world's most carbon *per capita* (Charfeddine et al. 2018). Yet its 2009 proposal to stage the 2022 Finals promised a 'carbon-neutral' event ("Qatar's World Cup" 2022), pretending that construction, desalination, and water use in an arid climate would be offset by future generations enjoying the facilities—despite gate receipts in the hundreds for local matches (Carbon Market Watch 2022; Raji 2022). The bid succeeded, for all the country's problematic vote-winning, labor abuses, disdain for human rights, and unsustainability as a venue, given limited natural resources, vast importation of workers to construct venues, abundant use of air-conditioning in stadia, and hundred and fifty daily flights for attendees (Abusin et al. 2020; Cabot 2022)—further confirmation that FIFA is, as the *New Yorker* put it, 'a rancid institution' (Blake 2022).

As soon as Qatar won the right to host, it 'funneled billions of dollars from … hydrocarbon revenues into infrastructure projects, supercharging the construction industry' and generating approximately four times the amount of carbon per capita as the US equivalent (Iskander 2021: 505, 516). Hotels were not ready in time, so apart from glorified containers and more luxurious accommodation, the organizers shamelessly used cruise ships as floating accommodation and short daily flights for tourists to better-prepared sites in neighboring states (Foxman and Nair 2022).

FIFA had claimed offsets would mitigate emissions, a problematic conceit: one would need to plant 600 million trees to counter Qatar 2022's climate criminality (Hometree 2022). Despite—perhaps thanks to—trademarking "Sustainable FIFA World Cup 2022™," the event's environmental despoliation was greater than any other World Cup or summer Olympics ("Qatar's World Cup" 2022). As usual, the USA provided the most international tourists of any country, accounting for 191,055 tons of CO_2 (Hometree 2022), an astounding 14.7 kt from private jets (Gössling et al. 2024). This was invisible in Qatar's *Greenhouse Gas Accounting Report Ex Post Assessment* (FIFA World Cup Qatar 2023).

Le Monde exposed the environmental propaganda as a 'mirage' (Mandard 2022). *Scientific American* termed the event a "climate catastrophe" (Boykoff 2022), 'the dirtiest in history' according to the conceptual artist Andrei Molodkin (2022), thanks to Qatar/FIFA's 'white elephants and green lies' (Auclair et al. 2024a) found Pablo Alabarces declaring 'la Copa Mundial de Qatar sea el hecho más escandaloso, más grotesco, desde que el fútbol se constituyó en una mercancia global' [the Qatar World Cup was the most scandalous, the most grotesque, since football became a global business] (Frieiro 2022). France's Notre Affaire à Tous [Everyone's Business] collective explained that:

> En présentant la Coupe du Monde comme neutre en carbone, la FIFA laisse donc les détentrices de billets—et plus largement les supporters—croire que le fait d'assister à un tel événement n'a aucun impact sur l'environnement, ce qui est clairement incompatible avec la part des émissions liées aux voyages internationaux dans les émissions globales de GES de la Coupe du Monde. [By presenting the World Cup as carbon neutral, FIFA leads ticket holders—and more broadly fans—to believe that attending such an event has no impact on the environment, which is clearly incompatible with the international travel that affects the Cup's greenhouse-gas emissions.] (2022: 14)

The Commission Suisse pour la Loyauté [Swiss Fairness Commission] received complaints about the organizers' duplicity from Notre Affaire à Tous in alliance with Carbon Market Watch (Belgium), Britain's New Weather Institute, Alliance Climatique (Switzerland), the Dutch Reclame Fossiettvrij, and Fossil Free Football. Unsurprisingly, FIFA argued that its efforts had been 'autentique et sincère' [real and sincere] (quoted in Danjon 2023). The Commission ruled that the Association had lied in claiming the Cup was the first 'fully carbon neutral' such event ("Swiss Commission" 2023). Neither this nor millions paid in bribes to officials via 'years of grubby backhanders and shady deals' (Blake and Calvert 2016) led to prosecutions; Swiss law protects FIFA from corruption charges because it is classified as an NGO (Blake 2022).

Meanwhile, Infantino and his courtiers ignored FIFA's own report covering Qatari crimes against nearly a million migrant workers who had made the 2022 Finals possible, thousands of whom had perished—a report the Association released almost a year after receiving it (FIFA Sub-Committee 2023). In place of following the recommendation to devote its 2022 profits to poorly paid and poorly treated guest workers who had made the site, and the loved ones of those who died doing so, FIFAcrats allocated the money to develop the sport ("FIFA Produces" 2024; MacInnes 2024). Meanwhile, Doha proudly announced that the Finals represented 'Qatar's IPO' (Initial Public Offering) (quoted in Malik 2025).

Qatar provided private jets for Infantino to swan around the globe, to the tune of more than 600,000 kilometers flown between 2021 and 2024 (Kunti et al. 2024); like Messi, Infantino is an élite polluter. He proudly announced the Saudi oil firm Aramco as a World Cup sponsor in 2024, praising its 'strong track record of supporting world-class events, but also a focus on developing grassroots sport initiatives' (quoted in "FIFA Announces" 2024). That decision raised questions about the Global North luxuriating in the Global South's internal inequalities. In Thomas Piketty's words:

> the Middle East is the most inegalitarian region of the world, in large part because its oil resources (which should be left in the ground) are concentrated in very thinly populated areas, where oligarchies accumulate unlimited financial reserves on the international markets, with the active support of the West, which is only too happy to be able to sell them weapons or recuperate part of the funds in their banking systems or their sports clubs. (2022: 225)

Away from host countries, consider the electricity demand created by media coverage. ESPN Deportes, Disney's Spanish-language US TV-sports channel, didn't have the rights to film the 2010 event, but dispatched twenty-five reporters to South Africa anyway to realize the promotional campaign '*90 minutos no son suficientes*' [90 minutes aren't enough], troping the duration of matches to indicate the importance of background and synoptic material (Miller 2010b).

2010 included a deal with YouTube whereby viewers across 17 languages and 120 nations posted their goal celebrations. There was no audit of the energy sources powering such exchanges (Sherter et al. 2024). We can say that UK power use during international matches surges by as much as 10% in the 'TV pick-up' or 'Half-Time Kettle Effect' (Selectra 2020) when people head for the kitchen and bathroom. The National Grid has correlated viewership and energy during England's tragi-comic departure from the 1990 World Cup to demonstrate pride in its ability to service unsustainable needs: 'A whopping 2800 MW at the end of the penalty shoot-out!' (2010 and 2020). Prior to the 2014 Finals in Brazil and the 2016 Euros football fans provided the National Grid's biggest custom (2014a; Miller 2018). For Brazil's matches in 2014, UK surges totaled 4348 megawatts, the equivalent of over a million and a half kettles being turned on at the same time (National Grid 2014b). The Qatar Finals drained reserves for Wales-England (900MW), England-Iran (819MW), and France-England (914MW) (Heynes 2022). The 2023 Women's World Cup Final stimulated a half-time surge of 500 MW (Mavrokefalidis 2023) and the 2024 Euros Final saw the English consume 1300 MW at half-time, as they opened refrigerators, flushed toilets, and boiled water (Green 2024).

When they weren't flushing or boiling water, much of what audiences saw and heard was commodity fetishism, AKA sponsorship. FIFA promises sponsors:

- The use of the Official Marks
- Exposure in and around the stadium, in all Official FIFA publications and on the official website,
- Acknowledgement of their support through an extensive FIFA World Cup™ sponsor recognition programme
- Ambush marketing protection
- Hospitality opportunities
- Direct advertising and promotional opportunities and preferential access to FIFA World Cup™ broadcast advertising (FIFA 2025)

Football's assiduous greenwashing of environmental miscreants has seen the Cup's principal sponsors include numerous firms in need of ecological cover. The 2010 Finals were Coca-Cola's biggest-ever promotion, just as reporters were exposing the firm's assiduous efforts to harness sporting "researchers" to its propaganda. For decades, Coke had clandestinely paid millions to exercise "scientists" to apportion the blame for rampant obesity on insufficient exercise rather than its own products. For the corporation and its cloistered chorines, 'science was not a truth above interests and politics; it was a commodity like, say, caffeine or high-fructose corn syrup, to be mixed and manufactured in ways that maximize profits and minimize government regulation' (Greenhalgh 2024: 7–9, 13). Hundreds of meretricious mercenaries took sugary dollars and published—as instructed—a multitude of pseudo-academic articles extolling the need for playing sports rather than eschewing fast food (Matos Serodio et al. 2016). When the whistle blew, Coke ceased paying supine scholars, but continued product placement made it the world's third biggest sponsor of sports after Nike and Adidas in 2024 (Ampere 2025: 14).

In 2023, the firm was named the world's worst plastic polluter for the sixth time in a row (Break Free From Plastic 2023; Cowger et al. 2024).

European consumers, DC activists, and Los Angeles County litigated it for fallacious advertising and promotion (Ding and Associated Press 2024; Johnson 2024). Meanwhile, Coke inconspicuously discarded a pledge to make a quarter of its packaging reusable by 2030 (Horton 2024), even as its reliance on plastics was scheduled to exceed 4.1 million metric tons that year, 40% more than in 2018 and 20% up on the 2023 figure, with large proportions headed for the ocean and other waterways (Oceana 2025).

A million plastic bottles of water are sold each minute and five trillion plastic bags come into circulation annually ("A New Way" 2025), mostly derived from fossil fuels. It's anticipated that by 2050, plastics will provide half the increase in fossil-fuel demand and be responsible for a quarter of the emissions keeping the world above its targeted annual increase in temperature (Karali et al. 2024; International Energy Agency 2018). The OECD explains the immensity of the problem:

> Globally, the annual production of plastics has doubled, soaring from 234 million tonnes (Mt) in 2000 to 460 Mt in 2019. Plastic waste has more than doubled, from 156 Mt in 2000 to 353 Mt in 2019. After taking into account losses during recycling, only 9% of plastic waste was ultimately recycled, while 19% was incinerated and almost 50% went to sanitary landfills. The remaining 22% was disposed of in uncontrolled dumpsites, burned in open pits or leaked into the environment. … Plastics have a significant carbon footprint, contributing 3.4% of global greenhouse gas emissions throughout their lifecycle … 90% coming from their production and conversion from fossil fuels. (2022: 14)

Plastics are 'choking' the planet. Less than a tenth are recycled, because the cost of producing new goods is cheaper (United Nations Environment Programme n.d.; Brigham 2022). And *The Lancet Countdown on Health and Plastics* argues that plastics account for over $1.5 trillion in economic losses due to illness and death (Landrigan et al. 2025). As the *Economist* acknowledges, 'it will often be more profitable for a business to dump costs, such as pollution, on to society than to bear them directly' ("The Fundamental" 2022).

Disreputable conduct by the Cup's sponsors is ongoing. As bushfires raged across the Australian summer of 2025, kangaroo survivors "ran" the risk of slaughter to become raw material for football boots vended by FIFA's friends ("He Outran" 2025). Computer manufacturer Lenovo paid millions when the Federal Trade Commission and 32 US states found it had clandestinely loaded software on laptops 'that compromised security protections in order to deliver ads to consumers' ("Lenovo" 2017). Hyundai reportedly used child labor to make motor cars. It has been censured by the ASA for promotions claiming that a new model 'purifies the air as it goes' and deploying Chelsea footballers to deceive people over the time needed to charge electric vehicles (advertisement quoted in "ASA Ruling" 2021; "ASA Ruling on Hyundai" 2023). Qatar Airways flies across the globe. Wanda is a vast property developer; when it failed to meet a scheduled payment to FIFA, the firm was swiftly ostracized—just part of its scandalous financial history (Cronin and Ross 2024). Mengniu tortures cows and has sold poisoned infant formula ("China's Lethal" 2019).

UK television coverage of the 2018 World Cup Finals featured product placement for a multitude of dangerous sponsors promoting alcohol, sugar beverages, and fast "food" (Ireland et al. 2024). Across ten European countries, almost three-quarters of football clubs benefit from alcohol sponsors (Purves et al. 2022), which deplete and pollute ground water. Budweiser has over-stimulated rice cultivation, Ireland's Advertising Standards Authority determined that the firm misled customers via claims to use renewable energy, and the UK Environment Agency criticized it for recycling failures (Tweedale 2018; "Complaints Decisions" 2024; "Budweiser Budvar" 2024). McDonald's suppliers cultivate grain to feed livestock at unsustainable levels. The parent company is so lily-livered that having forced numerous media outlets to settle out of court for daring to criticize it, it took out the biggest libel case in British history to punish almost-penniless activists who had exposed its malpractice. And McDonald's often causes *E. coli* outbreaks (Singh et al. 2025).

Gazprom, the extractive sponsor to end all extractive sponsors, has shared oligarchic profits with FIFA and engaged in greenwashing through its Football for Friendship scheme via the slogan "Earth is a ball. Football drives the planet!" (n.d.). The company presided over a video contest inviting the young to record "Small steps to save the planet" in an 'environmental online challenge' ("Football for Friendship" 2021), a core part of its efforts to 'communiquer une image positive de l'industriel climaticide' [create a positive image of industrial climaticide] (Correia 2022: 429).

After the 2022 Finals, Nike achieved remarkable levels of positive awareness among viewers (SponsorPulse 2023). Meanwhile, in addition to generations of the company killing kangaroos to capture their bodies

(Touma 2023), executive use of the firm's Gulfstream G650ERs emitted 20% more CO_2 in 2023 than 2015 as part of the quest to 'improve productivity and address security concerns for executives' (Davis et al. 2024). Nike remains in recovery from decades of scandals over sweatshop factories, performance-enhancing drugs, and sexist, destructive relations with athletes. Needless to say, it claims to be focused on 'sustainable solutions to help protect our planet' (Hart 2020; Collier and Hernandez 2022).

Conclusion

The 2026 Finals were expanded to include forty-eight nations, with fixtures in three massive countries—Canada, Mexico, and the USA—across four time zones, sixteen venues separated by several thousand kilometers (Paché 2023). A projected five and a half million spectators confronted a veritable atlas, with Mexico City its southernmost point, Vancouver its northern equivalent, Boston furthest to the east, and San Francisco to the west (Paché 2024). Air travel would be a logistical norm because of primitive rail infrastructure in what Radio France referred to as a 'cocktail très carboné' [a very carbonated cocktail] (Cannesant 2023). If the Qatari World Cup of 2022 emitted 6 million tons of CO_2, double FIFA's promise (Dauphin 2022), one can only imagine the two figures for North America—the fantastical and the real—from transport alone. The New Weather Institute with Scientists for Global Responsibility calculate that aviation 'emissions will rise by between 160% and 325% for each of the three tournaments in 2026, 2030 and 2034' in comparison with recent Finals (2025b).

In addition, the period scheduled—high summer—saw 96% of the US population experience extreme heat for a week or more in 2023, while forty-five cities had vastly increased average temperatures (Lockwood 2023b). The following year set numerous records due to hitherto anomalous patterns of drought and heavy rain (National Oceanic and Atmospheric Administration 2024). Various bio-meteorological measures, such as oxygen, water, thermal climate, and physical exertion, could see more than half the 2026 sites generating 'extreme heat stress' among footballers, with illness a serious consequence (Lindner-Cendrowska et al. 2024; Craig and Karabas 2024; Mullan et al. 2025).

Laughable US mass transit, poor facilities, and violent fans led *Foreign Policy* to declare 2026 "Heading for Disaster" (De Loera-Brust 2024).

For some, it promised to be 'the largest hospitality program in the history of sports'—if one could afford 'stadium suites and hospitality villages' and 'exclusive food and entertainment experiences,' which FIFA's hired help refers to as 'curated' (quoted in Thomas 2024). Everything will be fine, unless you're a Mexican citizen or an Iranian player looking to enter the country—or anyone who can't afford to have their lives 'curated.' Across eighteen US embassies and consulates in early 2025, people waited an average of 330 days to get visas (700 from Colombia) and confronted an Administration dedicated to repelling all boarders (Baxter with Wilkinson 2025). Human Rights Watch wrote to Infantino about the brutal and discriminatory acts of the US government toward visitors, contrasting such conduct with FIFA doctrine (Worden 2025). Amnesty International criticized stepped-up infringements of human rights and civil liberties (Amnistía Internacional 2025).

The environmental impacts in Mexico alone were horrifying. Guadalajara's Primavera reserve is close to its Akron stadium and home to pumas, an almost extinct species, golden eagles, migratory birds, and deer. Wildlife and plant conservation and water resources were put at severe risk due to Cup development and the presence of 50,000 spectators and 4000 cars. Mexico City and Monterrey (also next to a key biological corridor) were subject to similar strains (Pistolesi and Pelliccia 2024; Collins 2025), notably the Azteca's misuse of water resources (Woodford 2025).

In 2030, the hosts will be Morocco, Portugal, and Spain. They're not contiguous, though close-by, and Spain has imperial possessions on the North African coast. The event will be a relief from the absurdity of 2026 because of Iberian railway excellence, albeit with even more severe weather conditions: all three hosts had record maximum temperatures in the summer of 2024, close to 50 °C in some areas, and experienced 'climate-fueled drought, flooding and wildfires' (Carosella 2023). Saudi Arabia is the setting in 2034, a prize secured when the Saudis paid FIFA vast amounts through sponsorship from its supposedly private firm Aramco (Cannesant 2023), with all that implies in terms of replaying the Qatari disgrace/repaying Aramco's investment. That sponsorship is estimated to add about 30 million tCO2e during and after the 2026 Finals, thanks to increased sales for the fossil-fuel criminal (New Weather Institute with Scientists for Global Responsibility 2025b).

The notorious Crown Prince Mohammed bin Salman said:

> Sport used to participate in Saudi GDP [Gross Domestic Product] by 0.4%. Today it's 1.5%. So it's economic growth, it's jobs, it's a calendar, it's entertainment, it's tourism and you can see we are now ranked number one in the Middle East … [i]f sport washing is going to increase my GDP by way of 1%, then I will continue doing sport washing. ("Saudi Crown" 2023)

The clown prince and his devotees also wished to reshape the country's world standing: sponsoring twenty-one sports via 300 contracts was designed to divert attention from slaughtering Ethiopian migrants and feminists, *inter alios* ("Why the World" 2024; Kingdom of Saudi Arabia 2023; Reda and Viviez 2021; Lockwood 2023c; Roan 2023; "Time for LIV" 2024; Elsborg 2023). Focusing on 2024 and beyond, Play the Game found '910 Saudi sponsorships in sport' (Elsborg et al. 2024: 4).

The *Guardian* parodied 2034's bidding process:

> The candidates to stage the 2034 edition of the grandest and most lucrative entertainment event on the planet lineup as follows. Favourites: Saudi Arabia. Shortlisted: Saudi Arabia. Dark horses: Saudi Arabia. Chief sponsor of Fifa: Saudi Arabia. Only bid not eliminated by a gathering sense of inevitability: Saudi Arabia. (Ronay 2024)

Carbon Market Watch noted that the Association had put 'greed ahead of people and planet' and the Sport for Climate Action Network declared 2034 part of the Saudi project to ensure fossil-fuel's future, alongside disrupting environmental negotiations and ensuring ongoing national addictions to Aramco poisons. Climate Action Tracker found the state's attitudes to allocating a fair share of the impact of climate change 'CRITICALLY INSUFFICIENT' (2024). The message from FIFA was 'oil and gas stands for dynamism, prosperity and power, not the regressive, lethal pollution it is'.

Infantino celebrated the sites selected:

> The GREATEST SHOW ON EARTH will be organised by CA Canada, MX Mexico and the US United States in 2026—in North America. The next two editions of the FIFA World Cup are set to be hosted in Africa (MA Morocco) and Europe (PT Portugal and ES Spain)—with three celebratory matches played in South America (AR Argentina, PY Paraguay and UY Uruguay)—in 2030 and in Asia (SA Saudi Arabia) in 2034. Three editions, five continents and ten countries involved in staging matches in the tournament—that's making football truly global! (Infantino 2024)

The solipsistic man we encountered at the beginning of this chapter used pre-pubescent graphics to express his excitement. A wizard wheeze. Infantino went on to describe the Association's choices as 'uniting the entire world' (quoted in "FIFA President" 2025), his juvenilia matched only by his bombast.

Freedom House gave Saudi a human-rights score of eight out of a hundred (2024). Examination of the nation's death certificates indicated chronic failure to record etiology satisfactorily (Aljerian 2019). FIFA blithely ignored distinguished legal counsel's careful explanation that the Saudis infringed the Association's own human-rights policies. It preferred advice based on the would-be host's bizarre version of such doctrines, which of course exclude free speech, same-sex love, and trade unions (Pieth et al. 2024; Das 2024).

Tens of thousands of guest workers quickly perished in construction projects related to the Cup and "development" in general (*Kingdom Uncovered: Inside Saudi Arabia*). When Bangladeshi workers hired to build a Saudi resort went on strike to protest unpaid wages, they were sacked (Kunti and Selliaas 2025). Further abuse through eviction was matched by environmental destructiveness (Ames 2024; ALQST for Human Rights 2023; "Saudi Arabia/FIFA" 2024; Amnesty International 2024).

As two US Senators put it in their correspondence with Infantino:

> While the Saudi bid document pledges to engage with its own state agencies on these issues, the Saudis have failed to provide any mention of partnerships with Saudi human rights experts, human rights organizations, trade unions, or rights advisors to ensure basic standard. More concerningly, the Saudis have failed to address how they will uphold labor protections, press freedoms, non-discrimination and inclusion standards. Indeed, human rights organizations, experts, and governing bodies, like the United Nations, continue to sound the alarm on Saudi Arabia's deteriorating human rights conditions and their inability to monitor labor and human rights conditions on the ground due to Saudi's repressive and restrictive policies. (Wyden and Durbin 2024)

The African Regional Organisation of the International Trade Union Confederation protested to the Confederation of African Football about 'Saudi Arabia's egregious abuse and undermining of migrant workers' rights,' which it 'feared will worsen with its planned bid to host the 2034 FIFA World Cup' because 'migrant workers, especially Africans, are given the most crude and brutal human treatment' (Odigie 2024). The BWI produced a scathing investigation of Saudi behavior toward employees, from basic exploitation to forced labor (2024). Progressive Members of the European Parliament wrote to Infantino, noting the Saudis' 'repression of independent civil society and critical voices,' silencing of superstitions other than their own, imprisonment of dissidents and activists, risks to guest construction workers, and failure to protect queer people (Roba i Giner et al. 2024). Thirty other members from various parties wrote a second 'excoriating' critique (Walker 2024). FairSquare's report on migrant deaths was horrifying (2025a), its condemnation of FIFA's complicity shared with scholars, NGOs, and fans (2025b).

The Saudis' claim to host took as its slogan "Growing. Together" in a bid 'based primarily on a desire for growth' while of course guaranteeing 'sustainability' (Saudi Arabia 2024). The Aramco-and-friends plan was obvious: to replicate the *succès d'estime* of the 1978 Finals, hosted by an Argentine *junta* cutting a swathe through democracy under the loving hand of the USA. Those fascists engaged in social clearance prior to the Cup, removing potential protestors from areas where overseas reporters were to stay, reducing overt oppression during the tournament in such zones then re-intensifying it once the press corps had departed (Scharpf et al. 2023).

None of this halted FIFAcrats in their tracks; after all, they had awarded Saudi Arabia's bid a record grade. Human rights? 'medium risk.' Environment? 'low risk' (FIFA 2024: 94). Note the use of the word 'risk'—to reputation, not people or planet.

Putting personal or collective corruption to one side, another word matters here: growth. The political-economic priority remains selling the game in sizeable and affluent regions (Conn 2017). The Association is a transnational actor without democratic oversight, enjoying greater autonomy than most corporations—and fewer tax liabilities (Paché 2023; FairSquare 2024). A glance at its accounts shows how revenues have grown:

FIFA income

Years	*$Billion*
2011–14	5.15
2015–18	6.44
2019–22	7.57

Source: Based on Richter (2024)

At the end of each tournament, teams and their supporters may have memories and hopes founded on what they have seen and done—or not seen and done—during the Finals, per A.S. Byatt noting that the 'English press, and the English team, and the English supporters always collude to tell an improbable story about what is going to happen before a competition begins' (2008). But 'más allá del análisis deportivo, de organización, económico y de impacto social, llega la hora de revisar el impacto que un evento de estas proporciones tiene para el planeta' [after evaluating the event *qua* sport, and its socio-economic impact, comes the time to examine what an event of such proportions means for the Earth] (de la Cruz 2022).

For the World Cup is 'simultaneously the greatest sporting festival on the planet, and a tawdry commercial machine run at enormous human and environmental cost for the benefit of torturers, murderers, exploiters and the unquenchably greedy, scrutiny stymied by the sheer variety of outrage' (Wilson 2024). Such cynical use of sports helps draw young people toward both perilous consumption and tolerance of criminality (Brown 2016; Graham 2025).

Ecology only matters in football to the extent that governing associations, host countries, and the sport in question consider it a factor in their public image and their sponsors'. Instead of more greenwashing and pseudo-reform, football desperately needs a dramatic transformation to disrupt its environmental impact, corrupt rule, systematic human-rights offenses, and sponsorship by the animal-slavery addiction and money laundering of gambling and other miscreants.

FIFA has a 'fatal climate blind spot' (New Weather Institute with Scientists for Global Responsibility 2025b). Its carbon party must end. Abolish the World Cup.

Bibliography

NB—Links from US Science Have Been Removed By Its Government

"2025 Budget Notes: FIFA World Cup 2026 Toronto." (2025). City of Toronto https://www.toronto.ca/legdocs/mmis/2025/bu/bgrd/backgroundfile-252534.pdf.

"A New Way to Recycle Plastic is Here." (2025, April 16). *Economist* https://www.economist.com/business/2025/04/16/a-new-way-to-recycle-plastic-is-here.

"Almost Half the World Tuned in at Home to Watch 2010 FIFA World Cup South Africa™." (2011, July 20). FIFA www.fifa.com/worldcup/archive/southafrica2010/organisation/media/newsid=1473143/index.html.

"America's Gambling Boom Should be Celebrated, not Feared." (2024, December 7). *Economist* https://www.economist.com/leaders/2024/12/05/americas-gambling-boom-should-be-celebrated-not-feared.

"ASA Ruling on Hyundai Motor UK Ltd." (2021, June 9). Advertising Standards Authority [UK]. Complaint A21-1096716 Hyundai Motor UK Ltd https://www.asa.org.uk/rulings/hyundai-motor-uk-ltd-a21-1096716-hyundai-motor-uk-ltd.html.

"ASA Ruling on Hyundai Motor UK Ltd." (2023, June 28). Advertising Standards Authority [UK]. Complaint G22-1147624 Hyundai Motor UK Ltd https://www.asa.org.uk/rulings/hyundai-motor-uk-ltd-g22-1147624-hyundai-motor-uk-ltd.html.

"ASA Ruling on Repsol SA." (2023, October 18). Advertising Standards Authority [UK]. Complaint A23-1206254 Repsol SA https://www.asa.org.uk/rulings/repsol-sa-a23-1206254-repsol-sa.html.

"BP Ad: Carbon Footprint." (2003). https://www.youtube.com/watch?v=ywrZPypqSB4.

"Budweiser Budvar Pays Charity £414,000 for Recycling Failure." (2024, August 15). Environment Agency https://www.gov.uk/government/news/budweiser-budvar-pays-charity-414000-for-recycling-failure.

"Calculan costo ambiental del Mundial." (2014, June 24). *Revista Mundo Verde* http://www.revistamundoverde.net/noticias-ambientales/calcula-costo-ambiental-del-mundial.

"China's Lethal Milk Scandal Reverberates a Decade Later." (2019, January 22). *Straits Times* https://www.straitstimes.com/asia/east-asia/chinas-lethal-milk-scandal-reverberates-a-decade-later.

"Climate X Report Projects $800 Million in Climate-Related Losses for Fifa World Cup Venues by 2050." (2024, December 3). Climate X https://www.climate-x.com/articles/industry/climate-x-report-projects-800-million-in-climate-related-losses-for-fifa-world-cup-venues-by-2050.

"Complaints Decisions." (2024, January). Advertising Standards Authority [Ireland]. C&C Group—Budweiser Reference 41478 https://adstandards.ie/complaint/alcohol-35/.

"FIFA and UEFA Welcome Helsinki Report on Sport from the European Commission." (1999, December 4). FIFA http://www.fifa.com/news/y=1999/m=12/news=fifa-and-uefa-welcome-helsinki-reportsport-from-the-european-commissi-71582.html.

"FIFA Announces Deal with Saudi Oil Company Aramco." (2024, April 25). *BBC* https://www.bbc.com/sport/football/articles/c51n3ld43yro.

"FIFA President Infantino Defends Saudi World Cup, Trump Relationship." (2025, March 1). *RTL Today* https://today.rtl.lu/sport/international/a/2281240.html.

"FIFA Produces Whitewash of Saudi Arabia World Cup Bid and Ignores Own Report to Provide Compensation for Migrant Workers in Qatar." (2024, November 30). Amnesty International https://www.amnesty.org/en/latest/news/2024/11/global-fifa-produces-whitewash-of-saudi-arabia-world-cup-bid-and-ignores-own-report-to-provide-compensation-for-migrant-workers-in-qatar/.

"Football for Friendship Raises Environmental Awareness Whilst Breaking Records." (2021, July 23). *Euronews* [advertorial for Gazprom] https://www.euronews.com/business/2021/07/21/football-for-friendship-raises-environmental-awareness-whilst-breaking-records.

"He Outran the Flames and Lost His Mum. Now He Could be Shot." (2025, January 16). *Animals Australia* https://animalsaustralia.org/our-work/shooting-and-hunting/urge-adidas-to-stop-using-bushfire-victims-for-shoes/.

"How Sports Gambling Became Ubiquitous." (2024, December 5). *Economist* https://www.economist.com/finance-and-economics/2024/12/05/how-sports-gambling-became-ubiquitous.

"Lenovo Settles FTC Charges it Harmed Consumers With Preinstalled Software on its Laptops that Compromised Online Security." (2017, September 5). Federal Trade Commission https://www.ftc.gov/news-events/news/press-releases/2017/09/lenovo-settles-ftc-charges-it-harmed-consumers-preinstalled-software-its-laptops-compromised-online.

"Nearly Half of All Professional Football Stadiums Put at Risk by Climate Change." (2023, October 9). Zurich UK https://www.zurich.co.uk/media-centre/football-and-climate-change.

"Petronor y Repsol se convierten en los partners de energía del Athletic." (2024, April 19). Athletic Club de Bilbao https://www.athletic-club.eus/noticias/2024/04/19/petronor-y-repsol-se-convierten-en-los-partners-de-energia-del-athletic/.

"Premier League Agrees Record £6.7bn Domestic TV Rights Deal." (2023, December 4). *BBC* https://www.bbc.com/sport/football/67619756.

"Premier League Clubs' Flight Habits Revealed." (2025, May 28). Campaign for Better Transport https://bettertransport.org.uk/media/premier-league-clubs-flight-habits-revealed/.

"PSG Coach Sets Off Storm With Quip About Flying." (2022, September 6). https://www.youtube.com/watch?v=gRlhBqofTdM.

"PSG Face Backlash for Mocking Eco-Friendly Travel Plea." (2022, September 6). *France 24* https://www.france24.com/en/france/20220906-psg-face-backlash-for-mocking-eco-friendly-travel-plea.

"Qatar's World Cup Will Emit More CO2 Than Any Recent Sporting Event." (2022, December 1). *Economist* https://www.economist.com/graphic-detail/2022/12/01/qatars-world-cup-will-emit-more-co2-than-any-recent-sporting-event.

"Repsol patrocinará a la selección peruana tras haber causado el mayor desastre ecológico de su historia." (2023, September 15). *El Salto* https://www.elsaltodiario.com/greenwashing/repsol-patrocinara-seleccion-peruana-haber-causado-mayor-desastre-ecologico-historia.

"Saudi Arabia/FIFA: Organizations Demand a Credible Human Rights Assessment for 2034 World Cup Bid." (2024, October 28). Amnesty International https://www.amnesty.org/en/latest/news/2024/10/saudi-arabia-fifa-organizations-demand-a-credible-human-rights-assessment-for-2034-world-cup-bid/.

"Saudi Crown Prince Interview with Fox News." (2023). https://www.youtube.com/watch?v=w0NxI44yBDM.

"Session is Done: FIFA World Cup 2026 Gets $21m." (2024, March 11). The Chamber https://www.seattlechamber.com/news/2024/03/11/policy-advocacy/session-is-done-fifa-world-cup-2026-gets-21m/.

"Swiss Commission for Fairness Upholds Complaints Against FIFA." (2023, June 7). Commission Suisse pour la Loyauté https://www.faire-werbung.ch/de/swiss-commission-for-fairness-upholds-complaints-against-fifa/.

"The Fundamental Contradiction of ESG is Being Laid Bare." (2022, September 29). *Economist* https://www.economist.com/leaders/2022/09/29/the-fundamental-contradiction-of-esg-is-being-laid-bare.

"Why the World is Teeming with So Many New Sports Leagues." (2024, August 23). *Economist* https://www.economist.com/culture/2024/08/23/why-the-world-is-teeming-with-so-many-new-sports-leagues.

Abusin, Sana, Noora Lari, Salma Khaled, and Noor Al Emadi. (2020). "Effective Policies to Mitigate Food Waste in Qatar." *African Journal of Agricultural Research* 15, no. 3: 343–50.

Adorno, Theodor and Max Horkheimer. (1979). *Dialectic of the Enlightenment*. Trans. John Cumming. London: Verso.

Ahmed, Fathima, and Leon Pretorius. (2010). "Mega-Events and Environmental Impacts: The 2010 FIFA World Cup in South Africa." *Alternation* 17, no. 2: 274–96.

Alarcón, Adriana. (2025, June 12). "Mexico's World Cup 2026: Infrastructure Race Amid Investment Cuts." *Mexico Business News* https://mexicobusiness.news/infrastructure/news/mexicos-world-cup-2026-infrastructure-race-amid-investment-cuts.

Aljerian, Khaldoon. (2019). "Death Certificate Errors in One Saudi Arabian Hospital." *Death Studies* 43, no. 5: 311–15.

ALQST for Human Rights. (2023, February). *The Dark Side of Neom: Expropriation, Expulsion and Prosecution of the Region's Inhabitants* https://alqst.org/uploads/the-dark-side-of-neom-expropriation-expulsion-and-prosecution-en.pdf.

Althusser, Louis. (1971). *Lenin and Philosophy and Other Essays.* Trans. Ben Brewster. New York: Monthly Review Press.

Ames, Nick. (2024, October 28). "Saudi Arabia's World Cup Bid Gathers Pace—But at What Cost to Human Life?" *Guardian* https://www.theguardian.com/football/2024/oct/28/saudi-arabia-world-cup-2034-qatar-fifa-neom.

Amnesty International. (2024, November 11). *High Stakes Bids: Dangerously Flawed Human Rights Strategies for the 2030 and 2034 FIFA World Cups* https://www.amnesty.org/es/documents/ior10/8712/2024/en/.

Amnistía Internacional. (2025, June 11). *Copa Mundial: Cuando falta un año para su celebración, aumentan los ataques a los derechos* https://www.amnesty.org/es/latest/news/2025/06/world-cup-a-year-out-growing-attacks-on-rights/.

Ampere. (2025). *A Year of Sport 2024: Full Stream Ahead in 2025.* https://19957038.fs1.hubspotusercontent-na1.net/hubfs/19957038/2024%20in%20Sports-1.pdf?utm_medium=email&_hsenc=p2ANqtz-_dXEmIwhUZXNR8DZTyCGUGWDN8MjPoO5EcDUzk62EM2P1PCObUjAH9Y8GOaZ2ESRWnL-vWSERAUBeKBETg6joUvYAj0w&_hsmi=307835695&utm_content=307835695&utm_source=hs_automation.

Arsenal. (n.d.). https://www.arsenal.com/news/sustainability.

Ashoo, Sam. (2021, December 3). "Climate Crisis: How Can Football Make a Difference?" *Euronews* https://www.euronews.com/2021/10/28/the-climate-crisis-why-football-can-no-longer-hide.

Attali, Jacques. (2001). *Bruits: Essais sur l'économie politique de la musique*, 2nd ed.. Paris: Librairie Arthème Fayard/Presses Universitaires de France.

Auclair, Philippe, Andy Brown, Jack Kerr, Sam Kunti, and Steve Menary. (2024a, August 12). "White Elephants and Green Lies." *Josimar* https://josimarfootball.com/2024/08/12/white-elephants-and-green-lies/.

Auclair, Philippe, Andy Brown, Jack Kerr, Sam Kunti, and Steve Menary. (2024b, October 21). "Child's Play." *Josimar* https://josimarfootball.com/2024/10/21/childs-play/.

Auclair, Philippe. (2021, August 12). "The Trillion-Dollar Gambling Game." *Josimar* https://josimarfootball.com/2021/08/12/the-trillion-dollar-gambling-game/.

Auclair, Philippe. (2022, January 31). "Gambling with Lives." *Josimar* https://josimarfootball.com/2022/01/31/gambling-with-lives/.

Auclair, Philippe. (2023, June 23). "In the Name of Charity." *Josimar* https://josimarfootball.com/2023/06/23/in-the-name-of-charity/.

Auclair, Philippe. (2024, June 27). "Uefa's Lofty Environmental Ambitions and the Elephant in the Room." *Guardian* https://www.theguardian.com/football/article/2024/jun/27/uefa-lofty-environmental-ambitions-euro-2024-elephant-in-the-room.

Auclair, Philippe. (2025, July 29). "Plus ça change." *Josimar* https://josimarfootball.com/2025/07/29/plus-ca-change/.

Austin, Daniel. (2025, April 18). "From Beer to Betting—How Have Football Shirt Sponsors Changed?" *BBC* https://www.bbc.com/sport/football/articles/cge1lp0734jo.

Badiou, Alain. (2007). *The Century*. Trans. Alberto Toscano. Cambridge: Polity.

Barcelona FC. (n.d.). https://www.fcbarcelona.cat/ca/club/sostenibilitat.

Baudrillard, Jean. (1983). *In the Shadow of the Silent Majorities ... Or the End of the Social and Other Essays*. Trans. Paul Foss, Paul Patton, and John Johnston. New York: Semiotext(e).

Bautista-González, Manuel A. (2023, October 30). "The Environmental Footprint of Payment Cards." *CashEssentials* https://cashessentials.org/the-environmental-footprint-of-payment-cards/.

Baxter, Kevin with Tracy Wilkinson. (2025, January 29). "Visa Approval Crisis Threatens to Cost 2026 World Cup and L.A. Olympics Millions." *Los Angeles Times* https://www.latimes.com/sports/story/2025-01-29/sports-visas-delay-olympics-world-cup.

Beck, Ulrich. (2009). *World at Risk*. Trans. Ciaran Cronin. Cambridge: Polity.

Biel, Jonas, Vincent Reinke, Tobias Finger, and Arne Niemann. (2024). "No Longer Sidelined? Football Fandom, Belonging, and the Boundaries of Europe." *Journal of Contemporary European Studies* 32, no. 4: 1343–365.

Blake, Heidi and Jonathan Calvert. (2016). *The Ugly Game: The Qatari Plot to Buy the World Cup*. New York: Simon & Schuster.

Blake, Heidi. (2022, December 1). "Qatar's Tarnished World Cup." *New Yorker* https://www.newyorker.com/culture/cultural-comment/qatars-tarnished-world-cup.

Borges, Jorge Luis with Richard Burgin, Daniel Bourne, Stephen Cape, Charles Silver, and Gloria López Lecube. (2013). *The Last Interview and Other Conversations*. Trans. Kit Maude. New York: Melville House.

Boykoff, Jules. (2022, November 23). "The World Cup in Qatar is a Climate Catastrophe." *Scientific American* https://www.scientificamerican.com/article/the-world-cup-in-qatar-is-a-climate-catastrophe/.

Break Free From Plastic. (2023). *Brand Audit Report 2023* https://drive.google.com/file/d/1YFyfRv4mviZZXa8b1HdpucDX3WEwJzv/view.

Brecht, Bertolt. (1964). "Emphasis on Sport." *Brecht on Theatre*. Ed. John Willett. Basingstoke: Macmillan. 6–9.

Brigham, Katie. (2022, January 29). "How the Fossil Fuel Industry is Pushing the Plastics on the World." *CNBC* https://www.cnbc.com/2022/01/29/how-the-fossil-fuel-industry-is-pushing-plastics-on-the-world-.html.

British Screen Forum. (2021, April). *Business Briefing: High-End Television Drama Investment: An Update* https://britishscreenforum.co.uk/wp-content/uploads/2021/04/British-Screen-Forum-High-end-Television-Drama-An-Update.pdf.

Brown, Katherine. (2016). "Association Between Alcohol Sports Sponsorship and Consumption: A Systematic Review." *Alcohol and Alcoholism* 51, no. 6: 747–55.

Brown, Andy, Philippe Auclair, Jack Kerr, Steve Menary, and Samindra Kunti. (2025, February 6). "Strangelove: How the Premier League Can Enable Illegal Gambling and Money Laundering." *Josimar* https://josimarfootball.com/2025/02/06/strangelove-how-premier-league-learned-to-stop-worrying-and-embrace-illegal-gambling-and-money-laundering/.

Building and Wood Workers' International. (2024). *Migrant Worker Conditions in the Kingdom of Saudi Arabia: An Overview of Worker Abuses Linked with Forced Labour* https://www.diarioconstitucional.cl/wp-content/uploads/2024/06/Vea-texto-investigativo-de-la-denuncia.pdf.

Buzzoni, Lorenzo, Marta Portocarrero, and Chris Matthews. (2025, March 20). "Revealed: How Gambling Embedded Itself into European Football." *Investigate Europe* https://www.investigate-europe.eu/posts/revealed-how-gambling-betting-industry-sponsors-european-football.

Byatt, A. S. (2008, June 29). "Lie Back and Think of Europe." *Observer* https://www.theguardian.com/books/2008/jun/29/sportandleisure.review.

Cabot, Cyrielle. (2022, October 28). "Qatar: derrière "la neutralité carbone", la Coupe du monde du "greenwashing"." *France 24* https://www.france24.com/fr/sports/20221028-qatar-derri%C3%A8re-la-neutralit%C3%A9-carbone-la-coupe-du-monde-du-greenwashing.

Cabral, Amílcar. (1979). *Unity and Struggle: Speeches and Writings: Texts Selected by the PAIGC* [Partido Africano da Independência da Guiné e Cabo Verde]. Trans. Michael Wolfers. New York: Monthly Review Press.

Cannesant, Raphael. (2023, December 1). "Coupe du Monde de football: la FIFA se moque-t-elle des enjeux environnementaux?" *Radiofrance* https://www.radiofrance.fr/mouv/coupe-du-monde-de-football-la-fifa-se-moque-t-elle-des-enjeux-environnementaux-spoiler-oui-9717302.

Carbon Market Watch and éclaircies. (2024). *Going for Green: Assessing the Climate Strategy and Communication of the 2024 Paris Olympics* https://carbonmarketwatch.org/wp-content/uploads/2024/04/Going-for-green-Olympics-report.pdf.

Carbon Market Watch. (2022, May). *Poor Tackling: Yellow Card for 2022 FIFA World Cup's Carbon Neutrality Claim* https://carbonmarketwatch.org/wp-content/uploads/2022/05/Poor-tackling-Yellow-card-for-2022-FIFA-1.pdf.

Carosella, Vitas. (2023, October 6). "FIFA Announces 2030 and 2034 World Cup Plans Amongst Climate Concerns." *Forbes* https://www.forbes.com/sites/vitascarosella/2023/10/06/fifa-announces-2030-and-2034-world-cup-plans-amongst-climate-concerns/.

Charfeddine, Lanouar, Afnan Yousef Al-Malk, and Kholoud Al Korbi. (2018). "Is it Possible to Improve Environmental Quality Without Reducing Economic Growth: Evidence from the Qatar Economy." *Renewable and Sustainable Energy Reviews* 82, no. 1: 25–39.

Chandler, Raymond. (1949). *The Little Sister*. New York: Houghton Mifflin.

Collier, Eric and Anthony Hernandez. (2022, July 23). "The Nike Saga: Winning at All Costs." *Le Monde* https://www.lemonde.fr/en/summer-reads/article/2022/07/23/the-nike-saga-winning-at-all-costs_5991163_183.html.

Collins, Andrea, Calvin Jones, and Max Munday. (2009). "Assessing the Environmental Impacts of Mega Sporting Events: Two Options?" *Tourism Management* 30, no. 6: 828–37.

Collins, Andrea and Andrew Flynn. (2008). "Measuring the Environmental Sustainability of a Major Sporting Event: A Case Study of the FA Cup Final." *Tourism Economics* 14, no. 4: 751–68.

Collins, Marybeth. (2025, July 14). "World Cup Projects in Mexico Threaten Ecosystems. *Environment + Energy Leader* https://www.environmentenergyleader.com/stories/world-cup-projects-in-mexico-threaten-ecosystems,84079?.

Committee on Women, Children, Family Relations and Gender Equality. (2022, January 26). *Executive Summary Committee Report Senate Resolution No. 131* https://legacy.senate.gov.ph/lis/committee_rpt.aspx?congress=18&q=542.

Conn, David. (2017). *The Fall of the House of FIFA*. New York: Nation Books.

Cornelissen, Scarlett, Bob Urmilla, and Kamilla Swart. (2011). "Towards Redefining the Concept of Legacy in Relation to Sport Mega-Events: Insights from the 2010 FIFA World Cup." *Development Southern Africa* 28, no. 3: 307–18.

Correia, Mickaël. (2018). *Une histoire populaire du football*. Paris: Éditions La Découverte.

Correia, Mickaël. (2022). *Criminels climatiques*. Paris: Éditions La Découverte.

Cowger, Win, Kathryn A. Willis, Sybil Bullock, Katie Conlon, Jorge Emmanuel, Lisa M. Erdle, Marcus Eriksen, Trisia A. Farrelly, Britta Denise Hardesty, Kristinna Kerge, Natalie Li, Yedan Li, Adam Liebman, Neil Tangri, Martin Thiel, Patricia Villarrubia-Gómez, Tony R. Walker, and Mengjiao Wang. (2024). "Global Producer Responsibility for Plastic Pollution." *Science Advances* 10, no. 17 https://doi.org/10.1126/sciadv.adj8275.

Crabb, Lauren A. H. (2018). "Debating the Success of Carbon-Offsetting Projects at Sports Mega-Events. A Case from the 2014 FIFA World Cup." *Journal of Sustainable Forestry* 3, no. 2: 178–96.

Crafton, Adam. (2024, April 30). "NBC to Continue Push for Premier League Games in U.S., Says Leading Exec." *The Athletic* https://www.nytimes.com/athletic/5460003/2024/04/30/premier-league-games-united-states-nbc/.

Craig, Christopher and Ismail Karabas. (2024). "Forecasting Thermal Stress for Sports Tourists at the 2026 FIFA World Cup." *International Journal of Biometeorology* 68: 2731–742.

Cronin, Ben and Martin Ross. (2024, March 5). "EXCLUSIVE: FIFA Withdraws Wanda Sponsorship Rights Over Missed Payments." *SportBusiness* https://www.sportbusiness.com/news/exclusive-fifa-withdraws-wanda-sponsorship-rights-over-missed-payments/.

Cronin, Ben. (2024, September 16). "Qatar Airways Ties Down UCL in 6-Year, €500m Deal." *SportBusiness* https://www.sportbusiness.com/news/exclusive-qatar-airways-ties-down-ucl-in-6-year-e500m-deal/.

Cusce Nobre, Eduardo Alberto. (2017). "Preface." *Sports Mega-Events and Urban Legacies: The 2014 FIFA World Cup, Brazil.* Ed. Eduardo Alberto Cusce Nobre. Cham: Palgrave. v-vi.

Daddi, Tiberio, Francesco Rizzi, Gaia Pretner, Niccolò Todaro, Eleonora Annunziata, Marco Frey, and Fabio Iraldo. (2022). "Environmental Management of Sport Events: A Focus on European Professional Football." *Sport, Business and Management: An International Journal* 12, no. 2: 208–32.

daMatta, Roberto. (1991). *Carnivals, Rogues, and Heroes: An Interpretation of the Brazilian Dilemma.* Trans. John Dury. Notre Dame: University of Notre Dame Press.

Danjon, Alexis. (2023, June 7). "La FIFA a bien fait du greenwashing pour la Coupe du monde 2022." *L'Équipe* https://www.lequipe.fr/Football/Actualites/La-fifa-a-bien-fait-du-greenwashing-pour-la-coupe-du-monde-2022/1401281.

Das, Shanti. (2024, November 2). "'It's Created an Internal Shitstorm': Turmoil at UK Law Firm Accused of 'Whitewashing' Saudi World Cup Report." *Guardian* https://www.theguardian.com/world/2024/nov/02/top-uk-law-firm-accused-of-whitewashing-saudi-human-rights-report-for-world-cup-bid.

Dauphin, Sandy. (2022, November 8). "L'empreinte carbone du Mondial au Qatar serait près de deux fois supérieure à celle annoncée par la FIFA." *FranceInter* https://www.radiofrance.fr/franceinter/l-empreinte-carbone-du-mondial-au-qatar-serait-pres-de-deux-fois-superieure-a-celle-annoncee-par-la-fifa-8733673.

Davis, Rob, Agnel Philip, Alex Mierjeski, and Matthew Kish with Ryanne Mena and Jeff Frankl. (2024, August 13). "After Nike Leaders Promised Climate Action, Their Corporate Jets Kept Flying—and Polluting." *ProPublica* https://www.propublica.org/article/nike-corporate-jet-travel-carbon-emissions.

Death, Carl. (2011). "Greening the 2010 FIFA World Cup: Environmental Sustainability and the Mega-Event in South Africa." *Journal of Environmental Policy & Planning* 13, no. 2: 99–117.

de la Cruz, José Luis. (2022, December 20). "Mundial Catar 2022: la insostenibilidad ambiental de los macro eventos deportivos." *Público* https://blogs.publico.es/otrasmiradas/67095/mundial-catar-2022-la-insostenibilidad-ambiental-de-los-macro-eventos-deportivos/.

De Loera-Brust, Antonio. (2024, August 11). "The World Cup is Coming to America—And Heading for Disaster." *Foreign Policy* https://foreignpolicy.com/2024/08/11/world-cup-united-states-2026-germany-qatar-russia-olympics/.

Deloitte. (2025, January 23). *Deloitte Football Money League 2025* https://www.deloitte.com/uk/en/services/financial-advisory/analysis/deloitte-football-money-league.html?utm_source=substack&utm_medium=email.

de Menezes, Jack. (2015, November 28). "Arsenal Criticized for Taking 'Ridiculous, Ludicrous and Farcical' 14-Minute Plane Flight to Norwich." *Independent* https://www.independent.co.uk/sport/football/premier-league/arsenal-criticised-for-taking-ridiculous-14-minute-plane-flight-to-norwich-a6752486.html.

Ding, Jaimie and Associated Press. (2024, November 1). "Coca-Cola and Pepsi, Two of the World's Top Plastic Polluters, Just Got Slapped with a Lawsuit Alleging Exaggerated Recycling Claims and Downplayed Health Effects." *Fortune*https://fortune.com/2024/11/01/coca-cola-pepsi-plastic-pollution-lawsuit-los-angeles-county/.

Dmowski, Seweryn. (2015). "Football Sites of Memory in the Eastern Bloc 1945–1991." *European Football and Collective Memory.* Ed. Wolfram Pyta and Nils Havermann. Basingstoke: Palgrave Macmillan. 171–84.

Dosumu, Adekunle, Ian Colbeck, and Rachel Elizabeth Bragg. (2017). "Greenhouse Gas Emissions as a Result of Spectators Travelling to Football in England." *Nature* 7: 6986.

Doyle, Julie. (2011). *Mediating Climate Change.* Abingdon: Routledge.

Duval, Antoine and Ben Van Rompuy, eds. (2016). *The Legacy of Bosman: Revisiting the Relationship Between EU Law and Sport.* The Hague: Asser/ Cham: Springer.

Eagleton, Terry. (2010, June 15). "Football: A Dear Friend to Capitalism." *Guardian* https://www.theguardian.com/commentisfree/2010/jun/15/football-socialism-crack-cocaine-people.

Eisenstein, Sergei. (1987). *Nonindifferent Nature.* Trans. Herbert Marshall and Roberta Reader. New York: Cambridge University Press.

Elias, Norbert and Eric Dunning. (1986). *Quest for Excitement: Sport and Leisure in the Civilizing Process.* Oxford: Basil Blackwell.

Elsborg, Stanis and Karim Zidan with Nathalie Zahle and Lasse Kjær Nielsen. (2024). *Saudi Arabia's Grip on World Sport*. Play the Game https://www.playthegame.org/media/rxflxlpt/saudi-arabias-grip-on-world-sport.pdf.

Elsborg, Stanis. (2023). "The Power Players Behind Saudi Arabia's Sports Strategy." *Play the Game* https://www.playthegame.org/news/the-power-players-behind-saudi-arabias-sports-strategy/.

Engar, Aadil and Jacques du Toit. (2023). "Cape Town's 2010 FIFA World Cup Stadium Location and its Spatial and Environmental Justice Implications." *Sport Stadiums and Environmental Justice*. Ed. Timothy Kellison. London: Routledge. 158–71.

Estrada, Daniela. (2010, June 3). "World Cup 2010: Climate Change Fouls and Goals." *Guardian* www.guardian.co.uk/environment/2010/jun/03/climate-change-world-cup.

FairSquare. (2024, October). *Substitute: The Case for the External Reform of FIFA*. https://fairsq.org/wp-content/uploads/2024/10/FIFA_Substitute_Report_v2_Single-Pages.pdf.

FairSquare. (2025a, May 14). *Underlying Causes: Unexplained Migrant Worker Deaths in Saudi Arabia* https://fairsq.org/wp-content/uploads/2025/05/Fairsquare_Underlying-Issues-Report-May-2025.pdf.

FairSquare. (2025b, May 27). "Joint Statement on FIFA Misgovernance" https://fairsq.org/wp-content/uploads/2025/05/FIFA-Reforms-Statement-27-May-2025.pdf.

Fanon, Frantz. (2018). *Alienation and Freedom*. Trans. Steven Corcoran. Ed. Jean Khalfa and Robert J. C. Young. London: Bloomsbury.

Fédération Internationale des Associations de Footballeurs Professionnels. (2024). *Excessive Workload Demands: Player Performance, Recovery and Health* https://fifpro.org/media/ad4lyibk/en_pwm_men-s-annual-report-2024.pdf.

Fédération internationale de football association (FIFA) contre BZ, en présence de : Union royale belge des sociétés de football association ASBL (URBSFA), Sporting du Pays de Charleroi SA, Fédération internationale des footballeurs professionnels, Fédération internationale des footballeurs professionnels – Division Europe, Union nationale des footballeurs professionnels (UNFP). (2024, October 4). C-650/22 ECLI:EU:C:2024:824.

Fett, Matthias. (2020). "The Game Has Changed—A Systematic Approach to Classify FIFA World Cups." *International Journal of Sport Policy and Politics* 12, no. 3: 455–70.

FIFA Sub-Committee on Human Rights & Social Responsibility. (2023, December 22). *Report on Remedy and Legacy for Workers in the Context of the FIFA World Cup Qatar 2022* https://digitalhub.fifa.com/m/774a7297f704cebb/original/FIFA-Sub-Committee-on-Human-Rights-Social-Responsibility-Report.pdf.

FIFA. (2019). *2018 FIFA World Cup Russia™ Sustainability Report* https://img.fifa.com/image/upload/ya7pgcyslxpzlqmjkykg.pdf.

FIFA. (2024). *FIFA World Cup 2034™ Bid Evaluation Report* https://digitalhub.fifa.com/m/51d8a13714827f1d/original/FIFA-World-Cup-2034-Bid-Evaluation-Report.pdf.

FIFA. (2025). "Partners." *Inside FIFA* https://inside.fifa.com/tournament-organisation/partners.

FilmLA. (2019a). *2018 Television Report* https://filmla.com/wp-content/uploads/2019/02/2018_TV_Report_web.pdf.

FilmLA. (2019b). *2019 Television Report* https://filmla.com/wp-content/uploads/2019/10/2019_TV_Report-WEB.pdf.

Forbes, Don Nico. (2025, January 24). "Soccer Tackles Its Carbon Problem." *Wall Street Journal* https://www.wsj.com/articles/soccer-tackles-its-carbon-problem-efff5bcb?utm_source=substack&utm_medium=email.

Fossil Free Football, Game Changer Sponsorship Pledge, and Badvertising. (2024, October 7). *UEFA's €500m Qatar Airways Deal Pollutes Like a Coal-Fired Power Plant* https://www.fossilfreefootball.org/2024/10/07/publication-uefas-e500m-qatar-airways-deal-pollutes-like-a-coal-fired-power-plant/.

Foxman, Simone and Adveith Nair. (2022, November 18). "What Qatar Built for the Most Expensive World Cup Ever." *Bloomberg* https://www.bloomberg.com/graphics/2022-what-qatar-built-for-the-world-cup/?leadSource=uverify%20wall&embedded-checkout=true.

Freedom House. (2024). *Freedom in the World 2024: Saudi Arabia* https://freedomhouse.org/country/saudi-arabia/freedom-world/2024.

Frieiro, Leonardo. (2022, November 15). "El fútbol, una mercancía global: Una entrevista con Pablo Alabarces." *Jacobin* https://jacobinlat.com/2022/11/el-futbol-una-mercancia-global/.

GAe Engineering. (n.d). *Juventus Stadium* https://www.gae-engineering.com/project/juventus-stadium/.

Galeano, Eduardo. (1995). *El fútbol a sol y sombra y otros escritos.* Ciudad de México: Siglo XXI Editores.

Galeano, Eduardo. (2013). *Soccer in Sun and Shadow*, rev. version. Trans. Mark Field. New York: Nation Books.

Gambling Commission. (2025, May 16). "TGP Europe Leaves GB Market Following Commission Investigation." https://www.gamblingcommission.gov.uk/news/article/tgp-europe-leaves-gb-market-following-commission-investigation.

García Márquez, Gabriel. (1996). *Noticia de un secuestro.* Mexico City: Editorial Diana.

Garrahan, Matthew. (2025, March 2). "Football's Man in the Middle: The Premier League CEO on Keeping the Peace." *Financial Times* https://www.ft.com/content/ade619c9-7191-4fe2-8ae9-e14992ccc4b9.

Glanville, Brian. (2018, November 8). ""Infantino is as Rotten as Blatter"." *World Soccer* https://www.worldsoccer.com/world-soccer-latest/infantino-rotten-blatter-407740.

Global Web Index. (2019). *Sports Media Landscape* https://www.gwi.com/reports/sports-media-landscape.

Godson, Matilda. (2024, April 10). "The Uncomfortable Truth About Kangaroo Leather." *Sokito* https://www.sokito.com/live-coverage/the-uncomfortable-truth-about-kangaroo-leather/.

Gössling, Stefan, Andreas Humpe, and Jorge Cardoso Leitão. (2024). "Private Aviation is Making a Growing Contribution to Climate Change." *Communications Earth & Environment* 5. https://doi.org/10.1038/s43247-024-01775-z.

Graham, Christopher J. (2025). "Alcohol Normalisation in Europe: The Role of Alcohol Sports Sponsorship." *The Lancet Regional Health—Europe* 48 doi:https://doi.org/10.1016/j.lanepe.2024.101156.

Gramsci, Antonio. (1994). *Pre-Prison Writings.* Ed. Richard Bellamy. Trans. Virginia Cox. Cambridge: Cambridge University Press.

Gray, James. (2023, October 6). "Fifa 'Prioritising Profit Over People' with 2030 World Cup Plans to Fly Teams 5,000 Miles." *The i Paper* https://inews.co.uk/sport/football/fifa-prioritising-profit-over-people-world-cup-2667248?srsltid=AfmBOoosMSupyfcOTXv3EWkaz6JV9AfLvzBoPy0fsxLogDS878sIAzUC.

Green, Molly. (2024, July 16). "Energy Demand Saw Rapid 1300MW Pick Up During Euro 2024 Football Final Halftime." *Current* +- https://www.current-news.co.uk/energy-demand-saw-rapid-1300mw-pick-up-during-euro-2024-football-final-halftime/.

Greenhalgh, Susan. (2024). *Soda Science: Making the World Safe for Coca-Cola.* Chicago: University of Chicago Press.

Habtemariam, Dawit. (2024a, May 7). "World Cup Tickets Will Go to Corporations, Not Fans." *Skift* https://skift.com/2024/05/07/world-cup-tickets-will-go-to-corporations-not-fans/.

Hall, Stuart. (1998). "The Windrush Issue: *Postscript.*" *Soundings* 10: 188–92.

Halliday, Fred. (2009). "International Relations in a Post-Hegemonic Age." *International Affairs* 85, no. 1: 50.

Hare, David. (2015). *The Blue Touch Paper: A Memoir.* New York: WW Norton.

Harraghy, Daniel. (2024, August 16). "How the Premier League's Global Popularity is Driving its Revenue Growth." *SportsPro* https://www.sportspro-media.com/insights/opinions/premier-league-tv-rights-revenue-global-popularity-sponsorship-data-ampere/.

Hart, Matt. (2020). *Win at All Costs: Inside Nike Running and its Culture of Deception*. New York: Dey St.

Heller, Agnes. (1984). *Everyday Life*. Trans. G. L. Campbell. London: Routledge & Kegan Paul.

Heynes, George. (2022, December 14). "Avoiding a 'Messi' Situation: Managing Demand Pick-Ups During the FIFA World Cup 2022." *Current* +- https://www.current-news.co.uk/avoiding-a-messi-situation-managing-demand-pick-ups-during-the-fifa-world-cup-2022/.

Hibberd, James. (2021, April 16). "Amazon's 'The Lord of the Rings' to Cost $465M for Just One Season." *Hollywood Reporter* https://www.hollywoodreporter.com/tv/tv-news/amazons-lord-of-the-rings-cost-465-million-one-season-4167791/.

Hobsbawm, E. J. (1989). *The Age of Empire 1875–1914*. New York: Vintage Books.

Hodgson, Camilla with Tom Wilson. (2023, June 7). "UK Watchdog Bans Shell, Repsol and Petronas Greenwashing Ads." *Financial Times* https://www.ft.com/content/fec9e504-7006-495a-964a-ba2ded4df5ad.

Hometree. (2022). "How Much Will it Cost to Power the World Cup This Winter?" https://www.hometree.co.uk/articles/how-much-will-it-cost-to-power-the-world-cup-this-winter/?referrer=611041&awc=1701916780217206262d94f1cc551a7a7ddc83d0101c338&utmsource=AWIN&utmmedium=affiliate&utmcampaign=Sub+Networks&utmcontent=78888-Skimlinks.

Horkheimer, Max. (1973). *Critical Theory: Selected Essays*. Trans. Matthew J. O'Connell *et al.* New York: Continuum.

Horton, Helena. (2024, December 3). "Coca-Cola Accused of Quietly Dropping its 25% Reusable Packaging Target." *Guardian* https://www.theguardian.com/environment/2024/dec/03/coca-cola-accused-dropping-reusable-packaging-target.

Hua, Jing, Martin Lundqvist, Shanti Naidu, Mikael T. Ekvall, and Tommy Cedervall. (2024). "Environmental Risks of Breakdown Nanoplastics from Synthetic Football Fields." *Environmental Pollution* 347 doi:https://doi.org/10.1016/j.envpol.2024.123652.

Hughes, Matt. (2025, August 4). "Fifa Facing Multibillion-Pound Compensation Claim from Players." *Guardian* https://www.theguardian.com/football/2025/aug/04/fifa-multibillion-pound-compensation-claim-former-players-transfer-regulations?CMP=Share_iOSApp_Other.

Hyde-Clarke, Nathalie, Rune Ottosen, and Toby Miller. (2014). "Nation-Building and the FIFA World Cup, South Africa 2010." *African Football, Identity Politics and Global Media Narratives: The Legacy of the FIFA 2010 World Cup*. Ed. Chari Tenda and Mhiripiri Nhamo Anthony. London: Palgrave Macmillan. 15–28.

Infantino, Gianni. (2024). https://www.instagram.com/gianni_infantino/p/CzErr6Uo1QJ/?hl=en.

Ingle, Sean. (2022, November 19). "'I Feel Gay, Disabled … Like a Woman Too:': Infantino Makes Bizarre Attack on Critics." *Guardian* https://www.theguardian.com/football/2022/nov/19/fifa-gianni-infantino-world-cup-qatar.

Inteligência Esportiva Research Institute and Play the Game. (2023). *Research on the Sports Legacy of Brazilian Stadiums from the 2014 FIFA World Cup* https://www.playthegame.org/media/z3bkh5fh/research-on-the-sports-legacy-of-brazilian-stadiums-from-the-2014-fifa-world-cupupdated-10.pdf.

International Energy Agency. (2018). *The Future of Petrochemicals: Towards More Sustainable Plastics and Fertilisers* https://iea.blob.core.windows.net/assets/bee4ef3a-8876-4566-98cf-7a130c013805/TheFutureofPetrochemicals.pdf.

Ionescu, Diana. (2024, December 27). "Mexico City Raises Mobility, Public Works Budget Ahead of World Cup." *Planetizen* https://www.planetizen.com/news/2024/12/133448-mexico-city-raises-mobility-public-works-budget-ahead-world-cup.

Ireland, Robin, Magdalena Muc, Christopher Bunn, and Emma Boyland. (2024). "Marketing of Unhealthy Brands During the 2018 Fédération Internationale de Football Association (FIFA) World Cup UK Broadcasts—A Frequency Analysis." *Journal of Strategic Marketing* 32, no. 6: 762–77.

Iskander, Natasha. (2021). *Does Skill Make Us Human? Migrant Workers in 21st-Century Qatar and Beyond.* Princeton: Princeton University Press.

Johnson, Lamar. (2024, September 4). "Coca-Cola Must Face Greenwashing Lawsuit: D.C. Appeals Court." *ESG Dive* https://www.esgdive.com/news/d-c-appeals-court-rules-coca-cola-must-face-greenwashing-lawsuit/726071/.

Justice for Players. (n.d.). https://justiceforplayers.com/.

Kang, Jasnoor. (2020, January 31). "Lifecycle of a Soccer Ball." *StoryMaps* https://storymaps.arcgis.com/stories/ef946b742331425fb773a2d06f6b44f0.

Karali, Nihan, Nina Khanna, and Nihar Shah. (2024). *Climate Impact of Primary Plastic Production.* Energy Analysis & Environmental Impacts Division, Berkeley Lab https://live-lbl-eta-publications.pantheonsite.io/sites/default/files/2024-09/climateandplasticreportlbnl-2001585.pdf.

Kaufman, Mark. (2020). "The Carbon Footprint Sham: A 'Successful, Deceptive' PR Campaign." *Mashable* https://mashable.com/feature/carbon-footprint-pr-campaign-sham.

Kazim, Hasnain. (2010, March 16). "The Football Stitchers of Sialkot." *Spiegel International* https://www.spiegel.de/international/world/globalization-in-pakistan-the-football-stitchers-of-sialkot-a-683873.html.

Kennedy, Lindsey and Nathan Southern. (2024, August 16). "How Premier League Sponsors Run Human Trafficking Syndicates." *Seasons of Crime* https://www.seasonsofcrime.com/p/how-premier-league-sponsors-run-human.

Kingdom of Saudi Arabia. (2023). *Saudi Arabia: A Story of Transformation: Vision 2030* https://www.vision2030.gov.sa/media/jvllaxpl/story-of-transformation-2023.pdf.

Kipling, Rudyard. (1902). "The Islanders." https://www.kiplingsociety.co.uk/poem/poems_islanders.htm.

Konyn, Carol. (2024, April 2). "The Carbon Footprint of Football: Unveiling the Dark Side of the Game." *Earth* https://earth.org/the-carbon-footprint-of-football/.

Krauze, Enrique. (1994, June 28). "México en un balón." *El País* https://elpais.com/diario/1994/06/28/opinion/772754409850215.html.

Kunti, Sam and Andreas Selliaas. (2025, January 6). "The Zone of Interest." *Josimar* https://josimarfootball.com/2025/01/06/the-zone-of-interest/.

Kunti, Sam, Steve Menary, Andy Brown, Philippe Auclair, and Jack Kerr. (2024, September 12). "Fly Another Day." *Josimar* https://josimarfootball.com/2024/09/12/fly-another-day/.

Lamborelle, Aymone and Beatriz Rios. (2019, September 4). *Sustainability in Sports? What's Football's Carbon Footprint?* Euractiv https://www.euractiv.com/section/health-consumers/infographic/whats-footballs-carbon-footprint/.

Landrigan, Philip J., Sarah Dunlop, Marina Treskova, Hervé Raps, Christos Symeonides, Jane Muncke, Margaret Spring, John Stegeman, Bethanie Carney Almroth, Thomas C. Chiles, Maureen Cropper, Megan Deeney, Lizzie Fuller, Roland Geyer, Rachel Karasik, Tiza Mafira, Alexander Mangwiro, Denise Margaret Matias, Yannick Mulders, Yongjoon Park, Costas A. Velis, Roel Vermeulen, Martin Wagner, Zhanyung Wang, Ella M. Whitman, Tracey J. Woodruff, and Joacim Rocklöv. (2025). *The Lancet Countdown on Health and Plastics.* https://doi.org/10.1016/S0140-6736(25)01447-3.

Lennon, John. (1971, January 21). "Interview with Tariq Ali and Robin Blackburn." *Red Mole* https://www.beatlesinterviews.org/db1971.0121.beatles.html.

Lindner-Cendrowska, Katarzyna, Kamil Leziak, Peter Bröde, Dusan Fiala, and Marek Konefał. (2024). "Prospective Heat Stress Risk Assessment for Professional Soccer Players in the Context of the 2026 FIFA World Cup." *Scientific Reports* 14. https://doi.org/10.1038/s41598-024-77540-1.

Lockwood, David and Matt Hickson. (2023, October 24). "What Could More European Football Mean for Planet?" *BBC* https://www.bbc.com/sport/football/67159156.

Lockwood, David. (2023a, March 23). "Premier League Domestic Flights: BBC Sport Research Shows 81 Flights from 100 Games." *BBC* https://www.bbc.co.uk/sport/football/65017565.

Lockwood, David. (2023b, September 15). "World Cup 2026: Could Climate Crisis Impact the Men's Tournament?" *BBC* https://www.bbc.com/sport/football/66211285.

Lockwood, David. (2023c, November 2). "Saudi Arabia: New Research on Extent of Country's Investment in Sport." *BBC* https://www.bbc.com/sport/67295892.

Lockwood, David. (2024, July 28). "Which Premier League Clubs Fly the Most in Pre-Season?" *BBC* https://www.bbc.com/sport/football/articles/cm52013g52jo.

López-Castro, Fernanda. (2025, January 10). "Gobierno de Sheinbaum remodelará AICM y AIFA rumbo al Mundial 2026: estos serán los cambios." *infobae* https://www.infobae.com/mexico/2025/01/10/gobierno-de-sheinbaum-remodelara-aicm-y-aifa-rumbo-al-mundial-2026-estos-seran-los-cambios/.

Lukács, Georg. (1972). *History and Class Consciousness: Studies in Marxist Dialectics*. Trans. Rodney Livingstone. Cambridge, Mass.: MIT Press.

Lyjak, Adam. (2023, January 10). "The Finances Behind the 2022 World Cup." *Michigan Journal of Economics* https://sites.lsa.umich.edu/mje/2023/01/10/the-finances-behind-the-2022-world-cup/.

MacInnes, Paul. (2024, November 30). "Fifa Ignores Own Report Into Qatar World Cup Over Workers' Compensation." *Guardian* https://www.theguardian.com/football/2024/nov/30/fifa-ignores-own-report-into-qatar-world-cup-over-workers-compensation?CMP=Share_iOSApp_Other.

MacInnes, Paul. (2025, March 11). "Fifa Accused on Blocking Labour Inspection at 2026 World Cup Stadium." *Guardian* https://www.theguardian.com/football/2025/mar/11/fifa-accused-blocking-labour-inspection-azteca-stadium-2026-world-cup?CMP=Share_iOSApp_Other.

Maguire, Kieran and Christina Philippou. (2024, October 17). "Why America is Buying Up English Football—And What it Means for the Future of 'Soccer'." *The Conversation* https://theconversation.com/why-america-is-buying-up-english-football-and-what-it-means-for-the-future-of-soccer-240695.

Malhado, Acácia Cristina Mendes and Rainer Rothfuss. (2013). "Transporting 2014 FIFA World Cup to Sustainability: Exploring Residents' and Tourists' Attitudes and Behaviours." *Journal of Policy Research in Tourism, Leisure and Events* 5, no. 3: 252–69.

Malik, Nasrine. (2025, July 22). "The Go-Between: How Qatar Became the Global Capital of Diplomacy." *Guardian* https://www.theguardian.com/world/2025/jul/22/how-qatar-became-the-global-capital-of-diplomacy.

Mandard, Stéphane. (2022, November 18). "Coupe du monde 2022: Le mirage de la compensation carbone pour atteindre la «neutralité» promise par la FIFA." *Le Monde* https://www.lemonde.fr/football/article/2022/11/18/coupe-du-monde-2022-le-mirage-de-la-compensation-carbone-pour-atteindre-la-neutralite-promise-par-la-fifa_6150404_1616938.html.

Matos Serodio, Paulo, Martin Mckee, Deborah Cohen, and David Stuckler. (2016). "Coca-Cola—A Model of Transparency in Research Partnerships?" *European Journal of Public Health* 26, no. 1. https://doi.org/10.1093/eurpub/ckw170.021.

Mavrokefalidis, Dimitris. (2023, August 21). "Nation's Roar for Lionesses Ignites Power Surge." *Energy Live News* https://www.energylivenews.com/2023/08/21/nations-roar-for-lionesses-ignites-enormous-power-surge/.

McCarthy, Michael. (2009, December 8). "Football's Carbon Footprint Comes Under Fire." *Independent* https://www.independent.co.uk/environment/climate-change/footballs-carbon-footprint-comes-under-fire-1836035.html.

McKie, Robin, Michael Savage, and Philip Cornwall. (2019, May 11). "As English Fans Get Set to Cross Europe, Anger Rises at Football's Carbon Bootprint." *Guardian* https://www.theguardian.com/environment/2019/may/11/anger-carbon-bootprint-english-football-finals-champions-league-europa-league.

Miller, Toby, Geoffrey Lawrence, Jim McKay and David Rowe. (2001). *Globalization and Sport: Playing the World*. London: Sage Publications.

Miller, Toby. (2010a). *Television Studies: The Basics*. London: Routledge.

Miller, Toby. (2010b, June 4). "Soccer Conquers the World." *Chronicle of Higher Education*: B6–B9.

Miller, Toby. (2018). *Greenwashing Sport*. London: Routledge.

Mitchell, James. (2007). "Sustainable Soccer: How Green Projects at International Sporting Events Benefit the Fans, the Global Climate, and Local Populations." *Sustainable Development Law & Policy* 7, no. 2: 52, 84.

Molodkin, Andrei. (2022). *Líbero: The Dirtiest Cup* https://thedirtiestcup.com/.

Mullan, Donal, Iestyn Barr, Neil Brannigan, Nuala Flood, Oliver R. Gibson, Catherine Hambly, Alan T. Kennedy-Asser, Aimée C. Kielt, Tom Matthews, and Madeleine Orr. (2025). "Extreme Heat Risk and the Potential Implications for the Scheduling of Football Matches at the 2026 FIFA World Cup." *International Journal of Biometeorology* 69: 753–63.

Müller, Martin, David Gogishvili, and Sven Daniel Wolfe. (2022). "The Structural Deficit of the Olympics and the World Cup: Comparing Costs Against Revenues Over Time." *Environment and Planning A: Economy and Space* 54, no. 6: 1200–218.

Murphy, Andrew and Valentin Simon. (2021, April). *Private Jets: Can the Super-Rich Supercharge Zero-Emission Aviation?* Transport & Environment https://www.transportenvironment.org/wp-content/uploads/2021/05/202209privatejetsFINALwithaddendum.pdf.

National Grid. (2014a, May 30). National Grid Up for the World Cup www.nationalgrid.com/mediacentral/uk-press-releases/2014/national-grid-up-for-the-world-cup/.

National Grid. (2014b, July 11). World Cup Fans Switched on to Brazil https://media.nationalgrid.com/press-releases/archive/world-cup-fans-switched-on-to-brazil/.

National Oceanic and Atmospheric Administration. (2024, September 10). "U.S. Sweltered Through its 4th-Hottest Summer on Record." https://www.noaa.gov/news/us-sweltered-through-its-4th-hottest-summer-onrecord#:~:text=Meteorological%20summer%20(June%20through%20August,fourth%2Dhottest%20summer%20on%20record.

New Weather Institute with Scientists for Global Responsibility. (2025a, February 3). *Dirty Tackle: The Growing Carbon Footprint of Football* https://www.new-weather.org/wp-content/uploads/2025/01/Dirty_tackle_The_growing_carbon_footprint_of_football.pdf.

New Weather Institute with Scientists for Global Responsibility. (2025b, July). *FIFA's Climate Blind Spot: The Men's World Cup in a Warming World* https://www.newweather.org/wp-content/uploads/2025/07/FIFAs_climate_blind_spot.pdf.

Nielsen, Kristian S., Jan M. Bauer, Ramit Debnath, Charles A. Emogor, Sonja M. Geiger, Sakshi Ghai, Wencke Gwordz, and Ulf J. J. Hahnel. (2024). "Underestimation of Personal Carbon Footprint Inequality in Four Diverse Countries." *Nature Climate Change*. https://doi.org/10.1038/s41558-024-02130-y.

Notre Affaire à Tous. (2022, November 2). *Plainte devant le Jury de Déontologie Publicitaire* https://climatecasechart.com/wp-content/uploads/non-us-case-documents/2022/20221102_18433_complaint.pdf.

Oceana. (2025, March). *Coca-Cola's World with Waste: The Soft Drink Company's Enormous and Worsening Plastic Problem and How Reusable Packaging Can Solve it* https://oceana.org/wp-content/uploads/sites/18/2025/03/Coca-Colas-World-With-Waste-Oceana-Report-2025_reduced-size-1.pdf.

Odigie, Akhator Joel. (2024, October 15). Letter to Patrice Motsepe https://www.ituc-africa.org/IMG/pdf/2024_oct_letter_caf_eng.pdf.

Orwell, George. (1945, December). "The Sporting Spirit." *Tribune* https://orwell.ru/library/articles/spirit/english/e_spirit.

Orwell, George. (1949). *Nineteen Eighty-Four* https://files.libcom.org/files/1984.pdf.

Organisation for Economic Co-operation and Development. (2022). *Global Plastics Outlook: Economic Drivers, Environmental Impacts and Policy Options* https://www.oecd-ilibrary.org/docserver/de747aef-en.pdf?expires=1732255776&id=id&accname=guest&checksum=69258D4A3FF9BCA837D7EBCED4EAF6F5.

Oxfam International. (2024). *Carbon Inequality Kills: Why Curbing the Excessive Emissions of an Elite Few Can Create a Sustainable Planet for All* https://

oxfamilibrary.openrepository.com/bitstream/handle/10546/621656/bp-carbon-inequality-kills-281024-en.pdf?sequence=1.

Paché, Gilles. (2020). "Sustainability Challenges in Professional Football: The Destructive Effects of the Society of the Spectacle." *Journal of Sustainable Development* 13, no. 1: 85–96.

Paché, Gilles. (2023). "Le ballet des FIFA World Cups: d'une aberration à l'autre." *Revue de Management et de Stratégie*: 1–10

Paché, Gilles. (2024). "FIFA World Cups: An Own Goal Against Sustainability." *IUP Journal of Management Research* 24, no. 3: 51–58.

Paul Nkemngu, Acha-Anyi. (2012). "The 2010 FIFA World Cup as Sustainable Tourism: A Community Perspective." *Journal of Tourism & Hospitality* 1, no. 5. https://doi.org/10.4172/2167-0269.1000106.

Pieth, Mark, Stefan Wehrenberg, and Rodney Dixon. (2024, May 22). "Submission to FIFA Regarding Saudi Arabia's Bid to Host the Men's 2034 World Cup." https://wehrenberg-law.ch/media/documents/Legal-Submission-to-FIFA-22-May-2024.pdf.

Piketty, Thomas. (2022). *A Brief History of Equality*. Trans. Steve Rendall. Cambridge, Mass.: The Belknap Press of Harvard University Press.

Pirks, Natalie and Tom Grundy. (2024, September 27). "29,000 Gambling Ads in Premier League Weekend, Says Research." *BBC* https://www.bbc.com/sport/football/articles/c4g07v7x70lo.

Pistolesi, Alice and Monica Pelliccia. (2024, July 2). "Across Mexico, World Cup Infrastructure Threatens Biodiversity and Communities." *Mongabay* https://news.mongabay.com/2024/07/across-mexico-world-cup-infrastructure-threatens-biodiversity-and-communities/.

Poniatowska, Elena. (1995). *Nothing, Nobody: The Voices of the Mexico City Earthquake*. Trans. Aurora Camacho de Schmidt and Arthur Schmidt. Philadelphia: Temple University Press.

Premier League. (2025, February 13). Written Evidence to UK Culture, Media and Sport Committee https://committees.parliament.uk/writtenevidence/135100/pdf/.

Preston, Michelle. (2020, June). *Hell for Leather*. Vegetarian & Vegan Foundation https://cdn.viva.org.uk/wp-content/uploads/2020/06/Hell-for-Leather.pdf.

Purves, Richard I., Amber Morgan, and Nathan Critchlow. (2022, November 13). *Alcohol Sponsorship of Football: Examining the Nature of Sponsorship Relations for Professional Football Teams Across Countries with Varied Restrictions on Alcohol Marketing*. Scottish Health Action on Alcohol Problems https://shaap.org.uk/downloads/reports/465-alcohol-sponsorship-of-football-examining-the-nature-of-sponsorship-relations-for-professional-football-teams-across-countries-with-varied-restrictions-on-alcohol-marketing.html.

Raji, Khalid. (2022, November 17). "Qatar 2022: The Environmental Cost of the FIFA World Cup." *Earth* https://earth.org/qatar-2022/.

Reda, Ahmed and Laurent Viviez. (2021, September 15). *How Sports Events Can Transform the Saudi Arabian Economy.* EY https://www.ey.com/enpk/services/consulting/how-sports-events-can-transform-the-saudi-arabian-economy#:~:text=Sports'%20contribution%20to%20GDP%20has,sports%20services%20such%20as%20gyms.

Republic of South Africa, Department of Environmental Affairs and Tourism, Norwegian Agency for Development Cooperation, and Norwegian Embassy in South Africa. (2009). *Feasibility Study for a Carbon Neutral 2010 FIFA World Cup in South Africa* https://www.norway.org.za/NR/rdonlyres/3E6BB1B1FD2743E58F5B0BEFBAE7D958/114457/FeasibilityStudyforaCarbonNeutral2010FIFAWorldCup.pdf.

Richter, Felix. (2024, December 11). "The World Cup is FIFA's Quadrennial Cash Cow." *Statista* https://www.statista.com/chart/30976/fifa-annual-revenue/.

Roan, Dan. (2023, December 14). "Why is Saudi Arabia Spending so Much on Sport?" *BBC* https://www.bbc.com/sport/67713269.

Roan, Dan. (2024, September 14). "England 'Could Face Euros Ban' Over Regulator Plan." *BBC* https://www.bbc.com/sport/football/articles/c9wkjnvpy2ko.

Roba i Giner, Diana, Ana Miranda Paz, Cristina Guarda, Carolina Morace, David Cormand, Catarina Vieira, Pernando Barrena, Li Andersson, Rainer van Lanschot, Chloé Ridel, Estelle Ceulemans, erik Marquardt, Mélissa Camara, and Emma Fourreau. (2024, December 6). Letter to Gianni Infantino https://www.eunews.it/wp-content/uploads/2025/01/MEPs-on-FIFA-2034-World-Cup-in-Saudi-Arabia-and-2030-in-Morocco-Portugal-and-Spain.pdf.

Ronay, Barney. (2024, November 15). "Saudi Arabia's World Cup: How Close Could FIFA Get to Corporate Manslaughter?" *Guardian* https://www.theguardian.com/football/2024/nov/15/saudi-arabias-world-cup-how-close-could-fifa-get-to-corporate-manslaughter?CMP=Share_iOSApp_Other.

Rossi, Raffaello, Jamie Wheaton, Maria Moxey, and Edoardo Tozzi. (2023). *New Season, More Self-Regulation, More Marketing: The Prevalence of Gambling Adverts During the Opening Weekend of the English Premier League 2023/24.* University of Bristol https://www.bristol.ac.uk/media-library/sites/business-school/documents/BRISTOL-UNI-GAMBLING-Report2023-2.pdf.

Sant'Anna School of Advanced Studies, Institute of Management. (2020). *The Environmental Awareness and Behaviour of Professional Football Supporters: An Empirical Survey* https://lifetackle.eu/assets/files/LIFETACKLEReportonsupporterssurvey.pdf.

Sartre, Jean-Paul. (2004). *Critique of Dialectical Reason. Volume 1: Theory of Practical Ensembles.* Trans. Alan Sheridan-Smith. Ed. Jonathan Ree. London: Verso.

Saudi Arabia. (2024). *Vision & Legacy* https://saudi2034bid.com/vision-legacy/#tournament-concept.

Schaerlaeckens, Leander. (2025, March 12). "Gianni Infantino and Donald Trump Have Taken the 2026 World Cup for Themselves." *Guardian* https://www.theguardian.com/football/2025/mar/12/gianni-infantino-donald-trump-2026-world-cup.

Scharpf, Adam, Christian Gläßel, and Pearce Edwards. (2023). "International Sports Events and Repression in Autocracies: Evidence from the 1978 FIFA World Cup." *American Political Science Review* 117, no. 3: 909–26.

Selectra. (2020, January 22). *What Happens When Everyone in the UK Watches the World Cup at the Same Time?* https://selectra.co.uk/energy/news/world/world-cup-2018-electricity-surges.

Shakespeare, William. (1994). *Complete Works of William Shakespeare.* N.p.: World Library.

Sherter, Alain with Anne Marie Lee and Associated Press. (2024, September 24). "Visa Monopolizes Debit Cards, Hurting Consumers and Businesses, Justice Department Alleges." *CBS* https://www.cbsnews.com/news/visa-debit-cards-justice-department-lawsuit/.

Silic, Dario. (2022). "Is the FIFA World Cup in Qatar Profitable? *SSBM* https://www.ssbm.ch/is-the-fifa-world-cup-2022-in-qatar-profitable/.

Singh, Parminder, Ashok Kumar Balaraman, Rachana Mehta, and Sanjit Sah. (2025). "Urgent Need for Enhanced Food Safety Protocols in Fast-Food Supply Chains: Lessons from the Recent Multi-State *E. coli* Outbreak." *Infectious Diseases* 57, no. 1: 109–11.

Solidarity. (2001). "Paris May 1968." *Beneath the Paving Stones: Situationists and the Beach, May 1968.* Comp. Dark Star. Edinburgh: AK Press/Dark Star. 64–96.

Solnit, Rebecca. (2021, August 23). "Big Oil Coined 'Carbon Footprints' to Blame Us for Their Greed. Keep Them on the Hook." *Guardian* https://www.theguardian.com/commentisfree/2021/aug/23/big-oil-coined-carbon-footprints-to-blame-us-for-their-greed-keep-them-on-the-hook.

Spanne, Autumn and *The Daily Climate.* (2014, June 19). "Brazil World Cup Fails to Score Environmental Goals." *Scientific American* https://www.scientificamerican.com/article/brazil-world-cup-fails-to-score-environmental-goals/?WT.mcid=send-to-friend.

SponsorPulse. (2023). *FIFA World Cup 2022 Sponsor Impact Report. Market: USA* https://insights.sponsorpulse.com/hubfs/FIFA%20World%20Cup%202022%20Sponsor%20Impact%20Report.pdf?_gl=1*u5y7jp*_gcl_au*NDQyMjg2ODY5LjE3MzM1OTY5OTg.*_ga*OTM4Mzc5MTMzLjE3MzM1OTY5

OTg.*_ga_392WGBYBSG*MTczMzU5Njk5OC4xLjEuMTczMzU5NzAzOC4yMC4wLjA.

SportBusiness. (2024). *Global Media Report 2024* https://www.sportbusiness.com/global-media-report-2024/.

Stanton, John, David Lockwood, and Katie Gornall. (2021, November 12). "Premier League: Should Clubs Stop Flying to Domestic Matches for Environmental Reasons?" *BBC* https://www.bbc.co.uk/sport/football/59213173.

Sturrock, Leanne. (2018, July 12). "The Environmental Impact of the World Cup—Are FIFA Scoring Sustainability Goals?" *thegreatProjects* https://www.thegreatprojects.com/blog/environmental-impact-of-the-world-cup.

Subcomandante Marcos. (2017). *Professionals of Hope: The Selected Writings of Subcomandante Marcos*. New York: The Song Cave.

The Climate Coalition. (2018). *Game Changer: How Climate Change is Impacting Sports in the UK* https://www.theclimatecoalition.org/gamechanger.

Thomas, Ian. (2024, July 18). "Inside Evolving FIFA 2026 World Cup Plan to Cash in on the Rise of High-End Sports Tourism." *CNBC* https://www.cnbc.com/2024/07/18/fifa-2026-world-cup-plan-to-cash-in-on-high-end-sports-tourism.html.

Tóffano Pereira, Rodrigo Pinheiro, Viachaslau Filimonau, and Glaydston Mattos Ribeiro. (2019). "Score a Goal for Climate: Assessing the Carbon Footprint of Travel Patterns of the English Premier League Clubs." *Journal of Cleaner Production* 227: 167–77.

Touma, Rafqa. (2023, March 15). "Animal Activists and Commercial Industry at Odds After Nike Halts Use of Kangaroo Leather." *Guardian* https://www.theguardian.com/business/2023/mar/15/animal-activists-and-commercial-industry-at-odds-after-nike-halts-use-of-kangaroo-leather.

Tweedale, Alistair. (2018, November 21). "Fifa Talks a Good Game But its Drive for Sustainability Highlights Football's Hypocrisy." *Telegraph* https://www.telegraph.co.uk/football/2018/11/21/fifa-talks-good-game-drive-sustainability-highlights-footballs/.

UEFA. (n.d.). https://www.uefa.com/sustainability/environment/climate-advocacy/.

UEFA. (2020). "Venues Confirmed for EURO 2020." https://www.uefa.com/uefaeuro-2020/news/025e-0fac6d3ee9e4-85b1a76389ea-1000/.

UEFA. (2024). *Environmental, Social and Governance Report: Post-Event Assessment* https://www.efdn.org/wp-content/uploads/2024/11/uefa_euro_2024_esg_report_compressed.pdf.

UEFA. (2025, March 7). *The European Club Finance and Investment Landscape* https://ecfil.uefa.com/2024.

Ukhova, Daria, Virve Marionneau, Janne Nikkinen, and Heather Wardle. (2024). "Public Health Approaches to Gambling: A Global Review of Legislative Trends." *The Lancet Public Health* 9, no. 1: e57–67.

Vargas Llosa, Mario. (1997). *Making Waves*. Trans. and Ed. John King. New York: Farrar, Strauss and Giroux.

Vásquez, Alex. (2024, December 21). "Mexico City to Boost Mobility, Security Ahead of 2026 FIFA World Cup." *Business Standard* https://www.business-standard.com/world-news/mexico-city-to-boost-mobility-security-ahead-of-2026-fifa-world-cup-124122100164_1.html.

Vlessing, Etan. (2024, July 18). "CBS Sports Nabs U.S. TV Rights to English Football League Games." *Hollywood Reporter* https://www.hollywoodreporter.com/business/business-news/cbs-sports-english-football-league-1235952038/.

Walters, Geoff and Richard Tacon. (2011). *Corporate Social Responsibility in European Football: A Report Funded by the UEFA Research Grant Programme* http://www.sportbusinesscentre.com/wp-content/uploads/2012/08/CSR2.pdf.

Wardle, Heather, Louisa Degenhardt, Virve Marionneau, Gerda Reith, Charles Livingstone, Malcolm Sparrow, Lucy T. Tran, Blair Biggar, Christopher Bunn, Michael Farrell, Viktorija Kesaite, Vladimir Poznyak, Jianchao Quan, Jürgen Rehm, Angela Rintoul, Manoj Sharma, Jeremy Shiffman, Kristiana Siste, Daria Ukhova, Rachel Volberg, Joana Salifu Yendork, and Shekhar Saxena. (2024). *The Lancet Public Health Commission on Gambling* https://doi.org/10.1016/S2468-2667(24)00167-1.

Wartuny, Geffin. (2024). "Journal Review of Capitalism in European Soccer." *LATTE: A Journal of Language, Culture, and Technology* 1, no. 1: 17–23.

Weber, Max. (1985). ""Churches" and "Sects" in North America: An Ecclesiastical Socio-Political Sketch." Trans. Colin Loader. *Sociological Theory* 3, no. 1: 7–13.

Weinstein, Roy, Peter Quies, Ryan Smart, Jeremy De Gracia, and Joe Hale. (2024, June). *Projected Economic Impact of FIFA World Cup 26™ County of Los Angeles June-July 2026*. Micronomics Economic Research and Consulting https://static1.squarespace.com/static/61b53e492ea58d13b806ccb3/t/668c5701d897eb08bf8a2355/1720473346077/FWC_26_LA-Economic_Impact_Report-Micronomics%282024-06-25%29.pdf.

Wells, H. G. (1902). *Anticipations of the Reaction of Mechanical and Scientific Progress Upon Human Life and Thought*, 2nd ed.. London: Chapman & Hall.

Wilson, Jonathan. (2024, December 7). "Fifa's Infantino and Saudi Arabia 1, Football and Human Decency 0." *Guardian* https://www.theguardian.com/football/blog/2024/dec/07/fifa-infantino-saudi-arabia-world-cup-2034.

Wittgenstein, Ludwig. (1965). "I: A Lecture on Ethics." *The Philosophical Review* 74, no. 1: 3–12.

Woodford, Isabel. (2025, August 25). "'We Get Nothing': Residents Fight Back as Mexico Builds Big for 2026 World Cup." *Guardian* https://www.theguardian.com/global-development/2025/aug/25/we-get-nothing-residents-fight-back-as-mexico-builds-big-for-2026-world-cup?CMP=Share_iOSApp_Other.

Woods, Chelsea L. and Ashli Q. Stokes. (2019). "'For the Game, for the World': An Analysis of FIFA's CSR Initiatives." *Public Relations Inquiry* 8, no. 1: 49–85.

Worden, Minky. (2025, May 5). Letter to Gianni Infantino. Human Rights Watch https://www.hrw.org/news/2025/05/06/letter-fifa-re-human-rights-responsibilities-2026-world-cup.

Wyden, Ron and Richard Durbin. (2024, November 25). Letter to Gianni Infantino https://www.wyden.senate.gov/imo/media/doc/fifa_world_cup_2034_letter2.pdf.

Zipes, Jack. (2023). *Buried Treasures: The Power of Political Fairy Tales.* Princeton: Princeton University Press.

Zirin, Dave. (2011, June 8). "Let Them Play: Behind FIFA's Decision to Ban Iran's National Women's Team." *The Nation* www.thenation.com/blog/161228/let-them-play-behind-fifas-decision-ban-womens-team-iran.

CHAPTER 3

The Carbon Summer Olympics

Two and a half millennia ago, Pindar noted that 'the fame of the Olympic festivals shineth from afar' (1915: 13). But at the end of the pre-Christian era, Cicero deemed 'going to watch the Olympic Games, a most despicable motive'—at times of political crisis, '*impardonable*' (2008: 241). Juvenal wrote that everyday Romans '[c]ouldn't care less' when it came to democracy: 'Only two things really concern them: bread and the Games' (1998: 505). Conversely, Francis Bacon regretted the Olympics' demise, denouncing seventeenth-century sports' 'mercenary ostentation' (1863). The first critique was concerned with political apathy, the second commodification—an old story in constant need of retelling.

Per the Introduction, a preparation for politics, and disenchantment with it, has always indexed and been encouraged by the Olympics, with money and oil central—ancient Greece and Rome invested heavily in sports, almost to the level of their military budgets (Neils 1992) and Athens gave huge amounts of olive oil as prizes in the Panathenaic Games (Neils 1992). The fantasy of a once-amateur event, admired for some sort of purity and a distance from extractive industry, is misleading.

From the moment of their re-emergence, the modern Games were articulated to development and capitalism: the first three events featured trade fairs in Paris, St. Louis, and London. Kodak was an early sponsor, and many others followed. From 1984, sponsors were offered exclusivity within their industry—hence Lionel Ritchie rather than Michael Jackson

T. Miller, J. P. Carañana, *Global Sports Go Green—Or Do They?*, Global Culture and Sport Series,
https://doi.org/10.1007/978-3-032-14207-8_3

singing at the Los Angeles Games, because of their respective sugar-drink contracts (Echikson 1998; Clarey 1998; Boykoff 2016). Broadcasting rights and sponsorship have provided immense funding:

Olympic revenues

Years	*$Billion*
2004–08	3.8
2008–12	4.7
2012–16	5.2
2016–20	6.4
2020–24	7.7

Source: Based on Noble et al. (2025)

Given the money involved, it can come as no surprise that GDP is the key predictor of contemporary Olympic medal success ("Olympic Glory" 2021) or that the jamboree is hugely popular. The 2020 Tokyo Olympics drew audiences of over 3 billion, 93% watching on TV (International Olympic Committee 2021: 29). More than half the world's population "engaged" Paris 2024, to use the artless term of the day ("Olympics President" 2024). The event gained more spectators for NBC's streaming services than all previous Games combined. The conglomerate's Olympic revenue was $1.9 billion (Szalai 2024). Records were set by the BBC and Warner Bros. Discovery across Europe, where streaming, so-called social media, cable, satellite, and broadcast systems gathered well over 215 million viewers (Saleh 2024).

The modern Games are constitutively environmental, their very division between summer and winter contests climatic. As such, the natural world should be front and center. But many sports avoid or alter geography. Contestants in the pool are sedulously sheltered from salt and river water, basketballers securely shielded from weather. Skateboarders and bicycle-motocross riders rely on the evisceration of anything natural in their path. Our fellow-animals are largely absent, other than the enslaved horses of the pentathlon until their liberation from 2024, and equestrianism (one that stepped out of line in 2021 was slapped into submission by rider and trainer), skeet shooting's mimetic birds ("Tokyo Olympics: Germs" 2021; "Sin" 2014), and occasional mascots (*pace* the renowned "Animal Olympics" of Théâtre Sans Frontières and 動物オリムピック大会).

At its 1994 centennial conference, the IOC declared protecting the environment 'an essential component of Olympism' (n.d.-b). The Committee boasts that it 'supports potential hosts to think about how to achieve social, environmental and economic benefits that start long before, and continue long after, the events themselves' (n.d.-a) and the *Olympic Charter* proclaims a 'Movement' to 'encourage and support a responsible concern for environmental issues, to promote sustainable development in sport and to require that the Olympic Games are held accordingly' (International Olympic Committee 2024a: 14). A *Carbon Footprint Methodology* promises 'effective carbon reduction strategies' (International Olympic Committee 2018: 11) and the IOC offers "Climate Action Awards." In 2020, the Committee decreed the Games would become "Climate Positive" by 2030, meaning 'each organizing committee would be legally required to remove more carbon from the atmosphere than the games emit' (Foth 2025). The IOC itself promised an "Olympic Forest" in Africa ("Olympic Games to Become" 2020). But these policies are symbolic and reformist at best (Jiménez-García et al. 2020; Cain and Callan 2025). The numerous true believers in the value of playing host—tourism apparatchiks, policy boosters, "travel writers," corporate shills, and scholarly chorines—seem chronically unaware of the environmental implications.

This reactionary story will hardly surprise, given the Committee's inbred oligarchic composition: "royalty" from Liechtenstein, Monaco, Britain, Luxembourg, Qatar, Jordan, Bhutan, Saudi Arabia, Malaysia, Denmark, and the Netherlands, plus Belgian and British aristocrats and a film star (2025; Noble et al. 2025). As *Slate* magazine put it, '[w]hen I think of the IOC, I think of a bloated European man receiving a bag filled with money as the digestif to a five-course lunch served on someone else's yacht' (Peters 2021). Juan Antonio Samaranch, IOC president for decades, liked to be addressed as 'Your Excellency' and told the *New York Times* 'Franco did good things' (quoted in Cohen and Longman 1999: 26). Unsurprisingly, he adopted a 'dictatorial style' (Tomlinson 1996: 586). The corruption of the Samaranch regime was legendary: bid committees hiring sex workers to "service" IOC men, paying bribes, and putting money into scholarships for relatives of Committee members, *inter alia* (Lusetich 1999: 1).

Like aristocracies, plutocracies, and all carbon mega-events, the summer Games are mismanaged extravagances, characterized by 'a slate of ingrained negative externalities: overspending, intensified policing to protect the spectacle, displacement of marginalized populations, greenwashing, and corruption' (Boykoff 2025a). The cities involved have generally been wealthy, polluting societies allied to problematic sponsors (Miller 2018), with bribery and graft endemic (Zirin and Boykoff 2023). Between 1960 and 2024, their average costs were over $8 billion at 2022 prices—excluding general construction. Budget overruns were universal, beyond 100% in half of them (Budzier and Flyvbjerg 2024). Merely preparing a bid to host requires $50–150 million (McBride et al. 2024). Such massive misallocations of public and private resources have relied on two myths to seduce the gullible burghers of city after exploited city: reusable new infrastructure and ongoing tourism: 'escalating and consistently uncontrollable costs' are 'rationalised by the appeal to legacy, in realms from the economic to the environmental, participation to regeneration, diplomacy to public health, national pride to global harmony' (Tomlinson 2014: 139).

Olympic budget overruns

Year	*Percentage*
2004	49
2008	2
2012	76
2016	352
2020	128
2024	115

Source: based on McBride et al. (2024)

Social, economic, and environmental devastation has routinely accompanied the carbon Games, compromising any supposed accommodation to our climate crisis. The record shows that in the decades since the IOC announced commitments to the environment, the Games have become less, not more, sustainable: 'as the IOC increasingly boasted its own environmental sustainability bona fides, its follow-through actually weakened' (Boykoff 2025b). Before, during, and after the Olympics, host countries' carbon emissions increase, regardless of the size of their economy or population, due to an 'extended level of consumption of raw materials, worsening land and air traffic, draining local water supplies, and increasing waste disposal' (Ceccon et al. 2024). No host has met its ecological goals (Delaplace and Müller 2024).

To help explain why, this chapter focuses on the environmental impact of four cases over the last seven decades and the climate calumny of key participants.

History

The Games greenwash themselves, their sponsors, and their hosts. The process started and has continued with shunting the "wrong" populations away from public view. In addition to instituting deforestation, the 1936 Berlin Olympics uprooted untold numbers of people from their homes to create "necessary" infrastructure. That strategy seemed to have reached its apogee with the forced removal of hundreds of thousands for the 1988 Games in Seoul—until Beijing's Leninists cleared more than a million residents from their dwellings in advance of the 2008 event (Davis 2006; Ross 2009; Worden 2008, 2015). Much of this devastation has been conducted in the name of development, as this chapter shows, beginning with the 1964 and 1968 Olympics, which were promulgated as calling cards to the world from Japan and Mexico that in fact illustrated how the environment and human rights were playthings for the IOC and its state courtiers.

The 1964 Tokyo Games announced a fabulous new modernity emerging from the nation's wartime defeat, personified by Yoshinori Sakai. A student-athlete from the north-east of Hiroshima, he was born just hours after Yanquis dropped "Little Boy" on the city, killing 140,000 people (Tagsold 2010). Known as the "Atomic Bomb Boy," Sakai carried the Olympic torch on its last leg (Whiting 2021: 108), a human testimony to surviving US violence and growing up "after" Japanese fascism.

In the twenty years since the War,

> the popular image of Japan held in the West had improved only slightly: cartoon depictions had shifted from the wartime caricatures of Japan as a scheming nation of slavering beasts to an Occupation-era representation of Japanese as pet monkeys or children learning at the feet of their American educators. (Abel 2015: 140)

Consider these disturbing words from *Sports Illustrated* prior to the Games: '[t]wenty years ago the way for an American to see Tokyo was through his periscope or his bombsight. Now he marries Japanese girls at a rate of 2000 a year, and has learned that the best massage is the patter of tiny Oriental feet up and down the spine'. Meanwhile, de Gaulle had (allegedly) referred to the Japanese people as 'transistor radio salesmen' (quoted in Abel 2015: 154), while *Walk Don't Run* found Cary Grant desperately seeking cheap accommodation during the Games' housing crisis while on a mission to purchase transistors for his UK factory.

In stark contrast to those reductive tropes, by the mid-1960s, Japan was a net exporter of goods, its heavy industry and micro-electronic manufacture suddenly dominant forces in ship- and car-building and entertainment technology. The Olympics offered a vision in which the humiliation of defeat and occupation could be overdetermined by the cosmopolitan spectacle of a 'peaceful, friendly, productive, innovative, and modern' country (Tomizawa 2019: 14). The two weeks of display incarnated the country's changes audiovisually, pioneering color broadcasting and distributing the Olympics live via satellite ("The Evolution" 2010; "1964 Memories" 2014).

Such innovations put Japan on an equal footing with the winners of World War II and at the head of non-Western countries in terms of development. In addition, they set up the state as a welcome, trustworthy route to prosperity rather than a crazed pseudo-deity bent on destruction (Wilson 2012). No 'longer a militarist pariah shattered by war, the nation had successfully reinvented itself as a peaceful democracy on its way to becoming a world economic powerhouse' (Whiting 2021: 123) via the 'conversion of a war economy back into a peace economy, the reconstruction of destroyed economic potential, and … the integration of new technology' (Foucault 2008: 79).

Tokyo could no longer be regarded as 'a war-scarred, dilapidated, diseased, and polluted third-world megalopolis'. This suddenly amenable eastern outpost of international monopoly capital, a 'valuable partner of the Western powers,' combined futuristic modernity with distinctive tradition (Abel 2015: 140–41). The image of a dedicated, obsessive, and efficient people was articulated away from conflict, toward pacific prosperity. References to traditional culture were frequent, such as the きもの [kimono] worn by medal presenters, but excluded martial iconography, per the 侍 [samurai]. This would not be a modernity drawing on violence and conquest, unlike the 1930s and 1940s (Collins 2011: 2244). Rather, it would harmonize 'the way of thinking prior to modernization and that of modern Europe' (Foucault 2024: 48).

One of Robert Whiting's informants from the time said, 'I like the smell of freshly dried asphalt. … It means progress' (quoted in Whiting

2021: 82). Ten thousand buildings were constructed, including major apartment blocks, new metro lines and freeways, and a monorail connecting Tokyo airport to downtown (Whiting 2021). Tourists encountered 'state-of-the-art transportation whizzing between an upgraded airport and smart new hotels' (Zimmerman 2021). The 東海道新幹線 [Tōkaidō Shinkansen] high-speed rail link between the capital and Osaka, partly funded by a World Bank loan (Abel 2015: 163), was the first of its kind in the world, 'the putative crown jewel of the Olympic effort' (Whiting 2021: 90).

But Japan's transformation was achieved through violence, like all rapid, massive modernizations (Britain, the USA, the Soviet Union, China): the US military's munitions needs during the Korean War gave business to a World-War-II-enemy-turned-Cold-War ally, its geopolitical location crucial to proxy superpower struggles (Tomizawa 2019: 46). And the rebirth of 'a war-ravaged city into a major international capital,' seemingly overnight, concealed 'a dark side … environmental destruction and human misery' (Whiting 2014).

"Thanks" to the Olympics, slum-like conditions and a regrettable recent history gave way to space-age brutalism, a peace-loving monarchy, and dutiful athletes. The nation showed the TV world what that meant—without worrying about the ecological impact. 1964's flashy welcome to Tokyo modernity featured a façade of newness, technology, democracy, and efficiency—a neat, convenient lesson in how not to fall prey to socialism when emerging from fascism (Wilson 2012) and the 'chance to tell a fresh story' ("Tokyo 1964—Designing" 2021). Pictorial commemorations highlighted "Tokyo Reborn" and "Made in Japan" ("1964 Memories" 2014). Decades later, the official version celebrated a 'revolutionary cultural and design legacy' ("Tokyo 1964—Designing" 2021).

The city's western and central zones were selected for Olympic development, thereby protecting visitors' vulnerable noses from a sewage-laden Tokyo Bay. The unhoused, sex workers, and mobsters were invisibly exiled to undergo 'spiritual training' (Whiting 2021: 97–98), further Bowdlerizing daily life. As 'construction rose to a frenzied pitch, Tokyo residents put up black curtains to shut out the blinding all-night work lights and climbed into their futons wearing ear plugs to block out the incessant noise of the pile drivers and bulldozers'. The tramway system was virtually eviscerated in favor of freeways. Hundreds of thousands of animals were killed as part of a 'reconstruction' that destroyed many 'navigable waterways and put an end to what had been a vibrant, commercial river culture' (Whiting 2021: 131). Local birds and fish migrated to avoid contamination, while 'invasive species flourished'. Concrete and neon hid polluted rivers and tributaries. The organizers covered riverbanks with vinyl, mass-sprayed deodorants, and cavalierly dropped chemicals into the water to hide despoliation for the duration of the Games, even as contaminants crested via industrial runoff, toxic chemicals, and sewage. This would no longer be a 'water city'.

The Shinkansen also had a hidden history:

> The bullet train connected distant cities at unprecedented speed, but it also tore through a region dense with people, industry, agriculture, and history. Opening day was undoubtedly a more somber occasion for those who had been evicted from their homes, had been pressured into selling their land, or were coping with damage to their livelihoods, local environments, and communities. (Abel 2022: 1–2)

The Olympics' opening ceremony fetishized away the productive and destructive work that had manufactured its conditions of possibility. The closing corollary saw bursting fireworks freeing participants from orderly marching so they could bow, dance, and gently canter round the stadium. A hint emerged of the underpinnings to this shiny, blossom-laden modernity—fleetingly but immediately—in the remarkable official documentary 東京オリンピック [*Tokyo Olympiad*]. The film started with an impossibly bright, luminous sun—a reminder of both the nation's flag and the atomic bombs' terrible, prodigious heat. It then moved to black-and-white footage of buildings being demolished, fading into color imagery of workers and the Olympic stadium. Just a hint, in what the BBC has declared "The Greatest Film About Sport Ever Made" that 'captured the essence of a post-crisis nation'.

Beyond the hint, problematic factors were barely noticed in the illumination of a newly lit, seemingly eternal day that blasted wartime horrors into the past (Jorgenson and Clark 2016). For the uncritical visitor, 'everywhere you looked you saw a glistening new building' (Whiting 2021: 97–98), *The Times* rhapsodizing that Tokyo 'has emerged, as if from a chrysalis, to stand glitteringly ready', the *Guardian* gushing over its 'futuristic' tone. The Games 'epitomised and accelerated the country's remarkable modernisation and reentry to the international community after the second world war'. But really 1964 exemplified Japan's postwar 'disregard for the environment'.

Mexico City in 1968 was also conceived as a means of evoking the country's economic and social advances, encapsulating the state's project of self-formation under the sign of modernity. But unlike Japan's break with fascism and eschewal of socialism, the Mexican plan, for domestic consumption and beyond, was to showcase political stability dating back decades and ongoing development. Successful bids to stage the 1968 Games and 1970 men's World Cup represented 'a political and administrative coup' (Arbena 1991: 359), 'una suerte de reconocimiento internacional a su transformación en un país moderno o semimoderno' [a dream of international recognition of transformation into a modern—or semimodern—country] (Paz 1994).

The prominent architect Pedro Ramírez Vázquez chaired the nation's Olympic Committee. He hoped the world would 'forget an image of Mexico of a figure covered by a poncho and a sombrero sleeping soundly beneath the shadow of a tree.' In its place, a 'new international image of Mexico is being created this Olympic Year' (quoted in Castañeda 2010: 103). Deep tradition and youthful vigor would be partners in a paean to memory, economy, and the future, the country's pre-Hispanic heritage merging comfortably with mid-twentieth-century developmentalism—Mayan and Aztec cultures alongside a fast-growing GDP. The local *bourgeois* media repeatedly told people the country would be scrutinized by the world (Elias 2021).

Per Tokyo, the environment was something to be mastered. The nation's new Olympic facilities established micro-climates dedicated to dealing with Mexico City's high-altitude, earthquake-prone topography, hot temperatures, and sudden storms, while new technology meant the first-ever synthetic Olympic track was relatively impervious to heat and rain, and divers made their way to the platform via elevator (Underwood 1968).

An ideology known modestly as 'total design' (Castañeda 2014: 134, 152) highlighted a new 'confidence in the nation's material accomplishments and sense of cosmopolitan belonging' (Zolov 2004: 162). Plans included the release of doves to fly over imposing new freeways across the city prior to the Games, incarnating a spirit of harmony. The official slogans were "Mexico offers international peace to all countries" and "The Olympics of Peace" (quoted in de Anda Alanis 2021: 66). A cultural festival was convened to demonstrate the loving nature of the event (Paz

1994). In his annual report to Congress, Mexican President and alleged CIA informer (Kriza 2019: 83) Gustavo Díaz Ordaz announced, 'Nos complace señalar que México ha sido el primer país que, en la época moderna, devuelve a los juegos su profundo contenido humanista y cultural' [It's a source of pride that Mexico is the first country in modern times to fill the Games with humanistic, cultural content] (1968).

In reality, such qualities were on meagre display in a nation dominated by a repressive state apparatus flexing its power in the face of student unrest against both authoritarianism and propaganda created for the event that asserted the country's democracy and modernity (Castañeda 2010: 112–14; Poniatowska 1971). Díaz Ordaz warned that:

> quienes intentaron aprovecharlo con fines ideológicos y políticos y la de quienes se propusieron sembrar el desorden, la confusión y el encono, para impedir la atención y la solución de los problemas, con el fin de desprestigiar a México, aprovechando la enorme difusión que habrán de tener los encuentros atléticos y deportivos, e impedir acaso la celebración de los juegos Olímpicos [some people, driven by ideological and political ends, have sought to seed disorder, confusion, and animosity, to disrupt the examination and solution of social problems, with the aim of bringing Mexico into disrepute, jeopardizing our Olympic Games celebrations]. (1968)

The military mounted Operación Galeana, a concerted attempt to wipe out the student movement. Ten days before the opening ceremony, the state apparatus massacred, brutalized, and arrested hundreds, perhaps thousands of demonstrating students, workers and bystanders in Mexico City's Plaza de las Tres Culturas (the pre-Hispanic, colonial, and modern), a space next to the vast Tlatelolco housing projects that had been designed to juxtapose 'pre-Conquest Aztec ruins with colonial, and post-revolutionary architecture' (Beckhart Coppinger 2021: 431). The *bourgeois* media and the IOC sought to minimize the extent of the slaughter and the importance of the protest (Aviña 2007). Ironically, it had taken place just a year after the Tlatelolco Treaty for the Prohibition of Nuclear Weapons in Latin America and the Caribbean was signed (n.d.), to be followed by the 1974 Conference of Tlatelolco, where unindicted war criminal and US Secretary of State Henry Kissinger met his equivalents from across Latin America.

There is evidence that before the Tlatelolco massacre, IOC President Avery Brundage had secretly advised Díaz Ordaz that if student unrest were not quelled, the Games might not proceed (Rodda 2008: 19). In the week after Tlatelolco, Brundage told journalists, 'Si los Juegos Olímpicos tuvieran que ser interrumpidos cada vez que los políticos violan las leyes de la humanidad, jamás habría encuentros deportivos internacionales' [If the Olympic Games were interrupted every time politicians committed human rights violations, there would be no international sporting events] (quoted in "Brundage en Defensa" 1968). But while the IOC and the Mexican state were celebrating the opening ceremony as a sign of peaceful competition, student protestors were honoring their dead, tending to survivors' wounds, and nominating the event as a 'noche triste' [sad night] ("Tlateloloco" 2015).

The official film of the Games, *Olimpiada en México* [*The Olympics in Mexico*] did not deal with Tlatelolco. It depicted a happy Latin journey from colonization, with a voiceover describing the bizarre decision to re-enact Cristóbal Colón's voyage of "discovery" in the route taken by the Olympic flame: 'La antorcha sigue la ruta de Colón. América ya no es conquistada; sino simbólicamente encontrada' [The torch follows the path laid down by Christopher Columbus. This time America isn't being conquered, but symbolically encountered].

Impunity for the state and its servants persists (Aviña 2007) and the massacre remains ignored by the IOC. Below, you'll see the outcome of my search for a reference to Tlatelolco on the Committee's site (Fig. 3.1).

The official record avows that the 'Olympic Games Mexico City 1968 are remembered as one of the most iconic editions in history, and their legacy has become an integral part of the city's daily life' ("Mexico '68 Legacy" 2018). No, IOC hegemons. The legacy of those Olympics was blood-laden, its stain wiped from public view by a public-relations machinery steeped in deception, airbrushing the truth to foster 'a notable amnesia' (Nicholson-Smith 2008).

Mexico's truth commission on the Guerra Sucia (Dirty War) waged against the left from 1965 to 1990 revealed that well over a thousand people were shot in the back of the head and their bodies dropped in the ocean in a concerted state campaign of displacement, disappearance, imprisonment, torture, and murder. The commission's report nominated Tlatelolco as one of 'los momentos más infames que mostraron cómo el régimen de la época orquestó una letal represión contra diversos sectores de la población, entre ellos estudiantes, profesionistas, campesinos, obreros' [the infamous moments that illustrate how the regime of the time orchestrated the lethal oppression of diverse groups: students,

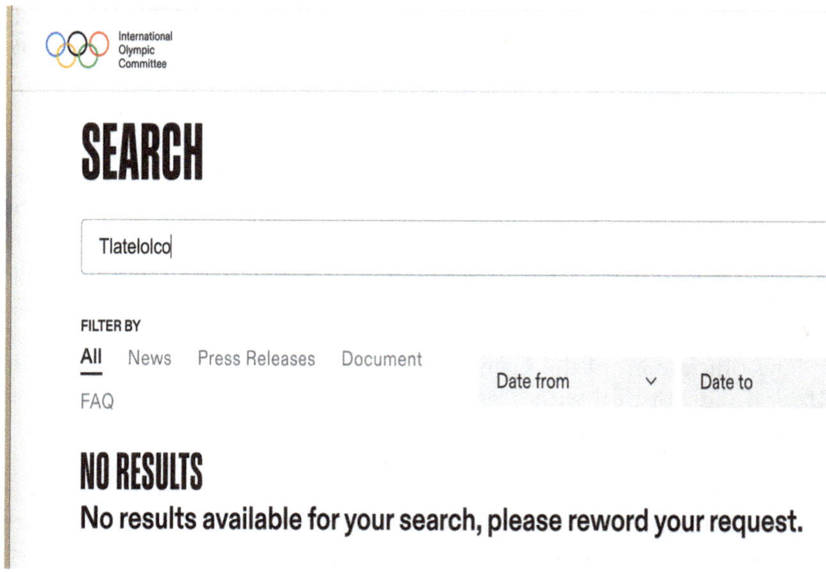

Fig. 3.1 (Source: https://www.olympics.com/ioc/overview)

professionals, peasants, and workers, among others]. It noted that Tlatelolco's firehouse became a key site for torturing and disappearing people. Some of that horror can be witnessed in *TLATELOLCO, las claves de la masacre* [*TLATELOLCO, The Keys to a Massacre*]. In 2024, Claudia Sheinbaum Pardo dedicated part of her first press conference as President to apologizing for the massacre on behalf of the government. Her mother had been among the protestors (Nájar 2024). We shall read more about this powerful legacy in Chap. 6.

In our own time, the desire to showcase modernity has been accompanied by the requirement to appear ecological as much as developmental, with disingenuous promises and predictably failed results. Athens in 2004 missed or abandoned almost every environmental target and did not produce a credible evaluation (WWF 2004). China promoted the 2008 Games as green, humane, and scientifically advanced. Needless to say, its opening ceremony instantiated two ideas, progress and unity, disavowing Taiwan and Hong Kong as discrete entities and highlighting a safe sameness. The weather was manipulated to ensure a dry night (Chen et al.

2012). And Beijing remained one of the most polluted places on earth (Dai and Chen 2014). Rio, which announced that the 2016 Olympics would turbocharge investment in the city's inadequate sewage systems, gave up on the project (Goldblatt 2016: 606; see also Huijuan et al. 2013). The hosts *did* manage to clear *favelas* that stood in their way (Rocha 2023) and line the pockets of Organizing Committee Chair Carlos Arthur Nuzman through money laundering and corruption (Zirin and Boykoff 2023).

The 2012 London Olympic Authority promised the "greenest ever" Games, a boast as laughable as it was predictable. The city needed forty venues over 500 hectares, with huge implications for construction, maintenance, transport, post-Olympic utilization, gentrification, consumption, pollution, noise, light, water, and sewage (Pitts and Liao 2013). Perfidious Albion's government guaranteed a utopia for local residents:

> The 2012 Games will make a huge and permanent impact on East London. They will create new homes, new public transport facilities, new business opportunities and new spaces in which to live, work and play. The 2012 Games will create a positive new image for the area, which will attract new investment to East London and the Thames Gateway more widely, and help accelerate both the pace and quality of its regeneration. But a transformation of an area is not just about new buildings and new business. It is about the people and the quality of the lives they lead, and involving them in planning and shaping the improvements that the London Games will bring to the area. 2012 will benefit residents and communities through more jobs, training and creating a cleaner, better and healthier environment for everyone to live in. (Department for Culture, Media and Sport and Department for Digital, Culture, Media & Sport 2008)

A decade and more later, a similarly duplicitous IOC took 2012 as an exemplary case of 'the power of the Olympic and Paralympic Games to deliver lasting benefits that can transform a community, its image and its infrastructure' ("Sport Takes" 2024).

East London, 2012's principal site, had been a manufacturing center from the middle of the nineteenth century until the 1960s. It was transformed by the 'environmental impacts of industrialization, post-industrial piecemeal development, spatial disconnection, and long-standing patterns of social deprivation' (Davis 2016: 425). Decaying substructure emptied many buildings, including warehouses. From the 1980s, cultural workers moved in. Prior to the Games, 'more than 700 artists' studios in an area of less than half a km2' nestled alongside light industry—auto-repair shops, printers, book distributors, and so on—and new immigrants (Marrer-Guillamón 2016: 125).

The area's prevailing national image was of impoverished migration, dirty warehouses, white adolescent party-going, and electric waste (abandoned appliances formed a 'fridge mountain'). That would be wiped away, along with less perilous forms of life:

> The fact that land had been given by covenant to the people of east London for allotments made no difference to the Olympic Delivery Board, the GLA [Greater London Authority] or ministers. One-hundred-year-old apple trees, charming allotment sheds, beautiful ranges of vegetables grown by a very ethnically mixed allotment community made a living picture of the best of British life. Manor Garden allotments could have been a showcase of a diverse, healthy London. Instead it was all destroyed to build a cement walkway. (Miller 2023)

The opening ceremony announced a purportedly open and diverse Britain, proud of its history, yet intrinsically and restlessly innovative. What did those fine sentiments amount to in practice? Almost $17 billion spent on the Games, equating to $1.6 million per athlete, leading to a 76% cost overrun (Budzier and Flyvbjerg 2024), and emitting more than 3.5 million tons of CO_2 (Short 2015) compared to Britain's annual average of 550 million. These astonishing levels were largely caused by flying thousands of athletes, bureaucrats, horses, journalists, and objects to and from the Games (Davis 2009; Samuel and Stubbs 2013; International Civil Aviation Organization 2016). And what happened to surges in the popular use of mass transit during the event? They quickly subsided to normal patterns, a typical difficulty with the fantasy that mega-events encourage sustainable transport.

The photographs below depict official versions of these processes (Fig. 3.2):

After the event, bureaucrats, scholars, and activists conducted very different evaluations. They divide neatly in two: apart from the relatively autonomous National Audit Office (2022), everything associated with the state, the IOC, and capital found nothing but fresh love and even fresher roses in mind-bendingly self-affirming paeans to London 2012 (University of East London 2015; London Assembly Plenary 2022). The *bourgeois* media concurred. *Le Monde* deemed the Games an 'exemplary' model for

Fig. 3.2 (Source: Toby Miller)

Fig. 3.2 (continued)

Fig. 3.2 (continued)

Fig. 3.2 (continued)

Fig. 3.2 (continued)

Fig. 3.2 (continued)

Paris in 2024 (Albert 2024). *Der Spiegel* celebrated their liveliness, efficiency, and Britishness (Volkery 2012). CNN declared a 'spectacular triumph for London,' discerning a 'mood of euphoria' that dispelled 'arguments over spiralling costs and an uncertain legacy' in favor of 'an open-minded and progressive games,' indexing a 'self-confident, inclusive society' at 'a cathartic and regenerative moment.' The city was destined to 'shine with luminosity in the collective memory' (Hooper 2012). *The Daily Beast* said, 'Take a Bow, London' (Jukes 2012), and the BBC called it a 'triumph ... a crucial catalyst in uniting the nation,' while the *Daily Mirror* and *Daily Mail* were redundantly grateful: 'Thanks a Million' were shared headlines. Content analysis of coverage across the globe found portrayals of London 'transforming itself in a wholly positive way,' with 'markedly increased salience of reporting of East London in general and Stratford in particular in the context of regeneration'.

Critical research on the ground told an entirely different story. Before the Games began, analysts discerned 'un estado de excepción: una suerte de paraíso capitalista, en el que la actividad económica está regulada por contratos exclusivos, toda actividad política está criminalizada, y los privilegios son protegidos policial y militarmente' [a state of exception: the dream of a capitalist paradise, with economic activity regulated by exclusive contracts, political activity criminalized, and privilege protected by the police and military] (Marrero-Guillamón 2011). And so it proved.

Workers seeking construction employment for the Games were excluded by political affiliation (Nelson 2012). State funds allocated to boil and wash soil, lay pipes, bury cables, and build roads for the Games served the interests of a cosmopolitan élite rather than working-class, immigrant, and artistic residents of the East End. An 'orgy of development' stimulated 'property speculation and social displacement.' Hackney Wick in particular, which artists had remade from a discarded light-manufacturing area into a vibrant, organic alternative to corporate life, saw locals promised 50% of new housing would be affordable by the poor. They got a fifth of that, as left-wing city and centrist national governments were succeeded by reactionaries bent on diverting wealth and comfort toward the élite. Qatari-owned developers ensured there would be little or no space for working people, who trapsed dispiritedly but determinedly off into less-costly areas, exiled by spiraling rents (Yeo and Miller 2017: 12, 15). Hackney Wick became a sizeable middle- and upper-class enclave, circled by zones of poverty where tens of thousands of households experienced grievous hardship (Foster 2022; Wainwright 2022; Jakar and Philippou 2025):

> the displacement of low-income residents living in affordable housing and of unskilled and semi-skilled workers employed in traditional manufacturing jobs prior to the Games ... made way for their replacement with mainly market housing and employment opportunities targeted at professional groups from outside. (Davis and Bernstock 2023: 213)

Putative post-Olympic job opportunities for residents were largely taken by Londoners from other neighborhoods. No serious attempt was made to reach out to an impoverished migrant population alienated from the city's major projects, a prejudice justified by the private sector as necessary because of the allegedly urgent requirement to fill positions based on existing skills. A primary labor market in West and North London, alongside the transnational class, left East Londoners in a dismal secondary spot (Vadiati 2020), with 'livelihood insecurity ... entrenched' (Woodcraft et al. 2024); as if it were ever meant to be otherwise. For this is typical. The Centre on Housing Rights and Evictions' survey of what happens to the popular classes during summer Games is shocking (2007; see also Doustaly and Zembri-Mary 2024). By 2024, Stratford housing prices averaged over £750,000 (Foxtons 2025).

Looking back on the Olympics from the vantage point of twelve years, Radio-télévision belge de la Communauté française [French Community Radio-Television Belgium] asked whether it had been 'un feu d'artifice ou un pétard mouillé ?' [a firecracker or a damp squib]. Ecological targets had not been met, participation in sports had not increased, and public goods had gone into corporate hands (Marlet 2024). For example, West Ham United Football Club came to use the London Stadium for home fixtures, converted from track-and-field with £320 million of public money (double the original estimate). Taxpayers proceeded to fund it to the tune of 20 million more each year to cover heating, cleaning, and maintenance. The team paid less than half those annual costs in rent (Roan 2022; Skóra 2024).

The real outcome of the 2012 conceit was East London's gentrification, undertaken as part of Britain's risible ongoing attempt to reorient its economic base toward the digital by creating 'a networked technology cluster running from Shoreditch to Stratford.' This hierarchical urban development welcomed museums, theaters, universities, and big business via a strategy that moved in on areas with an organic character created by local artists and communities, a fashionable subculture that unwittingly set the conditions of existence for property development, displacement, and the reproduction of inequality (Yeo and Miller 2017: 12, 15). The Games were a gift to the wealthy young ("The Legacy" 2022).

Toby worked in the area and lived in the former Athletes' Village for several years, in the course of which Shinjoung Yeo and he prepared a report on the experience of local artists. It drew the ire of an executive running the building where we worked, whose qualifications were time spent in the military and a right-wing think tank. His lackeys had called the police when some of us mounted a peaceful picket as strike action, in which we handed people crossing the line a pamphlet and their choice of flowers or candy. A while later, Toby lost his job, as did most of his colleagues (he has no reason to think these things were related) and, like so many others we had studied, could no longer afford to live there. Years later, the executive assured the world that the Olympics 'were a catalyst for sustained economic regeneration and economic development' (Poole 2024). But then he would, wouldn't he?

In 2020, the IOC signed up to that great capitalist oxymoron of our times, the UN's "Sustainable Development Goals" ("Tokyo 2020, 100" 2021) among other outlandish claims of environmental leadership. Tokyo 2020 was one more case of failed cost control: $12.9 billion, a fifth more than estimated, with a further $9.8 billion spent on associated projects (Mair and Duignan 2025: 109). Like London, it was putatively designed to be green, with 5000 medals constructed from electronic waste—discarded cellphones, consoles, and laptops (Nakagawa 2018). Adopting the slogan "Be better, together, for the planet and the people," Japan guaranteed a Games characterized by mass transit, renewable energy, and recycled rainwater ("Tokyo 2020, 100" 2021) among other outlandish claims of environmental leadership. Most events were held on Tokyo Bay islands atop landfill that had been dumped in preparation for 1964, to the point where a public-health crisis was identified the following year. The *Financial Times* applauded this as a 'scavenging mission accomplished'. The Olympic Channel's animation *Tomorrow's Leaves* (Yoshiyuki Momose 2021) brought together imaginary young athletes from five continents (some via wind or sea, others through gas-guzzlers) to compete, collaborate, survive, and ultimately restore the planet through a magical greening that occurred just as it was time to return to their homes, secure in the knowledge that the vegetation of the title had a future.

Those happy tales excluded the reality of increased policing in parks, the destruction of lively neighborhoods, ballooning rents, the takeover of small businesses by corporate chains, rampant corruption, brutal clearances of homeless folks, and the eviction of those in public housing (Suzuki et al. 2018; Mair and Duignan 2025: 109; Zirin and Boykoff 2023). Parks were razed and iconic landmarks ruined. The organizers pillaged Malaysian and Indonesian rainforests to build a stadium on land that had once been tree-lined. Fukushima was re-imagined for local and international propaganda: don't fear radiation—come home, or visit for the first time—it's safer than smoking (Zirin and Boykoff 2019; Rainforest Action Network, n.d.; McCurry and Associated Press 2023; Boykoff and Gaffney 2020; Ware 2021) (*à propos*, 1964 had featured an Olympic cigarette, facilitating massive increases in Japanese smoking and lung cancer) (Tomizawa 2016). Even before COVID-19 clouded preparations, the Japanese public showed little excitement about the Games by contrast with 1964, and less when forced to live under a state of emergency (Takanobu 2023).

Japan 'pours concrete with abandon, generates mountains of plastic waste … scours the world's oceans for fish' and 'burns a lot of coal'. Even local officials estimated 2.73 million tons of CO_2 would be emitted courtesy of the 2020 Games—more than many countries produce in a year ("A Greener Games" 2021), though banning foreign spectators because of the syndemia ultimately diminished that figure (The Tokyo Organising Committee 2021). But Olympic personnel alone were responsible for 35,365.5 tCO_2 emissions, courtesy of travel to Tokyo (Ito et al. 2022).

There were also questions about 2020's seasonal timing. The 1964 Games had been held in October to avoid 'intense midsummer heat and the September typhoon season' (Griggs 1963). But the IOC relies for 70% of its revenue on television; repeating such a calendar was unacceptable to NBC. And ever since poor ratings for the September 2000 in Sydney, NBC gets what NBC buys: $7.75 billion in control, if not intellect, competence, or artistry (Carter with Sandomir 2000; Branch and Rich 2021; Timms 2021; Noble et al. 2025). The network wanted live TV for US audiences during an otherwise fallow summer before the Fall's new drama series, baseball playoffs, and the NFL. Its 2025 deal insisted on 'tighter collaboration on content production, distribution and advertising' in return for $3 billion more to cover the Games through 2036 (Noble et al. 2025).

The Japanese Olympic Committee promised 'mild and sunny' weather for 2020, 'an ideal climate for athletes'. But during the 2018 summer, a thousand fatalities were declared 'The First Undeniable Climate Change Deaths' (Merino 2020). In 2019, the Games period saw Tokyo's daily maximum temperature average 92°F, with 80% humidity. Twenty thousand people were hospitalized nationwide (Branch and Rich 2021). Exertional heat causes strokes, organ failure, and mortality. Cyclists, triathletes, and runners are most at risk from sickness and injuries related to heat, but many other competitors suffer (Gamage et al. 2020; Hollander et al. 2021; Bandiera et al. 2024). An increased prevalence clearly correlates with climate change (O'Connor and DeGroot 2024; Bouchama et al. 2022), as the American College of Sports Medicine recognizes (Roberts et al. 2023).

In 2021, host broadcaster NHK's meteorologist Sayaka Mori noted that the weather was 'really torturing the Olympians and volunteers' (2021). The athletes' own testimony was equally damning—and vivid (Nagao et al. 2021). Dozens required urgent medical treatment due to the heat, most notably in the marathon and walking events (Sugawara et al. 2022), where they were 'fainting, leaving the competition in wheelchairs and asking what happens if they die while competing' (Lo and Jacobsen 2021). We have reached the point where the very notion of a Summer Olympics imperils the health of participants (Smith et al. 2014), with athletes falling ill and expiring in sporting venues (Wallace et al. 2019; Bernard et al. 2021). Olympic hegemons know all this. It's been exhaustively, kindly explained to them by numerous scholars (Bougault et al. 2025).

But the IOC has put responsibility onto the athletes, as if the impact on different sports and identities were not its doing, and failed to discipline the many federations that do next to nothing to care for competitors (Lei and Wang 2021; Santini and Henderson 2021). Meanwhile, reactionary scholars have celebrated a purported correlation between 'less dense air' produced by climate change and more world records (Wang et al. 2024), despite evidence to the contrary and the repugnance of privileging "performance" over health and sustainability (Brocherie et al. 2024).

Needless to say, NBC covered the 2021 'heat wave' (Siemaszko 2021) and typhoon as 'another hit of nature's power' ("Pandemic Olympics" 2021). Network hegemons complained that a 'drumbeat of negativity' dragged down TV ratings (Hsu 2021). But those numbers, and empty hotels and stadia, enshrined a void at the core of the movement's very being: a witless, mute, plastic, masked COVID-19 testimony to the IOC's absurdities. The anachronistic invocation of a year that had already passed to describe the Olympics (2020 in 2021) celebrated a pomp and arrogance so thoroughly articulated to marketing that it overdetermined the simplest chronological truth (McDonald 2020). Then again, as the *Los Angeles Times* put it, 'how else will they unload all that merchandise?'. Four things characterized these Games: 'overspending of public money ... greenwashing ... the militarization of the public sphere' (Boykoff 2021: 50), and 'commercial real-estate opportunism' (Zimmerman 2021).

Supposedly by contrast, Paris 2024's organizers proclaimed 'principles of moderation, innovation and boldness' (International Olympic Committee 2024c). There was even an official *Paris 2024 Social Charter* (2024). Event propaganda included a promise to halve London's carbon footprint ("Paris 2024 Presents" 2024) thanks to renewable energy, vegan food, carbon capture, reduced emissions—no air -conditioning—and avowedly "green" projects in Africa, Asia, and the Americas ("Paris 2024 Carbon" 2024). The Seine would be made swimmable after a century of pollution caused by agricultural runoff, consumer trash, sewage, and industrial waste (Moutiez 2024).

Paris Mayor Anne Hidalgo defended the Games somewhat indecorously in *Le Monde:* 'Fuck aux réacs, fuck à cette extrême droite' [Fuck the reactionaries, fuck the far right]. She also denounced the 'extrême gauche' [far left] (quoted in Bronner 2024). The *Guardian* went overboard (McMillan 2024), declaring 'local cynics' vanquished by 'unabashed joy' (Hurst 2024). The Australian Broadcasting Corporation announced that 'dalliances with controversy or outright negativity generally felt incredibly forced or opportunistic, and tended to exist mostly in the outer-spheres of social or bad-faith media' (Bilton 2024). Dear old *Sports Illustrated* used what could only have been magic to discover 'Parisians believe in art as something that should be experienced communally but absorbed individually. These Olympics were an expression of that' (Rosenberg 2024). We were told that unlike Beijing in 2008, 2024 was not a case of 'a sullied country using the Olympics to improve its image,' but 'a sullied Olympics ... using a country to decontaminate itself' (Wertheim 2023). The EU called Paris 2024 'economically viable' and a successful promotion of sustainability (Starcevic 2025).

The clamor reached the point where 'criticar los Juegos parece ahora mismo casi un sacrilegio en el país vecino' [criticizing the Games in the host country became almost sacrilegious] (Bonet 2024). *Libération* referred to 'une tyrannie de la joie officielle' [a tyranny of official joy] (Lambert 2024). Amidst such 'joy,' initial claims about carbon neutrality—and beyond—were quietly dropped (Carbon Market Watch and éclaircies 2024) along with some planned new metro lines ("Métro gratuit" 2024), while there were steep price hikes on mass transit (Henley 2024). Aquatic events were held across the country and in Tahiti, 15,000 kilometers away, with environmental impacts that offended local and international competitors alike ("JO-2024" 2023). A cruise ship, complete with restaurant, swimming pool, bars, and tattoo salon, housed surfers and executive hangers-on, who felt little need to spend time in Tahiti proper (Garot 2024). Such craft are ecological disasters (Transport & Environment 2024a, b). We'll read about them again in Chap. 4.

There was a 'Cara B' [another side] (Bonet 2024). Pollyannaish accounts of the Games involved a decision to 'leave out those who suffered' (Boykoff and Zirin 2024), from humans and other animals to the climate. For the organizers' greenwashing pomposity ran against history (Delmas 2017). It was no surprise that Carbon Market Watch and éclaircies found the Parisian ecological initiatives 'incomplete' and official reportage 'short of achieving transparency' (2024). Here is the briefest of summaries of the havoc wrought by the IOC and its French chorines:

> Destruction des jardins ouvriers d'Aubervilliers, de l'aire des vents à la Coumeuve, du bois de Taverny, des milieux boisés à Élancourt ou des roseaux à Vaires-sur-Marne, encerclement du groupe scolaire Anatole France par une bretelle d'autoroute, destruction des coraux à Tahiti, bétonisation à Marseille et autres méfaits encore : les JOP ont saccagé les espaces verts et naturels des banlieues parisiennes, Marseille et Tahiti [destruction of the workers' gardens of Aubervilliers, of the Aire des Vents in La Courneuve, of the Taverny woods, of the wooded areas of Élancourt and the reeds of Vaires-sur-Marne, encirclement of the Anatole France school by a motorway slip road, destruction of coral in Tahiti, concreting in Marseille and other misdeeds: the Olympics ravaged the green, natural spaces of Parisian suburbs, Marseille, and Tahiti]. (Saccage 2024b)

The Seine kept filling up with dangerous bacteria (Walker 2024), even on days when officials proclaimed it safe (Rouget 2024). The promise of no air-conditioning in athlete accommodation was broken: 2500 cooling units sated vulnerable Global Northerners, who had threatened to bring their own ("AC-Free" 2024); Australian Olympocrat Matt Carroll had complained, 'We're not going for a picnic.' It certainly wasn't some church outing for the big man, managing as he did on a salary of $440,000 (Osborne 2024). The *Los Angeles Times*' editorial board thundered that the absence of air-conditioning was 'unnecessarily austere and shortsighted'.

The Australians kept whining ("Australian Athletes" 2024) and a fellow Anglo moaner said, 'I need meat to perform' (quoted in Latham-Coyle 2024). Earlier paeans to an 'eco-friendly menu' duly fell silent (Morrison 2024), as vast numbers of dead animals were imported and food and waste mounted (Poinssot 2024). As is typical of sporting events, nothing was said about where and how the meat was obtained—the kind of slaughter involved (Sher and Fusco 2024). Six hundred and fifty tons of ice were on hand to treat human injuries, ten times the 2021 figure (Garot 2024), thanks to a febrile faith in unproven doctrines of "cryotherapy" (Racinais et al. 2024).

Many events were held on the capital's outskirts, in the Seine-Saint-Denis region. Per East London in 2012, it's a *département* marked by poverty, immigration, unemployment, and poor social services, policing, and school safety (Gilbert 2023; Institut national de la statistique et des études économiques 2023; Chaumier 2023). For the *Guardian*, this was 'France's poorest and most unloved banlieue' (Willsher 2024). In an appropriately Olympian adjective, the *Economist* called it 'scruffy' ("Paris's Stunning" 2024).

Condé Nast Traveler found the Games giving 'hope' to Seine-Saint-Denis (Tramuta 2024). But several infrastructure plans never came to fruition (Rédaction 2022). Some that did relied on exploiting undocumented migrant labor (Kristanadjaja 2022), with hundreds of workers injured, dozens gravely so (Fisné-Koch 2024). The neighborhood hosted the Athletes' Village, again per East London. The fact that beds were made of cardboard was a point of pride for organizers ("Ecológica" 2024) when they weren't busy evicting people sleeping rough in makeshift cardboard protection across the city or dealing with the *New York Post* mocking 'anti-sex beds' designed to discipline 'horny athletes' prone to 'orgies' (Court 2024). (Neighborhood sex workers were "moved on" [Gagnepain 2024].)

Forget putative Puritanism; this was gentrification through displacement (Hussain et al. 2024; Lindgaard 2024). Just a fifth of Village apartments were reserved for post-Games social housing, and the new pool came at the cost of destroying community gardens (Bernard 2023; Cadoret and Guitton-Boussion 2021). Porte de la Chappelle, the site of badminton and rhythmic gymnastics, witnessed unprecedented 'nettoyage social' [social cleansing] (Gable 2024). Twelve and a half thousand migrants and unhoused people were exiled for the duration (Bonet 2024; Janetsky 2024).

The state set new Parisian "standards" of surveillance and police harassment—45,000 police, 15,000 soldiers, agents from thirty-two additional counties, requirements that citizens flash quick-response codes in many streets, and drone and static cameras proliferating their doleful features across the city (Boykoff 2024; Le Foll 2024; Molinari 2024). Seine-Saint-Denis residents felt as though they were living behind a barricade (Boganda 2024). Interior Minister Gerald Darmanin, who had promised 'delincuencia zero' [zero criminality], declared 'une belle médaille d'or pour les forces de sécurité' [a beautiful gold medal for the security forces] (quoted in Molinari 2024; "Paris 2024: les forces" 2024). Their number included an officer who had shot and killed a young man just before the Games. He was rewarded with duty at the opening ceremony (Polloni 2024). After the Olympics, *Libération* published a video of police brutality (Hallisat and Courtois 2024). Critics looked back at the *Social Charter* as akin to a blueprint for gentrification and displacement, with enormous cost overruns financed by the people (Fisné-Koch 2024; Bonet Torra 2024; Morteaux 2024).

The 2024 Games straightforwardly repeated what many carbon sports have done for sixty years, accelerating destruction, pollution, expulsion, and speculation (Kokabi 2024). The Saccage collective's post-Olympics report posed uncomfortable questions:

> Comment savoir combien de personnes ont fui l'Île-de-France ? Combien ont dû renoncer à des déplacements ne pouvant plus payer un ticket de métro à 4 euros ? Combien de familles ont renoncé à prendre le frais dans le parc à côté de chez elles car celui-ci était occupé par une fan zone ? Combien de personnes ont passé l'été à étouffer dans leur appartement sans ouvrir les fenêtres, à cause du bruit incessant des sirènes de police et des hélicoptères qui tournaient toute la journée ? Combien de personnes n'ont pas bénéficié des soins nécessaires car bloquées par les interdictions de circuler ? Combien partiront de leur logement devenu trop cher ? [Have we any idea how many people left l'Île-de-France? How many had to forego travel because they

> couldn't afford a four euro metro ticket? How many families gave up on a stroll in the fresh air because their neighborhood park was transformed into a fan zone? How many people spent the summer in stifling heat, their windows shut, because police sirens and helicopters whirred all day long? How many didn't get the care they needed because of blocked traffic? And how many had to leave their homes due to skyrocketing rents?] (2024a)

As the event drew to a close, sustainability director Georgina Grenon proudly detailed her desire 'to turn these Games into a laboratory … an accelerator' (quoted in "Olympics Equipment" 2024); spurious science met *cliché* capitalism. Putative environmental successes were hailed by *Forbes* (Ménage 2024). The IOC's self-satisfied report concealed authorship, methods, calculations, setbacks, and failures, certain in this happy peroration: '[a]t Paris 2024, all roads led to sustainability' (International Olympic Committee 2024b). The organizers claimed they had reduced Rio and London's carbon footprints by 54.6%, at 1.59 million tons of CO_2 (*Paris 2024 Sustainability & Legacy Post-Games Report Summary* 2024). We find the work inscrutable, and the detail on journeys by spectators, athletes, horses, equipment, and hangers-on (AKA executives) minimal.

One can only imagine what awaits the unhoused of Toby's beloved, lost Los Angeles in 2028. They numbered over 45,000 in mid-2024, with an average monthly income of under $400 and rents escalating because construction had not kept up with population growth (Flaming et al. 2024; Byrne et al. 2021; United States Department of Housing and Urban Development 2023). That was six years after then-Mayor Eric Garcetti announced 'I'm confident that by the time the Olympics come, we can end homelessness on the streets of LA' (quoted in Zirin and Boykoff 2022).

With new powers from the Supreme Court to arrest the homeless, would the city revert to the social clearances of the 1984 Olympics, when the police adopted a policy of 'trying to sanitize the area,' sending mounted officers across Downtown to imprison people or require them to 'relocate' (quoted in Chandler 2018)? A 2024 Human Rights Watch report highlighted LA's 'Criminalization of Unhoused People' as 'Cruel and Ineffective.' In keeping with the social clearances that accompanied construction of the world's priciest stadium, SoFi, homeless folks and longstanding residents were already being shunted from public space when the Paris Games concluded, in order 'to make way for tourists and investors, despite promises to prioritize housing solutions over forced removals'

(Robbins 2024). A 'multipronged strategy of displacement' operated in a city 'greased for private profit' (Lee White 2022).

But fear not: Airbnb announced a partnership with the IOC to assist athletes ("AIRBNB" 2025) and the organizers, LA28, guaranteed 'employment opportunities, economic benefit and pride to people across Los Angeles' ("How Will" n.d.). Its head had been a brigadier general in the empire's failed colonization of Afghanistan, which he described as the 'most complex logistics operation since World War II' (quoted in Wharton 2024)—perfect preparation for the Olympics. Everything should be fine, as long as Angelin@s provided half a billion dollars in tax-free subvention demanded by the IOC for broadcasting and a wonderful euphemism, 'general assistance and transfer of knowledge' (Noble et al. 2025; International Olympic Committee 2024d: 15).

With tragic prescience, *The Nation* magazine published an article in 2022 entitled 'Organizers Don't Care That the 2028 Olympics Could Be on Fire' (Zirin and Boykoff 2022). Following the horrendous blast furnaces of 2025, Governor Gavin Newsom referred to the "Recovery Games" and the "Comeback Games" (quoted in Boykoff and Zirin 2025), very much per the absurd legitimation voiced for the 1986 World Cup in Mexico. Wiser voices worried 'about sports usurping resources and valuable attention that should be paid to rebuilding' (Wharton 2025). Opinion polling indicates most Angelin@s favored reconstructing the city after its terrible fires as their priority for public policy; a fifth chose the Games, mostly older men (NOlympics LA 2025); we are as we are.

The 1984 jamboree had brought some temporary environmental gifts: 'revolutionary real-time traffic information, an ambitious new strategy to make deliveries, a rapid-transit bus system, and a plea for employers to institute flex work hours' (Walker 2018). Their emulation and betterment would depend on making inroads into the city's profound, profoundly disturbing, car masochism and homeless sadism. Just as plans for improved mass transit were shelved, the organizers announced a flying-taxi service—not approved for US travel, but why bother with such niceties (Abbas Ahmadi 2025)?

The Queensland Government said the 2032 Olympics would prove 'a catalyst for positive change and improved infrastructure right across' the state and 'benefit all Queenslanders in the years leading up to the Games, and for years beyond' through job creation and 'multiple social and health benefits.' It knew this *per medio* conferring with '500 of Australia's best, brightest and boldest minds.' (My colleagues there didn't qualify.) The propaganda even undertook 'to remove more carbon from the atmosphere than the Games emits' (n.d.). But almost a decade before the 2032 event, housing was posing problems for the Brisbane Games. Prices and homeless numbers accelerated as workers were imported to undertake construction. Promises of athletes' villages transmogrified into affordable homes rang hollow (Martin et al. 2024). Although the 2032 Games were the first Olympics contractually obliged to be 'climate-positive,' that was "amended" (IOC 2022) with weasel words (Foth 2025). Meanwhile, controversy over customarily unreliable figures about costs to the public, and debates about new construction, were in full swing (McKenna 2024). Two former state Premiers concurred across party lines that the Games were a 'hot mess' (quoted in McGowan 2025). The usual *clichés* were in play.

And sponsorship? It accounts for 30% of Olympic revenue (Noble et al. 2025). The Games' commercial "partners" involved in these extravaganzas have included major polluters seeking to appear green: Visa, Dow, MacDonald's, Coca-Cola (which paid a reputed $3 billion in 2020, extending its sponsorship to 2032), Toyota, and General Electric (GE) (Samuel and Stubbs 2013; "Olympics Under Pressure" 2024). We have already encountered some of them.

Such companies typically claim a high moralism while dodging essential environmental standards; helpfully, the IOC 'issues no hard-and-fast requirement for Games hosts' (Müller 2015: 197). The Committee's dealings with sponsors are far from transparent, but the period 2017–21 saw it gain $2.3 billion from sixteen corporations in return for global marketing rights—double the price of the previous cycle (Germano et al. 2024). Leading sponsors for the 2020 Olympics included SMBC and Mizuho, major financiers of coal plants that used green rhetoric while refusing to make meaningful guarantees protecting the environment (Tzanelli 2023: 57).

We read about Nike in Chap. 2. It has been a regular participant, committing somewhere near a billion dollars in return for unprecedented product placement and sponsorship at the Paris Games (Reid 2024). The company promised 'athlete-informed, data-driven and visually unified' outfits ("A First" 2024). Its cast-offs went into Cambodian kilns that poisoned indentured labor (Kasztelan and Quashie-Idun 2022).

Dow is the Olympics' "Official Carbon Partner," guaranteeing the Games can offset emissions. Dow should know: converting fossil fuels into plastics is the firm's 'core business' (Glasson and Hutchins 2024). It chairs the Alliance to End Plastic Waste, which began in 2019 to great fanfare. Five years later, the Alliance had cleaned up the same amount of plastic as its executive committee—Chevron Phillips, Shell, ExxonMobil, Dow, and TotalEnergies—produced every forty-eight hours (Boren and Howard 2024).

The Political Economy Research Institute (PERI) placed Dow at number twelve in its *Toxic 100 Air Polluters Index* (2024), as the scandal continued over the 1984 Bhopal leakage of methyl isocyanate, which killed tens of thousands and affected half a million, a tragedy perpetrated by a company Dow subsequently bought (Eckerman 2005). Forty years on, people still felt the physiological and psychological impact of 'the world's deadliest industrial accident.' Court hearings into reparations were deemed illegitimate by the corporation. Its servants in the US "justice" system defeated civil claims (Sharma 2024: 2145; "Suffering" 2024; Ganguly et al. 2018).

PERI's *Index* put GE at thirty-sixth place (2024). The company's criminality is massive and long-standing ("GE" 2001; "General Electric" 2015; Greene 2024), such as setting records for dumping polychlorinated biphenyls into US waterways (Cray 2001; Betts 2006). To stave off democratic regulation, GE announced a policy of "ecomagination," a word/campaign that emerged from $90 million of product development. It featured advertisements of trees growing from smokestacks, a computer-generated elephant dancing in a rainforest, and a "clean" factory. The intention was to show the corporation 'addressing the problems of tomorrow, today' (quoted in Anderson 2005; also see Fisher 2005) even as GE continued to invest in unsustainable energy ("GE's Verona" n.d.).

One sponsor needed particular redemption in the run-up to the 2012 Olympics: BP was in environmental disgrace following the explosion of the Deepwater Horizon oil rig in 2010, which propelled 134 million gallons of oil into the Gulf rightly known as "of Mexico," across the coastlines of five US states. The firm was responsible for the deaths of 11 people, 160,000 sea turtles, 105,000 sea birds, 8 billion oysters, and half the dolphin population. Six years later, BP and its partners, which included Haliburton, were fined over $20 billion and paid another $50 billion in compensation and clean-up costs (National Oceanic and Atmospheric Administration 2017; Moore 2025). It took the company a long time, and considerable litigation, to acknowledge what had happened. Along the way, BP repeatedly invoked metaphors of war and death to describe the struggle to manage the disaster it had made (McClintock 2010).

London 2012 represented a veritable coming-out party for that toxifying, terrorizing corporation:

> At their recent AGM [Annual General Meeting], the BP board outlined how they had made a business case internally for their sponsorship of the Olympics, the costed returns for which included building and protecting their brand. Inside the industry this is understood as maintaining the "social license to operate." (England 2012)

In the USA, where the firm's image had been the worst of all Games sponsors, the company's publicity professed dedication to the nation and Olympic ideals (Young 2012). After weeks of billboards and commercials promoting this self-regard, BP became the second-most popular Games enabler, behind Visa (Levintova 2012). Just 18% of those polled were unaware that the company had sponsored the London absurdity. Of the 82% majority, well over a third believed BP sought 'a cleaner planet' (Reynolds 2012). That corporate camouflage led even mainstream commentators to call for an ethical overhaul:

> The relationship between the Olympics and some commercial sponsors and suppliers has not realised its potential for creating sustainable change. We recognise the complexity of contractual arrangements which can involve both London 2012 and the IOC but feel the opportunity to put relationships with sponsors and suppliers on a greener footing has not been fully grasped. These commercial arrangements need to be reviewed to ensure that future Games are able to deliver greater gains for sustainability, (Bioregional and WWF 2012)

All terribly polite. Critics on the left were more scathing:

> organizers created a new category of corporate sponsor: "sustainability partners." Six such sponsors—BP, BMW, BT, Cisco, EDF [Electricité de France] Energy, and GE—debuted as "sustainability partners." ... [T]he sponsorship program was simply a pay-to-play charade ... no environmental standards ... needed to be met in order to become a green corporate partner. (Boykoff and Mascarenhas 2016)

A decade later, PERI's *Index* ranked BP at number 84 (2024). Then the corporation terminated its commitments to a green transition as 'misguided,' cutting its low-carbon budget by £5 billion and increasing fossil-fuel investments by twice that figure, toward 'an unwavering focus on growing long-term shareholder value' (quoted in Ambrose 2025).

Though not as dramatic as BP's criminality, the Games' other 'sustainability partners' have poor records, before and since. BMW allegedly relied on forced labor in China to manufacture vehicles (Senate Finance 2024), was fined hundreds of millions of euros for colluding with so-called competitors to minimize innovations in emissions technology ("Antitrust" 2021), and breached advertising laws by claiming it sold 'Zero Emissions Cars' (quoted in "ASA Ruling on BMW" 2024). The UK's telecommunications regulator has repeatedly found BT guilty of deceiving and failing customers, and it presided over accounting fraud ("BT Fined £2.8m" 2024; "BT Fined 117.5m" 2024; Parodi 2019). Cisco knowingly provided the US Government with faulty surveillance machinery and is involved in lengthy litigation alleging its part in the torture of 法轮功 [Falun Gong] members in China (Goodwin 2019; *Doe v. Cisco Systems, Inc.* 2024). EDF has engaged in insider trading, manipulated markets, and denied indigenous Mexicans their rights ("French Energy" 2022; "Lawsuit Filed" 2024).

Unsurprisingly, a virtual podium of the worst environmental criminals among the corporations involved in Paris 2024 was crowded (New Weather Institute 2024). Lululemon Athletica sponsored the Canadian team. It makes clothing from animals and forests and has tens of thousands of workers and hundreds of stores in twenty-two countries. The firm's governing slogan, "Be Human, Be Well, Be Planet," led to the claim that '[o]ur products and actions avoid environmental harm and contribute to restoring a healthy planet' ("Our Sustainability" n.d.). But well over half its materials come from fossil fuels and are not biodegradable ("Lululemon: Canadian" 2024). Lululemon's catastrophic emissions poisoned air and water through pollution and waste treatment, releasing '1.2 million metric tonnes of CO_2e in 2022,' nearly 65% more than 2020 (Action Speaks Louder 2024: 4). The New Climate Institute rated the firm's record 'Shallow'. Meanwhile, its founder was busy insisting the road to profitability had been forged by clarifying that "certain customers" were unwelcome. As if in embarrassed reaction, Lululemon hosted workshops designed to 'resist capitalism ... and the ways violent colonialism has erased these histories to enforce consumerism'—this from a billion-dollar corporation notorious for white-supremacist personnel policies.

France Nature Environnement obtained a confidential document indicating Olympic sponsor Coca-Cola would distribute 18 million bottles of sugar drinks during the Parisian fortnight, more than half made from plastic ("JO 2024: Coca" 2024; Chrisafis 2024). (As noted in Chap. 2, the company had just been named the world's worst plastic polluter.) Fellow-sponsor Toyota provided Paris organizers with a fleet of unsustainable hydrogen-powered vehicles, even as it admitted to falsifying emissions and safety tests and was under investigation by the Australian Competition and Consumer Commission for misleadingly promoting itself as green ("Hydrogen Cars Risk" 2024; Liang 2024; Joshi 2024; Environmental Defenders Office 2023). Coke exited the ranks of sponsors, as did Panasonic and Bridgestone, casting doubt as they did so on whether athletic achievement was really the IOC's priority (Noble et al. 2025).

Visa was the sole currency for purchases (Germano et al. 2024) and Omega 2024's "Official Timekeeper." Mechanical watches are built to last half a century, but the luxury models created for particular events and moments are not, even as their makers claim commitments to sustainability in the light of critiques that they have been slow to address environmental problems thanks to gold and jewelry mining, animal imprisonment and slaughter, transport, water, and the labor process (Swithinbank 2022a, 2022b). WWF's ecological ranking of watch and jewelry corporations rated Omega as middling, with much to improve in transparency of information and water use (2023: 22, 37).

Louis Vuitton (LVMH is Vuitton's acronymous makeover luxury conglomerate—we'll use the names interchangeably) luxuriated in animal torture ("Les Jeux Olympiques de Paris 2024" 2024) while stocking élite lounges with Moët Hennessy and storing the Olympic torch in its checkerboard-design luggage—just a small part of the $1.3 billion from sponsors that enabled IOC revelers ("LVMH is Splurging" 2024). If you were a Games hegemon, this all worked rather well. And fear not: the Committee assured us it was determined to '[c]ontrol and limit the commercialisation' of the Games (2021: 23). Perhaps; but that did not preclude product placement, as LVMH, Coke, and Samsung disported their wares: the torch stopped at LVMH hotels, medals were presented on its trays, athletes photographed themselves with Samsung telephones provided by officials, and—by chance—wandered around the opening ceremony with gold-colored bottles of Coca-Cola water or LVMH clothing. The principal performers were dressed by Dior (Germano et al. 2024) and

NBC fetishized an interview about Moët Hennessy, Louis Vuitton, and sports as a grand reunion of luxury and athletics (Ginsky 2024).

The sign of Louis Vuitton marked a new moment in Olympic history: luxury and mass-market brands on the podia together (Raji 2024). That amounted to clandestine advertising, which the IOC had previously abjured; now, its chorines propagandized such marketing as a means of avoiding drains on the public purse (Dang and Spencer 2024). The *Wall Street Journal*, a willing servant of capital, announced that LVMH had 'turned the Paris Olympics and Paralympics into one long exercise in product placement' (Kostov and Robinson 2024). To quote Hall, the Games' sponsors are 'deep enemies of the Olympic ideal' (Jhally 2016: 341).

Conclusion

Six decades ago, Raymond Williams yearned for fantasy to be made real: 'Who has not a wish to see the world as the Olympic Games show it?' (2011: 24). Perhaps; but he was in poor company. Consider Hitler's welcome to the 1936 event:

> Sporting and chivalrous competition awakens the best human qualities. It does not sever, but on the contrary, unites the opponents in mutual understanding and reciprocal respect. It also helps to strengthen the bonds of peace between the nations. May the Olympic Flame therefore never be extinguished. (quoted in Organisationskomitee fiir die XI. Olympiade Berlin 1936 E.V. 1937: 6)

Sixty years later, Andrew Jennings reflected on these jamborees like this:

> The aristocrats who launched the Olympics were the supreme power in their age. Their dominance has passed to international capitalist enterprises, more powerful than many national governments. Acquiring the Olympics—and the spirit of passionate internationalism that comes with them—has been a sound move by big business. (1996: 296)

The Games are dedicated to exploiting athletes, their hosts, and the Earth, the IOC's public acknowledgment of our climate crisis matched only by its desire for growth (Glasson 2024). In *Jacobin*'s words:

> hosting the Olympics has come at a devastating cost for local residents who've experienced accelerated gentrification, displacement, privatization, municipal debt, environmental damage, and militarization of police alongside the implementation of permanently ramped-up surveillance regimes. (Lind Gustavussen 2023)

In 2016, eighty-year-old Tokyo resident 神野 浩平 [Kohei Jinno] was evicted in the name of Olympic construction. That repeated his experience of 1964, when his home and a nearby river were buried in concrete and surrounding vegetation sacrificed. This time, he was among hundreds of elderly people displaced from public housing, each one handed $1500 by the state to assist with moving costs that amounted to $9000. Kohei Jinno was left 'with a sad, lonely feeling' (quoted in Lies 2021).

Nothing could better express the ethical void at the very core of the Games (Jennings and Sambrook 2000): a simultaneously productive and destructive, unequal, and crisis-laden industrial modernity, disguised in a 'blaze of global publicity and bogus bonhomie' (Anderson 2004), whose 'benefits … accrue to the already privileged, whereas damaging impacts tend to fall on the most marginalized' (Talbot 2025: 132). Mr. Jinno saw that—twice—but was powerless to do anything about it.

Bibliography

NB—Links from US Science Have Been Removed By Its Government

"1964: Memories of the Tokyo Olympics." (2014). *Japan News* http://tokyo-olympics.the-japan-news.com/#titlePage.

"A First Look at the 2024 National and Federation Kits." (2024, April 12). Nike https://about.nike.com/en/newsroom/collections/2024-national-and-federation-kits.

"A Greener Games? Tokyo's Environmental Impact." (2021, July 17). *VOA* https://www.voanews.com/a/east-asia-pacificgreener-games-tokyos-environmental-impact/6208269.html.

"AC-Free Ambition for Paris Olympics Melts Away as Organisers Order 2,500 Cooling Units." (2024, July 2). *France 24* https://www.france24.com/en/live-news/20240702-no-ac-ambition-for-paris-olympic-village-melts-away.

"AIRBNB ATHLETE TRAVEL GRANT 2025." (2025, January 21). https://www.olympics.com/athlete365/offers/airbnb/airbnb-athlete-travel-grant-2025.

"Antitrust: Commission Fines Car Manufacturers €875 Million for Restricting Competition in Emission Cleaning for New Diesel Passenger Cars." (2021, July 8). European Commission https://ec.europa.eu/commission/presscorner/detail/en/ip_21_3581.

"ASA Ruling on BMW (UK) Ltd." (2024, February 7). Advertising Standards Authority [UK]. Complaint A23-1209400 BMW (UK) Ltd https://www.asa.org.uk/rulings/bmw%2D%2Duk%2D%2Dltd-a23-1209400-bmw%2D%2Duk%2D%2Dltd.html.

"Australian Athletes Benefit from Extra Comforts as Olympic Village Criticised." (2024, July 31). *New Daily* https://www.thenewdaily.com.au/sport/olympics-2024/2024/07/31/olympics-australians-village.

"Brundage en Defensa de las Olimpiadas." (1968, October 8). *El Informador*: 1, 6.

"BT Fined £17.5m for 999 Call-Handling Failures." (2024, July 22). Ofcom https://www.ofcom.org.uk/phones-and-broadband/telecoms-infrastructure/bt-fined-17.5m-for-999-call-handling-failures.

"BT Fined £2.8m Over Failures Affecting 1.1 Million EE and PlusNet Customers." (2024, May 22). Ofcom https://news.sky.com/story/bt-fined-2-8m-over-failures-affecting-1-1-million-ee-and-plusnet-customers-13141349.

"Ecológica y sostenible, así es la Villa Olímpica de París 2024." (2024, July 7). *TeleMadrid* https://www.telemadrid.es/deportes/Ecologica-y-sostenible-asi-es-la-Villa-Olimpica-de-Paris-2024-0-2684731547%2D%2D20240703090000.html.

"French Energy Regulator (CRE) Fines EDF with the First Sanction for Insider Trading Under REMIT, and EDF Trading Limited (EDFT) for Market Manipulation." (2022, May 10). European Union Agency for the Cooperation of Energy Regulators https://www.acer.europa.eu/news-and-events/news/french-energy-regulator-cre-fines-edf-first-sanction-insider-trading-under-remit-and-edf-trading-limited-edft-market-manipulation.

"GE: Decades of Misdeeds and Wrongdoing." (2001, July–August). *Multinational Monitor* 22, nos. 7 & 8 https://www.multinationalmonitor.org/mm2001/01july-august/julyaug01corp4.html.

"General Electric to Pay $2.25 Million for Violating Federal and State Environmental Laws in Waterford, New York." (2015, December 3). U.S. Department of Justice https://www.justice.gov/opa/pr/general-electric-pay-225-million-violating-federal-and-state-environmental-laws-waterford-new.

"How Will Los Angeles Benefit from the Games?" (n.d.). *LA28* https://la28.org/en/faqs/what-benefit-will-the-games-bring-to-los-angeles--.html.

"Hydrogen Cars Risk Derailing Green Credibility of Paris Olympics: Open Letter from Scientists Concerned that Hydrogen Cars are Misaligned with Net-Zero." (2024, July). Centre for Sustainable Road Freight https://www.csrf.ac.uk/wp-content/uploads/2024/07/Open-letter-from-scientists_-Hydrogen-cars_final.pdf.

"JO 2024: Coca et ses 10 millions de boissons en plastique, médaille d'or du green-washing." (2024, June 5). *France Nature Environnement* https://fne.asso.fr/actualites/jo-2024-coca-et-ses-10-millions-de-boissons-en-plastique-medaille-d-or-du-green-washing.
"JO-2024: Des surfeurs mobilisés pour dénoncer la tour controversée des juges à Tahiti." (2023, December 12). *France 24* https://www.france24.com/fr/info-en-continu/20231217-jo-2024-des-surfeurs-mobilis%C3%A9s-pour-d%C3%A9noncer-la-tour-controvers%C3%A9e-des-juges-%C3%A0-tahiti.
"Lawsuit Filed Against EDF for Alleged Failure to Respect French Duty of Vigilance Law Over Wind Project in Mexico." (2024, June 18). Business & Human Rights Resource Centre https://www.business-humanrights.org/en/latest-news/devoir-de-vigilance-edf-assign%C3%A9-en-justice-pour-ses-activit%C3%A9s-au-mexique/.
"Les Jeux Olympiques de Paris 2024 sont sponsorisés par la maltraitance animale." (2024, July 24). PETA France https://www.petafrance.com/actualites/les-jeux-olympiques-de-paris-2024-sont-sponsorises-par-la-maltraitance-animale/.
"Lululemon: Canadian Sportswear Brand Accused of Greenwashing Ahead of Olympic Opening Ceremony." (2024, July 25). *Euronews* https://www.euronews.com/green/2024/07/25/greenwashing-lululemon-hit-with-legal-complaint-before-olympic-opening-ceremony.
"LVMH is Splurging on the Olympics." (2024, July 25). *Economist* https://www.economist.com/business/2024/07/25/lvmh-is-splurging-on-the-olympics.
"Métro gratuit, cérémonie d'ouverture, environnement … Les promesses pas forcément tenues des Jeux olympiques de Paris 2024." (2024, July 20). *rtbf actus* https://www.rtbf.be/article/metro-gratuit-ceremonie-d-ouverture-environnement-les-promesses-pas-forcement-tenues-des-jeux-olympiques-de-paris-2024-11406393.
"Mexico '68 Legacy Gets Fresh Impetus on 50th Anniversary." (2018). International Olympic Committee https://olympics.com/ioc/news/mexico-68-legacy-gets-fresh-impetus-on-50th-anniversary.
"Olympic Glory Tends to Go Hand-in-Hand with Economic Clout." (2021, August 8). *Economist* https://www.economist.com/graphic-detail/2021/08/08/olympic-glory-tends-to-go-hand-in-hand-with-economic-clout.
"Olympics Equipment Heads for "Second Life" Post-Games." (2024, August 12). *Inside the Games* https://www.insidethegames.biz/articles/1147766/olympics-equipment-heads-second-life.
"Olympics President: Paris 2024 on Track to Reach 'More than Half the World's Population'." (2024, August 7). *IBC365* https://www.ibc.org/news/olympics-president-paris-2024-on-track-to-reach-more-than-half-the-worlds-population/11205.article.

"Olympics Under Pressure to Dump Coca-Cola Sponsorship After Nearly 100 Years Over Health and Environmental Concerns." (2024, August 7). *Fortune* https://fortune.com/europe/2024/08/07/olympics-under-pressure-to-dump-coca-cola-sponsorship-after-nearly-100-years-over-health-environmental-concerns-paris-nike/.

"Our Sustainability." (n.d.). Lululemon https://www.eu.lululemon.com/en-lu/c/community/about-us/our-sustainability.

"Pandemic Olympics Endured Heat, and Now a Typhoon's En Route." (2021, July 26). *NBC* https://www.nbcnews.com/news/world/pandemic-olympics-endured-heat-now-typhoon-s-en-route-n1274976.

"Paris 2024 Carbon Contribution Projects." (2024, June 28). Paris 2024 https://press.paris2024.org/assets/paris-2024-carbon-contribution-projects-2eda-7578a.html.

"Paris 2024: les forces de l'ordre, l'attraction inattendue des Jeux." (2024, August 8). *franceinfo* https://www.francetvinfo.fr/les-jeux-olympiques/paris-2024-les-forces-de-l-ordre-l-attraction-inattendue-des-jeux6714135.html.

"Paris 2024 Presents the Latest Advance in its Low Carbon Strategy." (2024, June 27). Paris 2024 https://press.paris2024.org/news/press-release-paris-2024-presents-the-latest-advance-in-its-low-carbon-strategy-175c-7578a.html.

"Paris's Stunning Vision for the Olympics Wins a Gold Medal." (2024, August 10). *Economist* https://www.economist.com/europe/2024/08/10/pariss-stunning-vision-for-the-olympics-wins-a-gold-medal.

"Sport Takes Prominent Place at World Urban Forum." (2024, November 11). International Olympic Committee https://olympics.com/ioc/news/sport-takes-prominent-place-at-world-urban-forum.

"Suffering from the Bhopal Disaster in India Continues, 40 Years on." (2024, December 1). *Economist* https://www.economist.com/asia/2024/12/01/suffering-from-the-bhopal-disaster-in-india-continues-40-years-on.

"The Evolution of TV: A Brief History of TV Technology in Japan." (2010). Sec. 12: "Color TV Broadcasts and the Tokyo Olympic Games." http://www.nhk.or.jp/strl/aboutstrl/evolution-of-tv-en/p12.html.

"The Legacy of the London Olympics." (2022, July 27). *Economist* https://www.economist.com/britain/2022/07/27/the-legacy-of-the-london-olympics.

"Tlatelolco, la testimonianza di Oriana Fallaci." (n.d.). *YouTube* https://www.youtube.com/watch?v=yW8q5i7mdgo.

"Tlatelolco, según los periódicos del 3 de octubre de 1968." (2015, October1). *Expansión* https://expansion.mx/nacional/2015/10/01/tlatelolco-segun-los-periodicos-del-3-de-octubre-de-1968.

"Tokyo Olympics: German Pentathlon Coach Thrown Out." (2021, August 7). *BBC* https://www.bbc.com/sport/olympics/58127366.

Abbas Ahmadi, Ali. (2025, May 15). "LA 2028 Olympics Plans Air Taxi Service for Spectators." *BBC* https://www.bbc.com/news/articles/cvgnp103zr9o.

Abel, Jessamyn R. (2015). *The International Minimum: Creativity and Contradiction in Japan's Global Engagement, 1933–1964.* Honolulu: University of Hawaii Press.

Abel, Jessamyn R. (2022). *Dream Super-Express: A Cultural History of the World's First Bullet Train.* Stanford: Stanford University Press.

Action Speaks Louder. (2024, April 18). *'Be Planet'? Toxic Pollution and Fossil Fuel Reliance in Lululemon's Supply Chain* https://speakslouder.org/report/be-planet-toxic-pollution-and-fossil-fuel-reliance-in-lululemons-supply-chain/.

Albert, Eric. (2024, January 5). "What Legacy for Paris 2024? The Exemplary Conversion of the London 2012 Olympic Park." *Le Monde* https://www.lemonde.fr/en/sports/article/2024/01/05/what-legacy-for-paris-2024-the-exemplary-conversion-of-the-london-2012-olympic-park_6402549_9.html.

Ambrose, Jillian. (2025, February 26). "BP Blames 'Misplaced' Faith in Green Transition for its Renewed Focus on Fossil Fuels." *Guardian* https://www.theguardian.com/business/2025/feb/26/bp-oil-and-gas-spending-green-energy-scale-back?CMP=Share_iOSApp_Other.

Anderson, Christopher. (2005). "Little Green Men." *Flow* 2, no. 8 http://flowtv.org/2005/07/advertising-ge-ecoimagination.

Anderson, Perry. (2004, June 3). "Stand-Off in Taiwan." *London Review of Books*: 12–17.

Arbena, Joseph L. (1991). "Sport, Development, and Mexican Nationalism, 1920–1970." *Journal of Sport History* 18, no. 3: 350–64.

Aviña, Rafael. (2007). *¡No queremos olimpiadas! ¡Queremos revolución! 1968: Una crónica desde la infancia y todas las películas sobre el 2 de octubre.* Mexico City: Palabra de Clio.

Bacon, Francis. (1863). *The Works.* Ed. James Spedding, Robert Leslie Ellis, and Douglas Denon Heath. Boston: Taggard and Thompson.

Bandiera, David, Sebastien Racinais, Frédéric Garrandes, Paolo Emilio Adami, Stéphane Bermon, Yannis P. Pitsiladis, and Antonio Tessitore. (2024). "Heat-Related Risk at Paris 2024: A Proposal for Classification and Review of International Federations Policies." *British Journal of Sports Medicine* 58: 860–69.

Beckhart Coppinger, Sarah. (2021). ""Loco Tlatelolco:" The Conflicting Past of this Historic Neighborhood, 1960–1964." *The Latin Americanist* 65, no. 4: 431–55.

Bernard, Paquito, Guillaume Chevance, Celia Kingsbury, Aurélie Baillot, Ahmed-Jérôme Romain, Virginie Molinier, Tegwen Gadais, and Kelsey N. Dancause. (2021). "Climate Change, Physical Activity and Sport: A Systematic Review." *Sports Medicine* 51, no. 5: 1041–059.

Bernard, Raphaël. (2023, August 25). "Seine-Saint-Denis: contre les Jeux Olympiques, Saccage 2024 fait son "Toxic Tour"." *Citoyens* https://94.citoyens.com/2023/seine-saint-denis-contre-les-jeux-olympiques-saccage-2024-fait-son-toxic-tour,25-08-2023.html.

Betts, Kellyn. (2006). "PBDEs and PCBs in Computers, Cars, and Homes." *Environmental Science & Technology* 40, no. 24: 7452.

Bilton, Dean. (2024, August 12). "The Paris Olympics, a Timely Celebration of Humanity, Taught Us About Joy, Hope and Each Other." *ABC* https://www.abc.net.au/news/2024-08-12/what-the-paris-olympics-taught-us-about-joy-hope-and-each-other/104211818.

Bioregional and WWF. (2012). *Towards a One Planet Olympics Revisited* http://assets.wwf.org.uk/downloads/towardsaoneplanetolympicsrevisited.pdf.

Boganda, Cyprien. (2024, August 9). "Saint-Denis à l'heure de Paris 2024: une fête sous surveillance." *l'Humanité* https://www.humanite.fr/societe/jeux-olympiques-paris-2024/saint-denis-a-lheure-de-paris-2024-une-fete-sous-surveillance.

Bonet, Enric. (2024, August 11). "La cara B de una fiesta olímpica "inclusiva" y "popular"." *El Salto* https://www.elsaltodiario.com/juegos-olimpicos/cara-b-una-fiesta-olimpica-inclusiva-popular.

Bonet Torra, Enric. (2024, August 16). "Francia afronta la resaca económica de los Juegos Olímpicos de París." *Público* https://www.publico.es/internacional/francia-afronta-resaca-economica-juegos-olimpicos-paris.html#md=modulo-portada-bloque:2col-t4;mm=mobile-medium.

Boren, Zach and Emma Howard. (2024, November 20). "Companies Behind Campaign to 'End Plastic Waste' Produced 1,000 Times More Plastic Than it Cleaned Up." *Unearthed* https://unearthed.greenpeace.org/2024/11/20/alliance-to-end-plastic-waste-oil-chemical-exxon-shell-total/.

Bouchama, Abderrezak, Bisher Abuyassin, Cynthia Lehe, Orlando Laitano, Ollie Jay, Francis G. O'Connor, and Lisa R. Leon. (2022). "Classic and Exertional Heatstroke." *Nature Reviews: Disease Primers* 8, no. 8. https://doi.org/10.1038/s41572-021-00334-6.

Bougault, Valerie, Christopher Carlsten, Paolo Emilio Adami, Nicola Sewry, Wolfgang Schobersberger, Torbjørn Soligard, Lars Engebretsen, Richard Budgett, Martin Schwellnus, and Ken Fitch. (2025). "Air Quality, Respiratory Health and Performance in Athletes: A Summary of the IOC Consensus Subgroup Narrative Review on 'Acute Respiratory Illness in Athletes." *British Journal of Sports Medicine* 59, no. 7: 480–90.

Boykoff, Jules and Christopher Gaffney. (2020). "The Tokyo 2020 Games and the End of Olympic History." *Capitalism Nature Socialism* 31, no. 2: 1–19.

Boykoff, Jules and Gilmar Mascarenhas. (2016). "The Olympics, Sustainability, and Greenwashing: The Rio 2016 Summer Games." *Capitalism Nature Socialism* 27, no. 2: 1–11.

Boykoff, Jules and Dave Zirin. (2024, August 9). "The Paris Olympics Has Launched a War on the Poor." *The Nation* https://www.thenation.com/article/world/paris-olympics-homeless

Boykoff, Jules and Dave Zirin. (2025, January 15). "The Olympics Aren't the Way to Rebuild Los Angeles." *The Nation* https://www.thenation.com/article/society/los-angeles-olympics-fires/.

Boykoff, Jules. (2016). *Power Games: A Political History of the Olympics*. London: Verso.

Boykoff, Jules. (2021). "Celebration Capitalism, Sports Mega-Events, and the Tokyo 2020 Olympics." *Journal of Sport and Gender Studies* 19: 50–66.

Boykoff, Jules. (2024). *What Are the Olympics For?* Bristol: Bristol University Press.

Boykoff, Jules. (2025a). "The Olympics in Crisis: Challenges, Reforms, and the Path Forward." *International Sports Law Journal* 25, no. 1: 29–48.

Boykoff, Jules. (2025b). "Greenwash Gold?: The Paris 2024 Olympics." *Capitalism Nature Socialism* 36, no. 2: 171–78.

Branch, John and Motoko Rich. (2021, July 30). "Tokyo Braces for the Hottest Olympics Ever." *New York Times* https://www.nytimes.com/2019/10/10/sports/tokyo-braces-for-the-hottest-olympics-ever.html.

Brocherie, Franck, Olivier Girard, Adèle Mornas, Janne Bouten, and Grégoire P. Millet. (2024). ""Citius, Altius, Fortius" in the Face of Global Warming: Not as Simple as it Seems." *Climate and Atmospheric Science* 7. https://doi.org/10.1038/s41612-024-00774-3.

Bronner, Luc. (2024, August 6). "Anne Hidalgo et les Jeux olympiques: « C'est pas complètement foutu, on peut être ensemble et être heureux ensemble! »." *Le Monde* https://www.lemonde.fr/politique/article/2024/08/06/anne-hidalgo-avec-les-jo-les-gens-se-disent-c-est-pas-completement-foutu-on-peut-etre-ensemble-et-on-peut-etre-heureux-ensemble6269386823448.html?lmdmedium=al&lmdcampaign=envoye-par-appli&lmdcreation=ios&lmdsource=default.

Budzier, Alexander and Bent Flyvbjerg. (2024). *The Oxford Olympics Study 2024: Are Cost and Cost Overrun at the Games Coming Down?* Said Business School Working Paper https://arxiv.org/pdf/2406.01714.

Byrne, Thomas H., Benjamin F. Henwood, and Anthony W. Orlando. (2021). "A Rising Tide Drowns Unstable Boats: How Inequality Creates Homelessness." *Annals of the American Academy of Political and Social Sciences* 693, no. 1: 28–45.

Cadoret, NnoMan and Justine Guitton-Boussion. (2021, September 6). "À Aubervilliers l'« écœurement » après la destruction des jardins ouvriers." *Reporterre* https://reporterre.net/A-Aubervilliers-l-ecoeurement-apres-la-destruction-des-jardins-ouvriers.

Cain, Alison and Michael Callan. (2025). "Principles in Practice? A Policy Review of the IOC's Environmental Sustainability Agenda." *Frontiers in Sports and Active Living* 7 https://doi.org/10.3889/fsport.2025.1511092.

Carbon Market Watch and éclaircies. (2024). *Going for Green: Assessing the Climate Strategy and Communication of the 2024 Paris Olympics* https://carbonmarketwatch.org/wp-content/uploads/2024/04/Going-for-green-Olympics-report.pdf.

Carter Bill with Richard Sandomir. (2000, September 20). "Sydney 2000: Television. NBC Ratings for Olympics Are Worst Ever." *New York Times* https://www.nytimes.com/2000/09/20/sports/sydney-2000-television-nbc-s-ratings-for-olympics-are-worst-ever.html.

Castañeda, Luis. (2010). "Beyond Tlatelolco: Design, Media, and Politics at Mexico '68." *Grey Room* 40: 100–26.

Castañeda, Luis. (2014). *Spectacular Mexico: Design, Propaganda, and the 1968 Olympics.* Minneapolis: University of Minnesota Press.

Ceccon, Angelique, Andrew Hargrove, and Jamie Sommer. (2024). "Do the Olympics Impact CO_2 Emissions? A Cross-National Analysis." *Global Transitions* 6: 241–48.

Centre on Housing Rights and Evictions. (2007). *Fair Play for Housing Rights: Mega-Events, Olympic Games and Housing Rights* https://www.ruig-gian.org/ressources/Report%20Fair%20Play%20FINAL%20FINAL%20070531.pdf.

Chandler, Jenna. (2018, July 12). "LA 'Sterilized' its Streets for the '84 Olympics—How Will it Treat the Homeless in 2028?" *Curbed* https://la.curbed.com/2018/7/12/17454676/los-angeles-olympics-homeless-police-militarization-security.

Chaumier, Morgan. (2023, November 30). "En Seine-Saint-Denis, « la promesse républicaine n'est pas tenue »." *l'Humanité* https://www.humanite.fr/politique/education/en-seine-saint-denis-la-promesse-republicaine-nest-pas-tenue.

Chen, Chwen Chwen, Cinzia Colapinto, and Qing Luo. (2012). "The 2008 Beijing Olympics Opening Ceremony: Visual Insights into China's Soft Power." *Visual Studies* 27, no. 2: 188–95.

Chrisafis, Angelique. (2024, August 7). "Greenwash Games? French Public Opinion Points Finger at Coca-Cola Over Olympics Plastic Waste." *Guardian* https://www.theguardian.com/sport/article/2024/aug/07/greenwash-games-french-public-points-finger-at-coca-cola-over-olympics-plastic-waste?CMP=ShareiOSAppOther.

Cicero. (2008). *Selected Letters.* Trans. P. G. Walsh. Oxford: Oxford University Press.

Clarey, Christopher. (1998, December 22). "Bribery Allegations Suddenly Raise International Olympic Committee's Visibility: Just Who Are the Members of the Arcane IOC?" *New York Times* http://www.nytimes.com/1998/12/22/

sports/bribery-allegations-suddenly-raise-international-olympic-committees.html.
Cohen, Roger and Jeré Longman. (1999, February 7). "Olympic Chief's Expansion Goals Left Little Zeal to Pursue Abuses." *New York Times*: 1, 26.
Collins, Sandra. (2011). "East Asian Olympic Desires: Identity on the Global Stage in the 1964 Tokyo, 1988 Seoul and 2008 Beijing Games." *International Journal of the History of Sport* 28, no. 16: 2240–260.
Court, Andrew. (2024, May 14). "'Anti-Sex' Beds Have Arrived at Paris Olympics—After Horny Athletes Admit to Orgies Amid Competition." *New YorkPost*https://nypost.com/2024/05/14/lifestyle/anti-sex-beds-have-arrived-at-paris-olympics-after-horny-athletes-admit-to-orgies/.
Cray, Charlie. (2001, July/August). "Toxics on the Hudson: The Saga of GE, PCBs and the Hudson River." *Multinational Monitor* 22, nos. 7/8 https://www.multinationalmonitor.org/mm2001/072001.
Dai, Xiaodong, and Guo-Ming Chen. (2014). "The Construction of National Image in the Media and the Management of Intercultural Conflicts." *The Handbook of Media and Mass Communication Theory*. Ed. Robert S. Fortner and P. Mark Fackler. Malden: Wiley-Blackwell. 708–24.
Dang, Sheila and Mimosa Spencer. (2024, August 12). "Paris Sets New Trend as the Product Placement Olympics." *Reuters* https://www.reuters.com/sports/olympics/paris-sets-new-trend-product-placement-olympics-2024-08-10/.
Davis, Juliet and Penny Bernstock. (2023). "From Inclusive Legacy Promises to Exclusive Realities: Planning, Design, and Displacement in Post-Olympic East London." *Urbanizing Suburbia Hyper-Gentrification, the Financialization of Housing, and the Remaking of the Outer European City*. Ed. Tahl Kaminer, Leonard Ma, and Helen Runting. Berlin: jovis Verlag. 212–34.
Davis, Juliet. (2016). "The Making and Remaking of Hackney Wick, 1870–2014: From Urban Edgeland to Olympic Fringe." *Planning Perspectives* 31, no. 3: 425–57.
Davis, Mike. (2006). *Planet of Slums*. London: Verso.
Davis, Tricia Holly. (2009, July 26). "A Mixed Record So Far on Environmental Issues." *Financial Times* https://www.ft.com/content/31503024-77eb-11de-9713-00144feabdc0.
de Anda Alanis, Enrique. (2021). "The 19th Olympic Games Celebrated in Mexico in 1968 and Their Architecture." *Das moderne Erbe der Olympischen Spiele: Historische Sportstätten zwischen Konservierung und Konversion/The Modern Heritage of the Olympic Games: Historic Sports Sites between Conservation and Conversion*. Ed. Sigrid Brandt, Jörg Haspel, Ralph Paschke, and John Ziesemer. Berlin: ICOMOS. 66–72.
Delaplace, Marie and Martin Müller. (2024, June 18). "Can We Really Measure the Ecological Footprint of the Olympics?" *Polytechnique Insights: A Review by the Institut Polytechnique de Paris* https://www.polytechnique-insights.com/

en/columns/planet/can-we-really-measure-the-ecological-footprint-of-the-olympics/.

Delmas, Aurélie. (2017, September 14). "Merci de l'avoir posée—Paris 2024: Jeux verts ou greenwashing?" *Libération* https://www.liberation.fr/france/2017/09/14/paris-2024-jeux-verts-ou-greenwashing1596104/.

Department for Culture, Media and Sport and Department for Digital, Culture, Media & Sport. (2008). *Our Promise for 2012: How the UK Will Benefit from the Olympic Games and Paralympic Games* https://www.gov.uk/government/publications/our-promise-for-2012-how-the-uk-will-benefit-from-the-olympic-games-and-paralympic-games.

Díaz Ordaz, Gustavo. (1968, September 1). *Informe*. Legislatura XLVII Año II Período Ordinario Fecha 19680901 Número de Diario 3. https://cronica.diputados.gob.mx/DDebates/47/2do/Ord/19680901-I.html.

Doe v. Cisco Systems, Inc. (2024). Human Rights Law Foundation https://hrlf.net/doe-vs-cisco-systems/.

Doustaly, Cécile and Geneviève Zembri-Mary. (2024). "Is Urban Planning Returning to the Past in Search of a Sustainable Future? Exploring the Six Paris and London Olympic Games." *Planning Perspectives* 39, no. 3: 675–700.

Echikson, William. (1998, February 9). "The Olympics: Governance: Making the Games Run on Time." *Business Week*: 66–68.

Eckerman, Ingrid. (2005). *The Bhopal Saga: Causes and Consequences of the World's Largest Industrial Disaster*. Hyderabad: Universities Press.

Elías, Alex. (2021). *Mexico City's Olympic Games: Citizenship and Nation Building, 1963–1968*. Cham: Palgrave.

England, Phil. (2012, April 25). "#London2012: An Olympian Exercise in Corporate Greenwashing." *Ceasefire* https://ceasefiremagazine.co.uk/london2012-olympian-exercise-corporate-greenwashing/.

Environmental Defenders Office. (2023, March 2). "Complaint About Misleading Advertising by Toyota Motor Corporation Australia Ltd." https://climatecasechart.com/wp-content/uploads/non-us-case-documents/2023/20230302_19330_complaint.pdf.

Fisher, Daniel. (2005, August 15). "GE Turns Green." *Forbes*: 80–85.

Fisné-Koch, Audrey. (2024, August 9). "Les Jeux olympiques de Paris ont-ils été exemplaires sur le plan social?" *Alternatives Economiques* https://www.alternatives-economiques.fr/jeux-olympiques-de-paris-ont-ete-exemplaires-plan-social/00112042?utmsource=emailing&utmmedium=email&utmcontent=09082024&utmcampaign=hebdo.

Flaming, Daniel, Gary L. Blasi, Anthony W. Orlando, and Sam Tsemberis. (2024, September 4). *Excelling for the 2028 Olympics: Restoring, Not Displacing, LA's Unsheltered Residents*. Economic Roundtable https://economicrt.org/publication/excelling-for-the-2028-olympics/.

Foster, Aurelia. (2022, July 20). "London Olympics: Residents 'Betrayed' Over Housing Promise." *BBC* https://www.bbc.co.uk/news/uk-england-london-62138346.

Foth, Marcus. (2025, March 30). "Brisbane 2032 is no Longer Legally Bound to be 'Climate Positive'. Will it Still Leave a Green Legacy?" *The Conversation* https://theconversation.com/brisbane-2032-is-no-longer-legally-bound-to-be-climate-positive-will-it-still-leave-a-green-legacy-246672.

Foucault, Michel. (2008). *The Birth of Biopolitics: Lectures at the Collège de France, 1978–79.* Trans. Graham Burchell. Ed. Michel Senellart. Houndmills: Palgrave Macmillan.

Foucault, Michel. (2024). *The Japan Lectures: A Transnational Critical Encounter.* London: Routledge.

Foxtons. (2025). https://www.foxtons.co.uk/living-in/hackney-wick.

Gable, Margaux. (2024, August 8). "A Paris, le nouveau visage Olympique de la Porte de la Chapelle accusé d'être un cache-misère." *Libération* https://www.liberation.fr/societe/ville/a-paris-le-nouveau-visage-olympique-de-la-porte-de-la-chapelle-accuse-detre-un-cache-misere-2024080 87TC6A3PFHJG7ZKF6IZTNS4HCME/.

Gagnepain, Névil. (2024, August 5). "Prostitution: les femmes du bois de Vincennes contraintes de déserter." *Mediapart* https://www.mediapart.fr/journal/france/050824/prostitution-les-femmes-du-bois-de-vincennes-contraintes-de-deserter.

Gamage, Prasanna, Lauren V. Fortington, and Carline F. Finch. (2020). "Epidemiology of Exertional Heat Illnesses in Organised Sports: A Systematic Review." *Journal of Science and Medicine in Sport* 23, no. 8: P701–09.

Ganguly, Bani Bandana, Shouvik Mandal, Nalok Banerjee, Nitin N. Kadam, and Rita Abbi. (2018). "Effect of Age at Exposure on Chromosome Abnormalities in MIC-Exposed Bhopal Population Detected 30 Years Post-Disaster." *Mutation Research* 809: 32–50.

Garot, Blandine. (2024, August 7). "JO Paris 2024: découvrez notre pódium du greenwashing." *novethic* https://www.novethic.fr/environnement/climat/jo-paris-2024-decouvrez-notre-podium-du-greenwashing.

Germano, Sara, Josh Noble, and Adrienne Klasa. (2024, August 1). "The Official Medal Selfie: Paris Olympics Steps Up Product Placement." *Financial Times* https://www.ft.com/content/4db45132-13b9-4304-8cb3-514ea921a318?FTCamp=engage/CAPI/website/Channel_muckrack//B2B.

Gilbert, Olivier. (2023). "The Olympic Games: Social and Environmental Innovation." *Field Actions Science Reports* 25: 20–23.

Ginsky, Kaya. (2024, August 21). "LVMH Puts Mark on Olympics as Luxury Brands Embrace Sports." *CNBC* https://www.cnbc.com/2024/08/11/lvmh-puts-mark-on-olympics-as-luxury-brands-embrace-sports.html.

Glasson, Ben and Brett Hutchins. (2024). "“Carbon Partners” and Collaborative Greenwashing: The Sustainability Partnership Between Dow Chemical and the Olympic Games." *Journal of Sport & Social Issues* 48, no. 5: 271–89.

Glasson, Ben. (2024). "Reality Offsets: Climate Meets Capitalism at the Olympic Games." *European Journal of Cultural Studies* 28, no. 1: 137–55.

Goldblatt, David. (2016). *The Games: A Global History of the Olympics.* New York: WW Norton.

Goodwin, Bill. (2019, August 1). "Cisco Pays $8.6m After Whistleblower Discloses Security Flaws in Video Surveillance System." *Computer Weekly* https://www.computerweekly.com/news/252467624/Cisco-pays-83m-after-whistleblower-discloses-security-flaws-in-video-surveillance-system.

Greene, Shayna. (2024, August 16). "General Electric Loses Bid for Quick Win in Contamination Suit." *Bloomberg Law* https://news.bloomberglaw.com/environment-and-energy/general-electric-loses-bid-for-quick-win-in-contamination-suit.

Griggs, Lee. (1963, October 28). "A Very Dry Run in Tokyo." *Sports Illustrated* https://web.archive.org/web/20140222142617/http://sportsillustrated.cnn.com/vault/article/magazine/MAG1075316/index.htm.

Hallisat, Ismaël and Caurentin Courtois. (2024, September 19). "Les images du déchaînement de violences subi par un homme dans un commissariat parisien." *Libération* https://www.liberation.fr/societe/police-justice/les-images-revoltantes-des-violences-subies-par-un-homme-dans-un-commissariat-parisien-20240919_Q2XURAOL7BDCTFWJH6LBF46RN4/.

Henley, Jon. (2024, July 12). "‘It Might Actually be Cool’: Wary Parisians Seduced by Prospect of Olympics." *Guardian* https://www.theguardian.com/sport/article/2024/jul/12/it-might-actually-be-cool-wary-parisians-seduced-by-prospect-of-olympics?CMP=ShareiOSAppOther.

Hollander, Karsten, Milan Klöwer, Andy Richardson, Laurent Navarro, Sébastien Racinais, Volker Scheer, Andrew Murray, Pedro Branco, Toomas Timpka, Astrid Junge, and Pascal Edouard. (2021). "Apparent Temperature and Heat-Related Illnesses During International Athletic Championships: A Prospective Cohort Study." *Scandinavian Journal of Medicine & Science in Sport* 31: 2092–102.

Hooper, Simon. (2012, August 12). "Olympics 2012: A Spectacular Triumph for London." *CNN* https://edition.cnn.com/2012/08/12/opinion/hooper-london-triumph/index.html.

Hsu, Tiffany. (2021, August 5). "NBC Tries to Salvage a Difficult Olympics." *New York Times* https://www.nytimes.com/2021/08/05/business/media/nbc-olympics-tv-ratings.html.

Huijuan, Cao, Hidemichi Fujii, and Shunsuke Managi. (2013). *Environmental Impact of the 2008 Beijing Olympic Games.* Economics Discussion Paper

2013-30, Kiel Institute for the World Economy https://www.economics-ejournal.org/economics/discussionpapers/2013-30.

Hurst, Alexander. (2024, August 12). "France Specialises in Glum Self-Doubt. Has This Joyful Olympics Finally Changed That?" *Guardian* https://www.theguardian.com/commentisfree/article/2024/aug/12/france-olympics-paris-city-cool?CMP=ShareiOSAppOther.

Hussain, Umer, Adam Ali, MacIntosh Ross, and Shakiba Moghadam. (2024, May 8). "Paris 2024 Olympics: How the Games Are Being Used to Marginalize the Most Vulnerable." *The Conversation* https://theconversation.com/paris-2024-olympics-how-the-games-are-being-used-to-marginalize-the-most-vulnerable-228855.

Institut national de la statistique et des études économiques. (2023, March 30). *Immigrés et descendants d'immigrés: Édition 2023* https://www.insee.fr/fr/statistiques/6793282?sommaire=6793391#:~:text=Les%20immigr%C3%A9s%20r%C3%A9sident%20fr%C3%A9quemment%20dans,Mayotte)%20(figure%201).

International Civil Aviation Organization. (2016). *ICAO Carbon Emissions Calculator* https://www.icao.int/environmental-protection/CarbonOffset/Pages/default.aspx.

International Olympic Committee. (n.d.-a). *Becoming an Olympic Games Host* https://olympics.com/ioc/becoming-an-olympic-games-host.

International Olympic Committee. (n.d.-b). "XII Olympic Congress Paris 1994" https://www.olympics.com/ioc/paris-1994-olympic-congress.

International Olympic Committee. (2018). *Carbon Footprint Methodology for the Olympic Games* https://library.olympic.org/Default/doc/SYRACUSE/184686/carbon-footprint-methodology-for-the-olympic-games-international-olympic-committee?lg=fr-FR.

International Olympic Committee. (2021). *IOC Marketing Report Tokyo 2020* https://stillmed.olympics.com/media/Documents/International-Olympic-Committee/IOC-Marketing-And-Broadcasting/IOC-Marketing-report-Tokyo-2020.pdf.

International Olympic Committee. (2022, with addendum). *Olympic Host Contract: Operational Requirements* https://stillmed.olympics.com/media/Documents/Olympic-Games/Brisbane-2032/Host-Contract/OHC-Operational-Requirements2022-addendum.pdf.

International Olympic Committee. (2024a, July 23). *Olympic Charter* https://stillmed.olympics.com/media/Document%20Library/OlympicOrg/General/EN-Olympic-Charter.pdf.

International Olympic Committee. (2024b). *Paris 2024: More Sustainable Games* https://olympics.com/ioc/paris-2024-sustainable-games.

International Olympic Committee. (2024c). *Olympic Games Paris 2024* https://www.olympics.com/en/olympic-games/paris-2024.

International Olympic Committee. (2024d, July). *Olympic Host Contract* https://stillmed.olympics.com/media/Documents/Olympic-Games/Salt-Lake-City-Utah-2034/Olympic-Host-Contract-Principles-2034.pdf.

International Olympic Committee. (2025). *IOC Members* https://www.olympics.com/ioc/members.

Ito, Eiji, James Higham, and Joseph M. Cheer. (2022). "Carbon Emission Reduction and the Tokyo 2020 Olympics." *Annals of Tourism Research Empirical Insights* 3, no. 2 https://doi.org/10.1016/j.annale.2022.100056.

Jakar, Gidon and Christina Philippou. (2025). "Legacy and Urban Regeneration: Long-Term Local Impact of the London 2012 Olympic Games." *European Planning Studies* 33, no. 2: 183–203.

Janetsky, Megan. (2024, July 26). "Migrants and Homeless People Are Cleared Out of Paris During the Olympics." *AP* https://apnews.com/article/olympics-2024-paris-migrant-camp-3ef2a08d8da1085148ed409dcb44d6f6.

Jennings, Andrew and Clare Sambrook. (2000). *The Great Olympic Swindle: When the World Wanted its Games Back*. New York: Simon & Schuster.

Jennings, Andrew. (1996). *The New Lords of the Rings: Olympic Corruption and How to Buy Gold Medals*. London: Pocket Books.

Jhally, Sut. (2016). "Stuart Hall: The Last Interview." *Cultural Studies* 30, no. 2: 332–45.

Jiménez-García, Mercedes, José Ruiz-Chico, Antonio Rafael Peña-Sánchez, and José Antonio López-Sánchez. (2020). "A Bibliometric Analysis of Sports Tourism and Sustainability (2002–2019)." *Sustainability* 12, no. 7: 2840–858.

Jorgenson, Andrew K. and Brett Clark. (2016). "The Temporal Stability and Developmental Differences in the Environmental Impacts of Militarism: The Treadmill of Destruction and Consumption-Based Carbon Emissions." *Sustainability Science* 26, no. 11: 505–14.

Joshi, Saumya. (2024, July 12). "Paris Olympics 2024: Why Are Scientists Urging the Olympics to Ditch Its Official Car?" *Oneindia* https://www.oneindia.com/sports/paris-olympics-2024-why-are-scientists-urging-the-olympics-to-ditch-its-official-car-011-3876999.html.

Jukes, Peter. (2012, August 13). "Take a Bow, London: The Olympics Were a Triumph." *The Daily Beast* https://www.thedailybeast.com/take-a-bow-london-the-olympics-were-a-triumph/.

Juvenal. (1998). *The Sixteen Satires*, 3rd ed. Trans. and Ed. Peter Green. Harmondsworth: Penguin.

Kasztelan, Marta and Sam Quashie-Idun. (2022, August 8). "Revealed: Garment Waste from Nike, Clarks and Other Leading Brands Burned to Fuel Toxic Kilns in Cambodia." *Unearthed* https://unearthed.greenpeace.org/2022/08/08/garment-waste-nike-clarks-cambodia-bonded-workers-toxic/.

Kokabi, Alexandre-Reza. (2024, July 23). "Les JO 2024 à Paris, une catastrophe écologique en Seine-Saint-Denis." *Reporterre* https://reporterre.net/Les-JO-2024-a-Paris-une-catastrophe-ecologique-en-Seine-Saint-Denis.

Kostov, Nick and Joshua Robinson. (2024, October 2). "LVMH Took Over the Paris Olympics. Now it's Snagged Formula One." *Wall Street Journal* https://www.wsj.com/business/retail/lvmh-took-over-the-paris-olympics-now-its-snagged-formula-one-b570c0f4.

Kristanadjaja, Gurvan. (2022, December 5). "Sans-papiers sur les chantiers de Paris 2024: «Les JO ne pourraient pas se faire sans nous»." *Libération* https://www.liberation.fr/societe/sans-papiers-sur-les-chantiers-les-jeux-olympiques-ne-pourraient-pas-se-faire-sans-nous-20221205310A3NFTMZBADLDM35V7OVI3TI/.

Kriza, Elisa. (2019). "Anti-Communism, Communism, and Anti-Interventionism in Narratives Surrounding the Student Massacre on Tlatelolco Square (Mexico, 1968)." *Bulletin of Latin American Research* 38, no. 1: 82–96.

Lambert, Stéphane. (2024, August 15). "JO de Paris 2024: pouvez-vous vibrer sans nous tyranniser?" *Libération* https://www.liberation.fr/idees-et-debats/tribunes/jo-de-paris-2024-pouvez-vous-vibrer-sans-nous-tyranniser-20240815GIL4Q6QHNNGBXNHRAZXEHOKZ7A/.

Latham-Coyle, Harry. (2024, August 6). "Adam Peaty Claims Athletes Found Worms in Food at Olympic Village: 'It's Just Not Good Enough'." *Independent* https://www.independent.co.uk/sport/olympics/adam-peaty-olympic-village-food-worms-paris-2024-b2591821.html.

Lee White, Lauren. (2022, August 19). "The 2028 L.A. Olympics Are Already Creating a Housing Disaster." *The New Republic* https://newrepublic.com/article/167453/2028-los-angeles-olympics-housing.

Le Foll, Clément. (2024, July 30). "En catimini, Saint-Denis achète un logiciel de surveillance algorithmique." *Mediapart* https://www.mediapart.fr/journal/france/300724/en-catimini-saint-denis-achete-un-logiciel-de-surveillance-algorithmique.

Lei, Tze-Huan and Faming Wang. (2021). "Looking Ahead of 2021 Tokyo Summer Olympic Games: How Does Humid Heat Affect Endurance Performance? Insight into Physiological Mechanism and Heat-Related Illness Prevention Strategies." *Journal of Thermal Biology* 99 https://www.sciencedirect.com/science/article/abs/pii/S0306456521001431.

Levintova, Hannah. (2012, August 11). "BP's Olympic Ads Seek to Erase Oil Spill from Memory." *Mother Jones* https://www.motherjones.com/politics/2012/08/bps-olympic-ads-are-helping-deepwater-horizon-go-away/.

Liang, Annabelle. (2024, June 4). "Toyota Raided as Safety Testing Scandal Grows." *BBC* https://www.bbc.com/news/articles/c1wwj1p2wdyo.

Lies, Elaine. (2021, July 1). "'Sad, Lonely Feeling': Tokyo Man Evicted Twice, 50 Years Apart, for Olympic Construction." *Reuters* https://www.reuters.com/

world/asia-pacific/sad-lonely-feeling-tokyo-man-evicted-twice-50-years-apart-olympic-construction-2021-07-01/.

Lindgaard, Jade. (2024). *Une ville face à la violence olympique*. Quimperlé: Éditions Divergences.

Lind Gustavussen, Mathilde. (2023, April 9). "The Gentrification Scheme Behind Los Angeles's "No-Build" Olympics." *Jacobin* https://jacobin.com/2023/04/los-angeles-olympics-2028-development-politics-gentrification.

Lo, Jo and Marlene Jacobsen. (2021, August 6). "'Hottest Games Ever': At the Tokyo Olympics, Elite Sport Met the Climate Crisis." *Climate Home News* https://www.climatechangenews.com/2021/08/06/hottest-games-ever-tokyo-olympics-elite-sport-met-climate-crisis/.

London Assembly Plenary. (2022, October 6). *Transcript of Agenda Item 3—Question and Answer Session: Legacy of the 2012 London Olympic Games and Paralympic Games* https://gla.moderngov.co.uk/documents/s102250/Minutes%20-%20Appendix%201%20-%20Transcript%20of%20Agenda%20Item%203.pdf.

Lusetich, Robert. (1999, January 11). "Games Elite Faces Purge Over Bribes." *Australian*: 1.

Mair, Judith and Mike Duignan. (2025). "How Does Hosting the Olympics Benefit Local Communities?" *Events and Society: Bridging Theory and Practice*. Ed. Mike Duignan. Abingdon: Routledge. 107–14.

Marlet, Pierre. (2024, July 23). "Les Jeux Olympiques de Londres 2012 : un feu d'artifice ou un pétard mouillé ?" *RTBF* https://www.rtbf.be/article/les-jo-de-londres-2012-du-feu-d-artifice-de-promesses-pour-les-britanniques-au-petard-mouille-du-brexit-11408173.

Marrero-Guillamón, Isaac. (2011). "London 2012: espacio de excepción." *urbe. Revista Brasileira de Gestão Urbana* 3, no. 2: 179–90.

Marrero-Guillamón, Isaac. (2016). "The Politics and Aesthetics of Assembling: (Un)building the Common in Hackney Wick, London." *Urban Cosmopolitics: Agencements, Assemblies, Atmospheres*. Ed. Anders Blok and Ignacio Farías. Abingdon: Routledge. 125–45.

Martin, Declan, Liss Ralston, and Tom Alves. (2024). *South East Queensland Displacement Monitoring: Baseline Report*. Australian Housing and Urban Research Institute for QShelter https://www.ahuri.edu.au/research/south-east-queensland-displacement-monitoring.

McBride, James, Noah Berman, and Melissa Manno. (2024, July 20). "The Economics of Hosting the Olympic Games." *Council on Foreign Relations* https://www.cfr.org/backgrounder/economics-hosting-olympic-games.

McClintock, Anne. (2010, June 24). "Militarizing the Gulf Oil Crisis." *CounterPunch* https://www.counterpunch.org/2010/06/24/militarizing-the-gulf-oil-crisis/.

McCurry, Justin and Associated Press. (2023, July 3). "'Tokyo Would Lose its Soul': Anger Over Plans to Redevelop Historic City Park." *Guardian* https://www.theguardian.com/world/2023/jul/01/tokyo-anger-plans-redevelop-historic-city-park-meiji-jingu-gaien?CMP=sharebtntw.

McDonald, Scott. (2020, March 25). "The Reason Why Olympics in 2021 Will Still be Called the 2020 Olympic Games." *Newsweek* https://www.newsweek.com/reason-why-olympics-2021-will-still-called-olympic-2020-games-1494333.

McGowan, Elliot. (2025, January 30). "Brisbane 2032 in "Hot Mess"." *inside the games* https://www.insidethegames.biz/articles/1151589/brisbane-2032-olympic-games-concerns.

McKenna, Kate. (2024, November 29). "Second Review into Olympic Infrastructure Launched, with 100 Days to Identify Priorities for 2032." *ABC* https://www.abc.net.au/news/2024-11-29/100-day-review-olympic-and-paralympic-infrastructure-brisbane/104665538.

McMillan, Stephen. (2024, August 12). "28 Hours Later: An Olympic Zombie in Paris After the Games Are Over." *Guardian* https://www.theguardian.com/sport/article/2024/aug/12/28-hours-later-an-olympic-zombie-in-paris-after-the-games-are-over?CMP=ShareiOSAppOther.

Ménage, Gaëlle. (2024, August 5). "Paris donne aux JO un bel environnement." *Forbes* https://www.forbes.fr/societe/issu-du-magazine-paris-donne-aux-jo-un-bel-environnement/.

Merino, Daniel. (2020, July 23). "The First Undeniable Climate Change Deaths." *Slate* https://slate.com/technology/2020/07/climate-change-deaths-japan-2018-heat-wave.html.

Miller, Sue. (2023, August 9). "Beware the Brutal Side of the Olympic Spirit." *Guardian*https://www.theguardian.com/lifeandstyle/2023/aug/09/beware-the-brutal-side-of-the-olympic-spirit.

Miller, Toby. (2018). *Greenwashing Sport*. London: Routledge.

Molinari, Letizia. (2024, August 8). "Sin tregua olímpica para los sectores populares." Trans. Florencia Oroz. *Jacobin* https://jacobinlat.com/2024/08/sin-tregua-olimpica-para-los-sectores-populares/.

Moore, Malcolm. (2025, April 21). "Will Going Back to Basics Restore BP's Fortunes?" *Financial Times* https://www.ft.com/content/ab60efc1-4abf-4e92-bd1b-b09046ced3f5?emailId=6520e7be-7310-4050-95aa-88ec4689ffc0&segmentId=9264b0f7-e7ac-8f9b-044f-c10729049333.

Mori, Sayaka. (2021). https://x.com/sayakasofiamori/status/1422856371981094919.

Morteaux, Alexandre. (2024, June 28). "Quel bilan économique et environnemental des Jeux olympiques?" *Alternatives Economiques* https://www.alternatives-economiques.fr/bilan-economique-environnemental-jeux-olympiques/00111854.

Moutiez, Julia. (2024, July 16). "Nager dans la Seine: Un loisir ancien qui refait surface à l'heure du réchauffement climatique." *The Conversation* https://theconversation.com/nager-dans-la-seine-un-loisir-ancien-qui-refait-surface-a-lheure-du-rechauffement-climatique-234776.

Müller, Martin. (2015). "(Im-)mobile Policies: Why Sustainability Went Wrong in the 2014 Olympics in Sochi." *European Urban and Regional Studies* 22, no. 2: 191–209.

Nagao, Futa, Marika Oi, and Suranjana Tewari. (2021, August 6). "Tokyo 2020: Olympic Athletes Struggle with Heat at Games." *BBC* https://www.bbc.com/news/av/world-asia-58110846.

Nájar, Alberto. (2024, October 2). "Qué pasó en la matanza de Tlatelolco hace 56 años por la que Claudia Sheinbaum pidió perdón en su primer día como presidenta de México." *BBC Mundo* https://www.bbc.com/mundo/articles/cnvde3j6q6vo.

Nakagawa, Satoko. (2018, September 2). "Criticism Grows Against Gov't 'Urban Mines' Plan for Tokyo 2020 Medals." *The Mainichi* https://mainichi.jp/english/articles/20180902/p2a/00m/0na/012000c.

National Audit Office. (2022). *Grassroots Participation in Sport and Physical Activity* https://www.nao.org.uk/wp-content/uploads/2022/07/Grassroots-participation-in-sport-and-physical-activity.pdf.

National Oceanic and Atmospheric Administration. (2017, April 20). "Deepwater Horizon Oil Spill Settlements: Where the Money Went." https://www.noaa.gov/explainers/deepwater-horizon-oil-spill-settlements-where-money-went.

Neils, Jennifer (1992). *Goddess and Polis: The Panathenaic Festival in Ancient Athens.* Princeton: Princeton University Press.

Nelson, Nigel. (2012, December 2). "Rob the Builder: Thousands of Construction Workers Were Denied Jobs as They Were Put on a Sinister Blacklist." *Daily Mirror* https://www.mirror.co.uk/news/uk-news/blacklist-thousands-of-construction-workers-denied-1469233.

New Weather Institute. (2024, July). *Polluters on the Podium: Which Sponsors of the Paris 2024 Olympics Are the Biggest Polluters?* https://www.newweather.org/wp-content/uploads/2024/07/PollutersonthePodiumJuly2024.pdf.

Nicholson-Smith, Donald. (2008, October 1). "Black Glove/White Glove: Revisiting Mexico's 1968." *openDemocracy* https://www.opendemocracy.net/en/black_glove_white_glove_revisiting_mexicos_1968/.

Noble, Josh with Paul Caruana Galizia, Michael Acton, and Chris Campbell. (2025, March 13). "Does the Olympic Business Model Still Work?" *Financial Times* https://www.ft.com/content/7e0e48eb-7c47-4028-9d87-589284efeacd?emailId=c94b36ed-2372-451c-bcd2-5fb0337e9107&segmentId=9264b0f7-e7ac-8f9b-044f-c10729049333.

NOlympics LA. (2025). *Nølympics LA Public Opinion Poll 2025* https://nolympicsla.com/polling/.

O'Connor, Francis G. and David W. DeGroot. (2024). "Heat-Related Illness in Athletes." *Journal of the American Medical Association* 332, no. 8: 664–65.

Organisationskomitee für die XI. Olympiade Berlin 1936 E.V. (1937). *The XIth Olympic Games, Berlin, 1936: Official Report*. Ed. Friedrich Richter. Berlin: Wilhelm Limpert.

Osborne, Mike. (2024, April 18). "Sending Australian Team to Paris Olympics Will Cost $25M+." *Sports Hounds* https://sportshounds.com.au/2024/04/18/sending-australian-team-to-paris-olympics-will-cost-25m/#:~:text=AOC%20chief%20executive%20Matt%20Carroll's,medallist%20will%20earn%20in%20Paris.

Paris 2024 Social Charter. (2024). https://press.paris2024.org/assets/paris-2024-social-charter-pdf-deac-7578a.html.

Parodi, Emilio. (2019, February 13). "Exclusive: BT Executives Knew of Accounting Fraud in Italy Unit—Prosecutors." *Reuters* https://www.reuters.com/article/us-bt-italy-exclusive/exclusive-bt-executives-knew-of-accounting-fraud-in-italy-unit-prosecutors-idUSKCN1Q216M/.

Paz, Octavio. (1994). *El laberinto de la soledad*. Mexico City: Fondo de Cultura Económica.

Peters, Justin. (2021, July 24). "Olympics Jerk Watch: The International Olympic Committee." *Slate* https://slate.com/culture/2021/07/international-olympic-committee-jerks-ioc-corruption-covid-tokyo.html.

Pindar. (1915). *The Odes of Pindar Including the Principal Fragments*. Trans. John Sandys. London: William Heinemann; New York: Macmillan.

Pitts, Adrian and Hanwen Liao. (2013). "An Assessment Technique for the Evaluation and Promotion of Sustainable Olympic Design and Urban Development." *Building Research and Information* 41, no. 6: 722–34.

Polloni, Camille. (2024, August 5). "Visé par une enquête pour meurtre, un policier fait du BMX à la cérémonie d'ouverture des JO." *Mediapart* https://www.mediapart.fr/journal/france/050824/vise-par-une-enquete-pour-meurtre-un-policier-fait-du-bmx-la-ceremonie-d-ouverture-des-jo.

Poinssot, Amélie. (2024, August 2). "Sur le village Olympique, le gaspillage alimentaire n'esta pas évité." *Mediapart* https://www.mediapart.fr/journal/economie-et-social/020824/sur-le-village-olympique-le-gaspillage-alimentaire-n-est-pas-evite.

Poniatowska, Elena. (1971). *La noche de Tlatelolco: Testimonios de historia oral*. Mexico City: Ediciones Era.

Poole, Gavin. (2024, September 13). "London's Olympics Have Given the City a Legacy to be Proud of." *Guardian* https://www.theguardian.com/sport/2024/sep/13/londons-olympics-have-given-the-city-a-legacy-to-be-proud-of?CMP=Share_iOSApp_Other.

Racinais, Sebastien, Mohammed Ihsan, Marie-Elaine Grant, Valentin Dablainville, Yohan Rousse, Wolfgang Schobersberger, Richard Budgett, and Lars

Engebretsen. (2024). "Ice Challenge in Recent Summer Olympic Games." *British Journal of Sports Medicine* 58, no. 17: 943–45.

Rainforest Action Network. (n.d.). "Make Top Ramen Deforestation Free." https://www.ran.org/campaign/demand-deforestation-free-instant-noodles/.

Raji, Lilian. (2024, July 28). "How Bernard Arnault and LVMH are Dominating the 2024 Paris Olympics." *Forbes* https://www.forbes.com/sites/lilianraji/2024/07/27/how-bernard-arnault-and-lvmh-are-dominating-the-2024-paris-olympics/.

Rédaction Île de France. (2022, August 4). "Paris 2024: ces infrastructures abandonnées et retardées en Seine-Saint-Denis." *Actu Seine-Saint-Denis* https://actu.fr/societe/carte-paris-2024-ces-infrastructures-abandonnees-et-retardees-en-seine-saint-denis52658423.html.

Reid, Helen. (2024, April 12). "Nike Bosses Plan 'Biggest' Olympics Spend as Marketing Ramps Up." *Reuters* https://www.reuters.com/business/retail-consumer/nike-bosses-plan-biggest-olympics-spend-marketing-ramps-up-2024-04-12/.

Reynolds, John. (2012, February 17). "BP's Brand Image Benefits from London 2012 Sponsorship, Claims Research." *Campaign* http://www.campaignlive.co.uk/article/1117665/bps-brand-image-benefits-london-2012-sponsorship-claims-research?srcsite=marketingmagazine.

Roan, Dan. (2022, July 26). "London 2012: Ten Years On, What Lessons Can We Learn from London's Legacy?" *BBC* https://www.bbc.co.uk/sport/olympics/62299772.

Roberts, William O., Lawrence E. Armstrong, Michael N. Sawka, Susan W. Yeargin, Yuval Heled, and Francis G. O'Connor. (2023). "ACSM Expert Consensus Statement on Exertional Heat Illness: Recognition, Management, and Return to Activity." *Current Sports Medicine Reports* 22, no. 4: 134–49.

Rocha, Claudio M. (2023). "Rio 2016 Olympic Legacy for Residents of Favelas: Revisiting the Case of Vila Autódromo Five Years Later." *Social Sciences* 12, no. 3 https://doi.org/10.3390/socsci12030166.

Rodda, John. (2008). "'Prensa, Prensa': A Journalist's Reflections on Mexico '68." *Bulletin of Latin American Research* 29, no. 1: 11–22.

Rosenberg, Michael. (2024, August 11). "Paris Olympics Perfectly Captured the Spirit of its Host City." *Sports Illustrated* https://www.si.com/olympics/paris-olympics-perfectly-captured-spirit-of-host-city.

Ross, Andrew. (2009). *Nice Work If You Can Get It: Life and Labor in Precarious Times.* New York: New York University Press.

Rouget, Antton. (2024, August 17). "La Seine était complètement souillée quand la minister Oudéa-Castéra s'y est baignée." *Mediapart* https://www.mediapart.fr/journal/france/170824/la-seine-etait-completement-souillee-quand-la-ministre-oudea-castera-s-y-est-baignee.

Saccage. (2024a, September). *Pourquoi faire un bilan?* https://saccage2024.noblogs.org/files/2024/09/Pourquoi-faire-un-bilan-.pdf.

Saccage. (2024b, October). *Les jeux olympiques et paralympiques de Paris 2024: Bilan des saccages. Chapitre 3: Le saccage écologique* https://saccage2024.noblogs.org/files/2024/10/Bilan-des-Saccages-2024-ECOLOGIQUE.pdf.

Saleh, Tariq. (2024, August 13). "Significant Paris 2024 Audiences for WBD, BBC, NBC." *Sportcal* https://www.sportcal.com/media/significant-paris-2024-audiences-for-wbd-bbc-nbc/.

Samuel, Shalini and Wendy Stubbs. (2013). "Green Olympics, Green Legacies? An Exploration of the Environmental Legacies of the Olympic Games." *International Review for the Sociology of Sport* 48, no. 4: 485–504.

Santini, Dominique and Holly Henderson. (2021). "The Winners and Losers in the Race to Environmental Sustainability: A Ranking of Summer Olympic International Federation Progress." *Emerald Open Research* 3, no. 12 https://doi.org/10.35241/emeraldopenres.14195.1.

Senate Finance Committee. (2024, May 20). *Insufficient Diligence: Car Makers Complicit with CCP Forced Labor* https://www.finance.senate.gov/imo/media/doc/insufficient_diligence_-_car_makers_complicit_with_ccp_forced_labor1.pdf.

Sharma, Dinesh C. (2024, November 30). "40 Years After Bhopal: Lingering Health Effects and No Closure." *The Lancet* 404: 2145–46.

Sher, Chloe and Caroline Fusco. (2024). "Sports and Sustainable Development: The Troubling Absence of Meat Sourcing Policies in the Sports Sector." *Frontiers in Sports and Active Living* 6 https://doi.org/10.3389/fspor.2024.1341810.

Short, John Rennie. (2015, August 3). "The Green and the Gold: Can We Soften the Environmental Impact of the Olympics?" *The Conversation* https://theconversation.com/the-green-and-the-gold-can-we-soften-the-environmental-impact-of-the-olympics-45530.

Siemaszko, Corky. (2021, July 24). "Heat Wave Hits Tokyo as Olympic Organizers Battle to Keep Covid Rates Down." *NBC* https://www.nbcnews.com/news/olympics/heat-wave-hits-tokyo-olympic-organizers-battle-keep-covid-rates-n1274899.

Sin, Ben. (2014, February 12). "All the Modern Olympic Mascots in One Adorable Infographic." *Sports Illustrated* https://www.si.com/extra-mustard/2014/02/12/olympics-mascots-infographic.

Skóra, Paulina. (2024, July 3). "England: West Ham Close to New London Stadium Lease Agreement? Property Value Declines." *StadiumDB* https://stadiumdb.com/news/2024/07/england_west_ham_close_to_new_london_stadium_lease_agreement_property_value_declines.

Smith, Kirk R., Alistair Woodward, Bruno Lemke, Matthias Otto, Cindy J. Chang, Anna A. Mance, John Balmes, and Tord Kjellstrom. (2014). "The Last Summer

Olympics? Climate Change, Health, and Work Outdoors." *The Lancet* 388, no. 10045: P642–44.

Starcevic, Seb. (2025, April 28). "EU Backs Olympic Bid for 2036, While London Wants the Games in 2040." *Politico* https://www.politico.eu/article/eu-olympics-bid-2036-london-2040-glenn-micallef/.

Sugawara, Makoto, Yoshiaki Manabe, Fumihiro Yamasawa, and Yuri Hosokawa. (2022). "Athlete Medical Services at the Marathon and Race Walking Events During Tokyo 2020 Olympics." *Frontiers in Sports and Active Living* 4 https://doi.org/10.3389/fspor.2022.872475.

Suzuki, Naofumi, Tetsuo Ogawa, and Nanako Inaba. (2018). "The Right to Adequate Housing: Evictions of the Homeless and the Elderly Caused by the 2020 Summer Olympics in Tokyo." *Leisure Studies* 37, no. 1: 89–96.

Swithinbank, Robin. (2022a, January 28). "Pressure Builds on the Watch Industry Over Green Claims." *Financial Times* https://www.ft.com/content/cca70499-291a-4b95-813f-253584446c33.

Swithinbank, Robin. (2022b, May 13). Is Sustainability Message From Watchmakers Mere Green-Washing?" *Watchpro* https://www.watchpro.com/is-sustainability-message-from-watchmakers-mere-green-washing/.

Szalai, Georg. (2024, October 31). "Peacock Loss Narrows to $436M as Revenue Rises, Subs Jump to 36M on Paris Olympics Boost." *Hollywood Reporter* https://www.hollywoodreporter.com/business/business-news/comcast-q3-earnings-peacock-loss-nbcuniversal-paris-olympics-1236047497/.

Tagsold, Christian. (2010). "Modernity, Space, and National Representation at the Tokyo Olympics 1964." *Urban History* 37, no. 2: 289–300.

Takanobu, Saito. (2023, January 20). *How the Public Perceived the Tokyo Olympic and Paralympic Games.* NHK Broadcasting Culture Research Institute https://www.nhk.or.jp/bunken/english/research/yoron/20230130_3.html.

Talbot, Adam. (2025). "Olympics for Whom? Winners and Losers of Mega-Events." *Events and Society: Bridging Theory and Practice.* Ed. Mike Duignan. Abingdon: Routledge. 131–38.

The Tokyo Organising Committee of the Olympic and Paralympic Games. (2021). *Update to the Sustainability Pre-Games Report* https://gtimg.tokyo2020.org/image/upload/production/cas22cdv09dc0h9yaf5t.pdf.

Timms, Aaron. (2021, August 2). "NBC Paid $7.75bn for its Olympic Rights … and We Got Televisual Vomit." *Guardian* https://www.theguardian.com/sport/2021/aug/02/nbc-olympic-coverage-peacock-replays-primetime?CMP=ShareiOSAppOther.

Tomizawa, Roy. (2016, February 3). "Olympia Cigarettes: How the 1964 Tokyo Olympics Led to Increased Cases of Lung Cancer in Japan Today." *The Olympians from 1964 to 2020* https://theolympians.co/2016/02/03/olympia-cigarettes-how-the-1964-tokyo-olympics-led-to-increased-cases-of-lung-cancer-in-japan-today/.

Tomizawa, Roy. (2019). *1964—The Greatest Year in the History of Japan: How the Tokyo Olympics Symbolized Japan's Miraculous Rise from the Ashes*. Carson City: Lioncrest.

Tomlinson, Alan. (1996). "Olympic Spectacle: Opening Ceremonies and Some Paradoxes of Globalization." *Media, Culture & Society* 18, no. 4: 583–602.

Tomlinson, Alan. (2014). "Olympic Legacies: Recurrent Rhetoric and Harsh Realities." *Contemporary Social Science* 9, no. 2: 137–58.

Tramuta, Lindsey. (2024, June 26). "How Paris Hopes the Summer Olympics Will Transform the City—For Good." *Condé Nast Traveler* https://www.cntraveler.com/story/paris-olympics-sustainability.

Tzanelli, Rodanthi. (2023). *Space, Mobility, and Crisis in Mega-Event Organisation: Tokyo Olympics 2020's Atmospheric Irradiation*. London: Routledge.

Underwood, John. (1968, September 30). "Games in Trouble." *Sports Illustrated* https://vault.si.com/vault/1968/09/30/games-in-trouble.

United States Department of Housing and Urban Development. (2023). *The 2023 Annual Homelessness Assessment Report (AHAR) to Congress* https://www.huduser.gov/portal/sites/default/files/pdf/2023-ahar-part-1.Pdf.

University of East London. (2015, December). *Olympic Games Impact Study—London 2012 Post-Games Report* https://www.uel.ac.uk/sites/default/files/post-games-ogi-report.pdf.

Vadiati, Niloufar. (2020). *The Employment Legacy of the 2012 Olympic Games*. Cham: Palgrave Macmillan.

Volkery, Carsten. (2012, August 13). "'Golden Games' Give Britain Much-Needed Boost." *Der Spiegel* https://www.spiegel.de/international/world/praise-for-closing-ceremony-of-the-london-olympic-games-a-849720.html.

Wainwright, Oliver. (2022, June 30). "'A Massive Betrayal': How London's Olympic Legacy Was Sold Out." *Guardian* https://www.theguardian.com/uk-news/2022/jun/30/a-massive-betrayal-how-londons-olympic-legacy-was-sold-out?CMP=ShareiOSAppOther&fbclid=IwAR3vG%E2%80%A6.

Walker, Alissa. (2018, June 7). "The Olympics Fixed LA's Traffic Problem—Can the 2028 Games Do it Permanently?" *Curbed* https://la.curbed.com/2018/6/7/17419270/olympics-2028-los-angeles-1984-traffic.

Wallace, Julianne Poock, Eric Wiedenman, and Robert J. McDermott. (2019). "Physical Activity and Climate Change: Clear and Present Danger?" *Health Behavior and Policy Review* 6, no. 5: 534–45.

Wang, Shixin, Tiexi Chen, Jing-Jia Luo, Meng Gao, Hongchao Zuo, Fenghua Ling, Jianlin Hu, Chaoxia Yuan, Yuanjian Yang, Lina Wang, Huaming Huang, Naiang Wang, Yaojun Li, and Toshio Yamagata. (2024). "Warming Climate is Helping Human Beings Run Faster, Jump Higher and Throw Farther Through Less Dense Air." *Climate and Atmospheric Science* 7. https://doi.org/10.1038/s41612-024-00635-z.

Ware, Alyn. (2021, July 29). "Tokyo's Games Are Harming the Nuclear Weapons Ban Movement." *The Nation* https://www.thenation.com/article/society/tokyo-olympics-nuclear-weapons/.

Wertheim, Jon. (2023, July 26). "Paris 2024 is One Year Away. Do the French Care?" *Sports Illustrated* https://www.si.com/olympics/2023/07/26/paris-2024-olympics-one-year-away-daily-cover.

Wharton, Dave. (2024, November 14). "L.A. Olympics Boss Reynold Hoover Uses Wartime Problem-Solving Skills to Plan 2028 Games." *Los Angeles Times* https://www.latimes.com/sports/olympics/story/2024-11-14/olympics-boss-is-treating-the-games-like-wartime.

Wharton, Dave. (2025, January 15). "Can Fire-Torn L.A. Handle the World Cup, Super Bowl and Summer Olympics?" *Los Angeles Times* https://www.latimes.com/sports/olympics/story/2025-01-15/la-fires-world-cup-super-bowl-2028-summer-olympics.

Whiting, Robert. (2014, October 24). "Negative Impact of 1964 Olympics Profound." *Japan Times* https://www.japantimes.co.jp/sports/2014/10/24/olympics/negative-impact-1964-olympics-profound/.

Whiting, Robert. (2021). *Tokyo Junkie: 60 Years of Bright Lights and Back Alleys…And Baseball.* Berkeley: Stone Bridge Press.

Willsher, Kim. (2024, August 10). "'We'll Just be Left with Memories': Can Olympic Success Help Transform the Poorest Suburbs in Paris?" *Guardian* https://www.theguardian.com/sport/article/2024/aug/10/well-just-be-left-with-memories-can-olympic-success-help-transform-the-poorest-suburbs-in-paris.

Wilson, Sandra. (2012). "Exhibiting a New Japan: The Tokyo Olympics of 1964 and Expo '70 in Osaka. *Historical Research* 85, no. 227: 159–78.

Williams, Raymond. (2011). Raymond Williams on Television: Selected Writings. Ed. Alan O'Connor. London: Routledge.

Woodcraft, Saffron, José Izcue-Gana, Georgios Melios, Nikolaos Tzivanakis, Elisabetta Pietristefani, Gillian Chan, Zaafir Hassan, and Jorge Pérez. (2024, July). *Prosperity in and Around London's Olympic Park: Findings from the Prosperity in East London 2021–2031 Longitudinal Study.* Institute for Global Prosperity, Procol UK, and London Prosperity Board https://discovery.ucl.ac.uk/id/eprint/10195859/1/Prosperity%20in%20Olympic%20Park%20and%20Fringes%20Bulletin,%20PieL%20Wave%201%20July%202024.pdf.

Worden, Minky, ed. (2008). *China's Great Leap: The Beijing Games and Olympian Human Rights Challenges.* New York: Seven Stories Press.

Worden, Minky. (2015). "Raising the Bar: Mega-Sporting Events and Human Rights." *Human Rights Watch* https://www.hrw.org/world-report/2015/country-chapters/global-1.

WWF. (2004, July 16). *Environmental Assessment of the Athens 2004 Olympic Games* https://wwf.panda.org/wwf_news/?14215/Environmental-Assessment-of-the-Athens-2004-Olympic-Games.

WWF. (2023). *Time for Change: Demanding More Transparency and Responsibility in the Watch and Jewellery sector. Sustainability Rating and Industry Report 2023* https://www.wwf.ch/sites/default/files/doc-2023-11/_WWF_Watch%20and%20Jewellery%20Report%202023_Final.pdf.

Yeo, ShinJoung and Toby Miller. (2017). *Artists in Tech Cities.* SPACE https://spacestudios.org.uk/wp-content/uploads/2021/07/4501702tech-cities8web-2021.pdf.

Young, Noel. (2012, August 9). "BP Olympic Commercials Cheer on Team USA ... and Reputation Rises." *The Drum* https://www.the-drum.com/news/2012/08/09/bp-olympic-commercials-backing-team-usa-sees-its-reputation-rise.

Zimmerman, Max. (2021, June 3). "How the 1964 Olympic Games Changed Tokyo Forever." *Bloomberg* https://www.bloomberg.com/news/features/2021-06-03/how-the-tokyo-olympics-in-1964-changed-japan-s-capital-forever.

Zirin, Dave and Jules Boykoff. (2019, July 23). "These Women Have Lost Their Homes to the Olympics in Tokyo—Twice." *The Nation* https://www.the-nation.com/article/archive/tokyo-olympics-displacement/.

Zirin, Dave and Jules Boykoff. (2022, July 19). "Organizers Don't Care That the 2028 Olympics Could be on Fire." *The Nation* https://www.thenation.com/article/society/la-olympics-climate-change/.

Zirin, Dave and Jules Boykoff. (2023, June 27). "The Paris Olympics Are Developing a Familiar Stench." *The Nation* https://www.thenation.com/article/world/paris-olympics-corruption/.

Zolov, Eric. (2004). "Showcasing the 'Land of Tomorrow': Mexico and the 1968 Olympics." *The Americas* 61, no. 2: 159–88.

CHAPTER 4

The Carbon America's Cup

What is it about sail? People bobbing up and down in water. Christopher Cross sang that 'if the wind is right you can find the joy of innocence again' and *Sail-World* argued that the 'sea has always been and remains a mystery as attractive as it is impenetrable, a mystery that for most human beings is an ineffable fascination. The sea attracts because unknown, mysterious, wild, endless, for the thousands of secrets that it evokes, from the past and the future, invoking people's deepest feelings.' An 'un-measurable greatness reminds us of how small we are in front of Nature, like a much-needed dose of humility to calm the waves of ego and vanity that plague us' (Global Solo Challenge 2022).

Maybe, but humility is not the term associated with the America's Cup. Unlike the World Cup or the Olympics, the America's Cup of sailing is a corporate competition between private yacht clubs of the rich world. But *like* those events, its nationalistic overtones attract and legitimize government funds, conflating private and public interests (Sturm et al. 2021):

> During the 1987 America's Cup in Fremantle, much media attention was centered on items like computer assisted design of the yachts, satellite-navigation systems, hand-held weather pagers, the effectiveness of gennakers and other sails, the advantages of titanium-constructed apparatuses, and the superiority of Stars and Stripes' plastic-covered hull, which was designed on the basis of information garnered from private industries and the

T. Miller, J. P. Carañana, *Global Sports Go Green—Or Do They?*, Global Culture and Sport Series,
https://doi.org/10.1007/978-3-032-14207-8_4

> American government's aerospace, naval, and defense industries. (McKay and Huber 1992: 209)

Sports Illustrated called the Cup 'the greatest single international trophy in sport' (Mitchell 1958). Critics suggest it is won on land rather than sea, via espionage, litigation, design, technology, and media rhetoric; the race even smiles on overt collusion between competitors, per the managed capitalism of US sports (Riggs 1986; Ross 2023; Coll 2024; Mullen 2023). It's always been a nasty affair, as one would expect of hegemonic, technological masculinity, often overdetermined by struggles over people, machinery, and intellectual property that feature an 'armada of lawyers'. In the 2020s, teams of 150 included just a dozen sailors (Bell with Agini 2024) in what was essentially 'una competición entre ingenieros en medio del mar' [a competition between engineers in the middle of the ocean] (Burgaya 2024).

The Cup began in 1851 as a contest between British and US yacht squadrons, at the very end of "the age of sail" (Gaynor 2013). The first modern international sporting event, it was designed as 'a showcase of maritime supremacy, a demonstration of just how invincibly Britannia ruled the waves'. From that point on, the event pitted the previous winner (the USA for 130 years) against a challenger. The list of key former contestants is a rollcall of robber barons and oligarchs: JP Morgan, the Vanderbilts, Thomas Lipton, Ted Turner, and Marcel Bich (Black 2020: 313). They 'would probably no more have thought of going forward of the mast than … going to the galley and offering to help peel potatoes' (Mitchell 1958). A regatta for the famous and wealthy, the Cup helped form 'class identity for the world's rich' (Connell 1984: 432).

Technology was always a feature, from the yachts themselves to media milestones. The first US experience of wireless allowed radio's hero Marconi to cover the 1899 Cup for the *New York Herald* in Morse Code (Douglas 1987: 3–28). Underwater telegraphy pioneered transmission of facsimile photographs of the 1920 race to the *Daily Mirror* (Coopersmith 2015: 50–51).

1851 had seen the New Yorkers offer a wager of £10,000 (1.7 million in 2025 prices). No one took up the bet. The winners received $25,000, over a million today (Kidd 2017). Even seven decades ago, the contest was renowned for its 'astronomically mounting costs' (Mitchell 1958). But something elemental happened in the 1980s: '[a]fter 132 years as the personal dalliance of the rich,' competitors 'turned to Madison Avenue sages for marketing and to habitues of Wall Street for money'. Old money was no longer enough to fund the events; '[h]uge egos' became the norm.

Alan Bond, the soon-to-be-disgraced mogul whose team won in 1983, described the event pithily as 'where very successful men come to be with other very successful men' (quoted in "Fremantle Prepares" 1985). In Tony Kornheiser's words, 'they're not worried about which bus to take to work'. Michael Fay, a 1980s syndicate head, said, 'I didn't come here to win the America's Cup … I'm here to win the America's Cup industry' (quoted in Barrass 2015). Escalating costs reached their apogee in the period from 2013, via 'faster boats, course boundaries, and broadcast enhancements' ("37th America's Cup" 2024). In 2024, the price to compete was £120 million (Bell with Agini 2024). Hence the *Otago Daily Times*, from the country that held the Cup in the 2020s, referring to the competition as 'rich, white men playing with fancy toys' ("America's Cup Line" 2021). The *Irish Independent* asked how:

> a yacht race, which few normal people have even contemplated watching at any point in their lives, can be so big? By what metric is this calculated? Is it the "spin-offs"? All that fine wealth the plutocrats will be generating and ploughing back into the community as is their wont? (Lynch 2021)

To answer that query, we'll look into the event's sponsors, notably LVMH, prior to studying how subvention and greenwashing characterized the thirty-seventh Cup, staged in Barcelona in 2024. Here's an early hint: the 'metric' should engage social and environmental squalor.

Since 1983, the winner of a contest between clubs to challenge the holder has received a prize named after whichever sponsor handed over the most cash (Ruta and Manzoni 2015). Vichy servant and Nazi beneficiary Louis Vuitton (Willsher 2004) was followed by notorious racist Prada ("Prada Blackface" 2020) before naming rights returned to Louis (John and Jackson 2011). Prada continued to offer Cup sneakers, customized by color, initials, and material—a steal in 2025 at €720 to 970 a pair ("Shoes" n.d.). Or perhaps sir would like Prada's K-Way jacket for €290, available from the America's Cup store as an item that 'epitomizes modern performance wear,' satisfies 'the highest standards of durability and environmental consciousness,' and 'represents the dynamic spirit of the America's Cup.' A further option was the official timekeeper Omega's "Seamaster Diver 300M America's Cup Edition," which retailed for a mere €7200 to the doubtless discerning customer ("The Timekeeper's" n.d.). One could also purchase LEGO's racing yacht for grown-ups to assemble, using some of the 'billions of non-biodegradable bricks' the company creates each year (Baraniuk 2023), many of which wash up on beaches after despoiling the oceans and are thought to take between a century and a millennium to break up (Turner et al. 2020).

Apart from its dubious political history, LVMH is a leading figure in unsustainable fashion, and the world's largest luxury-goods merchandiser, while KnowTheChain, a key advocacy and research group, ranked it fiftieth out of sixty-five global brands in sub-contacting slavers across international supply chains (2023). In the weeks prior to the 2024 regatta, a Milan court appointed a commissioner to run a firm owned by the company that was reportedly overseeing illegal working conditions (Parodi 2024). Meanwhile, the corporation was modestly claiming the title of 'Artisan de toutes les victoires' [Artisan of all victories] ("LVMH est partenaire" 2024). It boasted of a natural connection to the Cup thanks to 'a shared love of the sea and dedication to victory,' even announcing itself as the event's 'guardian' (quoted in Reed 2024). The *Robb Report*, a magazine dedicated to luxury goods and well-heeled readers, highlighted the nexus of LVMH and the regatta as 'el evento del año' [the event of the year] because it 'personifica el vigor contemporáneo y la maestría artesanal de la maison' [personifies the house's contemporary vigor and artisanal mastery] in a way that 'comparte importantes valores' [shares important values] of newness and tradition in design, engineering, and logistics (Ramírez 2025: 71–72).

Vuitton's greenwashing work on behalf of others, as well as itself, was indexed in an early trial race in Saudi Arabia, part of that state's already-noted use of sports to distract from human-rights and ecological abuses. The 2024 Cup *Press Kit* announced that since its foundation in 1854, Louis Vuitton had been 'aiming for the finest quality in the respect of biodiversity' (Louis Vuitton 37th America's Cup 2024a). Apart from being poorly written, this was rather a vanguardist, not to say anachronistic, achievement, given the term "biodiversity" was coined in the 1980s (Sarkar 2021). It also ignored LVMH's infringement of animal rights, noted in Chap. 3, without the merest acknowledgment, self-reproach, or change on the part of this disgraceful company (Guitton-Boussion 2024).

Earlier that year, the firm had forbidden its workers from speaking to such decidedly unthreatening French and Anglo publications as *La Lettre*,

Puck, *Miss Tweed*, *Glitz Paris*, and *Le Canard enchaîné*, as well as the progressive paper *Mediapart* (Prarthana 2024). All were accused by the firm's head, Bernard Arnault, of hiring 'journalistes peu scrupuleux' [unscrupulous journalists] (quoted in Abzouz and Zerouali 2024).

Perhaps the old boy was responding to accusations in courts and papers that he laundered money, trafficked political influence, and installed his wealth offshore to avoid or evade taxes (Jolly 2023; Faye 2023; de Villepin 2023). Or was Arnault obsessively repeating the tactics he had deployed against the current-affairs program *Complément d'enquête* [*Under Investigation*] a decade before (Pocréaux 2024a)—attempts to steal a copy of *Merci, Patron*, which documented the firm's inhumane exploitation of the international division of labor? Or was the model Arnault's lawsuit against *Libération* for headlining his application for Belgian citizenship, supposedly to avoid paying French taxes, 'Casse-toi, riche con' [Fuck off, you rich cunt] (Le Vaillant and Service Médias 2012)? The paper was of course troping Arnault's close friend, former President and convicted criminal Nicolas Sarkozy, saying to a farmer who declined to shake his hand, 'Casse-toi alors, pauvre [peasant] con!' Reporters sans Frontières catalogued these repeated attacks on journalism as standard practice for LVMH and called for Arnault to be held to judicial account ("Procès" 2024).

Just after the thirty-seventh America's Cup drew to a close, France's ex-intelligence head Bernard Squarcini and nine others were accused of undertaking secret surveillance for the corporation, misusing government resources to do so. But don't imagine LVMH was on trial—three years earlier, it had paid prosecutors €10 million over the matter, 'without any admission of guilt.' Arnault testified in 2024, saying 'I am here as a witness, a simple witness.' As the trial unfolded, allegations were made that the indicted men had infiltrated Ruffin's progressive magazine *Fakir* on his behalf. The big man attacked attorneys' and journalists' supposed '*l'idéologie socialo-gauchiste*' [socialist-leftist ideology], modestly asserting that he had done more for France than the party wronged by the espionage (Arnault quoted in "LVMH Chief" 2024 and Lazard 2024; Pocréaux 2024b).

The papers Arnault owns headlined the story with his mocking the plaintiff (de Semeville 2024; Boutry 2024). Conversely, *Mediapart* and *Fakir* offered more, shall we say, comprehensive accounts of the testimony ("Dossier" 2024):

> Au bout d'une heure d'audition de Bernard Arnault, on peut le prendre par tous les sens le patron de LVMH n'était au courant de rien, ne savait rien, ne se souvient de rien quant à l'espionnage de Fakir et Ruffin. C'est son numéro 2 (aujourd'hui décédé) qui gérait tout ça, sa secrétaire (et pas lui) qui était stressée, et les éléments factuels qui disent le contraire, il ne s'en souvient pas. [After an hour listening to Bernard Arnault, we can say this much: the boss of LVMH was not aware of anything, knew nothing, and remembers nothing about spying on *Fakir* and Ruffin. It was his number 2 (now deceased) who managed all of that, his secretary (not he) who was suffering from stress, and he couldn't recollect any facts that might contradict this version of events.] (Pocréaux 2024b)

Squarcini was convicted of using state funds on behalf of the conglomerate (Jabkhiro and Spencer 2025).

In addition to LVMH, the variety of America's Cup sponsors is considerable; in descending order of importance/payment, one assumes, 'Title Partner, Official Timekeeper, Global Partners, Exclusive Supplier, Official Suppliers, Host Partners' ("Partners" n.d.), the latter government entities. The corporate polluters involved include Coca-Cola. Below you'll see a bus-stop billboard in Madrid promoting Coke, complete with quasi-obligatory shot of a young white woman with her mouth posed just so and the mark of the Cup in the bottom left-hand corner (Fig. 4.1).

Other notorious carbon emitters sponsoring the event included Emirates Airlines, Puig (fashion, perfume, and cosmetics), Cupra (automobiles), Yanmar (diesel machinery), and McConachy (superyachts), inter alia. The Cup holders were Emirates Team New Zealand, the corporate/statist title of the Royal New Zealand Yacht Squadron. "Emirates Team New Zealand" was sponsored by Toyota and Omega, Catalunya's own beer, Estrella Damm, and the luxury cruise line Explora (Bell with Agini 2024).

Like LVMH, several sponsors needed greenwashing in 2024: Emirates Airlines devoted much of the year to defending itself against an exposé that it had endangered workers (Noëth 2024), Cupra's cars were in recovery from a significant period rated as the least-reliable automobiles by UK motorists ("Britain's Most" 2024), and Yanmar paid the California Air Resources Board over $16 million for violating emission controls (*Yanmar Power* 2024). We've already read of brewing's environmental impact and will encounter superyachts and cruise ships in some detail later.

Fig. 4.1 (Source: Toby Miller)

Subvention, Subvention

The billionaires competing for the Cup have long had their hands extended in the quest for public money to finance their juvenilia. Massive cost overruns are generally matched by bloated expectations of popular economic benefits (Fresh Info 2021). Needless to say, private- and public-sector bureaucrats and academic chorines line up to provide inelegantly described, if tantalizing, fantasies about the Cup's 'leveraging legacies' (Wright and Barron 2021), available to those with the 'organizational capabilities to plan and execute a legacy by exploiting a large-scale sports event' (Ruta and Manzoni 2015: 232). The elect use these skills to explore "Value Creation and Stakeholder Management" (Jacopin 2006–07). Such unfortunate expressions of mandarin managerialism aim to facilitate the noble search for 'sites portuaires délaissés' [neglected port zones] (Lecardane 2008: 110) by dodging 'the threat of overthrow by discontented individuals' (Lassalle 2014).

It's worth re-reading those words. They signify service to state and capital through denotative and connotative references to work, administration, violence, and control, dismissing everyday residential life and democratic activism along the way. We might deride such sentiments as silly, but they can have material impacts on our planetary future. At least 2022 found the holder unable to find a city in its own country of New Zealand/Aotearoa ready to take on the burdensome, ill-advised commitment of hosting (Canela 2022).

Many cities have been seduced by boosters' seemingly fecund but ultimately sterile promises of urban consolidation and growth *per medio* the America's Cup. Western Australia's Maritime Museum avowed that 'Fremantle became the center of the universe' for five days in 1987 as 'restaurants bustled with patrons from around the world' in the fundamental transmogrification of 'a quiet port town into a dynamic, cosmopolitan destination' (Meager and WA Maritime Museum 2023). Decades later, the Port de Barcelona declared the 2024 regatta 'an event with a great capillarity'—a Deleuzian moment ("The Benefits" 2024).

A convention that cupholders stage the next regatta in their own marina was initially violated in Spain and Bermuda, opening the way to competitions for playing host. Such commodification reached new levels with the creation of America's Cup Management, which oversaw the choice of site for 2007 and ran two regattas. València paid the firm €24 million, even as the municipality was entering a period of intense socio-economic distress (Lassalle 2014: 43; Rius-Ulldemolins et al. 2021: 40).

The city's late mayor, Rita Barberà, made the clichéd claim that the Cup would 'poner València en el mapa' [put València on the map] at 'ningún coste para el erario público' [no cost to the public purse] (quoted in Canela 2022). This was per the "[Santiago] Calatrava Model," named for the notorious architect who became accustomed to working with *carte blanche* from city burghers, sans normal public-works oversight (Rius-Ulldemolins et al. 2021: 32). The organizers decided to expand the event, 'allowing the partners to get a better, earlier and more regular return for their investments' and purportedly ensuring València would become 'a city of knowledge' (quoted in Jacopin 2006–07: 64). They had succumbed to the magic of carbon mega-event mega-promises as part of the administration's response to the de-industrialization and political-economic centralization that accompanied Spain's accession to membership of the EU (Rius-Ulldemolins et al. 2021).

The reality? Staging the Cup led to sizeable ongoing municipal debt. Barberà's successors woke up to public expenditure of €370 million or 2.7 billion, depending on one's source, to the point where the debt, €180 or 500 million, may never be acquitted. The city also inherited empty and underused buildings constructed for ephemeral regatta purposes (Ruta and Manzoni 2015; Canela 2022; Vicente 2023; Rius-Ulldemolins et al. 2021: 40–41).

Despite that absurdity, Barcelona was willing to follow the example of its southerly coastal neighbor, as part of both cities' all-too-similar, all-too-familiar governmental-commercial vision of prosperity. City functionaries and the *bourgeois* media emphasized that '1.200 millones y 19.000 puestos de trabajo' [€1.2 billion and 19,000 jobs] would allegedly be created by playing host to the Cup in 2024, equating to 0.5% of Catalunya's GDP (Alcalá; 2024; "Copa América: Barcelona espera" 2024).

Such propaganda was very much per the 2012 Olympics: an ever-welcome urban transformation would 'fomentar la innovación, impulsando la transformación del sector logístico y abriendo las puertas a *start ups*' [foment innovation, dynamize logistics, and open doors to startups] ("El Port" 2024). Even the purportedly progressive paper *Público* ran an advertorial on behalf of Port de Barcelona celebrating the Cup's infrastructural, environmental, and technological legacy ("El Port" 2024).

The city's bid relied on a remarkable political geography: two rivers, ten beaches, a busy port, headquarters of the World Ocean Council, experience hosting a World Cup, an Olympics, expos—and ample taxpayer support. Through a public-private partnership, Barcelona handed over perhaps €70 million to capture the 2024 regatta in the expectation of 'un retorno de 1.000 millones' [a billion in return]. Identifying the public entities that contributed the people's revenue to this folly is difficult. The Administración General del Estado, Generalitat de Catalunya, Ayuntamiento de Barcelona, Autoritat Portuària de Catalunya, Consorci de Turisme de Barcelona, Diputació de Barcelona, Cambra de Comerç, and Barcelona Global provided a mixture of "investment," loans, guarantees, contracts, and free services (López 2022). The Català outlet *Directa* made a valiant effort to sort through the thicket of obfuscation to discern roughly who paid for what (J Rodríguez 2024b). We can say with certainty that public subvention gained the right to host the pseudo-private America's Cup without a plebiscite to decide on bidding, or paying for, the privilege. And that even the port authority acknowledged the following year that its finances were in debt by €3.5 million thanks to the regatta (Rosa 2025).

As the months turned into weeks prior to the event itself, additional demands were made of Barcelona's municipal authorities, such as guaranteeing loans for the organizers. La Fundació Barcelona Capital Nàutica [The Barcelona Nautical Foundation] was declared a team of 'salvavidas' [lifesavers] because it handed over more than €6 million from the government and €4 million from the Executive (Carnicé 2024a). The Fundació modestly nominated Barcelona a 'capital dels oceans' [capital of the oceans], a nationalistic hub for electrifying boats and decarbonizing technology: '100% electric, 100% català' ("Descarbonització" n.d.). When LVMH used Barcelona for a catwalk preamble to the Cup, the city declined to share details of any subvention on the grounds of 'el secreto empresarial' [business confidentiality] (Congostrina 2024a). Boasts of a billion euros in revenue for the city were made yet again, with little to back them up (Burgen 2024). Every spurious utterance, premise, and promise seemed to be available, with no boosterish *cliché* left unspoken. After the fact, it emerged that direct public subvention may have amounted to a €100 million (Burgaya 2024) (Fig. 4.2).

The photo above, of dark-suited, windswept "authorities" with embarrassing thumbs, grinning for the camera to celebrate "Barcelona's" Cup, is unstylish at best. Not very Vuitton or Prada—not even terribly LEGO, on your knees in the living room on a rainy Sunday. The picture is unlikely to generate confidence or allure—rather, it suggests primary-process public policy, where decisions are made to allocate funds or raise thumbs in sudden yet orchestrated hope and folly, via what the *British Medical*

Fig. 4.2 (Source: https://x.com/kikoP9/status/1509097921542819853)

Journal calls 'policy-based evidence' masquerading as 'evidence-based policy' (Marmot 2004).

The then-President of the Generalitat, Pere Aragonès, avowed that the Cup would be 'la mejor de la historia' [the best in history], just as Barcelona in 1992 represented what he termed 'un antes y un después en el olimpismo' [a before-and-after in Olympic history] (quoted in López 2022). Conversely, the distinguished urbanist and politician Jordi Borja described the legacy of 1992 as 'un virus maligno: la ciudad-negocio' [a malignant virus: the business city] that 'no siempre se ha sabido resistir a la tentación de venderla por piezas al mejor postor' [can't resist the temptation to sell out to the highest bidder] (Borja et al. 2003: 56).

The fantasy that the Cup would both index and animate the region's economy produced huge optimism in the *bourgeois* media (Martínez 2024). But claims about multiplier effects were soon brought into question for their methodology, transparency, rigor, and, shall we say, provenance. A million tourists, each spending several hundred euros, would be required to generate the revenue proclaimed. Those calculations excluded externalized costs, such as environmental and social destructiveness (D Rodríguez 2022). And although the region's ecology was frequently mentioned by the mainstream press and among the oligarchy, there was no practical articulation to relevant policies (Díaz-Pont 2023).

2020s' Barcelona was hardly a time to throw taxpayers' money around, with average gross annual salary below €34,000 and a gendered differential of more than 17% (Oficina Municipal de Dades and Gerència Municipal 2024). The event's absurd subvention of billionaires amounted to a socialism for the rich that used local people's taxes to pollute everyday life.

None of this bothered Spain's monarch, Felipe VI, who opened the Barcelona event in a flurry of publicity that featured his life-long passion for sailing and mucking about with the international ruling class. It was noted that the big man dressed casually, arrived punctually astride a hydrogen-powered vessel, and proclaimed the Cup a triumph in three languages (Estirado 2024). Felipe issued a Royal Decree declaring that 'la regata está considerada, por su duración y relevancia social, como uno de los acontecimientos deportivos con mayor repercusión mediática, y genera un importante impacto económico positivo en la región donde se celebra' [given its history and social importance, the regatta is considered one of the sporting events with the greatest media impact, responsible for generating positive economic outcomes wherever it is held] (Ministerio de Trabajo y Economía Social 2024). So that's alright then. The monarch knows—in three tongues.

Tourists attracted by the Cup numbered in the thousands, not millions (Burgaya 2024). There were persistent 'grumblings from investors, backers and sponsors about the lack of airtime, limited viewers and other issues related to the lack of global interest' ("America's Cup Update" 2024). As a consequence of the event's 'historic inaccessibility and unfamiliarity' (Black 2020: 314), television ratings were minimal—a common issue for the regatta, given its technicality and lack of arresting drama. Alec McHoul's account of his experiences in Western Australia during the 1987 Cup explains that restricted codes worked to convene yachting spectators, leaving others puzzled and bored:

> The place was full of yachties watching TV screens, especially in pubs and at countless street parties. People were commenting, loudly, on the races in Gage Roads on the coast out there on the edge of the Indian Ocean. All I could see were a few boats sailing on the sea. I was, to all intents and purposes, not a member. But they could see actually socially-accomplished practices happening; practices to which I was utterly blind. (2015: 116)

Less charitably, the *Los Angeles Times* quipped '[s]ome reporters bring novels to read while they're covering the race; others actually write novels during the race'. When Roger Vaughan, a participant and fan since 1958, tuned in to the thirty-seventh iteration, he 'came away yawning.' The sailors were virtually invisible, so dominant were technology and scientific management (2024).

The small sample of Spanish tourism attracted to the regatta was in a certain sense an artifact of design, per FIFA's 2026 ticketing methods. Mateu Hernández of Turisme de Barcelona spearheaded efforts to change the profile of visitors. As he put it, the 'people who follow the America's Cup love the sea and have plenty of disposable income. ... We need to improve the quality of those who visit' and 'push away low chaos and massive tourism' (quoted in Burgen 2024 and Mas de Xaxàs 2024).

Public space shrank and became corporate, as wealthy tourists improving 'the quality of those who visit' sent rental and food prices skyrocketing and displaced long-standing small businesses. Apartments of thirty square meters suddenly fetched €6000 a month in rent (Sacristán 2024; Vicente et al. 2024) in the quest for what New Zealand/Aotearoa referred to as 'high-worth visitors' ("America's Cup Line" 2021).

This was in tune with sponsors' needs. For example, marketing research brings into doubt LVMH's image with the general public (Rodríguez-Santos et al. 2008). But campaigns for brands of that kind are aimed at very particular customers, via an elaborately confected air of sophistication; the goal is not broad-based popularity, but the custom of global oligarchs. Paradoxically, these corporations love both the élite lights of narcissistic day and the clandestine shadows of capitalist skullduggery, depending on their needs of the moment.

In addition to landlubbers, international regattas such as the Cup attract sizeable numbers of wealthy "sailing" spectators to the surrounds of the competition. Marina Port Vell was the 'socio oficial de servicios de superyates de la regata' [official partner for superyacht services during the regatta]. The America's Cup Superyacht program promised participants a view of 'the spectacular race from the comfort of your own superyacht, on the water, from a unique vantage point' (Marina Port Vell, n.d.). This was not only about spectatorship, food, and entertainment. It also guaranteed 'socialising and networking with key players, officials, and teams behind the scenes' ("America's Cup: Your Winning" n.d.), per the parthenogenesis of the international ruling class. *The Robb Report* devoted six pages of glossy photography and admiring prose to noteworthy moments of just such togetherness (Ramírez 2025: 74–80). This of course assumed a separateness from the great unwashed. The *Financial Times* has eulogized one superyacht where '[s]taff access is by hidden doors and tight staircases, so guests need never unexpectedly meet the help' (Elder 2025).

The Barcelona story was disastrous for local sailing businesses. Basic mooring fees increased from €1000 to €15,000 a month. The regatta's crewing assistants were all foreigners on location (Burgen 2024), their salaries banked and taxes paid thousands of airmiles away. Most event "jobs" went to volunteers. One reflected memorably on her experience:

> From the moment I touched down in Barcelona from the U.S., I was captivated by its energy and charm. The city is a unique blend of old-world European elegance and modern vibrancy. Waking up every day to views of the Mediterranean Sea, walking through the narrow streets of the Gothic Quarter, or sipping a cafe con leche in a local plaza—it has all felt like living in a postcard. (Shanahan 2024)

Being deemed akin to a postcard may not seem the kindest simile to actual residents, but volunteerism was part of ensuring the sailing oligarchy paid very little for their fun, as the organizers joyfully recounted:

> Key to the success of the event were the 2,100 volunteers who worked tirelessly to deliver excellence at every touchpoint and who epitomised the spirit not only of Barcelona as a city, Catalonia as a region, and Spain as a country, but of the America's Cup itself. They were the lifeblood of the event, known as the 'green wave' for their stand-out uniforms, but their contribution was immense and the sense of gratitude from the organisers is unquantifiable. (Louis Vuitton 37th America's Cup 2024b)

We could all do with some free labor, the better to experience the 'spirit' of a place in a way that is, indeed, 'unquantifiable'—unless one calculates the savings from not paying local people living wages to enable a jolly jamboree.

Ethnography of Tourism and the CUP

In 2024, over 93.8 million people visited Spain, a record, as were the 10.9 million in August alone (Instituto Nacional de Estadística 2025). The tourism industry is worth more than €12 billion annually, accounting for 14% of GDP and 150,000 jobs nationwide. Well over two-thirds of the population welcomes visitors, but there are differences of opinion between the country as a whole and the big cities, which bear the brunt of the industry and its drawbacks (YouGov 2024). A third of Spaniards lack sufficient income or wealth for a week's holiday away from home (Cañada et al. 2024: 2) and those in popular destinations feel 'as though our only role is to make the place seem more authentic' in an era 'of corporate colonialism' (Burgen 2025).

The tens of millions of people who come to Barcelona each year (tens of millions flew into the city in 2024) are drains on its 1.7 million permanent residents. Municipal budgets bear a particularly intense burden (Goodman 2025; Mas de Xaxàs 2024). The OECD notes the complexities, not only because of tourists, but voluntary and refugee migrants as well (2018). Their percentage of the population has increased monumentally since 2000, to the point where over a third of Barcelona residents were born in 180 countries ("One of Every Four" 2024).

What was the city like as it prepared for the Cup? Toby will always be a 'guiri' [foreigner] in Spain, because of how he looks and sounds. He hasn't experienced hostility over this difference: he's polite; he appears to be what he is—shabby, fallen middle class—and can speak Castellano and read Català and Valencià. This is part of what he saw on a field trip to Barcelona from his then-home in Madrid (Fig. 4.3):

The picture above represents one of the first signs he encountered en route to the site of the Cup—in English, not Català, Valencià, Gallego, Euskera, or Castellano. Its message was clear: not being drunk, drugged, aggressive, narcissistic, and childish in public guaranteed a warm welcome (the primary referent would have been the British in Spain, on holiday or retired, who are notorious, rightly or wrongly, for their monolingualism and distaste for integration [O'Reilly 2000]). They are the people most likely to visit Spain—18.4 million did so in 2024 (Instituto Nacional de Estadística 2025). Survey research finds them the least popular visitors among tourism workers, because

> No hablan otros idiomas que no sea el suyo, no suelen disfrutar con las comidas y bebidas locales, son propensos a beber demasiado alcohol, al desorden, al destrozo de las habitaciones y siempre se quejan de los hoteles. También lideran la clasificación de los peor vestidos y de los que menos propinas dejan. [They speak no language other than their own, don't want to try local food or drink, consume too much alcohol, behave poorly, and damage their hotel rooms. They are also badly dressed and terrible tippers.] ("Los japoneses" 2014)

Fig. 4.3 (Source: Toby Miller)

The "welcome" message to determinedly monolingual Anglos that Toby read wouldn't have been any kind of salve to multilingual Catalans unable to afford rental prices due to tourists using accommodation that didn't contribute to the tax base, in a conjuncture when successive governments had failed to expand housing stock. In 2024, sixty-three families competed for each rental property, with the number of available homes down 75% on 2019 ("Five Years" 2024).

As he strolled toward the shore, the signage grew more cosmopolitan. Unlike the Anglo-only welcoming sign in the city center, the marina that hosted the "America's Cup Experience" provided multilingual information guiding visitors to the "Official Divulgation Center," where they could pay to learn about the ruling-class regatta's history, then pay some more to "ride" the "Simulator." With virtual-reality glasses strapped to their heads, seat belts across their waists, and artificial turbulence coursing through their bodies, they were virtually immersed in, yes, a *simulacrum* of high-speed yachting life; Plato and Baudrillard rode among them. Perhaps it goes without saying that the "Simulator," which cost €155,000, was a "gift" from an unwitting public, which also paid for media equipment (J Rodríguez 2024a). This reminds us of the "Flow Rider," a twelve-meter surf simulator on cruise ships that forms part of their successful drive to attract young people, who can ride the simulator while photos make it seem as though they are in the ocean (Bachelor and Skopoleti 2024) (Fig. 4.4).

Fig. 4.4 (Source: Toby Miller)

Fig. 4.4 (continued)

As if joining a religious cult—or enchanted by the ideologically inbred technological masculinity of northern California—several hortatory signs included product placement for corporations. The most telling read "THERE IS NO SECOND," troping an explanation of sailing reportedly given to the British monarch during the inaugural race (Wheatley 2024).

That hyper-competitive, capitalistic notion of mocking anyone but winners is emblematic of sports' simultaneously transcendent and imprisoning qualities. Such figures of speech cross-validate sports, businesses, and nations via myths of justice, representativeness, and upward mobility that idealize political, economic, military, and social life. They distort conflict then re-signify it as part of martial masculinity and prowess, a double-declutching exemplified by the Cup's 'global syndicates … represented and marketed under the banner of national flags' (John and Jackson 2011: 402).

After simulating privilege and learning that there *was* no second, Toby emerged into the "maremagnum," a sizeable new shopping mall that guaranteed "NATURAL ATTITUDE," among numerous corporate delights. This was all part of the formal, corporate, governmentalized economy, a standard in urban renovation over the last four decades, modeled on US exemplars (Fig. 4.5).

Whether or not the authorities managed to fine people who bought blankets, the informal sector occupied space and time. The pictures above illustrate both signage prohibiting any truck with street vendors *and* their presence, evidence of the tenacity with which these workers go about their business. The corporate world was on hand to sell things "legitimately." New Zealanders/Aotearoans (of course "Aotearoa" was an almost-invisible word in Catalunya) may have lamented the fact that their status as "DEFENDERS OF THE AMERICA'S CUP" did not include being anywhere near the event, while the team's status as the property of the "Emirates" was assured and very public.

As tourists and taxpayers continued on their not-very-merry way, turning back toward the city center they could look up and see a different dominant figure fifty meters above them, from another age that is still very much with us: Colón. The monument was built as part of Barcelona's 1888 Exposición Universal to commemorate an "explorer" paid by the Spanish crown to facilitate what became the colonization of the Americas, per his path mentioned in Chap. 3. Over the last few years, its correlatives in the Americas have been beheaded, defaced, destroyed, and dragged through city streets in massive popular protests against their incarnations

Fig. 4.5 (Source: Toby Miller)

Fig. 4.5 (continued)

Fig. 4.5 (continued)

Fig. 4.5 (continued)

of imperialism, colonialism, slaughter, dispossession, conversion, and slavery (Fig. 4.6).

Despite great controversy in Catalunya, the statue remains, a reminder of a pre-capitalist colonial epoch and its fetishization centuries later. Progressives demand the statue's removal as a symbol of colonialism and racism, particularly during the Fiesta Nacional de España [National Day of Spain] or Día de la Hispanidad [Hispanicity day] (celebrated as Fiesta or Día de la Raza [Feast or Day of the People] from 1918 to 1958) to commemorate the "discovery" of the Americas on October 12 1492 ("Barcelona: manifestantes" 2020). Naturally, the far right demonstrates

Fig. 4.6 (Source: Toby Miller)

around the statue to vindicate what it deems the greatest civilizing mission in history, exhibiting Nazi and Francoist signs and chanting "Sieg Heil" and "Viva Franco" (Regué 2020). At the October event to celebrate the Cup's finale, this ghastly relic was highlighted by drones drawing it on beach skies (I Rodríguez 2024b).

Environment

Toby's encounter with socialism for the wealthy was inevitably colored by the adjacency of the natural and built environments, nowhere better captured than in shipping. The multimillionaire team owners at the 2024 regatta had distinguished themselves by making money in such exploitative, polluting industries as tourism, pharmaceuticals, petroleum, pesticides, and real estate ("Multimilionaris" 2024). INEOS, the UK's failed challenger, is owned by an expatriate Brexiteer. One of Britain's principal emitters of greenhouse gases, it exports pesticides to the Global South that are illegal in the EU. A fan of hydraulic fracturing, the firm has released 'toxic chemicals, leaks, fires, and explosions' and engaged in sustained greenwashing. As the Cup began, INEOS faced a legal challenge to its environmental record (J Rodríguez 2024b; F Harvey 2020; "INEOS Wants" 2022; Food and Water Europe 2017; ClientEarth 2024; Ahmed and Pooler 2020).

Not to worry: Turisme de Barcelona confidently declared the Cup would express and stimulate the city's drive toward 'la economía azul que estamos defendiendo' [the blue economy we're supporting]. Port de Barcelona claimed the potential to 'strengthen awareness for environmental conservation' and 'regeneration of the ecosystem,' making the city 'a world reference in sustainable industry' ("The Benefits" 2024).

The World Bank (2017) and the OECD (2016) see the blue economy as a route to sustainable growth for small islands and coastal countries. Barcelona Activa is a true believer, expressing great pride that the city's "Global Blue Economy Ranking" provides 'a great vector of job creation, economic and social development of the city and the backbone of its coast' ("Barcelona *Blue*" n.d.). Barcelona Capital Nàutica asserted that "its" event would be the finest ever, opening up sea life to the world and decarbonizing sail (2024). The *bourgeois* media were on hand to tell us that the Cup's grand legacy would include heightened public awareness of the ocean's importance (Ayuso 2024; "La iniciativa" 2024). *Mundo Deportivo* avowed that it had reconnected citizenry and sea ("La Copa América" 2024).

Most of our planet is oceanic, a rich ecosystem giving life to perhaps 2 million species (Armstrong 2022: 1) that provide 'nature-based solutions' to the anthropogenic crisis (Riisager-Simonsen et al. 2022). But they are assaulted by corporations and governments disinterested and uninterested in animal rights, ecological harm, and socio-economic inequality (Armstrong 2022). The story is all too familiar: a magic elixir will blend incommensurate desires through enlightened capitalism, thanks to purportedly sustainable fish slaughter, mineral extraction, waste disposal, and marine bioprospecting. But there can be only one winner when the prize is 'a lucrative new frontier' that inevitably, inexorably involves 'ocean grabbing,' dispossession, pollution, and monetary and governmental inequalities (Bennett et al. 2021). The capitalist ocean economy prefers growth to justice.

The Cup's environmental heritage is dubious at best. Apart from the ecological impact of travel, numerous human-made changes to harbors have altered nature's labor. Toby recalls the dramatic transformation of Fremantle's beaches with the addition of groynes in preparation for the 1987 farce (homonymously, accidentally troped by Morley Safer on *60 Minutes* when he announced the town had been 'tarted up'). Failed real-estate efforts saw some Perth beaches 'painted green to add garden suburb appeal' (Martin et al. 2017) while Bermuda dredged the ocean floor for landfill to develop a village catering to competitors prior to the 2017 event (Jones 2015).

In 2019, Barcelona visitors' carbon footprint was approximately '9.6MtCO_2eq/year,' or '96.9 kg CO_2eq/visitor' each day, mostly because of aviation (Rico et al. 2019). The year before 2024's regatta, Plataforma No a la Copa Amèrica calculated that the total direct volume of emissions linked to the teams competing would reach 66,600 tons of CO_2, with approximately 100,000 tons more deriving from visitors and spectators ("La plataforma No" 2023). Meanwhile, the city's drinking water relied on reservoirs. Climate change had seen a major loss of rainfall, felt all the more keenly with vast temporary residency. Extreme shortages had accompanied two years of drought prior to the Cup. Diminished water quality increased disease (Forero-Ortiz et al. 2020; Benítez-Cano et al. 2024).

Cruise ships were once rarely seen during the event. The grateful burghers of Fremantle, more used to the brutal export of live sheep, were delighted by the announcement that five cruise vessels would turn up for the 1987 regatta ("Fremantle Prepares" 1985). By contrast, Barcelona is occupied by more cruise ships than anywhere else in Europe (Perdiguero and Sanz 2020). The industry is growing, from 13 million passengers worldwide in 2004 to 32 million twenty years later—all the while undertaking, of course, to 'protect and preserve our planet ... advancing an ambitious sustainability agenda ... sailing to a better future' (Cruise Lines International Association 2024: 3, 13, 26).

The International Council on Clean Transformation explains that 'even the most efficient cruise ships emit more CO_2 per passenger kilometer (CO_2/pax-km) than a passenger jet. The industry's overall environmental toll is grievous, with appalling implications for our fellow-animals' lifeworld, especially cetacean feeding and communication (Rolland et al.

2012; Tsujii et al. 2018; Williams et al. 2022), per the devastating impact of the 2021 regatta (Pine et al. 2024). The ships are also problematic in terms of their sewage treatment (Friends of the Earth 2024), responsible for a quarter of the world's oceanic waste. The industry falls quiet about emitting more sulfur oxides each year than a billion automobiles (Dijkstra and Simon 2023).

Cruise ships pollute Barcelona more than any other destination (Dijkstra and Simon 2023: 3). By the time of the Cup, 'más de 520 los cruceros … han atacado en la ciudad con 2.372.000 pasajeros (110.000 más que a finales de septiembre de 2023)' [over 520 ships had docked in the city, with 2,372,000 passengers (110,000 more than the same period in 2023)], far beyond the 1995 figure of 233,389 (Friends of the Earth 2024). The port now has seven terminals dedicated to these craft (Goodman 2025). Their polluting work done, the ships generally leave Barcelona within twenty-four hours of arrival, adding little to the local economy (Congostrina 2024b).

As noted earlier, Explora sponsored "Emirates Team New Zealand." The firm is part of the huge container-ship conglomerate MSC, Europe's dominant shipping polluter. MSC rates poorly on emissions and has a history of dumping garbage overboard and using problematic fuel (Transport & Environment 2024; Ship It Zero 2024; Friends of the Earth 2024; Clark and Wittels 2023). But don't let that worry you. Explora has promised its multi-billion-dollar fleet will 'redefine the ocean experience for a new generation of discerning travellers … a state characterised by calmness, unity, elegance and joy, encouraging inner and outer exploration and discovery' (Williamson 2024).

Barcelona's Cup waters were also laden with 'high sails of the moored mega yachts owned by all the millionaires and billionaires who had flown in for the occasion' (Bull 2024)—243 in all. Such craft have doubled in number this century. They are key sites of super-pollution by the world's oligarchy. On standby, each one generates about 7000 tons of CO_2 a year through propulsion, pools, helicopters, air conditioning, and workers (Oxfam International 2024: 10).

Conclusion

While organizers speak of a putative 'paradigm shift' in economic growth that comes with hosting the Cup, skeptics deride this as a mere 'drop in the ocean' next to the adverse effects local people encounter through gentrification (Macbeth et al. 2012). But that doesn't stop functionaries and chorines. The 2024 farce had barely concluded when the victor 'provided insight into the early stages of a vision of continued growth' of 'one of the unique properties in international sport.' Barcelona was served notice when the New Zealanders/Aotearoans announced plans to 'super charge the 38th [and next] America's Cup with increased global footprint, more teams and increased overall expansion opportunities,' noting that it 'would be difficult to expand the number of teams based on the available infrastructure space in Barcelona' ("America's Cup: Dalton" 2024). The message was clear: this is private, this is property, this is hidden public subvention, this is clandestine environmental despoliation; in short, the inevitable—and often overt—dislocation of lives. It meant invitations for more taxpayers' money around the world to pay for destructive oligarchic pleasures.

Grant Dalton, who ran the 2024 regatta for the "Emirates Team New Zealand," insisted that this 'most prestigious trophy in our sport' not 'disappear into a mire of commercialism' (quoted in Bell with Agini 2024). Per desired changes in the class composition of local tourism, and the profile of luxury-goods sponsors, his was no call against commodification per se, but a desire for growth twinned with the maintenance of an élite oligarchic pastime; as *SuperyachtNews* charmingly put it, 'a purist's sport' ("America's Cup Update" 2024). After all, he labeled Barcelona's mayor Jaume Collboni and his administration communist (Soler 2025). Needless to say, Dalton had rejected his own country as the 2024 host because the state had only promised $100 million in support ("America's Cup Line" 2021).

Sporting bureaucrats decreed the thirty-seventh Cup an unmitigated triumph for Catalunya, claiming it had 'servido para impulsar proyectos de descarbonización náutica; ha transformado urbanísticamente el litoral de Barcelona; ha democratizado la vela entre los más jóvenes' [stimulated nautical decarbonization, transformed Barcelona's urban beaches, and democratized sailing to include young people] (Buesa 2024). No evidence was provided for these grandiose claims. A trio of reports from the Universitat de Barcelona was predictably sanguine, with critical voices essentially absent; nor were the methods up to date (Suriñach and Vayá 2025a, b, c; Quiggin 2023).

Despite declaring the event an indubitable success, the municipal authorities immediately announced the city would not bid for the next absurdity—any attempt would have to be privately financed, given the expense and 2024's disappointing shortfall in visitors. Barcelona's capitalist class pressured the government to reconsider the decision (Suñe 2024). But fear not—the city had its eyes set on hosting golf's Ryder Cup and a leg of the Tour de France (Mercader 2024; Carnicé 2024b).

Meanwhile, València, incredibly, looked to prepare a new bid to host the Cup, at a projected cost of €200 million, echoing its profligacy of 2007 and 2010 (Lidón and Atienza 2024; Ferrando 2024; Arufe 2025). That looked like an extension of 'El modelo de la Comunidad Valenciana … un decálogo extractivista de "toma el dinero y corre", aún a costa de destruir absolutamente el interés de esta ciudad y de su litoral' [the València Community model … the ten extractivist commandments of "take the money and run," to the cost of the city and its beaches] (Burgaya 2024). It shared what by now should be a failed fantasy in full retirement: 'la promoción del turismo náutico por su "amplia transversalidad" con otros sectores' [the promotion of nautical tourism because of its impact across the economy] ("La Generalitat" 2024). This was "Déjà vu 2007" from reactionary forces. Carlos Mazón, President of the Generalitat, tried to outmaneuver any leftist opposition to such an extravagance: 'Que no pongan palos en las ruedas, que se callen y nos dejen trabajar' [Don't put obstacles in our way; just shut up and let us get on with the job] (Aguar 2024).

Why? Because this kind of sailing is 'an elite sport [used] as a gentrifying development plan' (Rius-Ulldemolins et al. 2021: 40), repeatedly invoked—and celebrated as such—by lustful developers and their obedient servants, less so by everyday citizens (Veitch 2013), as we shall see in Chap. 6. Ultimately, Napoli won the dubious distinction of hosting the next regatta. The decision drew the immediate wrath of US and UK teams for a lack of transparency over the deal struck between Royal New Zealand Yacht Club and the Italian state, apart from a €5 million entry fee (Wilson 2025; Gladwell 2025). Needless to say, the event would be 'a symbol of innovation, pride and opportunity for southern Italy' because 'in true Cup fashion, the message was as much about legacy as it was about racing' (Chisholm 2025). This would supposedly secure 'Italy's maritime heritage, soft-power credentials and strategic influence in the Mediterranean' and €690 million for Neapolitans ("Naples to Host" 2025). Perhaps, but the 'legacy' would undoubtedly mean riches for the wealthy, displacement of the poor, and public subvention from the shadows.

The *Independent* welcomed the thirty-seventh event as 'a surreal reimagining of Formula 1, with shades of Star Wars, set on the high seas'.

The *Economist* proclaimed that "The America's Cup is Becoming Formula One on the Water," more a 'high-octane rush' than 'a ponderous regatta' (2017). They were using apt figures of speech: the Port de Barcelona's proud point of comparison to the carbon event's new technologies was Formula One ("The Benefits" 2024) and the Mercedes team assisted INEOS, which said 'the America's Cup is like F1 on water' (quoted in McRea 2024)—ironically, in the very year that Formula One announced Madrid would be the new host of Spain's second Grand Prix (Richards 2024). Red Bull and Alpine were other racing teams affiliated with Cup contestants (Bell with Angini 2024), in the hope of 'emulating or even surpassing the levels of Formula 1 auto racing in terms of technology, style and frankly, largesse' (Cesare 2024).

We almost forgot, a few weeks after the absurd regatta, that great boating philanthropist Arnault was 'entré à l'Académie des sciences morales et politiques parmi les «immortels»' [included in the Academy of Moral and Political Sciences as one of the "immortals"] (Blin 2024). Who could possibly argue against his elevation? It would be churlish, no?

The carbon America's Cup is unsustainable in every sense—for nature, for residents of Catalunya, for everyone else. Like all mega-events that waste public resources and turn the globe into a polluting playground for affluent people with too much time and too many superyachts on their hands, this gift of labor should be dedicated to more legitimate ends. Economy, society, and environment demand and deserve nothing less.

Bibliography

NB—Links from US Science Have Been Removed By Its Government

"37th America's Cup Was Amazingly Boring." (2024, December 1). *Scuttlebutt Sailing News* https://www.sailingscuttlebutt.com/2024/12/01/37th-americas-cup-was-amazingly-boring/.

"America's Cup Line in the Sand." (2021, June 19). *Otago Daily Times* https://www.odt.co.nz/opinion/editorial/america%E2%80%99s-cup-line-sand.

"America's Cup Update." (2024, October 7). *SuperyachtNews* https://www.superyachtnews.com/operations/americas-cup-update-.

"Barcelona: manifestantes piden quitar estatua de Colón." (2020, June 14). *DW* https://www.dw.com/es/barcelona-manifestantes-piden-quitar-estatua-de-col%C3%B3n/a-53805920.

"Britain's Most Reliable Cars 2024 (Plus the Ones That'll Let You Down)." (2024, November 11). *What Car?* https://www.whatcar.com/news/reliability-survey-most-reliable-cars/n26137.

"Copa América: Barcelona espera 2,5 millones de turistas y un impacto de más de mil millones." (2024, April 2). *Palco23* https://www.palco23.com/competiciones/copa-america-barcelona-espera-25-millones-de-turistas-y-un-impacto-de-mas-de-mil-millones.

"Dossier L'affaire Squarcini." (2024, November 29). *Mediapart* https://www.mediapart.fr/journal/france/dossier/l-affaire-squarcini.

"El Port de Barcelona impulsa la innovación en el mundo portuario." (2024, August 1). *Público* https://www.publico.es/sociedad/port-barcelona-impulsa-innovacion-mundo-portuario.html#md=modulo-portada-fi.

"Five Years of Bad Decisions in Spain's Rental Market." (2024, October 29). *Idealista/news* https://www.idealista.com/en/news/property-for-rent-in-spain/2024/10/29/820581-five-years-of-bad-decisions-in-spain-s-rental-market.

"Fremantle Prepares for Invasion: Australians Earmark $960 Million in Improvements for the 1987 America's Cup Race." (1985, May 5). *Los Angeles Times* https://www.latimes.com/archives/la-xpm-1985-05-05-sp-8715-story.html.

"INEOS Wants to Drill UK Fracking Test Site in Attempt to Show it is Safe." (2022, April 11). *Guardian* https://www.theguardian.com/environment/2022/apr/11/ineos-wants-to-drill-uk-fracking-test-site-in-attempt-to-show-it-is-safe-jim-ratcliffe#:~:text=2%20years%20old-,Ineos%20wants%20to%20drill%20UK%20fracking%20test%20site,to%20show%20it%20is%20safe&text=The%20chemicals%20and%20energy%20company,wrestles%20with%20high%20energy%20prices.

"La Copa América reconecta de nuevo a los barceloneses con el mar." (2024, August 21). *Mundo Deportivo* https://www.mundodeportivo.com/vela/20240821/1002301649/copa-america-reconecta-nuevo-barceloneses-mar.html.

"La Generalitat refuerza la promoción del turismo náutico por su "amplia transversalidad" con otros sectores." (2024, October 23). *ValenciaPlaza* https://valenciaplaza.com/generalitat-refuerza-promocion-turismo-nautico-amplia-transversalidad-otros-sectores.

"La iniciativa del Ayuntamiento de Barcelona 'Viu la Vela', gran legado de la Copa América." (2024, September 25). *Mundo Deportivo* https://www.mundodeportivo.com/vela/20240925/1002321946/iniciativa-ayuntamiento-barcelona-viu-vela-gran-legado-copa-america.html.

"La plataforma No a la Copa Amèrica anuncia les primeres mobilitzacions." (2023, September 6). *Fosbury* https://fosbury.cat/fosbudiary/vela/la-plataforma-no-a-la-copa-america-anuncia-les-primeres-mobilitzacions/.

"Los japoneses, los mejores turistas. Los peores, los británicos." (2014, August 26). *CapitalMadrid* https://www.capitalmadrid.com/2014/8/26/35025/los-japoneses-los-mejores-turistas-los-peores-los-britanicos.html.

"LVMH Chief Executive Testifies in Trial of Former French Spy Agency Chief." (2024, November 28). *Guardian* https://www.theguardian.com/world/2024/nov/28/lvmh-chief-executive-testifies-trial-french-spy-agency-arnault-squarcini?CMP=Share_iOSApp_Other.

"LVMH est partenaire premium des jeux olympiques de Paris 2024." (2024). LVMH https://www.lvmh.com/fr/lvmh-et-les-jeux-olympiques-et-paralympiques-de-paris-2024.

"Multimilionaris a tota vela." (2024, September 9). *Directa* https://directa.cat/multimilionaris-a-tota-vela/.

"Naples to Host America's Cup 2027 in Strategic Soft-Power Move." (2025, May 15). *Decode39* https://decode39.com/10787/naples-to-host-americas-cup-2027-in-strategic-soft-power-move/.

"One of Every Four People in Barcelona is Now a Foreign National." (2024, June 11). *Barcelona International Welcome* https://www.barcelona.cat/internationalwelcome/en/news/one-of-every-four-people-in-barcelona-is-now-a-foreign-national-1408362.

"Prada Blackface Scandal Results in Major Settlement." (2020, February 5). Center for Constitutional Rights https://ccrjustice.org/home/press-center/press-releases/prada-blackface-scandal-results-major-settlement.

"Procès "Squarcini" : la responsabilité du fondateur de LVMH doit être clairement établie quant à l'espionnage de la rédaction de *Fakir*." (2024, November 27). Reporters sans Frontières https://rsf.org/fr/proc%C3%A8s-squarcini-la-responsabilit%C3%A9-du-fondateur-de-lvmh-doit-%C3%AAtre-clairement-%C3%A9tablie-quant-%C3%A0-l

"Shoes." (n.d.). Prada https://www.prada.com/eu/en/mens/shoes/americas-cup/c/10297EU.

"The Benefits of the America's Cup." (2024). *Port de Barcelona* https://www.portdebarcelona.cat/en/port-vell/americas-cup-barcelona-2024/benefits-americas-cup.

"The Timekeeper's Watch." (n.d.). Omega https://www.omegawatches.com/watches/seamaster/diver-300-m/america-s-cup-edition/product.

Abzouz, Yunnes and Khedidja Zerouali. (2024, September 18). "Liberté de la presse : LVMH voit rouge et met Mediapart sur liste noire." *Mediapart* https://www.mediapart.fr/journal/economie-et-social/180924/liberte-de-la-presse-lvmh-voit-rouge-et-met-mediapart-sur-liste-noire.

Aguar, Ximo. (2024, October 29). "Déjà vu 2007: Mazón redobla la presión contra el PSPV y el Gobierno por la Copa América." *ValenciaPlaza* https://valenciaplaza.com/deja-vu-2007-mazon-redobla-la-presion-contra-el-pspv-y-el-gobierno-por-la-copa-america.

Ahmed, Muriad and Michael Pooler. (2020, March 15). "INEOS: Why Jim Ratcliffe is Mixing Petrochemicals and Sports." *Financial Times* https://www.ft.com/content/31b5e7e0-5c76-11ea-b0ab-339c2307bcd4.

Alcalá, César. (2024, July 15). "1.200 millones y 19.000 puestos de trabajo: los beneficios para Barcelona de acoger la Copa América." *El Debate* https://www.eldebate.com/espana/cataluna/barcelona/20240715/1200-millones-19000-puestos-trabajo-beneficios-barcelona-acoger-copa-america212766.html.

Armstrong, Chris. (2022). *A Blue New Deal: Why We Need a New Politics for the Ocean*. New Haven: Yale University Press.

Arufe, Fernando. (2025, April 20). "Valencia tiene infraestructuras, experiencia y empresas preparadas para recuperar la America's Cup." *Economía Digital* https://www.economiadigital.es/valencia/economia/valencia-infraestructuras-empresas-recuperar-americas-cup.html.

Ayuso, Glòria. (2024, September 25). "Barcelona celebra la gran fiesta de la vela con una marea de jóvenes navegantes." *El Periódico* https://www.elperiodico.com/es/blau/copa-america/20240925/fiesta-viu-la-vela-barcelona-deportes-acuaticos-108562226.

Bachelor, Lois and Clea Skopoleti. (2024, November 26). "Selfies and Surf Simulators: The Young Cruisers Driving Boom in Sea Holidays." *Guardian* https://www.theguardian.com/environment/2024/nov/26/selfies-and-surf-simulators-the-young-cruisers-driving-boom-in-sea-holidays.

Baraniuk, Chris. (2023, September 30). "Lego is a Company Haunted by its Own Plastic." *Wired* https://www.wired.com/story/lego-haunted-by-its-own-plastic/.

Barcelona FC. (n.d.). https://www.fcbarcelona.cat/ca/club/sostenibilitat.

Barcelona Capital Nàutica. (2024, March). *Por la vela y por el mar: Memoria de actividades 2023* https://www.barcelonacapitalnautica.org/wp-content/uploads/2024/03/FBCNMemoria2023-CAST.pdf.

Barrass, Tony. (2015, June 9). "WA's Cup Ran Over with Champagne." *West Australian* https://thewest.com.au/news/australia/was-cup-ran-over-with-champagne-ng-ya-118643.

Bell, Amy with Samuel Agini. (2024, December 3). "America's Cup Fights to Boost Audience While Preserving Heritage." *Financial Times* https://www.ft.com/content/e6848cc8-0c10-429e-86ec-2ae4bd4e4773.

Benítez-Cano, Daniela, Patricia González-Marín, Anna Gómez-Gutiérrez, Marc Marí-Dell'Olmo, and Laura Oliveras. (2024). "Association of Drought Conditions and Heavy Rainfalls with the Quality of Drinking Water in Barcelona (2010–2022)." *Journal of Exposure Science & Environmental Epidemiology* 34: 175–83.

Bennett, Nathan James, Jessica Blythe, Carole Sandrine White, and Cecilia Campero. (2021). "Blue Growth and Blue Justice: Ten Risks and Solutions for the Ocean Economy." *Marine Policy* 125 https://www.sciencedirect.com/science/article/abs/pii/S0308597X20310381.

Black, David. (2020). "Sailing." *Routledge Handbook of Global Sport*. Ed. John Nauright and Sara Zipp. Abingdon: Routledge. 312–19.

Blin, Simon. (2024, December 2). "Bernard Arnault élu à l'Académie: le patron qui veut être partout, tout le temps, éternellement." *Libération* https://www.liberation.fr/idees-et-debats/opinions/bernard-arnault-elu-a-lacademie-le-patron-qui-veut-etre-partout-tout-le-temps-eternellement-20241202_LI6M6QQJD5AR7DXN22JQDUSLN4/.

Borja, Jordi with Majda Drnda, Mariela Iglesias, Mirela Fiori, and Zaida Muxí. (2003). *La ciudad conquistada*. Madrid: Alianza Editorial.

Boutry, Timothée. (2024, November 28). "Procès Squarcini : « François Ruffin est bien meilleur sur le plan cinématographique que politique », réagit Bernard Arnault." *Les Parisiens* https://www.leparisien.fr/faits-divers/proces-squarcini-francois-ruffin-est-bien-meilleur-sur-le-plan-cinematographique-que-politique-reagit-bernard-arnault-28-11-2024-3KNLKOAWKBFALPYFAB6ZBX4F5E.php.

Buesa, Cristina. (2024, October 25). "El deporte catalán también pide "reflexionar" sobre la decisión de Barcelona de no repetir la Copa América de vela." *El Periódico* https://www.elperiodico.com/es/blau/copa-america/20241025/ufec-deporte-catalan-pide-reflexionar-decision-no-repetir-copa-america-de-vela-110411271.

Bull, Andy. (2024, October 20). "New Zealand Win America's Cup as Britain's Challenge Falls Short." *Guardian* https://www.theguardian.com/sport/2024/oct/19/new-zealand-win-americas-cup-britain.

Burgaya, Josep. (2024, October 26). "Copa América en Barcelona: el engaño, solo una vez." *Metropoli* https://metropoliabierta.elespanol.com/opinion/20241026/copa-america-en-barcelona-el-engano-solo-una-vez/896540341_13.html.

Burgen, Stephen. (2024, August 26). "'It's About Money': Doubt Over America's Cup Benefits in Tourist Hotspot Barcelona." *Guardian* https://www.theguardian.com/world/article/2024/aug/26/barcelona-americas-cup-overtourism-super-rich-local-economy.

Burgen, Stephen. (2025, May 20). "Thinking of a Trip to Barcelona This Summer? Beware—Here's What You'll Find." *Guardian* https://www.theguardian.com/commentisfree/2025/may/20/barcelona-summer-tourism-locals-city.

Cañada, Ernest, Carla Izcara, Bastien Montovert, and Mariciana Zorz. (2024). *Proposals for the Design of Public Policies on Popular Tourism* Alba Sud Policy Brief #3 https://www.albasud.org/publ/docs/135.en.pdf.

Canela, Joan. (2022, February 28). "La Copa América no volverá a Valècia tras la negativa de Ayuntamiento y Generalitat de pagar el evento." *Público* https://www.publico.es/deportes/copa-america-copa-america-no-volvera-valencia-negativa-ayuntamiento-generalitat-pagar-evento.html.

Carnicé, Alba. (2024a, July 21). "Barcelona permite a la Fundación de la Copa América pedir créditos de hasta dos millones de euros." *Metropoli* https://metropoliabierta.elespanol.com/economia/20240721/barcelona-permite-la-fundacion-de-copa-america-pedir-creditos-hasta-dos-millones-euros/8696631060.html.

Carnicé, Alba. (2024b, October 23). "Barcelona también descarta repetir la Copa América en 2026: "Es económicamente insostenible"." *Metropoli* https://metropoliabierta.elespanol.com/informacion-municipal/20241023/barcelona-tambien-descarta-repetir-la-copa-america-en-es-economicamente-insostenible/895660434_0.html.

Cesare, Benjamin. (2024, November–December). "The Amazing America's Cup." *WindCheck* https://www.windcheckmagazine.com/article/the-amazing-americas-cup/.

Chisholm, Justin. (2025, May 27). "Italian Prime Minister Giorgia Meloni Welcomes America's Cup in Rome." *Yacht Racing Life* https://www.yachtracinglife.com/italian-prime-minister-giorgia-meloni-welcomes-americas-cup-in-rome/?ref=yacht-racing-life-daily-newsletter.

Clark, Aaron and Jack Wittels. (2023, July 31). "The Invisible Climate Impact of a Cruise Ship." *Bloomberg* https://www.bloomberg.com/news/articles/2023-07-31/cruise-ship-methane-leak-shows-hidden-climate-impact.

ClientEarth. (2024, September 17). "Environmental Groups are Once Again Taking INEOS to Court." https://www.clientearth.org/latest/press-office/press/environmental-groups-are-once-again-taking-ineos-to-court/.

Coll, Bernat. (2024, July 11). "Copa del América de vela: Los espías toman la costa de Barcelona." *El País* https://elpais.com/espana/catalunya/2024-07-11/copa-del-america-de-vela-los-espias-toman-la-costa-de-barcelona.html.

Congostrina, Alfonso L. (2024a, July 16). "Barcelona argumenta que no puede informar sobre los ingresos por el desfile de Louis Vuitton por cuestiones de "confidencialidad"." *El País* https://elpais.com/espana/catalunya/2024-07-16/barcelona-se-escuda-en-el-secreto-empresarial-para-no-revelar-cuanto-cobro-de-louis-vuitton-por-el-desfile-en-el-park-guell.html.

Congostrina, Alfonso L. (2024b, September 30). "Dos de cada tres cruceros que atracan en Barcelona permanecen menos de doce horas en la ciudad." *El País* https://elpais.com/espana/catalunya/2024-09-30/dos-de-cada-tres-cruceros-que-atracan-en-barcelona-permanecen-menos-de-doce-horas-en-la-ciudad.html.

Connell, R. W. (1984). "Class Formation on a World Scale." *Review (Fernand Braudel Center)* 7, no. 3: 407–40.

Coopersmith, Jonathan. (2015). *Faxed: The Rise and Fall of the Fax Machine.* Baltimore: Johns Hopkins University Press.

Cruise Lines International. (2024, May). *State of the Cruise Industry Report* https://cruising.org/-/media/clia-media/research/2024/2024-state-of-the-cruise-industry-report_updated-050824_web.ashx.

de Senneville, Valérie. (2024, November 28). "Procès Squarcini : François Ruffin « est bien meilleur cinéaste que politique », estime Bernard Arnault." *Les Echos* https://www.lesechos.fr/politique-societe/societe/proces-squarcini-francois-ruffin-est-bien-meilleur-cineaste-que-politique-estime-bernard-arnault-2134679.

de Villepin, Paul with Pauline de Saint Remy. (2023, August 25). "World's Richest Man lobbied Macron to Ditch Prime Minister." *Politico* https://www.politico.eu/article/emmanuel-macron-bernard-arnault-elisabeth-borne-gerald-darmanin-meet-the-worlds-richest-man-favorite-politician/.

Díaz-Pont, Joana. (2023). "Visibilising the Climate in Social Policies in Barcelona: Connections in the Urban Context." *Critical Social Policy* 43, no. 1: 119–39.

Dijkstra, Constance and Valentin Simon. (2023). *The Return of the Cruise: How Luxury Cruises Are Polluting Europe's Cities.* Transport & Environment https://www.transportenvironment.org/uploads/files/2023-Cruise-ship-study_2024-05-07-143832_lapq.pdf.

Douglas, Susan J. (1987). *Inventing American Broadcasting 1899–1922.* Baltimore: The Johns Hopkins University Press.

Elder, Bryce. (2025, April 8). "Yacht Envy: A Visit to England's Half-a-Billion Boatyard." *Financial Times* https://www.ft.com/content/aba8786e-7d9c-4270-9e56-61945995e62b?emailId=bb331970-754c-44a4-9404-832459231fda&segmentId=9264b0f7-e7ac-8f9b-044f-c10729049333.

Estirado, Laura. (2024, August 29). "Felipe VI inaugural la Copa América 2024: "Barcelona revive el espíritu de los JJOO del 92"." *El Periódico* https://www.elperiodico.com/es/estilo-y-famosos/20240829/rey-felipe-bienvenida-copa-america-vela-museu-maritim-salvador-illa-jaume-collboni-107491405.

Faye, Olivier. (2023, September 28). "Les interrogations de Tracfin sur une transaction entre des sociétés de l'oligarque russe Nikolaï Sarkisov et de Bernard Arnault." *Le Monde* https://www.lemonde.fr/societe/article/2023/09/28/les-interrogations-de-tracfin-sur-une-transaction-entre-des-societes-de-l-oligarque-russe-nikolai-sarkisov-et-de-bernard-arnault_6191354_3224.html.

Ferrando, Ramón. (2024, October 24). "Samaranch: "Es una buena idea invertir en la Copa del América en València"." *El Periódico* https://www.elperiodico.com/es/blau/copa-america/20241024/samaranch-buena-idea-invertir-copa-america-vela-110311523.

Food and Water Europe. (2017, November). *INEOS' Chequered Environmental Track Record in Europe* https://www.foodandwatereurope.org/wp-content/uploads/2017/11/FoodandWaterEuropeReport IneosChequeredRecordinEuropeNov2017.pdf.

Forero-Ortiz, Edwar, Eduardo Martínez-Gomariz, and Robert Monjo. (2020). "Climate Change Implications for Water Availability: A Case Study of Barcelona City." *Sustainability* 12, no. 5 https://doi.org/10.3390/su12051779.

Fresh Info. (2021). *36th America's Cup Impact Evaluation: Final Report* https://www.majorevents.govt.nz/dmsdocument/15674-36th-americas-cup-impact-evaluation-report.

Friends of the Earth. (2024, November). *2024 Cruise Ship Report Card* https://foe.org/wp-content/uploads/2024/11/CruiseShipReportCard_2024_final.pdf.

Gaynor, Jennifer L. (2013). "Ages of Sail, Ocean Basins, and Southeast Asia." *Journal of World History* 24, no. 2: 309–33.

Gladwell, Richard. (2025, May 22). "America's Cup: The Blitz of Statements Answered with Protocol Release." *Sail-World* https://www.sail-world.com/news/286349/Americas-Cup-The-Blitz-of-Statements.

Global Solo Challenge. (2022, May 7). "Why Do People Become so Passionate About Ocean Sailing?" *Sail-World* https://www.sail-world.com/news/248757/What-is-the-passion-for-ocean-sailing.

Goodman, Al. (2025, March 22). "Barcelona Finally Turned on its Crowds of Tourists. Now the City Faces a Major Problem." *CNN* https://edition.cnn.com/travel/barcelona-reckons-with-overtourism-summer-2025?cid=external-feeds_iluminar_flipboard.

Guitton-Boussion, Justine. (2024, August 7). "LVMH, Decathlon : ces sponsors des JO qui maltraitent les animaux." *Reporterre* https://reporterre.net/LVMH-Decathlon-ces-sponsors-des-JO-qui-maltraitent-les-animaux.

Harvey, Fiona. (2020, September 10). "Toxic Pesticides Banned for EU Use Exported from UK." *Guardian* https://www.theguardian.com/environment/2020/sep/10/toxic-pesticides-banned-for-eu-use-exported-from-uk.

Instituto Nacional de Estadística. (2025, February 3). *Estadística de Movimientos Turísticos en Fronteras* https://www.ine.es/dyngs/Prensa/FRONTUR1224.htm.

Jabkhiro, Juliette and Mimosa Spencer. (2025, March 7). "France's Former Spy Chief Found Guilty of Using Public Resources to Aid LVMH." *Reuters* https://www.reuters.com/world/europe/frances-former-spy-chief-found-guilty-using-public-resources-aid-lvmh-2025-03-07/.

Jacopin, Tanguy. (2006–07). "America's Cup: Value Creation and Stakeholder Management." *Finance & The Common Good/Bien Commun* 26: 64–70.

John, Alistair and Steve J. Jackson. (2011). "Call Me Loyal: Globalization, Corporate Nationalism and the America's Cup." *International Review for the Sociology of Sport* 46, no. 4: 399–417.

Jolly, Jasper. (2023, September 29). "Transactions Involving Bernard Arnault Investigated Over Suspected Money Laundering." *Guardian* https://www.theguardian.com/business/2023/sep/29/bernard-arnault-lvmh-nikolai-sarkisov.

Jones, Simon. (2015, August 11). "America's Cup Dispute Set for Court." *Royal Gazette* https://www.royalgazette.com/event/news/article/20150810/americas-cup-dispute-set-for-court/.

Kidd, Patrick. (2017, July 26). "America's Cup: The Rising Cost of Sailing's Ultimate Prize." *Boat International* https://www.boatinternational.com/yachts/americas-cup/americas-cup-the-rising-cost-of-sailings-ultimate-prize%2D%2D31985.

Lassalle, Frederic. (2014). "America's Cup 2007: Sailing, Glitter and Power." *Sport, Business and Management: An International Journal* 4, no. 1: 38–51.

Lazard, Violette. (2024, November 28). "« J'ai fait plus pour la France que François Ruffin ! », au procès Squarcini, le témoignage du milliardaire Bernard Arnault." *Le Nouvel Obs* https://www.nouvelobs.com/justice/20241128.OBS97111/j-ai-fait-plus-pour-la-france-que-francois-ruffin-au-proces-squarcini-le-temoignage-du-milliardaire-bernard-arnault.html.

Lecardane, Renzo. (2008). "Le front de mer de Valencia et l'héritage de l'America's Cup 2007." *Méditerranée* 111: 109–14.

Le Vaillant, Luc and Service Médias. (2012, September 10). "Le «Riche con !» provoque du buzz et une plainte." *Libération* https://www.liberation.fr/medias/2012/09/10/le-riche-con-provoque-du-buzz-et-une-plainte_845366/.

Lidón, Inma and Héctor Atienza. (2024, October 22). "Los 90 días que tiene Valencia para traer la Copa América de nuevo "a casa"." *El Mundo* https://www.elmundo.es/comunidad-valenciana/2024/10/22/6717b56be4d4d8e5028b4591.html.

López, Estela. (2022, March 29). "Barcelona capta la Copa América de Vela 2024 con 70 millones." *El Economista* https://www.eleconomista.es/catalunya/noticias/11692411/03/22/Barcelona-capta-la-Copa-America-de-Vela-2024-con-70-millones.html.

Louis Vuitton 37th America's Cup. (2024a). *Press Kit* https://www.americascup.com/press-kit.

Louis Vuitton 37th America's Cup. (2024b, October 22). "Moltes Gràcies Barcelona—The Jewel in the Sun." https://www.americascup.com/news/3759_MOLTES-GRACIES-BARCELONA-THE-JEWEL-IN-THE-SUN.

Lynch, Declan. (2021, August 1). "Our Ruling Class is at Ease with Itself and Just Sailing By." *Irish Independent* https://www.independent.ie/opinion/comment/our-ruling-class-is-at-ease-with-itself-and-just-sailing-by/40707762.html.

Macbeth, Jim, John Selwood, and Sarah Veitch. (2012). "Paradigm Shift or a Drop in the Ocean? The America's Cup Impact on Fremantle." *Tourism Geographies: An International Journal of Tourism Space, Place and Environment* 14, no. 1: 162–82.

Marmot, Michael G. (2004). "Evidence Based Policy or Policy Based Evidence?" *British Medical Journal* 328, no. 7445: 906–07.

Martin, Daniel, Jason Macarlino, Felix Joensson, and Kate Hislop. (2017, June 15). "How Did the America's Cup Instigate Perth's Green Sand Dunes?" *Foreground* https://www.foreground.com.au/planning-policy/perth-and-americas-cup/.

Martínez, David García. (2024, May 7). "El ABC de la America's Cup: Origen, equipos, formato e impacto económico." *Palco23* https://www.palco23.com/competiciones/americas-cup-la-competicion-que-se-reta-en-los-mares-desde-1851.

Mas de Xaxàs, Xavier. (2024, August 19). "My Barcelona is Being Destroyed by Mass Tourism—But Kicking Visitors Out Isn't the Answer." *Guardian* https://www.theguardian.com/commentisfree/article/2024/aug/19/barcelona-mass-tourism-visitors-city-industry?CMP=ShareiOSAppOther.

McHoul, Alec. (2015). "The Humdrum." *The Routledge Companion to Global Popular Culture*. Ed. Toby Miller. New York: Routledge. 112–18.

McKay, Jim and Debbie Huber. (1992). "Anchoring Media Images of Technology and Sport." *Women's Studies International Forum* 15, no. 2: 205–18.

McRae, Donald. (2024, September 27). "Ben Ainslie: 'We Were Having a Tussle and Then I Saw a Guy Pull a Knife Out'." *Guardian* https://www.theguardian.com/sport/2024/sep/27/ben-ainslie-sailing-americas-cup-interview.

Meager, Dani and WA Maritime Museum. (2023, November 5). "The Transformative Impact of the America's Cup on Fremantle." *Perth Now* https://www.perthnow.com.au/lifestyle/culture/the-transformative-impact-of-the-americas-cup-on-fremantle%2D%2Dc-12455806.

Mercader, Aleix. (2024, October 24). "Barcelona Global solicita una reunión "urgente" para intentar salvar la Copa América." *El Economista* https://www.eleconomista.es/actualidad/noticias/13050484/10/24/barcelona-global-solicita-una-reunion-urgente-para-intentar-salvar-la-copa-america.html.

Ministerio de Trabajo y Economía Social. (2024, July 24). *Real Decreto 712/2024, de 23 de julio, por el que se desarrollan las bonificaciones de cuotas aplicables a las actividades relacionadas con la organización y celebración de la XXXVII edición de la Copa América en la ciudad de Barcelona* https://www.boe.es/diario_boe/txt.php?id=BOE-A-2024-15206.

Mitchell, Carleton. (1958, May 12). "Defenders of the Cup." *Sports Illustrated* https://vault.si.com/vault/1958/05/12/defenders-of-the-cup.

Mullen, Tom. (2023, September 19). "Why Crowdsourcing Industrial Espionage is Economical for America's Cup." *Forbes* https://www.forbes.com/sites/tmullen/2023/09/19/why-crowdsourcing-industrial-espionage-is-economical-for-americas-cup/.

Noëth, Bart. (2024, April 21). "Emirates Airline's Staff Speak Out (Anonymously) After UAE Records Heaviest Rainfall in 75 Years." *Aviation24* https://www.aviation24.be/airlines/emirates-airline/staff-speak-out-anonymously-after-uae-records-heaviest-rainfall-in-75-years/.

Oficina Municipal de Dades and Gerència Municipal. (2024). *Els salaris a Barcelona. 2022* https://bcnroc.ajuntament.barcelona.cat/jspui/handle/11703/134076.

O'Reilly, Karen. (2000). *The British on the Costa Del Sol.* London: Routledge.

Organisation for Economic Co-operation and Development. (2016). *The Ocean Economy in 2030* https://www.oecd.org/en/publications/the-ocean-economy-in-20309789264251724-en.html.

Organisation for Economic Co-operation and Development. (2018). *Working Together for Local Integration of Migrants and Refugees in Barcelona* https://doi.org/10.1787/9789264304062-en.

Oxfam International. (2024). *Carbon Inequality Kills: Why Curbing the Excessive Emissions of an Elite Few Can Create a Sustainable Planet for All* https://oxfamilibrary.openrepository.com/bitstream/handle/10546/621656/bp-carbon-inequality-kills-281024-en.pdf?sequence=1.

Parodi, Emilio. (2024, June 11). "Exclusive: Italian Prosecutors Probing Supply Chain of Around a Dozen Fashion Brands." *Reuters* https://www.reuters.com/markets/europe/italian-prosecutors-probing-supply-chain-around-dozen-more-fashion-brands-source-2024-06-11/.

Perdiguero, Jordi and Alex Sanz. (2020). "Cruise Activity and Pollution: The Case of Barcelona." *Transportation Research Part D* 78 https://doi.org/10.1016/j.trd.2019.11.010.

Pine, Matthew K., Emily Hague, Anna Kebke, Lauren McWhinnie, and Charlotte R. Findlay. (2024). "Not so Silent Spectators: How Spectator Vessels at International Sailing Regattas Alter Marine Soundscapes." *Marine Pollution Bulletin* 202 https://doi.org/10.1016/j.marpolbul.2024.116309.

Pocréaux, Cyril. (2024a, November 21). "Les barbouzeries de Bernard Arnault : en immersion au procès du Squale." *Fakir* https://fakirpresse.info/les-barbouzeries-1-les-sbires-de-lvmh-a-la-barre.

Pocréaux, Cyril. (2024b, November 28). "Les barbouzeries de Bernard Arnault : en immersion au procès du Squale Partie 4." *Fakir* https://fakirpresse.info/les-barbouzeries-4-bernard-arnault-a-la-barre.

Prarthana, Prakasah. (2024, September 25). "Bernard Arnault Sent a Memo Banning LVMH Staff from Talking to Certain News Outlets, Sparking Backlash Among French Journalists—Including from Publications He Owns." *Fortune* https://fortune.com/europe/2024/09/25/bernard-arnault-memo-lvmh-staff-banning-french-media-news-outlets/.

Quiggin, John. (2023, November 30). "What is the Actual Economic Benefit of the Grand Prix?" *Crikey* https://www.crikey.com.au/2023/11/30/grand-prix-2023-victorian-economic-benefit-ernst-young/.

Ramírez, Sheila. (2025). "Encuentro triunfal." *Robb Report: Best of the Best*: 70–80.

Reed, Dave. (2024, October 17). "Legends and Louis Vuitton Get Their Due at America's Cup Hall of Fame Night in Barcelona." *Sailing World* https://

www.sailingworld.com/racing/legends-and-louis-vuitton-get-their-due-at-americas-cup-hall-of-fame-night-in-barcelona/.

Regué, Júlia. (2020, October 12). "12-O en BCN: Ofrenda a Colón, bendición a la virgen del Pilar y 'Cara al sol'." *El Periódico* https://www.elperiodico.com/es/politica/20201012/12-octubre-barcelona-colon-cara-sol-8152690.

Richards, Giles. (2024, January 23). "F1 Reveals Madrid to Replace Barcelona as Host of Spanish Grand Prix From 2026." *Guardian* https://www.theguardian.com/sport/2024/jan/23/f1-reveals-madrid-to-replace-barcelona-as-host-of-spanish-grand-prix-from-2026.

Rico, Anna, Julia Martínez-Blanco, Marc Montlleó, Gustavo Rodríguez, Nuno Tavares, Albert Arias, and Jordi Oliver-Sol. (2019). "Carbon Footprint of Tourism in Barcelona." *Tourism Management* 70: 491–504.

Riggs, Doug. (1986). *Keelhauled: Unsportsmanlike Conduct and the America's Cup*. Newport: Seven Seas Press.

Riisager-Simonsen, Christian, Gianna Fabi, Luc van Hoof, Noél Holmgren, Giovanna Marino, and Dennis Lisbjerg. (2022). "Marine Nature-Based Solutions: Where Societal Challenges and Ecosystem Requirements Meet the Potential of Our Oceans." *Marine Policy* 144 https://www.sciencedirect.com/science/article/pii/S0308597X22002457#bib1.

Rius-Ulldemolins, Joaquim, Francisco Torres Pérez, and Gil-Manuel Hernàndez. (2021). "Regional Entrepreneurialism, 'Creative City' and Strategic Projects in València." *The Hidden Side of the Creative City: Culture Instrumentalization, Political Control and Social Reproduction in Valencia*. Ed. Joaquim Rius-Ulldemolins, Juan Arturo Rubio-Arostegui, and Verònica Gisbert-Gracia. Cham: Palgrave Macmillan. 25–58.

Rodríguez, David. (2022, April 4). "Escepticismo entre los expertos sobre el impacto económico de eventos como el Mobile o la Copa América para Catalunya." *Público* https://www.publico.es/sociedad/escepticismo-expertos-impacto-economico-eventos-mobile-copa-america-catalunya.html.

Rodríguez, Jesús. (2024a, September 17). "La inversió pública oculta de la Copa Amèrica." *Directa* https://directa.cat/la-inversio-publica-oculta-de-la-copa-america/.

Rodríguez, Jesús. (2024b, October 11). "Un espectacle de drons per dopar les xifres d'assistència a la Copa Amèrica." *Directa* https://directa.cat/un-espectacle-de-drons-per-dopar-les-xifres-dassistencia-a-la-copa-america-de-vela/.

Rodríguez-Santos, Carmen, Stephane Ganassali, Francesco Casarin, Paoloa Cerchiello, Gunnar Mau, and Astrid Siebels. (2008). "How Perceived Image of a Sponsorship Event Influences on the Brand Perceptions: An International Research of the America's Cup and Louis Vuitton." *International Journal of Management Studies* 10, no. 3: 261–81.

Rolland, Rosalind M., Susan E. Parks, Kathleen E. Hunt, Manuel Castellote, Peter J. Corkeron, Douglas P. Nowacek, Samuel K. Wasser, and Scott D. Kraus.

(2012). "Evidence That Ship Noise Increases Stress in Right Whales." *Proceedings of the Royal Society B: Biological Sciences* 279: 2363–368.

Rosa, Anton. (2025, February 18). "El Puerto reconoce que la Copa América dejó un agujero de 3,5 millones." *Tót Barcelona* https://www.totbarcelona.cat/es/economia-es/el-puerto-reconoce-que-la-copa-america-dejo-un-deficit-de-35-millones-592069/.

Ross, Hamish. (2023). "Throwing a Lifeline to an Old Sport: A Modern Review of the America's Cup Deed of Gift." *International Sports Law Journal* 23: 240–58.

Ruta, Dino and Beatrice Manzoni. (2015). "City Capacity Building—Preparing to Exploit the Legacy of a Large-Scale Sports Event: The Case of Valencia and the 2007 America's Cup." *Routledge Handbook of Sport and Legacy: Meeting the Challenge of Major Sports Events.* Ed. Richard Holt and Dino Ruta. Asst. Ed. James Painter. Abingdon: Routledge. 231–44.

Sacristán, Claudia. (2024, July 14). ""No puedo ir ni a la playa de mi ciudad": los barceloneses, contra de la Copa América de Vela." *El Confidencial* https://www.elconfidencial.com/espana/cataluna/2024-07-14/no-puedo-ir-playa-ciudad-barceloneses-copa-america_3919767/.

Sarkar, Sahotra. (2021). "Origin of the Term *Biodiversity*." *BioScience* 71, no. 9: 893.

Shanahan, Abigail. (2024, October 21). "Experiencing the Louis Vuitton 37th America's Cup as a Team Family Member and Volunteer." https://www.americascup.com/news/3772_EXPERIENCING-THE-LOUIS-VUITTON-37TH-AMERICA-S-CUP-AS-A-TEAM-FAMILY-MEMBER-AND-VOLUNTEER.

Ship It Zero. (2024). *Ship It Zero Shipping Decarbonization Report Card* https://shipitzero.org/report-card/.

Soler, Jaume. (2025, March 10). "Grant Dalton llama "comunista" al alcalde de Barcelona." *Mundo Deportivo* https://www.mundodeportivo.com/vela/20250310/1002423002/grant-dalton-llama-comunista-alcalde-barcelona.html.

Sturm, Damion, Tom Kavanagh, and Robert E. Rinehart. (2021). "They Are Not 'Team New Zealand' or the 'New Zealand' Warriors! An Exploration of Pseudonationalism in New Zealand Sporting Franchises." *Sport in Society* 24, no. 11: 2019–035.

Suñe, Ramón. (2024, October 24). "La patronal Foment del Treball luchará por retener la Copa del América en Barcelona." *La Vanguardia* https://www.lavanguardia.com/local/barcelona/20241024/10048731/patronal-foment-treball-luchara-retener-copa-america-barcelona.html.

Suriñach, Jordi and Esther Vayá, eds. (2025a, March). *Impacte economic de la 37a Copa América de vela, Barcelona 2024* https://www.barcelonacapitalnautica.org/wp-content/uploads/2025/03/Report-1-Impacte-Economic_.pdf.

Suriñach, Jordi and Esther Vayá, eds. (2025b, March). *Valoració social de la 37a Copa Amèrica de vela, Barcelona 2024* https://www.barcelonacapitalnautica.org/wp-content/uploads/2025/03/Report-2-Impacte-social_.pdf.

Suriñach, Jordi and Esther Vayá, eds. (2025c, March). *Anàlisi qualitativa de la 37a Copa Amèrica de vela, Barcelona 2024* https://www.barcelonacapitalnautica.org/wp-content/uploads/2025/03/Report-3-Llegat_.pdf.

Transport & Environment. (2024, November 28). *How Much Europe's Ships Emitted in 2023* https://www.transportenvironment.org/articles/shipping-2023-emissions.

Turner, Andrew, Rob Arnold, and Tracey Williams. (2020). "Weathering and Persistence of Plastic in the Marine Environment: Lessons from LEGO." *Environmental Pollution* 262 https://doi.org/10.1016/j.envpol.2020.114299.

Tsujii, Koki, Tomonari Akamatsu, Ryosuke Okamoto, Kyoichi Mori, Yoko Mitani, and Naoya Umeda. (2018). "Change in Singing Behavior of Humpback Whales Caused by Shipping Noise." *PLoS ONE* 13, no. 10 https://doi.org/10.1371/journal.pone.0204112.

Vaughan, Roger. (2024, August 25). "Not Only Boring, it is Dehumanizing." *Scuttlebutt Sailing News* https://www.sailingscuttlebutt.com/2024/08/25/not-only-boring-it-is-dehumanizing/.

Veitch, Sarah. (2013). "Mythology as Hallmark Event Legacy: The Endurance of America's Cup Mythology in the City of Fremantle." *Journal of Sport & Tourism* 18, no. 2: 67–80.

Vicente, Sandra, Victòria Oliveres, and Pau Rodríguez. (2024, August 28). "El alquiler de temporada alcanza niveles inéditos en la Barceloneta coincidiendo con la Copa América." *El Diario* https://www.eldiario.es/catalunya/alquiler-temporada-alcanza-niveles-ineditos-barceloneta-coincidiendo-copa-america111591525.html.

Vicente, Sandra. (2023, October 10). "65 entidades se unen contra la Copa América de Barcelona: "Es la síntesis del expolio contra la ciudad"." *El Diario* https://www.eldiario.es/catalunya/65-entidades-unen-copa-america-barcelona-sintesis-expolio-ciudad_1_10585988.html.

Wheatley, Magnus. (2024). *There is No Second: The Definitive Account of the First Race in 1851 for What Would Become 'América's Cup'.* Newport: Seahorse Publishers.

Williams, Terri M., Susanna B. Blackwell, Outi Tervo, Eva Garde, Mikkel-Holger S. Sinding, Beau Richter, and Mads Peter Heide-Jørgensen. (2022). "Physiological Responses of Narwhals to Anthropogenic Noise: A Case Study with Seismic Airguns and Vessel Traffic in the Arctic." *Functional Ecology* 36: 2251–266.

Williamson, Jeannine. (2024, May 22). "Explora Journeys Showcases European Luxury at Sea in its Second 'Space Ship'." *Riviera* https://www.rivieramm.

com/news-content-hub/news-content-hub/explora-journeys-showcases-european-luxury-at-sea-in-its-second-space-ship-80851.

Willsher, Kim. (2004, June 3). "Louis Vuitton's Links with Vichy Regime Exposed." *Guardian* https://www.theguardian.com/world/2004/jun/03/france.secondworldwar.

Wilson, Joseph. (2025, May 22). "An America's Cup Without an American Boat? American Magic Says it May Not Sail in Naples." *AP* https://apnews.com/article/americas-cup-sailing-american-magic-b16c447cef426b2a209385c0ea8af460.

World Bank. (2017). *The Potential of the Blue Economy: Increasing Long-Term Benefits for Small Island Developing States and Coastal Least Developed Countries* https://openknowledge.worldbank.org/server/api/core/bitstreams/cee24b6c-2e2f-5579-b1a4-457011419425/content.

Wright, Richard Keith and Christopher Barron. (2021). ""It's Coming Home!" Leveraging Legacies in the City of Sails." *Place Event Marketing in the Asia Pacific Region*. Ed. Waldemar Cudny. London: Routledge. 143–61.

Yanmar Power Technology Co., Ltd. Settlement Agreement Notices of Violation: EPES-2021-C00572 & EPES-2022-C00572. (2024, March). California Air Resources Board https://ww2.arb.ca.gov/sites/default/files/2024-03/yanmar_power_technology_co_ltd_sa.pdf.

YouGov. (2024, August 7). *Impacto del turismo en la población residente* https://commercial.yougov.com/rs/464-VHH-988/images/YouGov-ES-Impacto-del-turismo-en-la-poblaci%C3%B3n-residente.pdf?version=0&utmmedium=email&utmsource=whitepaper&utmcampaign=WP-2024-08-ES-Turismofobia&mkttok=NDY0LVZISC05ODgAAAGV0KHp6pY9nLK2eypE1LjDUCM7U-KUYc8pGpfQswGs1FldEhCPFnn-XFQ3VvlOvZBVGRHPsWn3DSxSS9LtlwM3BGeICv50Mx4Mk5HTI2wimk.

CHAPTER 5

Formula One Carbon

Sports are increasingly dominated by technological innovation designed to corral the environment. For Luxemburg, '[e]ach new step in the perfection of productive technology is at the same time a step in the subjugation of physical nature' (2014: 173). Eco put it this way: 'Disneyland tells us that technology can give us more reality than nature can' (1987: 44). Such accounts construe the environment as in need of mastery, its conquest ideological and physical. This chapter goes to the heart of such projects, per Adorno's definition of modernity: the 'technical domination of nature' (1975).

That subjugation can play havoc with justice. For example—perhaps the most *telling* example—J. Robert Oppenheimer talked about the instrumental rationality animating some of the scientists who created the atomic bomb. Once they saw what was feasible, the device's impact lost intellectual, political, and emotional significance for them, overtaken by what he labeled its 'technically sweet' quality (Polenberg 2002: 47). (It should be noted that Oppenheimer blocked the Szilard Petition [1945] from reaching the Administration. Signed by dozens of his distinguished colleagues, it urged the government not to use the bomb against Japan. Not everyone had fallen prey to the 'technically sweet.')

Applied science can attract the reverence previously reserved to deities, mixed with a decontextualized, deracinated pleasure. It 'ravishes [people] into admiration' (Hume 2007: 19) as a 'technological sublime' bestows

T. Miller, J. P. Carañana, *Global Sports Go Green—Or Do They?*, Global Culture and Sport Series,
https://doi.org/10.1007/978-3-032-14207-8_5

quasi-sacred power onto machinery (Marx 1964: 96). The contemporary dominance of that sublime is attributable to eighteenth- and nineteenth-century industrial and consumer inventions, when it was thought people were governed by electrical impulses and telegraphy a physical manifestation of human intellect, matching the essence of humanity to labor. In the early twentieth century, radio waves were said to move across "the ether," a mystical substance that could contact the dead and cure cancer. The inter-war period saw claims that the human "sensorium" had been comprehensively re-trained by technology (Miller 1998).

This tendency was supercharged via the successful political-economic provision of food, power, water, and communications after World War II. In 1963, future UK Prime Minister Harold Wilson famously guaranteed technology would ensure 'undreamed of living standards and the possibility of leisure on an unbelievable scale … a Britain that is going to be forged in the white heat of this revolution' (1963: 2–3, 7). In 2025, his successor Keir Starmer, artless and clumsy as ever, pledged himself to a plan that 'mainlines A[rtificial] [I]intelligence into the veins of this enterprising nation' ("Prime Minister" 2025).

Ever since the 1960s, we have seen what began as the awe, admiration, and domination that people experienced when they looked at massive engineering feats, such as public works, re-directed onto technologies of consumer entertainment and the digital (Nye 1994). That strange enchantment, that *cliché* dalliance, began with overlapping and successive gadgets: transistor radios, color TV, cassettes, video tapes, wireless communication, cellphones, tablets, high definition, mobile computing, and virtual reality. Something similar happens with the mysterious seeming immateriality of invisible goods: so-called big data, digital currency, and artificial intelligence. Across those developments, one constant has been the fantasy that each new technology's supposedly innate capacity would endow its users/its used with transcendence. The banality of repetition has not disempowered this tendency, because of the interests it dutifully serves and the cult of newness it reproduces (Ogan et al. 2009).

Via high-concept design, the machinery of work, contact, and fun mixes the sublime—the awesome, the ineffable, the uncontrollable, the powerful—with the beautiful, the approachable, the attractive, the pliant, the soothing. Philosophical aesthetics has typically opposed them (Eagleton 1990), but cultural technology draws on futurist precepts from the early twentieth century for aesthetic and ideological inspiration as technological powers 'accelerate at a dizzying pace, carried forward to a blind tomorrow'. A 'physical object invested with political and metaphysical Ideality' could be transformed into something people 'grasped and panted and cried for' (Marx 1964: 206, 208).

Consider the young and the reckless in *The Great Gatsby*, 'careless people' who 'smashed up things and creatures and then retreated back into their money or their vast carelessness or whatever it was that kept them together, and let other people clean up the mess they had made' (Fitzgerald 1925: 306–07) or Silicon Valley's tiresome imprecation to "move fast and break things." Juliet Schor called this a 'materiality paradox': the greater the frenzy to buy goods for their transcendent or immaterial cultural meaning, the greater the use of material resources (2010: 40–41). Per the *Financial Times*, the arrogant, errant second Trump administration specializes in quick movements that shatter worlds: 'Carelessness, in Fitzgerald's vision, is more than negligence—it is a way of wielding power. It's the certainty that the world exists to absorb your damage' (Churchwell 2025).

That tendency is exemplified by some compelling history directly relevant to this chapter. In 1909, Filippo Tommaso Marinetti laid out futurism's foundational manifesto on the front page of *Le Figaro*. It fetishized danger, courage, energy, masculinity, aggression, velocity, and war, denouncing feminism and established institutions of knowledge. Automotive power was a central component. The manifesto's fourth and fifth points are crucial to the mythology of motorsport:

> 4. We affirm that the beauty of the world has been enriched by a new form of beauty: the beauty of speed. A racing car with a hood that glistens with large pipes resembling a serpent with explosive breath … a roaring automobile that seems to ride on grapeshot …
>
> 5. We intend to hymn man at the steering wheel, the ideal axis of which intersects the earth, itself hurled ahead in its own race along the path of its orbit. (2009a: 51)

Seven years later, thrilled by what he called 'our great liberating war' (World War I), Marinetti announced a 'religion-morality of speed' that would supplant conventional religions by accepting 'sensuality' and raging against 'slowness, memory, analysis, rest, and habit.' His love of '[h]uman energy, multiplied a hundredfold by velocity' was an exaltation of automobility (2009b: 224). It reads remarkably like a mixture of the peerless arrogance of petrol heads with contemporary ideologues of software and hardware, juxtaposing the "good" and the "bad" via this binarism:

> Velocity = synthesis of all forms of courage in action. Aggressive and martial. Slowness = analysis of all forms of stagnant prudence. Passive and pacifist. Velocity = disdain for obstacles, desire for the new and unexplored. Slowness = arrest, ecstasy, immobile adoration of obstacles, nostalgia for the already seen, idealization of tiredness and repose, pessimism about the unexplored. (2009b: 225)

Marinetti included what could be a mantra for contemporary motorsports: 'The Inebriation of great speeds in cars is simply the joy of feeling oneself merged with the only *divinity*' (2009b: 226). That technological sublime, adjacent to an emergent Italian fascism (but also to pre-Stalinist Soviet art [Harte 2009]), relied then—as it does today—on a '[f]etishism which attaches itself to the products of labour' once these objects are in the hands of consumers, who lust after them as if they were 'independent beings' (Marx 1987).

The tendency to desire newness repeats and repeats. Eighty years ago, Brecht wrote, 'I stood on the hill, and there I saw the Old approaching, it came as the New' (2018: 2229). Ever since futurism, willing drivers have entered, enjoyed, destroyed, and transformed nature, something marked again and again in motorsport, from tourism numbers to specialty magazines to a discourse of transcendence through exposure to nature—via recreational and sport-utility/all-terrain vehicles.

This chapter turns to the technological sublime's acme in carbon sports: motor-racing. A world of despoiling drivers, engines, roads, and runways, this 'theoretically unsustainable sport' (Bodie 2011: 493) sees 'speed, engine size, engine power and aerodynamic efficiency … widely discussed by participants, commentators and fans.' Less time is dedicated to the correlative 'dependence on the physical resources of planet Earth' (Dingle 2009: 81). Following a historical overview, we'll look into its bizarre environmental claims and the greenwashing service it provides sponsors.

History

Daredevilry as much as technology defined Formula One in its first decades. The press and the sport itself romanticized the presence behind the wheel of such aristocratic names as Dutch Count Carel Godin de

Beaufort, who paid for his own cars and drove barefoot, the Spanish nobleman Alfonso, Marquis de Portago, Switzerland's Baron Emmanuel "Toulo" de Graffenried, the German Count Wolfgang von Trips, and Prince Bira of Siam; and the "romantically" fatal crashes of the Counts, the Marquis, beer scion and Old Etonian Piers Courage, and farmer Jim Clark. As Virilio put it, '[t]he exploitation of high speeds … [was] a sport reserved for dandy-warriors, a fantasy allowed to people who are otherwise useless' (1991: 102). Formula One lionizes them as "Gentlemen Drivers" by contrast with the careerists of today (2022).

Incarnating that fantasy of the talented amateur, the *bourgeois* media built up the bravado of debonair posh-boys behind the wheel, concentrating on the drunken and sexual 'off-track hijinks' of a Mike Hawthorn ("The Gay Cavalier") in the 1950s ("Mike Hawthorn" 1959) or a James Hunt ("Hunt the Shunt" [crash]) in the 1970s, who wore a racing-uniform patch reading "Sex: Breakfast of Champions" (Keegan 2024) and whose antics were fetishized by his team owners post-mortem. They rhapsodized a time when Enzo Ferrari claimed 'marriage slowed his drivers down' (quoted in Williams 2021). The *Mail on Sunday* still headlines those days, per 'James Hunt Was an F1 Playboy Who Bedded 5000 Women' (Graham 2013) and Stirling Moss comparing the present with the past: 'When the race finishes, instead of chasing girls like they did in my day, now they go and say "thanks" to Vodafone.' A very partial nostalgia valorizes a time 'When Sex Was Safe & Racing Dangerous' ("When Playboys" 2011; "When Sex" 2015; "Racing Legends" 2012). Giles Richards explains:

> Long a bastion of male heterosexuality, Formula One is a sport that for years positively revelled in all that went with it. The drivers living a glamorous high-life alongside beautiful women was sold to the extent that it became intrinsically linked with … gladiatorial characters, of the race-fast and party-hard variety. (2024c)

One gets a fictional flavor of these *macho* hijinks in John Frankenheimer's *Grand Prix* (1966). Super Panavision 70 mm cameras shot derring-do, betrayal, death, and sex across famous venues. As one does, four lusty chaps competed for hearts and trophies, each with his own wee melodrama. Their cars were piloted by actual champions, including Clark, Juan Fangio, and Jack Brabham ("10 Great" 2023).

Those days always had another side. Consider the dutiful *Peronista* Fangio, who learned to drive watching buses; or Brabham, the grim businessman and subsequent eponym of a similarly grim Perth suburb. They were some distance from the smiling antics of dashing imperialists. And there was systematic exclusion. The commercialization and professionalization of motorsport marginalized women, while the occasional successes of Global Southerners, royalty, driver-manufacturers, and small firms were soon displaced by European corporate triumphs (Matthews and Pike 2016).

Per the desire of FIFAcrats and Catalan mavens to appeal to the "right" kind of tourist—a globe-trotting oligarchy—the not-so-good burghers of the world crave hosting Formula One in order to appear 'cosmopolitan' (Lowes 2018), perhaps understanding all too well that the event is 'a true microcosm of the world economy' (Lefebvre and Roult 2011: 330).

Fantasies of effective 'place marketing' run riot. Races supposedly attract tourism and investment through a sporting showcase:

> Motorsport mega[-]events, particularly those conducted on city street circuits, have become a preferred method of image making for many of the world's major cities. Urban leaders widely regard hosting a city-based motorsport event as a successful event-tourism strategy, one that can boost the image of cities involved; contribute to the local economy; and, especially, cultivate a "winner" image for the politicians who champion such spectacles. (Tranter and Lowes 2009: 152)

This despite the fact that, as we saw with other mega-carbon sports, the benefits to host cities are minimal—if any—compared to outlays and environmental costs (Gezici and Er 2014). Developing a Formula One circuit from scratch means buying land and creating infrastructure. It can cost over $270 million, with $18.5 million to follow in annual maintenance. Then there are payments: competition owner, Liberty Media, derives about a third of its annual income from 'race promotion revenue'. The fees it charges Grand Prix hosts vary wildly. In 2023, Azerbaijan, Saudi Arabia, Qatar, and Bahrain all paid in excess of $50 million; Britain, Spain, Austria, Italy, Japan, Brazil, Belgium, and Monaco below $30 million ("Fees Paid" 2024).

Table 5.1 Liberty media: formula one revenue

Year	*$Billion*
2017	1.78
2018	1.83
2019	2.02
2020	1.15
2021	2.14
2022	2.57
2023	3.21
2024	3.40

Sources: Based on "Total Revenue" 2024 and Agini 2025

Unlike other sites, Liberty operates the new Las Vegas race itself, but not purely at its own cost. For example, in 2024, the local Convention and Visitors Authority spent $1.7 million on $11,000 suites for airlines and other corporations, at a time when demand for hotel rooms and tickets was weak (Habtemariam 2024).

Economists at Large showed that the 2012 Australian Grand Prix lost the state of Victoria A$60.6 million (Campbell 2013). When the race became the "Formula 1® Rolex Australian Grand Prix," it allegedly gave 'a boost of over $3.3 billion a year to our economy,' encouraging Melbourne, the host city, to dub itself the 'events capital of Australia'. There was no evidence to back up either claim. The purported multiplier effects from all that expenditure had one referent—consultancy firms reiterating discredited *nostra* from a previous century (Quiggin 2023). Nor was there reference to the more than a hundred million in public money allocated to that year's contest, or a billion in the period since the government had announced the event would not cost taxpayers a sou (McCubbing and Samios 2024). Predictably, revenue amounted to less than half of all costs (Gordon 2022). Alternative calculations were difficult to muster: the Australian Grand Prix Organisation decreed that disclosing spectatorship data would pose a 'national security risk' (quoted in Logan 2024). From 2025, LVMH assumed the right to name the event after itself (Brittle 2025).

One thing is clear: *bourgeois* economics, fanboy scholarship, and the specialist media ignore the competition's ecological crimes, emphasizing "efficiency," growth, and putatively benign influences on consumer motoring while denying the insanity of car-based transportation and the lessons of environmental materialism (Mourào 2017; Henry et al. 2007;

Gibbs et al. 2010). For reactionary Formula One scholars dedicate their professional lives to the kind of administrative research mentioned in the Introduction. From a veritable cascade of examples, consider these recent instances: game theory to make pitstops more efficient (Aguad and Thraves 2024); cost-benefit analysis of the labor process (Parker et al. 2024); econometric studies into attracting spectators (Gasparetto et al. 2024); data on securing perennial capitalist sponsorship (Jensen et al. 2024); focus groups dedicated to making TV viewers devoted (Shah and Williams 2024); and proposals for 'universities to adapt their hospitality and business curricula and prepare students for careers in this profitable field,' especially 'VIP client management' (Fu 2024). Thus shall we know them.

And there is plenty for them to do, as the price for racing teams to participate has become enormous. Some have spent half a billion dollars a year, principally on hybrid power systems, chassis, labor, and travel (Cave and Miller 2015; Coch 2022; Tovey 2014). A hundred and fifty have gone bankrupt. In 2022, regulations set limits on each team's annual expenditure at $140 million, with further reductions planned. It remains unclear what that will eventually mean for drivers' salaries (then up to 55 million), thousands of other employees (executives "earning" 10 million), and engine prices (averaging 25 million) (Bell 2022). The transformation immediately benefited established teams with sunken investments in plant and equipment and efficient and effective logistics. By 2024, the average "value" of a team was $1.88 billion, up 276% on 2019.

A major change in management facilitated this controlled growth. The sport was once administered by a wealthy coterie of elderly white men. Their internal dynamics, passion for power, adoration of fossil fuel, and untold wealth animated a massive expansion of Formula One around the world (Richard Williams 2015) via a structural and indexical homology between the extractive and sporting industries. They exploited the brief, fragile career of lead drivers *contra* the lengthy, powerful impact of the contest's environmental despoliation (Silk et al. 2005).

Since Liberty Media Group's takeover in 2017, Formula One has become more corporate, driven by buttoned-down, sober executives who know how to "go green"—in image terms. It goes without saying that as 'business opportunities surface in emerging countries' (Lefebvre and Roult 2011: 330), Liberty has invited newly wealthy nations to act as hosts (Bustad and Andrews 2023). Among newcomers, Bahrain held races from 2004, Abu Dhabi 2009, Qatar and Saudi Arabia 2021, and Miami and Las Vegas 2022 and 2023. Future plans envisage "heritage" events such as Monaco, Silverstone, and Monza guaranteed slots, with flexibility for potential expansion elsewhere.

Quite apart from the ideological promotion of automobility, simply situating the events internationally makes it impossible to reduce emissions significantly (Lim 2022): there are now twenty-four races a year (Foye 2024; Richards 2023b). Clustering and regionalizing the calendar is discussed as a means of minimizing globe-trotting; the actual calendar suggests such conversations are rather token (Smith 2023). The 2023 schedule might have been designed to maximize flying between destinations, with 'teams and drivers going from Europe to Canada and then back to Europe' (Brinsford 2023).

Per FIFA and the IOC, motorsport's governor, the Fédération Internationale de l'Automobile (FIA), is an international élite of technocrats, business leeches, and *macho* men (Nichols and Savage 2017) run by the infamous Emirati Mohammed ben Sulayem (Richards 2023c, 2024d). Scandal after scandal has afflicted a running joke of potentially fraudulent deals, clownish chiefs, and lack of accountability and transparency (Mosley 2015; Richards 2023c, 2024b,e).

Meanwhile, the sport's glamour has become less risky—still captive to hegemonic masculinity, albeit in a more careful contemporary idiom, as opposed to the libertine imagery of the past, though the media continue to stereotype drivers, per British journalists' xenophobic abuse of Michael Schumacher (Painter 2024). Despite the changes, Formula One remains the creature of '[i]nternational playboys, Machiavellian billionaires, humble heroes, racing-world royalty, overachieving underdogs, aging has-beens, [and] hotheaded bullies' (Battan 2022).

This is the preserve of oligarchic offspring engaging in the parthenogenesis of gendered, racial, and financial privilege (Boxall-Legge 2020; Morgan and Scarlett 2021; Kew 2023). Gender also colors the competition, via a destructive homology: 'active' men are deemed 'active' drivers (Sturm 2022: 116–17), with women excluded from most seats of power. There are sexist wage disparities (Sylt 2019) and propaganda continues to depict the likes of 'Naomi Campbell and Heidi Klum hanging off the arms of the team bosses' (Reid 2008), though the job of "grid girl," long-performed by young women who promoted sponsors to the public and cheered on drivers, has ended (Tippett 2019). Just 1% of the competition's 40,000 employees in 2021 were African-descended (Onwurah

2021). Seven-time champion Lewis Hamilton waited many years to speak out on these topics, for fear of being labeled an 'angry black man' (quoted in Mehta 2023).

A further problem, most relevant to this investigation, derives from a desire we have seen before: Formula One's lust for growth. The sport involves cars driving round and round and round for rather a long time. Now that accidents are rare—and serious or mortal injury rarer—this does not make for a distinctive screen spectacle. One, somewhat severe, critique argues that:

> the races are mostly snooze fests in which drivers are merely technicians controlled by legions of engineers who dictate how to drive the car. The rules forbid carrying enough fuel to race flat out for an entire race and so there are titanic battles in the first few laps. Then the drivers dial it back for the majority of the remaining time so they can conserve fuel, conserve tires, and conserve their engines and transmissions. It is stage managed, like professional wrestling or roller derby. (Hanley 2019)

Absent a return to cavalier masculinity, something different, with a dramatic narrative, was needed. The Netflix series *Formula 1: Drive to Survive* (2018–) duly attracted new audiences to television coverage of Formula One races (50% of the total) (Lawrence 2021), particularly affluent young *Gringos* and female viewers (YouGov 2023; Smith 2021; "Driven to Watch" 2022). The program has clearly coincided with the sport's development: despite the COVID-19 syndemic, Formula One's overall receipts increased from $2136 billion to $2573 billion in 2022, via additional spectators and new TV contracts (Brown 2023). 2023 saw 6.5 million attend races, up 9% on the previous year, and a total of 1.6 billion viewers and 77 million followers on "social" media. Drivers were urged to open Twitch accounts and market videogames, which proved immensely popular with the desired public (McIntyre 2023; Orlove 2024). Operating income was $492 million, up a hundred million in a year ("Liberty Media" 2025).

In 2024, *Drive to Survive* featured among the ten most-popular Netflix programs in sixty-one markets (STOP. and Formula Money 2024: 4). It chronicled 'grand melodramas and intricate microdynamics … a new kind of broken fourth wall between the world of sports and entertainment' (Battan 2022). The idea drew on myths of heroism, per the era of a Hawthorn, a Clark, or a Hunt (Soble and Lowes 2024) but refashioned for a different audience: ‘imagine the *Real Housewives*, if the housewives were driving around at 300 kilometers an hour, and if occasionally one of the housewives caught on fire’ (Pinkerton 2021). *Forbes* magazine called it ‘the world’s glitziest infomercial’ (Robinson and Clegg 2024).

There has long been a raft of fiction by women that thrives on Formula One romantic betrayal and high-octane you-know-what (“The Ultimate List” 2024). Many of the covers feature chiseled bare chests and stubbled charm—beautiful souls in need of curation by the love of a “good woman.” The bedhopping antics of yore had been dedicated to the triangulated desire of straight men for one another, mediated through contests for women’s affection. *Drive to Survive* touched implicitly on female desire in ways that Formula One had hitherto avoided. Needless to say, Netflix—and Twitch—steered away from actually featuring women (who were allotted 2.65% of speaking time in sixty episodes) or making environmental statements (Roper 2024; Coleman 2023).

The year 2025 saw the release of *F1* (Joseph Kosinski), a big-screen bracket connecting the guy-revelry of *Grand Prix* to the sober norms of decades later. Brad Pitt’s character incarnated the presence of the past in the present. Authenticity was again granted, via the participation of dozens of Formula One drivers and personnel—plus the inevitable product placement, as desired by Hollywood, the FIA, and Liberty.

We’re Green—Dammit!

This was all apiece with an uncritical futurism, a technological sublime that characterizes the *bourgeois* media’s dual *foci* on corporate composites of metal, plastic, and rubber, and the still-infantile antics of many drivers (Hogan 2015; Glendenning 2016). “Glamour” was often ascribed to the cars as something beyond our means and comprehension, with environmentalism unappealing. Olivia Solon explained it this way:

> Sustainability and Formula 1 are sexless bedfellows. Those in the sustainability corner accuse FI of being a gas-glugging, tyre-burning, component-churning anachronism, but petrolheads say that “treehuggers” and sustainability rules take the fun out of the sport. (2011)

Formula One is responsible for more than 250,000 tons of CO_2 emissions annually, the same as 55,000 automobiles: 0.3% from racing, 45% the air, sea, and road transport of cars, and 27.7% thanks to the movement of workers, promoters, partners, executive hangers-on, and "mobile-hospitality facilities." Each team flies 160,000 kilometers a year to test cars and compete and numerous past and present drivers own private jets. Formula One's annual electricity account would power 45,000 US homes, while the cars' carbon-fiber chassis defy easy recycling (Scott 2013; Black 2010; McLaren Racing 2021: 10; King 2013; Hanley 2019; "Seven Incredible" 2023; Zerrenner 2019; Foye 2024).

The sport's popularity—over a hundred million people watched its 2021 finale on television (Ruiz 2022)—alone makes for deleterious environmental impacts, with streaming a key culprit. Selling media rights accounts for a third of Liberty's Formula One income. The company is in the vanguard of sports providing their own coverage. In the search for 'direct billing relationships,' it screens races live to subscribers in the Netherlands, Mexico, and Brazil and is particularly proud of attracting young people, whom it stoops to call 'data-comfortable'.

In many cases, the food prepared for events necessitates not only prodigious travel and electronic despoliation but immense waste. The 2022 race in Miami saw 90,000 pounds of leftovers, at a time when over a million residents of South Florida confronted food insecurity. Formula One's answer was to invite volunteers to share unused food or ditch it in landfill, which was turned into a bizarre sign of virtue (Coleman 2024). The events often impinge upon or destroy birdlife, waterways, trees, and parks (Tranter and Lowes 2009). The short-lived Korean race saw a thousand acres of rice paddies cleared, and the Dutch jamboree adversely affected sand dunes (Sturm 2023). The Australian Grand Prix benefits from the state discarding with a range of environmental and planning laws for approximately three months each year, detracting from public use of green space (Lowes 2004). Much is made of the circuit being inside a park that includes a 'lake populated with elegant, languid red-billed, black swans' (Richards 2025). Little is said of the threats to their wellbeing, despite evidence of heightened corticosterone levels as more and more people flock to the site. Noise from the Grand Prix correlates with increased levels of stress among these birds (Payne et al. 2012).

Most significantly, the sport's air travel is not credible. Even if Formula One teams minimized other aspects of their carbon footprint, transport would remain a problem. The real issue is their cavalier international flying, principally the repeated movement of heavy machinery across the globe (Richards 2023a). *À propos*, taking the public for fools, in 2024,

Formula One announced its venue for the second Spanish Grand Prix from 2026 as a sound ecological choice because Madrid's raceway was only a few kilometers from the airport (quoted in Richards 2024a).

Essentially,

> motorsport organizations might attempt to reframe their image as 'green' or environmentally friendly, it is difficult to disguise the direct, negative environmental impacts of the sport on energy use, local pollution, and greenhouse gas creation. (Tranter and Lowes 2009: 163)

In 2013, the UN Environment Programme warned the FIA that it needed to make motorsport more sustainable or face regulation (quoted in Allen 2014). In 2014, teams grudgingly accepted energy-efficient fuel systems that had been proposed years earlier (Mosley 2015). The sport's chorines argued that, far from its outmoded image as a 'gas-guzzling behemoth,' this once 'gaudy polluter' had become 'an environmental science lab' whose vehicles attained 52% thermal efficiency, compared to the 32% of commuter cars. We were told that the *cliché* of a 'fossil-fuelled dinosaur ... is outdated and irrelevant' (Richards 2021). Boosters claimed that Formula One provided a 'high-speed research and development laboratory for road cars' (Allen 2013). This was allegedly because competition between manufacturers for a luxury activity's most advanced engineering stimulated ever-greater fuel efficiency, passed on to everyday business and domestic motoring and supposedly diminishing the latter's carbon footprint (Scott 2013; King 2013; Allen 2014)—a trickle-down model of petroleum. Consider this from *Triple Pundit*:

> Fossil fuels simply won't last forever, so why should motorsport be free to use the Earth's resources at will? Formula One's advantage here is in having the world's finest minds under its considerable command; with the best technicians and engineers at work we can all hope for a better, carbon[-]free future. (Morris 2012)

These potentially positive externalities have led the way in conventional press reporting of Formula One and the environment (Elliott 2014). It has even been argued that '[c]yclists are miles behind Formula 1 in the environmental race' due to the massive impact of travel on events such as the Tour de France, weighed against the 'cutting-edge technology' that saw Formula One use a third of the fuel in 2014 compared to the previous year with no loss in performance (Pickford 2014). *Forbes* magazine pondered whether it might be "The World's Most Sustainable Sport" (Sylt 2015) and the *Financial Times* lauded 'the high-performance engineers … the backbone' of the competition, for whom 'the climate challenge is an irresistible problem to solve'.

Energy efficiency may be improving, thus reducing costs (Mytton 2020), but the Jevons Paradox is at work: when prices for goods and services fall, their use and expenditure increase (1866). And questions persist over the suitability, validity, and applicability of these technologies and fuels, both for motorsport and for commuter use (O'Shea et al. 2020). For while various FIA road-safety measures have been adopted by international agencies, attempts to reduce sulfur in fuel and limited efforts by the Fédération in 1995 to make Formula One carbon-neutral, generated minimal interest (Mosley 2015). Later publicity would focus on its acquisition of forests as a perceived effort to offset carbon emissions.

The Fédération's Environment and Sustainability Commission is headed by former Mexican President Felipe Calderón, notorious during his *sexenio* for environmental despoliation, rampant corruption, cosmic inefficiency, and grotesque militarization. FIA's *Environmental Strategy 2020–2030* promised rapid progress toward carbon neutrality in 2030 through reduced emissions and carbon-credit capitalism, per targets set at the Paris Conference of the Parties in 2015 (n.d.; also see Formula 1 n.d.). All this was to be done while maintaining 'relevance and leadership.' Calderón announced 'una nueva era que renueva un fuerte compromiso con el medio ambiente' [a new era that confirms our strong commitment to the environment] (quoted in "Felipe Calderón" 2022).

In 2019, Formula One laid claim to a history of 'technologies and innovations that have positively contributed to society and helped to combat carbon emissions.' It announced the possibility of a net-zero carbon hybrid power unit. The "Countdown to Zero" program involved reforms to car design, fuel systems, power units, and team travel. "Positive Race Print" claimed to reduce waste, augment low-carbon transportation, and support biodiversity (Formula 1 2019; Öztopcu 2024). Such greenwashing spoke disingenuously of the sport's past, denying a record of environmental despoliation, waste, and excess while denoting a yet "greener" future in ways that were supposedly credible—to which Greenpeace retorted that the promised transformations were worth little or nothing ("Expert Slams" 2023).

The photo below, of hegemonic officials proudly announcing Madrid's capture of a Grand Prix, displayed the usual group of gray functionaries (Fig. 5.1):

The event faced the typical problem of raising sufficient private money to fund the absurdity, confounding Community of Madrid President Isabel Diaz Ayuso (depicted above), who had promised there would be no drain on the public purse (Alcivar 2024).

The administration's planning document seeking comments on construction plans for the jamboree's proposed circuit was released amidst holiday season, with a month for comment—minimal time to study and document environmental implications, such as noise and petroleum pollution. Despite the dubious timing, serious concern was expressed over the din of engines, especially as some dwellings were just forty meters from the circuit, and the effect on hundreds of trees, which would be transplanted or destroyed. The mayoralty proposed temporary suspension of codes restricting noise to eighty decibels, a favor also accorded to Real Madrid when "needed" (Bogado 2025; "El Grand Prix" 2025; "Report Questions" 2025; Gutiérrez Cañas 2025). As I wrote the previous

Fig. 5.1 (Source: https://x.com/V9_PODCASTF1/status/1839383959333130410)

sentence, I thought of the deafening sounds screaming into my eardrums during the 1970 Race of Champions at Brands Hatch.

The usual unsubstantiated promises were made: 'un evento de gran magnitud para la ciudad, con inversiones millonarias y expectativas de posicionar a la capital como un referente en el automovilismo mundial' [an immensely important event for the city, with millions in investment and the prospect of positioning the city as a reference point in world motorsport] (Domingo Rodríguez 2025) plus tens of thousands of new jobs and half a million tourists adding 0.2% to the region's GDP. The Socialists sang along to the hymn sheet: the event 'puede convertirse en el motor de crecimiento para nuestra región' [could stimulate growth in the region] despite wanting details of the impact on public finances ("El PSOE" 2025). Some provisional construction costs were released: €137,130,388, via a shadowy private-public alliance (Antich 2025). Meanwhile, València's disused racetrack, built at taxpayers' expense for €98 million, housed desperate migrants in *chabolas* [shanty towns]. Locals raced "informally" late at night (Bono 2021; Castelló 2024).

Greenwashing Others

As we have seen, growth is a carbon mega-event watchword, part of an endless quest to multiply multinational corporate and state sponsors. Formula One 'corrals competition into a small closed shop and has become a honeyspot for sponsors' (Ames 2025), providing them with something beyond an Olympics or a World Cup, because it occurs every year, every month, and across the globe, rather than every four years, for two to three weeks, in one region (Blitz 2013; Dewhirst 2023). The events gain millions more in average sponsorship than any US sport (SponsorUnited 2025), providing Liberty with about a fifth of its racing income ("Liberty Media" 2025).

Rolex is a key collaborator. We've already seen the damage done to the environment by luxury watches—and Rolexes can retail for $38,000 each, leading to a scandal when Peruvian President Dina Boluarte was spotted wearing one she claimed was a cheap simulation. "Rolexgate" became an international neologism. That controversy matched the simultaneous revelation that the firm's founder, Hans Wilsdorf, had been a Nazi collaborator. Fortunately, *Le Temps* was on hand to counter such stories—a newspaper owned by a subsidiary of the Hans Wilsdorf Foundation . The firm showed its commitment to the environment by prescribing a 'nature ride' for hundreds of executives, confirming the seriousness of its "Perpetual Planet Initiative." This was just part of an answer to serial crises: harassment of employees by the company's managerial class; being fined over €90 million by France's anti-trust Autorité de la concurrence [Competition Authority] for preventing licensed distributors from selling its merchandise online across a decade of restrictive trade practice that included removing a former partner from the network without explanation and experiencing a poor environmental and human-rights rating from WWF (2023: 22, 37–40).

In 2016, Heineken signed on as a sponsor for $200 million, to the outrage of public-health advocates (GMM 2016). The company has a "colorful" past, with breweries in dozens of countries, including anti-democratic fiefdoms, in keeping with its 1960s support of white supremacists in southern Africa and long-standing relationships with oligarchs. Heineken benefits from monopolies across the continent—and special conditions for white expatriate employees, unlike the thousands of young black women working for below minimum wage in bar-side promotions, who suffer high levels of sexually transmitted disease (van Beemen 2019). The firm came under fire for subcontracting slavers (Paraguassu and Teixeira 2023) not long after being fined £2 million in Britain for 'seriously and repeatedly breaching the legally binding Pubs Code over three years' by obliging tenants to sell its own products (Pubs Code Adjudicator 2020). It was one of a thousand corporations that promised to depart Russia after the Ukrainian invasion, but failed to do so until publicly shamed (Kennedy 2023). Meanwhile, Heineken contributed to a horrific water shortage in Mexico (A Rodríguez 2022) and cut down a huge British orchard, destroying a natural bird habitat while avowing that it was 'incredibly important that we act responsibly and sustainably at all times' (quoted in Price and Bryer 2024). All this occurred just as Charlene de Carvalho-Heineken, the eponym's multi-billionaire heiress, was exposed as "sheltering" hundreds of millions in tax liabilities ("Heineken Heiress" 2024).

From 2025, the sport encountered an old friend of ours at:

> the crossroads of the LVMH Group's values of creativity and excellence with Formula 1's innovation and high-performance ... [a] partnership will bring together the best of these two worlds and provide unparalleled experiences combining thrilling sport and elegant art-de-vivre, wheel-to-wheel

> racing and time-tested craftmanship, for enthusiasts, fans and clients. LVMH and its teams will share their exceptional savoir-faire and entrepreneurial spirit with the world of Formula 1 through hospitality, bespoke activations, limited editions and outstanding content. ("Formula 1 and LVMH" 2024)

Formula One was set to receive a billion euros from the luxury-goods firm over the next decade (Klasa with Germano 2024; "LVMH Clinches" 2024). Per the America's Cup, the merchandising target was an international élite, with the conglomerate's 'stable of 75 brands packaged together in an effort to saturate almost every aspect of the sport' (Selleck 2024) because the market for luxury items, which had exploded with the addition of billionaires across the USA, China, and Britain, had suffered a downturn ("Why Louis" 2024).

Motorsports in general receive more money from environmental miscreants than anyone—billions—imbuing them with positive images derived from racing's putatively pleasurable connotations (New Weather Institute 2024; Irimia 2022). Formula One is of course most deeply associated with the extractive industries, as well as airlines and cellphone and car manufacturers, even as it secrets the environmental harm inflicted by such partners and sponsors. 'Global Partners' include Emirates Airlines, crypto.com, DHL, Aramco, and Pirelli. BBS, Ferrari, LIQUI MOLY, and Amazon are the somewhat-lesser "Official Partners." Aramco's Saudi owners avow that 'our global team is dedicated to creating impact in all that we do' to 'promote stability and long-term growth around the world.' DHL provides 'multi-modal transport … using land, sea and air freight, depending on the race calendar' ("Formula 1 Partners" n.d.).

Some sponsors are greenwashed for their allegedly positive focus on biofuels. Shell (Ferrari), Petronas and INEOS (Mercedes), and Aramco are "partners" in innovation and technology, regardless of the ecological damage they cause—perhaps because of it (Millington et al. 2022; Rayman 2023). And it almost feels supererogatory to list Shell's extraordinary record of greenwashing and climate malfeasance (ClientEarth n.d.). But just to mention a major issue from 2023–24: the firm was trying to sell its Nigerian subsidiary without mentioning or compensating the country for decades of 'grievous human rights abuses related to oil spills which have polluted the environment, contaminated drinking water and poisoned agricultural land, fisheries and people' ("Nigeria" 2024).

Crypto.com charges people to trade cryptocurrency on their behalf via a telephone application. It favors folks buying its own brand, which is kept beyond scrutiny (Witt 2022). DHL has been interrogated for labor abuses in the USA (Hornbeck 2023), received a poor rating from Carbon Market Watch and the New Climate Institute in terms of transparency and integrity (Day et al. 2023: 84), and was forced to adjust misleading claims about sustainable delivery following action by the Netherlands' Authority for Consumers and Markets ("DHL and PostNL" 2024). Petronas, a creature of the Malaysian state, was under investigation for corruption as its Formula One participation developed (Muhamed Radhi 2023). We have already considered the horror of INEOS and read about Emirates Airlines.

And Amazon? Founder Jeff Bezos' three-minute space jaunt in 2021 earned him more attention from NBC, CBS, and ABC than the networks' combined 2020 coverage of climate change (Macdonald 2021). How did this person make his fortune? Through a wildly successful strategy of waste: to lose money initially but become dominant through over-production, over-distribution, over-consumption, and over-destruction (Quiggin 2022: 179). The firm came to sell most imaginable commodities other than animals, growing into a 'giant conglomerate with a staggering number of industries in which it's the number one, two, or three player' (Dana Mattioli quoted in McCordick 2024)—almost 'monopolizing the qualities of daily life' (D Harvey 2020: 49).

Amazon's massive capacity has seen it build many warehouses in areas subject to dangerous weather, leading to casualties from climate change, and in predominantly working-class communities of color, where diesel trucks, planes, and trains contaminate the necessities of everyday life (Calma and Schiffer 2021). The company's role in polluting ports across the West Coast is part of an air-quality-crushing supply chain from China that has caused tens of thousands of deaths (Robertson 2021; Anenberg et al. 2019). Like DHL, Amazon gets a poor rating from Carbon Market Watch and the New Climate Institute for transparency and integrity (Day et al. 2023: 74). Every hour of every day, the firm dumps the equivalent of a vanload of plastic in the ocean. In 2020, perhaps 23.5 million pounds of its waste ended up in oceans, rivers, and lakes. Resident creatures mistook these toxic items for foodstuffs (Oceana 2021).

Needless to say, the corporation makes extravagant commitments to saving the climate. Its much-touted *Climate Pledge* has been exposed again and again as concerted greenwashing (STAND.earth 2024; Amazon Employees for Climate Justice 2024). Attempts to uncover how Amazon functions have led to viciously intimidatory, mendacious, and manipulative campaigns against workers and reporters (Breland 2021). The final word goes to Britain's ITV network. What it unearthed also happens in France and Germany:

> Amazon is destroying millions of items of unsold stock every year, products that are often new and unused. ... Undercover filming from inside Amazon's Dunfermline warehouse reveals the sheer scale of the waste: Smart TVs, laptops, drones, hairdryers, top of the range headphones, computer drives, books galore, thousands of sealed face masks—all sorted into boxes marked "destroy". ... An ex-employee ... told us: "From a Friday to a Friday our target was to generally destroy 130,000 items a week." (Pallot 2021)

The price Formula One pays for accepting the dirty money of these winning enterprises, notably Aramco?

> Formula 1 has deployed resources in Brussels to lobby politicians on issues vital to Saudi interests ... the racing group has used closed-door meetings to urge top EU officials to backtrack on climate-friendly plans in favour of 'e-fuels' that will help keep the internal combustion engine alive—and an autocratic monarchy in power. ("Trojan" 2023)

The overt oxymoron, that some of the world's most significant global polluters brandish green credentials and environmental solutions, points to empty CSR campaigns that aim to generate positive publicity (Miller et al. 2023), with 'Formula One seemingly engaging in virtue signalling and moral grandstanding without meaningfully tackling any of these issues or enacting any real change' (Sturm 2023: 177).

Tobacco is perhaps the best example. In 2007, Honda provided an "earth dream"-inspired livery of the planet on a black space background, devoid of any notable sponsors. This was allegedly done to raise awareness of environmental issues, with team boss Nick Fry stating that:

> Climate change is probably the single biggest issue facing the global community and FI is not immune from it. On the contrary, we believe that FI with its huge global profile and cutting[-]edge technology can play an important role in not only highlighting the issues but also playing our part in developing solutions. (quoted in "Honda Unveils" 2007)

The purportedly "green" desires underpinning that approach were lampooned when fans and the *bourgeois* media queried whether Honda's environmental focus was designed to divert attention from its loss of tobacco sponsorship.

The devastating effects of smoking on public health are well known. It also has a negative environmental impact, via the industry's deforestation, cultivation, curing, processing, manufacturing, distributing, and dumping waste, while smokers magically do not appear to regard dropping four and a half trillion poisonous objects on the ground each year as having an impact on anyone or thing. If they last to the age of fifty, each of these people needs three times the water of a non-smoker and leaves a carbon footprint that could only be offset by planting over a hundred trees: their addiction's annual fossil-fuel emissions amount to more than a decade of domestic electricity use (Ti et al. 2024; Karaman 2019; Houghton et al. 2018; Hendlin and Bialous 2020; Lee et al. 2016; Aráujo and Costa 2019; Zafaeiridou et al. 2018).

Greenwashing is a key priority for nicotine drug-dealers. A 1989 Formula One event saw the addiction they craved to develop on view almost 6000 times during television coverage (Blum 1991). The best-known company involved with Formula One has been Marlboro, part of Phillip Morris International (PMI), which adorned McLaren's livery for a quarter of a century and Ferrari's for a decade (Irimia 2022). Once Tony Blair took power in Britain, he met Bernie Ecclestone, a major Labour Party donor who then owned Formula One; the competition was soon exempted from legislation banning tobacco advertising (Stevenson and Agencies 2008). When cigarette commercials were nevertheless largely exiled from television worldwide in 2006, the peddlers turned to covert motorsport promotions as part of their restless search to manufacture and sustain (for a time) users (Reid 2015) via what is aptly referred to in medical research as smokescreen marketing (Barker et al. 2019).

Philip Morris' subliminal product placement has been noteworthy ever since. In place of brand names, the company deployed barcodes and a campaign as part of the search to transform smoking into a "safe" addiction, signaled with—hey presto—tropes of its usual logo. British American Tobacco (BAT) turned to a company slogan after being barred from naming itself (Dewhirst and Hunter 2002). Formula One team McLaren boasted that their 2019 'global partnership' aimed 'to deliver the world's tobacco and nicotine consumers a better tomorrow' (quoted in Mitchell 2019). On renewal in 2024, the team referred to this arrangement as a key commitment to diversity, equality, and inclusion (Brittle 2024).

The old image of 'rugged masculinity, independence, heroism, and adventure' had been displaced by 'transformation, progress, open dialogue, teamwork, innovation, technology, and science' (Dewhirst et al.

2023: 1844). Each was dedicated to fabricating addicts; or was it? BAT insisted ‘our partnership has nothing to do with BAT’s cigarette brands’ (quoted in Grez 2020), PMI that it had ‘nada que ver’ [nothing to do with] Netflix and Formula One’s tobacco product placement (quoted in “La serie” 2023). The FIA declared itself resolutely opposed to tobacco advertising, but powerless to control sponsorship contracts entered into by race teams (“Formule 1” 2023).

Formula One delivers over a billion minutes of streamed footage with nicotine content per year (STOP. 2022: 4–5). The Australian Communications and Media Authority found repeated instances of smokescreen tobacco advertising during television coverage of the Melbourne Grand Prix (Australian Communications and Media Authority 2021) and *Drive to Survive* has featured tobacco in half its opening sequences (STOP. 2022: 5). Researchers examining 400 minutes of the show’s fourth season found ‘just over a third (35.8%) contained images from cigarette companies, meaning that Netflix has likely screened a billion minutes of free tobacco branding since 2019’ (Assunta and Chamberlain 2024: 281). Such product placement landed the competition $105 million in 2021, $4.5 billion lifetime (STOP. and FM 2021). Millions of children aged eight to twelve follow the sport. They are in the sights of motorsport and corporate nicotine’s smoky alliance (STOP. and Formula Money 2024) in a project the World Health Organization calls *Hooking the Next Generation* (2024).

Official sponsors are carefully curated to exclude those drug dealers, though alcohol is included along with other industries responsible for planetary destructiveness, whether directly (such as the automotive sector) or indirectly (finance, for example).

Table 5.2 Formula one sponsors’ 2024 contributions

Industry	*Percentage*
Computing	19.0
Fashion	16.2
Automotive	13.4
Engineering	4.7
Alcohol	4.0
Finance	2.8
Airlines	2.5

Source: Based on “A Selection” 2024

Conclusion

Formula One embodies dual desires for growth and mobility in a fetishistic blend of economic planning and *bourgeois* liberty. On the one hand, regularity, reliability, and control of production, distribution, and consumption are incarnate in the competition's teams. On the other, their drivers exemplify neoliberal subjects, ever ready for adventure. Together with fans, these three actors exhibit the technological sublime.

The *New York Times* hailed the sport's supposed ecological breakthroughs in an op-ed by an author whose coffee-table history of Formula One retailed for $1341 (Spurgeon 2010). It came complete with '*Complimentary white gloves and a signature canvas tote bag*' ("*Formula 1: The Impossible Collection*" 2025); fitting for a sport whose not-very-private slogan is "Race on Sunday; Sell on Monday"—advertising on cars, tracks, and networks makes for quick sales to enthusiasts and improves sponsors' public image (Ang 2011).

For here's the real deal: motorsport's true formula, successful or not, has always been to go 'where the money was, washing its hands of moral considerations' (Williams 2020). It has done so in the spirit of the crypto-fascist Marinetti, exemplified in Madrid's promise for 2026: 'UN ENCUENTRO CON LA VELOCIDAD Y LA EMOCIÓN' [AN ENCOUNTER WITH SPEED AND FEELING] (n.d.). The disgraceful, but telling, link between the sport and militarism was only checked when air-force flybys prior to races were prohibited due to their environmental impact (Benson 2022).

Forbes magazine queried "Can Formula 1 be Entertaining and Sustainable?". The *Economist* asked, 'What's not to love about a billionaire's party?' ("A Problem" 2023). The answers were simple. They weren't the response of boosters and chorines, for whom this carbon sport sits at the 'apex of aerodynamics, skill and strategy' (Arafat et al. 2024). The appropriate reply? Formula One is forever 'espousing greenwashing rhetoric that ignores the environmental impact and ecological incongruency of global travel, transport, motorsport and car cultures' (Sturm 2023: 171)—what Jane Jacobs termed the 'attrition of automobiles' (1961: 903).

Carbon racing has a venal past, a venal present, and a venal future—if it is permitted to continue. The sport's self-justifications are absurd, its guarantees hypocritical, its partnerships unforgiveable, its destructiveness indubitable.

Bibliography

NB—Links from US Science Have Been Removed By Its Government

"10 Great Motor Racing Films." (2023, December 14). BFI https://www.bfi.org.uk/lists/10-great-motor-racing-films.

"A Problem at the Heart of Formula One." (2023, December 15). *Economist* https://www.economist.com/culture/2023/12/15/a-problem-at-the-heart-of-formula-one.

"A Selection of Formula 1 Sponsorship Trends." (2024, July 15). caytoo https://caytoo.co.uk/article/formula-1-sponsorship-trends/.

"DHL and PostNL Adjust Their Sustainability Claims Following an ACM Action." (2024, October 3). Authority for Consumers and Markets https://www.acm.nl/en/publications/dhl-and-postnl-adjust-their-sustainability-claims-following-acm-action.

"Driven to Watch: How a Sports Docuseries Drove U.S. Fans to Formula 1." (2022, May). *Nielsen* https://www.nielsen.com/insights/2022/driven-to-watch-how-a-sports-docuseries-drove-u-s-fans-to-formula-1/.

"El Grand Prix de Fórmula 1 en Madrid afectará a 729 árboles y el Ayuntamiento tendrá que suspender los estándares de calidad acústica." (2025, February 6). *Madrid On* https://www.madridon.com/2025/02/06/el-grand-prix-de-formula-1-en-madrid-afectara-a-729-arboles-y-el-ayuntamiento-tendra-que-suspender-los-estandares-de-calidad-acustica/.

"El PSOE pide explicaciones sobre los costes del Gran Premio: "Madrid merece la F1, pero sobre todo merece rigor"." (2025, February 8). *infobae* https://www.infobae.com/espana/agencias/2025/02/05/el-psoe-pide-explicaciones-sobre-los-costes-del-gran-premio-madrid-merece-la-f1-pero-sobre-todo-merece-rigor/.

"Expert Slams Climate Greenwashing in F1." (2023, August 23). *GrandPrix* https://www.grandprix.com/news/expert-slams-climate-greenwashing-in-f1.html.

"Fees Paid to Host Formula One Races in 2023, by Circuit." (2024, April 16). Statista Research Department https://www.statista.com/statistics/1360331/formula-1-hosting-fees/.

"Felipe Calderón dio su primer informe como miembro de la FIA, máximo órgano de automovilismo." (2022, March 11). *Infobae* https://www.infobae.com/america/deportes/2022/03/11/felipe-calderon-dio-su-primer-informe-como-miembro-de-la-fia-maximo-organo-de-automovilismo/.

"Formula 1 and LVMH Announce Historic 10-Year Global Partnership." (2024, January 10). LVMH https://www.lvmh.com/en/news-lvmh/formula-1-and-lvmh-announce-historic-10-year-global-partnership.

"Formula 1 Partners." (n.d.). Formula 1 https://www.formula1.com/en/information/formula-1-partners.2shz7bIVhGgS3ovP69N3Rs.

"*Formula 1: The Impossible Collection.*" (2025). Assouline https://www.assouline.com/products/formula-1-the-impossible-collection?srsltid=AfmBOorZbZ5vx-rDoptmqhBAlnmXcObosmhHEdu35-NhXtNEWBkVBXS4.

"Formule 1 : la série Netflix 'Formula 1 : Pilotes de leur destin' critiquée pour sa promotion du tabac." (2023, March 9). *Le Figaro* https://www.lefigaro.fr/flash-sport/f1-la-serie-netflix-formula-1-pilotes-de-leur-destin-critiquee-pour-sa-promotion-du-tabac-20230308.

"Heineken Heiress Finds New Way to Avoid Dutch Taxes, Sheltering Billions in Tax Havens: NRC." (2024, April 4). *NL Times* https://nltimes.nl/2024/04/04/heineken-heiress-finds-new-way-avoid-dutch-taxes-sheltering-billions-tax-havens-nrc.

"Honda Unveil New Livery." (2007, February 22). *Eurosport* https://www.eurosport.com/formula-1/season/2007/honda-unveil-new-liverysto1098918/story.shtml.

"La serie de Netflix sobre la Fórmula 1 recibe criticas [*sic.*] por promocionar el tabaco." (2023, March 8). *Life and Style* https://lifeandstyle.expansion.mx/entretenimiento/2023/03/08/la-serie-de-netflix-sobre-la-formula-1-recibe-criticas-por-promocionar-el-tabaco.

"Liberty Media Corporation Reports Fourth Quarter and Year End 2024 Financial Results." (2025, February 27). Liberty Media https://www.libertymedia.com/investors/news-events/press-releases/detail/557/liberty-media-corporation-reports-fourth-quarter-and-year.

"LVMH Clinches 10-Year Sponsorship Deal with Formula 1." (2024, October 2). *Reuters* https://www.reuters.com/business/lvmh-strikes-sponsorship-deal-with-formula-1-2024-10-02/.

"Mike Hawthorn Killed." (1959). *Pathé News* https://www.youtube.com/watch?v=SyiJQ5_CbXM.

"Nigeria: Shell Must be Held Fully Accountable for Human Rights Harms Before Being Allowed to Sell its Niger Delta Business." (2024, May 14). Amnesty International https://www.amnesty.org/en/latest/news/2024/05/nigeria-shell-must-be-held-fully-accountable-for-human-rights-harms-before-being-allowed-to-sell-its-niger-delta-business/.

"Prime Minister Sets Out Blueprint to Turbocharge AI." (2025, January 13). UK Government https://www.gov.uk/government/news/prime-minister-sets-out-blueprint-to-turbocharge-ai.

"Racing Legends—Stirling Moss." (2012). https://www.youtube.com/watch?v=UtRWhPQVAY8&t=1466s.

"Report Questions Environmental Impact of Madrid Grand Prix." (2025, February 7). *GrandPrix247* https://www.grandprix247.com/2025/02/07/report-questions-environmental-impact-of-madrid-grand-prix/.

"Seven Incredible Private Jets Owned by F1 Drivers: Lewis Hamilton, Max Verstappen and More." (2023, June 15). *PlanetF1* https://www.planetf1.com/features/f1-private-jets-owned-lewis-hamilton-max-verstappen.

"Szilard Petition." (1945). *Atomic Heritage Foundation* https://ahf.nuclearmuseum.org/ahf/key-documents/szilard-petition/.

"The Ultimate List of Formula One Romance Novels." (2024, August 3). *Ana's Attic* https://anasattic.com/formula-one-romance-novels/.

"Total Revenue of the Formula One Group From 2017 to 2022." (2024, April 10). Statista Research Department https://www.statista.com/statistics/1137226/formula-one-revenue/.

"Trojan Horsepower: Formula 1 and the Saudi Oil Lobby." (2023, December 7). *SourceMaterial* https://www.source-material.org/saudi-aramco-formula-1-fossil-fuel-lobby/.

"UN ENCUENTRO CON LA VELOCIDAD Y LA EMOCIÓN." (n.d.). IFEMA https://www.ifema.es/f1.

"When Playboys Ruled the World: Barry Sheene and James Hunt." (2011). https://www.youtube.com/watch?v=-4UXfyxgZsk.

"When Sex Was Safe & Racing Was Dangerous." (2015). https://www.youtube.com/watch?v=GGP5FEIMza4.

"Why Louis Vuitton is Struggling but Hermès is Not." (2024, December 16). *Economist* https://www.economist.com/business/2024/12/16/why-louis-vuitton-is-struggling-but-hermes-is-not.

Adorno, Theodor W. (1975) "Culture Industry Reconsidered." Trans. Anson G. Rabinbach. *New German Critique* 7: 12–19.

Agini, Samuel. (2025, March 15). "Formula 1 Owner Bets on Brad Pitt's Pulling Power to Shift into US Mainstream." *Financial Times* https://www.ft.com/content/61b0156f-0025-4f2b-8037-e7477c2e92c1?utm_source=semafor.

Aguad, Felipe and Charles Thraves. (2024). "Optimizing Pit Stop Strategies in Formula 1 with Dynamic Programming and Game Theory." *European Journal of Operational Research* 319, no. 3: 908–19.

Alcívar, Estíbaliz. (2024, September 25). "GP de España en Madrid sin inversores: Ninguna empresa quiere asumir el "riesgo" de organizar el evento de F1." *Car and Driver* https://www.caranddriver.com/es/formula-1/a62364002/gp-espana-madrid-ningun-inversor-riesgo-organizar-f1/.

Allen, James. (2013, November 1). "Hybrid Power Fits with Greener Agenda." *Financial Times*: 2.

Allen, James. (2014, March 14). "F1 Sponsors and Supporters Drive the Shift to Sustainability." *Financial Times* https://www.ft.com/content/ceed5302-a446-11e3-9cb0-00144feab7de.

Amazon Employees for Climate Justice. (2024). *Burns Trust: The Amazon Unsustainability Report* https://static1.squarespace.com/static/65681f099d7c3d48feb86a5f/t/668ebf702516716ca72bbf98/1720631157044/unsustainability-report.pdf.

Ames, Nick. (2025, February 14). "Ineos and Ratcliffe's Sporting Empire Risks Atrophy with Horizons Narrowing." *Guardian* https://www.theguardian.com/football/2025/feb/14/ineos-jim-ratcliffe-manchester-united-sporting-empire-ben-ainslie?CMP=Share_iOSApp_Other.

Anenberg, Susan, Joshua Miller, Daven Henze, and Ray Minares. (2019). *A Global Snapshot of the Air Pollution-Related Health Impacts of Transportation Sector Emissions in 2010 and 2015.* International Council on Clean Transportation/Climate & Clean Air Coalition https://theicct.org/sites/default/files/publications/Global_health_impacts_transport_emissions_2010-2015_20190226.pdf.

Ang, Ulysses. (2011, January 19). "Race on Sunday, Sell on Monday." *The Philippine Star* http://www.philstar.com/motoring/649035/race-sunday-sell-monday.

Antich, José. (2025, February 7). "Problemas para la Fórmula 1 en Madrid." (2025, February 7). *El Nacional* https://www.elnacional.cat/es/editorial/problemas-formula-1-madrid-jose-antich_1360055_102.html.

Arafat, Yasir, Muhammad Rizwan Azhar, and Waqas Uzair. (2024, June 24). "Formula One is Moving Towards Hybrid Engines and Renewable Fuel. Major Environmental Progress or Just 'Greenwashing'? *The Conversation* https://theconversation.com/formula-one-is-moving-towards-hybrid-engines-and-renewable-fuel-major-environmental-progress-or-just-greenwashing-232375#:~:text=According%20to%20a%20report%20from,the%20cars%20on%20the%20track.

Araújo, Maria Cristina B. and Monica F. Costa. (2019). "A Critical Review of the Issue of Cigarette Butt Pollution in Coastal Environments." *Environmental Research* 172: 137–49.

Assunta, Mary and Phil Chamberlain. (2024). "Ensuring a Safe, Tobacco Free Future for the Young: Protecting Children from Tobacco Industry Interference." *Tobacco Control* 33, no. 3: 281–82.

Australian Communications and Media Authority. (2021, December 13). *Investigation Report no. BI-607* https://www.acma.gov.au/publications/2022-02/report/investigation-report-bi-607.

Barker, Alexander B., Magdalena Opazo Breton, Rachael L. Murray, Bruce Grant-Braham, and John Britton. (2019). "Exposure to 'Smokescreen' Marketing During the 2018 Formula 1 Championship." *Tobacco Control* 28: e154–55.

Battan, Carrie. (2022, March 11). "How 'Drive to Survive' Remade Formula 1." *New Yorker* https://www.newyorker.com/culture/culture-desk/how-drive-to-survive-remade-formula-1.

Bell, Sebastien. (2022, February 20). "How Can an F1 Team Spend $450 Million in a Year and Still Make a Profit?" *CARSCOOPS* https://www.carscoops.com/2022/02/how-does-an-f1-team-spend-450-million-in-a-year-and-still-make-a-profit/.

Benson, Andrew. (2022, January 21). "Formula 1 Bans Military Air Displays at Grands Prix." *BBC* https://www.bbc.com/sport/formula1/60082961.

Black, Richard. (2010, June 30). "Formula One Embarks on Carbon-Cutting Drive." *BBC* http://www.bbc.com/news/10456984.

Blitz, Roger. (2013, March 15). "New Season, New Faces—The Drama Continues." *Financial Times*: 1, 3.

Blum, Alan. (1991). "The Marlboro Grand Prix—Circumvention of the Television Ban on Tobacco Advertising." *New England Journal of Medicine* 324: 913–17.

Bodie, Matthew T. (2011). "NASCAR Green: The Problem of Sustainability in Corporations and Corporate Law." *Wake Forest Law Review* 46, no. 3: 491–522.

Bogado, Ignacio. (2025, February 2). "Madrid evaluá reubicar residentes y ajustes del trazado para suavizar el impacto del GP de F1." *GP* https://grandepremio.com/es/f1/madrid-evalua-reubicar-residentes-ajustes-trazado-f1-2026/.

Bono, Ferran. (2021, August 9). "The Shanty Town on Valencia's Abandoned Formula 1 Circuit." *El País* https://english.elpais.com/society/2021-08-09/the-shanty-town-on-valencias-abandoned-formula-1-circuit.html.

Boxall-Legge, Jake. (2020, June 30). "How Mercedes Can Address the Lack of Minority Representation." *Motorsport* https://www.motorsport.com/f1/news/how-mercedes-can-address-the-lack-of-minority-representation-in-f1/4816804/.

Brecht, Bertolt. (2018). *The Collected Poems of Bertolt Brecht*. Trans. and Ed. Tom Kuhn and David Constantine with Charlotte Ryland. New York: Liveright Publishing.

Breland, Ali. (2021, June 25). "How Amazon Bullies, Manipulates, and Lies to Reporters." *Mother Jones* https://www.motherjones.com/politics/2021/06/amazon-journalists-pr-tactics/.

Brinsford, James. (2023, January 18). "Formula 1 Carbon Footprint Slammed After 2023 Calendar Release." *Newsweek* https://www.newsweek.com/formula-one-carbon-footprint-slammed-2023-calendar-1774806.

Brittle, Cian. (2024, April 8). "McLaren Racing Agrees Multi-Year BAT Sponsorship Extension." *Blackbook Motorsport* https://www.blackbookmotorsport.com/news/mclaren-racing-f1-formula-e-bat-british-american-tobacco-vuse/.

Brittle, Cian. (2025, January 30). "Louis Vuitton Officially Joins F1 as Part of LVMH Deal." *SportsPro* https://www.sportspro.com/news/f1-louis-vuitton-lvmh-sponsorship-january-2025/.

Brown, Maury. (2023, March 29). "Inside The Numbers That Show Formula 1's Popularity and Financial Growth." *Forbes* https://www.forbes.com/

sites/maurybrown/2023/03/29/inside-the-numbers-that-show-formula-1s-popularity-and-financial-growth/.
Bustad, Jacob J. and David L. Andrews. (2023). "The Circus Comes to Town: Formula 1, Globalization, and the Uber-Sport Spectacle." *Lives in the Fast Lane: Essays on the History and Politics of Motor Racing*. Ed. Damion Sturm, Stephen Wagg, and David L. Andrews. Cham: Palgrave Macmillan. 595–617.
Calma, Justine and Zoe Schiffer. (2021, May 25). "Amazon Workers Demand Company Quit Polluting Near Communities of Color." *The Verge* https://www.theverge.com/2021/5/25/22453516/amazon-workers-pollution-petition-letter-demand.
Campbell, Roderick. (2013). *Blowout! A Cost Benefit Analysis of the Australian Grand Prix*. Economists at Large http://www.ecolarge.com/wp-content/uploads/2013/03/Blowout-A-cost-benefit-analysis-of-the-Australian-Grand-Prix-Ecolarge-Final.pdf.
Castelló, Carlos Navarro. (2024, September 17). "Carreras illegals en el abandonado circuito urbano de Fórmula 1 de València." *El Diario* https://www.eldiario.es/comunitat-valenciana/valencia/carreras-ilegales-abandonado-circuito-urbano-formula-1-valencia_1_11656041.html.
Cave, Andrew and Alex Miller. (2015, July 22). "Passion Drives Sponsorship of Formula One." *Telegraph* http://www.telegraph.co.uk/investing/business-of-sport/formula-one-sponsorship/.
Churchwell, Sarah. (2025, April 4). "How Gatsby Foretold Trump's America." *Financial Times* https://www.ft.com/content/dd6f1a32-398e-47f2-85a3-41120c952554?emailId=6e35221e-171a-48ce-9e8d-bdede0f9e5f0&segmentId=9264b0f7-e7ac-8f9b-044f-c10729049333.
ClientEarth. (n.d.). *Greenwashing Files* https://www.clientearth.org/projects/the-greenwashing-files/.
Coch, Matt. (2022, February 23). "FI Teams Battle Development-Cost Cap Equation." *Speed Cafe* https://speedcafe.com/f1-teams-battle-development-cost-cap-equation/.
Coleman, Madeline. (2023, February 25). "'Drive to Survive' Season Five: Our Top Five Takeaways." *The Athletic* https://theathletic.com/4251298/2023/02/25/drive-to-survive-season-5-takeaways-f1/.
Coleman, Madeline. (2024, May 9). "How F1's Miami GP's Food Surplus Feeds the City and Fights Climate Change." *The Athletic* https://www.nytimes.com/athletic/5482029/2024/05/09/f1-miami-gp-food-rescue/.
Day, Thomas, Silke Mooldijk, Frederic Hans, Sybrig Smit, Eduardo Posada, Reena Skribbe, Santiago Woollands, Harry Fearnehough, Takeshi Kuramochi, Carsten Warnecke, Aki Kachi, and Niklas Höhne with Carbon Market Watch. (2023, February). *Corporate Climate Responsibility Monitor 2023: Assessing the Transparency and Integrity of Companies' Emission Reduction*

and Net-Zero Targets. Carbon Market Watch and the New Climate Institute. https://carbonmarketwatch.org/wp-content/uploads/2023/02/CorporateClimateResponsibilityMonitor2023.pdf.

Dewhirst, Timothy, Wonkyong Beth Lee, and Lauren Czaplicki. (2023). "Philip Morris International's Formula 1 Sponsorship-Linked Marketing: Transformation from Marlboro to Mission Winnow." *Nicotine and Tobacco Research* 25: 1838–846.

Dewhirst, Timothy and A. Hunter. (2002). "Tobacco Sponsorship of Formula One and CART Auto Racing: Tobacco Brand Exposure and Enhanced Symbolic Imagery Through Co-Sponsors' Third Party Advertising." *Tobacco Control* 11: 146–50.

Dewhirst, Tim. (2023). "The Geopolitics of Money Versus Morals: Location, Location, Location of the Formula 1 Race Calendar." *The Geopolitical Economy of Sport: Power, Politics, Money, and the State*. Ed. Simon Chadwick, Paul Widdop, and Michael M. Goldman. Abingdon: Routledge. 235–41.

Dingle, Gregory. (2009). "Sustaining the Race: A Review of Literature Pertaining to the Environmental Sustainability of Motorsport." *International Journal of Sports Marketing and Sponsorship* 11, no. 1: 75–91.

Domingo Rodríguez, Diego. (2025, January 9). "El Gran Premio de Fórmula 1 en Madrid: entusiasmo, críticas y debate sobre su impacto." *El Diario de Madrid* https://www.eldiariodemadrid.es/articulo/sociedad/gran-premio-formula-1-madrid-entusiasmo-criticas-debate-impacto/20250109162501087476.html.

Eagleton, Terry. (1990). *The Ideology of the Aesthetic*. Oxford: Blackwell.

Eco, Umberto. (1987). *Travels in Hyperreality*. Trans. William Weaver. London: Picador.

Elliott, Larry. (2014, March 23). "How F1 and Champagne Might Help Us to Solve Global Warming." *Guardian* http://www.theguardian.com/business/2014/mar/23/solve-global-warming-pension-champagne-formula-1.

Fédération Internationale de l'Automobile. (n.d.). *FIA Environmental Strategy 2020–2030* https://www.fia.com/sites/default/files/fiaenvironmentalstrategyv4web.pdf.

Fitzgerald, F. Scott. (1925). *The Great Gatsby*. New York: Scribner's.

Formula 1. (2019). "Formula 1 Sustainability Strategy." https://corp.formula1.com/sustainability/.

Formula 1. (2022). "More Efficient, Less Fuel, and Carbon Net Zero—7 Things You Need to Know about the 2026 F1 Engine Regulations." https://www.formula1.com/en/latest/article.more-efficient-less-fuel-and-carbon-net-zero-7-things-you-need-to-know-about.ZhtzvU3cPCv8QO7jtFxQR.html.

Foye, Lauren. (2024, July 9). "Racing Towards Net Zero: How F1 is Going Green in Response to Criticism Around its Impact on the Environment." *Zero Carbon Academy* https://www.zerocarbonacademy.com/posts/racing-towards-net-zero-how-f1-is-going-green-in-response-to-criticism-around-its-impact-on-the-environment.

Fu, Rachel. (2024, February 14). "Revving Up Tourism: Formula One and Other Big Events Look Set to Drive Growth in the Hospitality Industry." *The Conversation* https://theconversation.com/revving-up-tourism-formula-one-and-other-big-events-look-set-to-drive-growth-in-the-hospitality-industry-221706.

Gasparetto, Thadeu, Marina Orlova, and Anton Vernikovskiy. (2024). "Same, Same but Different: Analyzing Uncertainty of Outcome in Formula One Races." *Managing Sport and Leisure* 29, no. 4: 651–65.

Gezici, Ferhan and Serra Er. (2014). "What Has Been Left After Hosting the Formula 1 Grand Prix in Istanbul?" *Cities* 41, Part A: 44–53.

Gibbs, Carole, Meredith L. Gore, Edmund F. McGarrell, and Louie Rivers III. (2010). "Introducing Conservation Criminology: Towards Interdisciplinary Scholarship on Environmental Crimes and Risks." *British Journal of Criminology* 50, no. 1: 124–44.

Glendenning, Barry. (2016, October 9). "The Lewis Hamilton Enigma is Quite Why He Thinks He's a Mysterious Riddle." *Guardian* https://www.theguardian.com/sport/blog/2016/oct/09/lewis-hamilton-national-treasure-paranoid-f1.

GMM. (2016, June 16). "Critics Lash out at F1 Over New Heineken Sponsorship." *Autoweek* https://www.autoweek.com/racing/formula-1/a1848126/critics-lash-out-f1-over-new-major-beer-sponsor-deal/.

Gordon, Josh. (2022, December 21). "Grand Prix Costs Victoria Half a Billion Dollars Over 10 Years." *The Age* https://www.theage.com.au/national/victoria/grand-prix-costs-victoria-half-a-billion-dollars-over-10-years-20221221-p5c831.html.

Graham, Caroline. (2013, May 27). "To You, James Hunt Was an F1 Playboy Who Bedded 5000 Women. To Me, He Was Dad—Who Doted on Me … and His 300 Budgies." *Mail on Sunday* https://www.dailymail.co.uk/femail/article-2331043/James-Hunt-F1-playboy-bedded-5-000-women-To-Tom-Hunt-Dad%2D%2Ddoted%2D%2D300-budgies.html.

Grez, Matias. (2020, December 11). "Some Motorsport Teams Remain Addicted to Tobacco Company Sponsorship Deals, Despite Tobacco Causing 8 Million Deaths Each Year." *CNN* https://edition.cnn.com/2020/12/11/motorsport/formula-one-tobacco-sponsorship-deals-spt-itl/index.html.

Gutiérrez Cañas, Marta. (2025, February 13). "Adiós a la Ley del Ruido para el Gran Premio de F1 en Madrid." *La Razón* https://www.larazon.es/madrid/adios-ley-ruido-gran-premio-formula-uno-madrid-p7m_2025021367add7e5417ec20001f86058.html.

Habtemariam, Dawit. (2024, August 16). "Las Vegas Tourism Agency to Spend $1.7 Million on Formula 1 Tickets." *Skift* https://skift.com/2024/08/16/las-vegas-tourism-agency-to-spend-1-7-million-on-formula-1-tickets/.

Hanley, Steve. (2019, April 30). "The Gigantic Carbon Footprint of Formula One." *Clean Technica* https://cleantechnica.com/2019/04/30/the-gigantic-carbon-footprint-of-formula-one/.

Harte, Tim. (2009). *Fast Forward: The Aesthetics and Ideology of Speed in Russian Avant-Garde Culture, 1910–1930.* Madison: University of Wisconsin Press.

Harvey, David. (2020). *The Anti-Capitalist Chronicles.* Ed. Jordan T. Camp and Chris Caruso. London: Pluto Press.

Hendlin, Yogi Hale and Stella A. Bialous. (2020). "The Environmental Externalities of Tobacco Manufacturing: A Review of Tobacco Industry Reporting." *Ambio: A Journal of Environment and Society* 49, no. 1: 17–34.

Henry, Nick, Tim Angus, Mark Jenkins, and Chris Aylett. (2007). *Motorsport Going Global: The Challenges Facing the World's Motorsport Industry.* Cham: Palgrave Macmillan.

Hogan, Michael. (2015, January 11). "Suzi Perry on F1 Technology, Her Love of Apple and Useful Emojis." *Guardian* http://www.theguardian.com/technology/2015/jan/11/suzi-perry-tv-presenter-motor-sport-favourite-technology.

Hornbeck, Kenton. (2023, November 1). "DHL Faces New Labor Controversy: Unfair Practices Allegations Escalate." *LINK nky* https://linknky.com/business/2023/11/01/dhl-faces-new-labor-controversy-unfair-practices-allegations-escalate/.

Houghton, Frank, Sharon Houghton, Diane O'Doherty, Derek McInerney, and Bruce Duncan. (2018). "'Greenwashing' Tobacco Products Through Ecological and Social/Equity Labelling: A Potential Threat to Tobacco Control." *Tobacco Prevention and Cessation* 4: 37.

Hume, David. (2007). *Dialogues Concerning Natural Religion and Other Writings.* Ed. Dorothy Coleman. Cambridge: Cambridge University Press.

Irimia, Silvian. (2022, March 11). "Best Sponsors in Formula One History." *Autoevolution* https://www.autoevolution.com/news/best-sponsors-in-formula-one-history-183650.html.

Jacobs, Jane. (1961). *The Death and Life of American Cities.* New York: Vintage.

Jensen, Jonathan A., Joe B. Cobbs, Alex Mazer, and B. David Tyler. (2024). "Analyzing Brand Strategy on an International Scale: The Sponsorship Performance Cycle in Formula One Racing." *Journal of International Marketing* 32, no. 3: 23–42.

Karaman, M. Ihsan. (2019). "Smoking as an Environmental Health Problem." *International Journal of Human and Health Sciences* 3, no. 3: 123–26.

Keegan, Brendan. (2024, November 1). "Racing's Role in Shaping Modern Masculinity: Redefining Strength and Vulnerability." *Rolling Stone* https://www.rollingstone.com/culture-council/articles/racings-shaping-modern-masculinity-redefining-strength-vulnerability-1235148485/.

Kennedy, Niamh. (2023, August 25). "Heineken Sells Russian Business for $1 as it Completes Exit." *CNN* https://edition.cnn.com/2023/08/25/business/heineken-exits-russia/index.html.

Kew, Matt. (2023, December 2). "Hamilton: F1 Diversity Not Moved "Anywhere Near as Much as it Should Have"." *Motorsport* https://www.motorsport.com/f1/news/hamilton-f1-diversity-not-moved-anywhere-near-as-much-as-it-should-have-/10554313/.

King, Ed. (2013, March 11). "Formula One: The Petrol Heads Driving the Green Economy." *Responding to Climate Change* https://www.rtcc.org/2013/03/11/formula-one-the-petrolheads-driving-the-green-economy/.

Klasa, Adrienne with Sara Germano. (2024, October 2). "LVMH Strikes Sponsorship Deal with Formula 1." *Financial Times* https://www.ft.com/content/5bee32c7-3305-474e-a85b-93ff7a5e333f.

Lawrence, Andrew. (2021, December 17). "'Big Egos, Power Struggles, Stunning Betrayals': How Netflix's *Drive to Survive* Turned Americans into F1 Fans." *Guardian* https://www.theguardian.com/media/2021/dec/17/netflixs-drive-to-survive-americans-f1-fans.

Lee, Kelley, Natalia Carrillo Botero, and Thomas Novotny. (2016). "'Manage and Mitigate Punitive Regulatory Measures, Enhance the Corporate Image, Influence Public Policy': Industry Efforts to Shape Understanding of Tobacco-Attributable Deforestation." *Globalization and Health* 12. https://doi.org/10.1186/s12992-016-0192-6.

Lefebvre, Sylvain and Romain Roult. (2011). "Formula One's New Urban Economies." *Cities* 28, no. 4: 330–39.

Lim, Vanessa. (2022, February 10). "Race to Reduce Singapore's F1's Carbon Footprint a Good Start, but Experts are Mixed on its Impact." *Channel News Asia* https://www.channelnewsasia.com/singapore/f1-singapore-formula-one-carbon-footprint-environment-2487536.

Logan, Joan. (2024, March 2). "Melbourne's F1 Grand Prix a Financial Car Crash for Victoria." *Pearls and Irritations* https://johnmenadue.com/melbournes-f1-grand-prix-a-financial-car-crash-for-victoria/.

Lowes, Mark. (2004). "Neoliberal Power Politics and the Controversial Siting of the Australian Grand Prix Motorsport Event in a Public Park." *Loisir et Société/Society and Leisure* 27, no. 1: 69–88.

Lowes, Mark. (2018). "Toward a Conceptual Understanding of Formula One Motorsport and Local Cosmopolitanism Discourse in Urban Placemarketing Strategies." *Communication & Sport* 6, no. 2: 203–18.

Luxemburg, Rosa. (2014). *The Complete Works of Rosa Luxemburg. Volume I: Economic Writings*. London: Verso.

Macdonald, Ted. (2021, April 13). "In 2020, Broadcast TV Networks Mentioned Climate Solutions in Less Than 30% of Overall Climate Change Coverage."

Media Matters https://www.mediamatters.org/broadcast-networks/2020-broadcast-tv-networks-mentioned-climate-solutions-less-30-overall-climate.

Marinetti, F. T. (2009a). "The Founding and Manifesto of Futurism." *Futurism: An Anthology*. Ed. Lawrence Rainey, Christine Poggi, and Laura Wittman. New Haven: Yale University Press. 49–53.

Marinetti, F. T. (2009b). "The New Religion-Morality of Speed." *Futurism: An Anthology*. Ed. Lawrence Rainey, Christine Poggi, and Laura Wittman. New Haven: Yale University Press. 224–29.

Marx, Karl. (1987). *Capital: Vol. 1: A Critical Analysis of Capitalist Production*, 3rd ed. Ed. Friedrich Engels. Trans. Samuel Moore and Edward Aveling. New York: International Publishers.

Marx, Leo. (1964). *The Machine in the Garden: Technology and the Pastoral Ideal in America*. New York: Oxford University Press.

Matthews, Jordan J. K. and Elizabeth C. J. Pike. (2016). "'What On Earth Are They Doing in a Racing Car?': Towards an Understanding of Women in Motorsport." *International Journal of the History of Sport* 33, no. 13: 1532–550.

McCordick, Jack. (2024, April 23). "Inside the Brutal Business Practices of Amazon—And How It Became "Too Toxic to Touch"." *Vanity Fair* https://www.vanityfair.com/news/story/inside-amazon-business-practices.

McCubbing, Gus and Zoe Samios. (2024, November 28). "Grand Prix Has Cost Victoria Nearly $700m in the Last Decade." *Australian Financial Review* https://www.afr.com/companies/sport/grand-prix-has-cost-victoria-nearly-700m-in-the-last-decade-20241128-p5kua5.

McIntyre, Isaac. (2023, October 26). "Lando Norris Sets New Twitch Record After F1 Cancels Aus Grand Prix." *Dexerto* https://www.dexerto.com/entertainment/lando-norris-twitch-record-f1-cancels-aus-grand-prix-1341664/.

McLaren Racing. (2021, April 14). *Sustainability at McLaren Racing* https://static-cdn.mclaren.com/static/pdf/SustainabilityatMcLarenRacing14April2021.pdf.

Mehta, Amar. (2023, March 25). "Lewis Hamilton: Diversity in Formula 1 About Behind Scenes." *Sky Sports* https://www.skysports.com/f1/news/12433/12842284/lewis-hamilton-diversity-in-formula-1-about-behind-scenes-i-felt-like-lone-ranger.

Miller, Toby, Brett Hutchins, Libby Lester, and Richard Maxwell. (2023). "Formula One and the Insanity of Car-Based Transportation." *The History and Politics of Motor Racing: Lives in the Fast Lane*. Ed. Damion Sturm, Stephen Wagg, and David L. Andrews. Cham: Palgrave Macmillan. 733–60.

Miller, Toby. (1998). *Technologies of Truth: Cultural Citizenship and the Popular Media*. Minneapolis: University of Minnesota Press.

Millington, Rob, Audrey R. Giles, Nicolien van Luijk, and Lyndsay M. C. Hayhurst. (2022). "Sport for Sustainability? The Extractives Industry,

Sport, and Sustainable Development." *Journal of Sport and Social Issues* 46, no. 3: 293–317.

Mitchell, Scott. (2019, February 11). "Former BAR Team Owner BAT Back into Formula 1 with McLaren Deal." *Autosport* https://www.autosport.com/f1/news/former-bar-team-owner-bat-back-into-formula-1-with-mclaren-deal-5283606/5283606/.

Morgan, Rhys and Yohanes Scarlett. (2021, July). *Accelerating Change: Improving Representation of Black People in UK Motorsport.* The Hamilton Commission and the Royal Academy of Engineering https://static1.squarespace.com/static/5f29736c8982c82f61df71e0/t/60edd33a6f118478735acbbc/1626198854176/THC+-+Accelerating+Change+-+July+2021.pdf.

Morris, Alex. (2012, August 28). "Formula One's Environmental Initiatives." *Triple Pundit* https://www.triplepundit.com/podium/formula-environmental-initiatives/.

Mosley, Max. (2015). *Formula One and Beyond: The Autobiography.* London: Simon and Schuster.

Mourão, Paulo. (2017). *The Economics of Motorsports: The Case of Formula One.* Cham: Palgrave Macmillan.

Muhamad, Susana. (2023, December 13). Speech https://x.com/susanamuhamad/status/1735049874540744800?ref_src=twsrc%5Etfw%7Ctwcamp%5Etweetembed%7Ctwterm%5E1735049874540744800%7Ctwgr%5Ec0e274aa9c8c0603b25025d40ea0c902c7a15c7e%7Ctwcon%5Es1_&ref_url=https%3A%2F%2Fwww.climatechangenews.com%2F2023%2F12%2F13%2Fdubai-deal-ministers-and-observers-react-to-the-uae-consensus%2F.

Mytton, David. (2020). "Hiding Greenhouse Gas Emissions in the Cloud." *Nature Climate Change* 10: 700–01.

New Weather Institute. (2024, September 18). *Dirty Money: How Fossil Fuel Sponsors Are Polluting Sport* https://www.newweather.org/wp-content/uploads/2024/09/Dirty-Money-How-Fossil-Fuel-Sponors-Pollute-Sport-18-09-2024.pdf.

Nichols, Georgia and Mike Savage. (2017). "A Social Analysis of an Elite Constellation: The Case of Formula 1." *Theory, Culture & Society* 34, nos. 5–6: 201–25.

Nye, David E. (1994). *American Technological Sublime.* Cambridge, Mass.: MIT Press.

Oceana. (2021). *Exposed: Amazon's Enormous and Rapidly Growing Plastic Pollution Problem* https://oceana.org/reports/amazon-report-2021/.

Ogan, Christine L. Manaf Bashir, Lindita Camaj, Yunjuan Luo, Brian Gaddie, Rosemary Pennington, Sonia Rana, and Mohammed Salih. (2009). "Development Communication: The State of Research in an Era of ICTs and Globalization." *Gazette* 71, no. 8: 655–70.

Onwurah, Chi. (2021, July 16). "Lewis Hamilton is Right About Diversity. But the Issue Goes Way Beyond Motorsport." *Guardian* https://www.theguardian.com/commentisfree/2021/jul/16/lewis-hamilton-motorsport-diversity-f1-commission-black-female-minorities.

Orlove, Raphael. (2024, April 22). "F1 in its Streamer Era." *Road & Track* https://www.roadandtrack.com/car-culture/a60013306/f1-is-for-the-streamers/.

O'Shea, Michelle, Neil Perry, and Sarah Duffy. (2020, January 15). "Formula 1 Says It's Going Carbon Neutral but Fans Must Demand Greater Detail on How." *The Conversation* https://theconversation.com/formula-1-says-its-going-carbon-neutral-but-fans-must-demand-greater-detail-on-how-127328.

Öztopcu, Asli. (2024). "Environmental Sustainability in Sport: Formula 1's SDGs Responsiveness." *Environment, Development and Sustainability* 26: 31075–1097.

Painter, Corinne. (2024). "Reconstructing British Identity: Formula One, Michael Schumacher and the British Press at the Turn of the Century." *Language and Intercultural Communication* 24, no. 2: 77–89.

Pallot, Richard. (2021, June 22). "Amazon Destroying Millions of Items of Unsold Stock in One of its UK Warehouses Every Year, ITV News Investigation Finds." *ITV* https://www.itv.com/news/2021-06-21/amazon-destroying-millions-of-items-of-unsold-stock-in-one-of-its-uk-warehouses-every-year-itv-news-investigation-finds.

Paraguassu, Lisandra and Fabio Teixeira. (2023, October 6). "Heineken-Owned Brewery Added to Brazil 'Slave Labor' List." *Reuters* https://www.reuters.com/business/retail-consumer/heineken-owned-brewery-added-brazil-slave-labor-list-2023-10-06/.

Parker, Andrew, Alessandro Lomi, and Paola Zappa. (2024). "Effect of Time Pressure on Informal Advice Relations Across Organizational Units: Evidence from a Study of Collaboration Within a Formula One Racing Team." *Organization Studies* 45, no. 10: 1441–465.

Payne, Catherine J., Tim S. Jessop, Patrick-Jean Guay, Michele Johnstone, Megan Feore, and Raoul A. Mulder. (2012). "Population, Behavioural and Physiological Responses of an Urban Population of Black Swans to an Intense Annual Noise Event." *PLOS One* https://doi.org/10.1371/journal.pone.0045014.

Pickford, N. (2014, July 14). "Cyclists are Miles Behind Formula 1 in the Environmental Race." *Hull Daily Mail* https://www.hulldailymail.co.uk/Cyclists-miles-Formula-1-environmental-race/story-21460343-detail/story.html.

Pinkerton, Byrd. (2021, September 14). "One Good Thing: I Didn't Watch Any Sports. Then I Watched Netflix's Drive to Survive." *Vox* https://www.vox.com/22652554/drive-to-survive-f1-racing-netflix.

Polenberg, Richard, ed. (2002). *In the Matter of J. Robert Oppenheimer*. Ithaca: Cornell University Press.

Price, Ben and Gareth Bryer. (2024, April 27). "Brewing Giant Heineken Fells Orchard to Sell Land." *BBC* https://www.bbc.com/news/articles/cz5d6mj5gp8o.

Pubs Code Adjudicator. (2020, October 15). *PCA Investigation into Star Pubs & Bars* https://www.gov.uk/government/publications/pca-investigation-into-star-pubs-bars.

Quiggin, John. (2022). *Economics in Two Lessons: Why Markets Work So Well, and Why They Can Fail So Badly*. Princeton: Princeton University Press.

Quiggin, John. (2023, November 30). "What is the Actual Economic Benefit of the Grand Prix?" *Crikey* https://www.crikey.com.au/2023/11/30/grand-prix-2023-victorian-economic-benefit-ernst-young/.

Rayman, Josh. (2023). "The End of Oil? Formula One's Changing Face." *The Geopolitical Economy of Sport: Power, Politics, Money, and the State*. Ed. Simon Chadwick, Paul Widdop, and Michael M. Goldman. London: Routledge. 242–49.

Reid, Caroline. (2008, May 3). "For Formula One, Sex Sells; But Not the Way Max Likes It." *The Spectator* https://www.spectator.co.uk/article/for-formula-one-sex-sells-but-not-the-way-max-likes-it/.

Reid, Caroline. (2015, March 19). "20 Brands That Defined F1." *Raconteur* https://www.raconteur.net/20-brands-that-defined-f1/.

Richards, Giles. (2021, November 26). "Climate Emergency Accelerates F1's Efforts to Clean Up Its Image." *Guardian* https://www.theguardian.com/sport/2021/nov/26/climate-emergency-accelerates-f1-efforts-to-clean-up-image.

Richards, Giles. (2023a, May 21). "After the Flood, Storms Lie Ahead for Formula One in Race to Hit Carbon Zero." *Guardian* https://www.theguardian.com/sport/2023/may/21/after-the-flood-storms-lie-ahead-for-formula-one-in-race-to-hit-carbon-zero.

Richards, Giles. (2023b, July 5). "Next F1 Season Will be Longest Calendar of 24 Races in Sport's History." *Guardian* https://www.theguardian.com/sport/2023/jul/05/next-f1-season-will-be-longest-calendar-of-24-races-in-sports-history.

Richards, Giles. (2023c, December 30). "Ben Sulayem's Mishaps and Meddling Leave F1 Asking if FIA is Fit for Purpose." *Guardian* https://www.theguardian.com/sport/blog/2023/dec/30/ben-sulayems-mishaps-and-meddling-leave-f1-asking-if-fia-is-fit-for-purpose.

Richards, Giles. (2024a, January 23). "F1 Reveals Madrid to Replace Barcelona as Host of Spanish Grand Prix From 2026." *Guardian* https://www.theguardian.com/sport/2024/jan/23/f1-reveals-madrid-to-replace-barcelona-as-host-of-spanish-grand-prix-from-2026.

Richards, Giles. (2024b, March 22). "FIA Unable to Escape Scrutiny with Indictments Overshadowing Season." *Guardian* https://www.theguardian.com/sport/2024/mar/22/fia-formula-one-f1-christian-horner-susie-wolff-lewis-hamilton.

Richards, Giles. (2024c, July 16). "Ralf Schumacher Coming Out Will Have a Positive Impact on 'Evolving' F1." *Guardian* https://www.theguardian.com/sport/article/2024/jul/16/ralf-schumacher-formula-one-impact.

Richards, Giles. (2024d, September 19). "Lewis Hamilton Unhappy with 'Racial Element' of FIA President's Comments." *Guardian* https://www.theguardian.com/sport/2024/sep/19/lewis-hamilton-unhappy-with-racial-element-of-fia-presidents-comments.

Richards, Giles. (2024e, November 28). "F1 'Running Out of People' Says Sacked Race Steward in Rebuke to FIA President." *Guardian* https://www.theguardian.com/sport/2024/nov/28/f1-running-out-of-people-says-sacked-race-steward-in-rebuke-to-fia-president.

Richards, Giles. (2025, March 14). "Sun, Fun and a Favourite Son: Melbourne Makes a Full-Throttle Return to the Top of the F1 Calendar." *Guardian* https://www.theguardian.com/sport/2025/mar/14/melbourne-formula-one-f1-grand-prix-2025-guide-albert-park.

Robertson, Angeline. (2021). *Shady Routes: How Big Retail and Their Carriers Pollute Along Key Ocean Shipping Corridors.* STAND.earth/Pacific Environment https://shipitzero.org/wp-content/uploads/2021/11/FINAL-Ship-it-Zero-Campaign-Report.pdf.

Robinson, Joshua and Jonathan Clegg. (2024, November 12). "How 'Drive To Survive' Turned F1 Into The Real Housewives of Monte Carlo." *Forbes* https://www.forbes.com/sites/forbesdigitalcovers/2024/03/08/the-formula-book-excerpt-drive-to-survive-f1-real-housewives-of-monte-carlo-max-verstappen-christian-horner/.

Rodríguez, Alejandra. (2022, June 27). "Heineken responde a AMLO: Se dice abierta al diálogo por escasez de agua." *El Financiero* https://www.elfinanciero.com.mx/empresas/2022/06/27/heineken-responde-a-amlo-se-dice-abierta-al-dialogo-por-escasez-de-agua/.

Roper, Gemma. (2024, February 27). "Drive to Survive: Claire Williams Helps Make Season Six the Second-Best for Women's Speaking Time." *Females in Motorsport* https://www.femalesinmotorsport.com/post/drive-to-survive-claire-williams-helps-make-season-six-the-second-best-for-women-s-speaking-time.

Ruiz, José Luis. (2022, February 19). "El subidón de la Fórmula 1." *Marca* https://www.marca.com/motor/formula1/2022/02/19/6210af9bca4741ee4f8b45bd.html.

Schor, Juliet B. (2010). *Plenitude: The New Economics of True Wealth.* New York: Penguin.

Scott, Mike. (2013, August 5). "Could Innovation in Formula One Drive Sustainable Technology?" *Guardian* https://www.theguardian.com/sustainable-business/innovation-formula-one-sustainable-technology.

Selleck, Emily. (2024, October 7). "What We Can Expect from LVMH's Billion-Dollar F1 Deal." *Motorsport* https://www.motorsport.com/Culture/news/formula-1-lvmh-sponsorship/10660515/.

Shah, Dhwanil and Anthony Lamont Williams. (2024). ""Drive to Survive" Drives New Fans to Formula 1? Studying Viewer Experiences of a Sports Documentary and its Influence on the Sport's Fandom." *Communication & Sport* https://doi.org/10.1177/21674795241280209.

Silk, Michael, David L. Andrews, and Cheryl L. Cole, eds. (2005). *Sport and Corporate Nationalisms.* Oxford: Berg.

Smith, Luke. (2023, February 11). "How Do F1's Sustainability Goals Fit with a 23-Race Calendar?" *Motorsport* https://us.motorsport.com/f1/news/how-do-f1s-sustainability-goals-fit-with-a-23-race-calendar/10430282/

Smith, Luke. (2021, July 16). "'Drive to Survive' on Netflix has Ignited Formula 1." *New York Times* https://www.nytimes.com/2021/07/16/sports/autoracing/drive-to-survive-netflix-formula-one.html.

Soble, Caroline and Mark Lowes. (2024). "Narrative Storytelling as a Fan Conversion Tool in the Netflix Docuseries *Drive to Survive*." *Communication & Sport* https://doi.org/10.1177/21674795241238158.

Solon, Olivia. (2011, February 14). "F1 Needs to Have Sex with Sustainability." *Wired* https://www.wired.co.uk/news/archive/2011-02/14/f1-sustainability.

SponsorUnited. (2025, May). *Formula 1 Report: 2024 & 2025* https://cdn.prod.website-files.com/646f83ebeb2de03609e8a7bb/68373a6fcd198d8281b706e9_SponsorUnited%20Formula%201%20Turnkey%20Report%202024-25.pdf?utm_campaign=13471865-2025%20%E2%80%94%20F1%202024-25&utm_medium=email&_hsenc=p2ANqtz-_55IyPiB3upYBxOfQm91WD2PWyLzs32eUxdy5J44B3Tcim51raS18ha7x3tmggdPZrwLFLXTuBm-sExs-1n6DNIdPPhg&_hsmi=363846356&utm_content=363846356&utm_source=hs_automation.

Spurgeon, Brad. (2010, June 30). "The Formula One Environment Initiative." *New York Times* https://archive.nytimes.com/rendezvous.blogs.nytimes.com/2010/06/30/the-formula-one-environment-initiative/.

STAND.earth. (2024, September). *Prime Polluter.* STAND.earth, Pacific Environment, Clean Mobility Collective, and Ship It Zero https://stand.earth/wp-content/uploads/2024/09/PRIME_POLLUTER-09.2024.pdf.

Stevenson, Rachel and Agencies. (2008, October 12). "Blair Intervened Over F1 Tobacco Ban Exemption, Documents Show." *Guardian* https://www.theguardian.com/politics/2008/oct/12/tonyblair-labour.

STOP. and FM. (2021). *Driving Addiction: Tobacco Sponsorship in Formula One, 2021* https://exposetobacco.org/wp-content/uploads/TobaccoSponsorshipFormula-One-2021.pdf.

STOP. (2022). *Driving Addiction: F1, Netflix and Cigarette Company Advertising* https://exposetobacco.org/wp-content/uploads/F1-Netflix-Driving-Addiction.pdf.

STOP. and Formula Money. (2024). *Driving Addiction: A Race for Future Generations. How F1 and Big Tobacco Marketing Align to Recruit Young Consumers* https://exposetobacco.org/wp-content/uploads/Race-For-Future-Generations.pdf.

Sturm, Damion. (2022). "The Formula One Paradox: Macho Male Racers and Ornamental Glamour 'Girls'." *Sports, Gender and Mega-Events*. Ed. Katherine Dashper. Leeds: Emerald Publishing. 113–30.

Sturm, Damion. (2023). "Processes of Greenwashing, Virtue Signaling and Sportwashing in Contemporary Formula One: Formula Façade?" *The Future of Motorsports: Business, Politics and Society*. Ed. Hans Erik Næss and Simon Chadwick. Abingdon: Routledge. 167–82.

Sylt, Christian. (2015, August 25). "Is Formula One the World's Most Sustainable Sport?" *Forbes* https://www.forbes.com/sites/csylt/2015/08/25/is-formula-one-the-worlds-most-sustainable-sport/?sh=5357d4d02952.

Sylt, Christian. (2019, April 23). "Formula One Gender Pay Gap Revealed to be Above UK Average at All but One British Team." *Independent* https://www.independent.co.uk/f1/formula-one-gender-pay-gap-statistics-standings-uk-motorsport-a8882946.html.

Ti, Jinsong, Yu Zheng, Weidong Duan, Haobin Zhao, Yanqing Qin, Guangting Yin, Liangwen Xie, Wenshuai Dong, Xiaochong Lu, and Zhaopeng Song. (2024). "Carbon Footprint of Tobacco Production in China Through Life-Cycle-Assessment: Regional Compositions, Spatiotemporal Changes and Driving Factors." *Ecological Indicators* 165 https://doi.org/10.1016/j.ecolind.2024.112216.

Tippett, Anna. (2019, June 24). "Formula 1: There Can Be No Equality in Sport while Women's Bodies are Used for Promotions." *The Conversation* https://theconversation.com/formula-1-there-can-be-no-equality-in-sport-while-womens-bodies-are-used-for-promotions-117443.

Tovey, Alan. (2014, November 1). "Formula One's Vast Costs Are Driving Small Teams to Ruin." *Telegraph* http://www.telegraph.co.uk/sport/motorsport/formulaone/11203136/Formula-Ones-vast-costs-are-driving-small-teams-to-ruin.html.

Tranter, Paul J. and Mark Lowes. (2009). "Life in the Fast Lane: Environmental, Economic, and Public Health Outcomes of Motorsport Spectacles in Australia." *Journal of Sport and Social Issues* 33, no. 2: 150–68.

van Beemen, Olivier. (2019). *Heineken in Africa: A Multinational Unleashed*. London: Hurst.

Virilio, Paul. (1991). *The Aesthetics of Disappearance*. Trans. Philip Beitchman. N.p.: Semiotext(e).

Williams, Richard. (2015, October 23). "F1 is in Deep Trouble and Bernie Ecclestone is Not the Man to Save it." *Guardian* https://www.theguardian.com/sport/blog/2015/oct/23/f1-bernie-ecclestone-max-mosley.

Williams, Richard. (2021, August 29). "Louise King Obituary." *Guardian* https://www.theguardian.com/stage/2021/aug/29/louise-king-obituary.

Witt, Stephen. (2022, May 19). "Crypto.com Spent Its Way to the Top. Then the Market Crashed. Now What?" *GQ* https://www.gq.com/story/crypto-dot-com-explained.

World Health Organization. (2024). *Hooking the Next Generation: How the Tobacco Industry Captures Young Customers* https://escholarship.org/content/qt97d8k4xz/qt97d8k4xz.pdf.

WWF. (2023). *Time for Change: Demanding More Transparency and Responsibility in the Watch and Jewellery sector. Sustainability Rating and Industry Report 2023* https://www.wwf.ch/sites/default/files/doc-2023-11/_WWF_Watch%20and%20Jewellery%20Report%202023_Final.pdf.

YouGov. (2023). *The Global Sports Media Landscape: Sports Whitepaper 2023* https://commercial.yougov.com/rs/464-VHH-988/images/YouGov-Sports-Whitepaper-v4.pdf.

Zafeiridou, Maria, Nicholas S. Hopkinson, and Nikolaos Voulvoulis. (2018). "Cigarette Smoking: An Assessment of Tobacco's Global Environmental Footprint Across its Entire Supply Chain." *Environmental Science & Technology* 52, no. 15: 8087–094.

Zerrenner, Kate. (2019, November 22). "Fasten Your Seat Belts: Formula 1 Racing to the Carbon Neutral Finish Line." *Triple Pundit* https://www.triplepundit.com/story/2019/fasten-your-seat-belts-formula-1-racing-carbon-neutral-finish-line/85706.

CHAPTER 6

Refusing and Resisting Carbon Sports

From Los Angeles to London, people introduce themselves as "activists." We're often puzzled by this. Their daily tasks, means of financial support, and socio-cultural theories and practices are left undisclosed by the term. The unspoken assumption is that 'an activist must "live the issue", demonstrate relentless dedication, and contribute a sustained effort to duly merit the label' through a comprehensive 'alignment between personal identity and collective identity' (Bobel 2007). This leads to critiques of environmental movements, for example, as prone to 'Machiavellianism, psychopathy, narcissism … antihierarchical aggression, [and] anticonventionalism' (Zacher 2024), rather grandly diagnosed as 'pathological narcissistic grandiosity' (Krispenz and Bertrams 2024). Such concerns were dominant in the heyday of functionalist sociology described in the Introduction, when activism was associated with maladaptive personalities. The success of social movements for civil rights, feminism, and disabled people has diminished the currency of that perspective, but not in psychology (Goodwin and Jasper 2015).

Activism is often associated with the left, as broadly defined. It has typically referred to volunteer and professional campaigners and analysts struggling against racism, misogyny, war, inequality, injustice, and climate change. These diverse topics relate to divisions and concerns deriving from various social transformations of the past half-century. Much of the investment once placed in the working class as the next great progressive agent

T. Miller, J. P. Carañana, *Global Sports Go Green—Or Do They?*, Global Culture and Sport Series,
https://doi.org/10.1007/978-3-032-14207-8_6

of history after the *bourgeoisie* assumed "science" made proletarian uprising and eventual hegemony inevitable. But the breakup of the notion of a single force of change, the discrediting of state socialism, and the *faux* identification of rationality with only some categories of people militated against progressive investments in traditional labor and encouraged the intellectual skepticism of social movements and post-modernism alike—a simultaneous split from the certainties of a united oppositional tendency and grandiloquent forms of thought. Activism assumes people 'can be, must be reasoned with' at a time of crisis that affects us all. There is no requirement for either 'humble supplication or threats of armed insurrection.' Instead, persuasion can alter the 'social imaginary' of values and desires (Taylor 2004: 27–28) via synecdoche.

Activists who oppose racism, nationalism, militarism, colonialism, and sexism are not generally fans of objectivity, because they believe passionately in enlarging personal and collective freedom, equality, and social justice. Dubious about knowledge for its own sake, yet dismissive of research undertaken for business or the state, most are united on one front—the need to diminish damage to the planet—and a single truth—the science of climate change. In this instance, they are on the side of reason (de Coninck 2021). When very conventional entities such as the UN, refereed journals, and the professoriate at fancy schools say something that proves human-generated climate change is authentic, they applaud. Conversely, they deride coin-operated "skeptics" within think tanks and the *bourgeois* media and those who decry expertise. "Big" science is "in" when activism pits an oppositional, yet scholarly based, perspective against the reactionary refusal or conservative pseudo-incorporation of environmental critique (Shenker 2023).

Right-wing activism has quite the lineage, too. Consider US anti-immigrant marches of the 1840s, Ku Klux Klan terrorism, 1970s opposition to busing school pupils to integrate education, or attacks on abortion (Goodwin and Jasper 2015: 3). Today, such people draw on formerly progressive tactics of outraged skepticism against voices of expertise and authority: problematizing climate science on so-called social media, merrily rejecting state institutions as legitimate representatives of the people, and embracing the value of spectacle (Falkenberg et al. 2022; Caiani 2024). Like "our" activism, 'rightist movements originate with movement entrepreneurs, frame their messages, respond to external political opportunities, forge collective identity, develop strategies and tactics, and serve as a source of vision and voice' (Blee and Creasap 2010: 271).

Progressives generally use online fora to forward ideas and organize non-violent action—reactionaries to prioritize disinformation, aggressiveness, and links with fellow-travelers in the *bourgeois* media (Freelon et al. 2020). The right also engages in carefully calculated and crafted greenwashing, designed to obscure and legitimize climate criminality or blame pollution on immigration, among other targets of its ire. In the 2022 French presidential elections, Marine Le Pen adopted a platform of nationalist ecology, positioning herself as pro-animal welfare via proposed bans on halal and kosher meat (Mandard 2022a). Conservative activists utilize environmental discourse against diversity and argue on behalf of hunting but also conserving wildlife and sustaining rural tradition (Lovato 2004; Sexton et al. 2005). Some reactionaries argue that 'activists do not only represent a threat of attack to a sports organization, but can constitute an intangible asset' when coopted to legitimize social clearance and environmental destructiveness (Jacopin 2006–07: 68).

So there are lots of contradictions confronting this chapter as it looks beyond greenwashing efforts within carbon sports to consider resistance from progressive grassroots, academic, and athlete concerns. Following a survey of environmental social movements, we'll look at their impact on carbon mega-events.

Why Activism?

In pondering direct action by pro-environmental movements, it's easy to fall into celebratory or critical responses. The celebratory view maintains that a Cartesian distinction between hearts and minds is not sustainable, a sense of humor is crucial to avoid the image of environmentalists as finger-wagging scolds, corporate capital must be opposed in public, the media's need for vibrant textuality can be twinned with serious discussion to alert people who are not activists, and a wave of anti-élite sentiment is cresting (Shenker 2023). Governing-as-usual is unable to handle 'global, complex, interconnected issues that do not respect geographic borders or electoral timetables' (Venzon and Cahen-Salvador 2019). Climate change will not be seriously addressed by conventional party politics. Instead, 'participative democracy, horizontality, inclusiveness, and direct action' (Fians 2022) are called for by those 'impatient of waiting for the authorities to act on their behalf' (Carter 2010: 19–20) or who wish to punish the guilty themselves. Only non-violent direct action, 'putting your body on the line,' can 'accelerate the global transformations needed to avert environmental catastrophe' (McCarthy 2021).

Based on historical research about the public support needed to create significant political change, Extinction Rebellion takes it as a matter of writ/destiny that 'we need the involvement of 3.5% of the population' (Chenoweth 2021: 114–15; "Act Now" n.d.). Hence groups believing that anti-climate change gatherings of a few hundred thousand in a population of 8 billion signify "the" people's message (Giacomini and Turner 2015), in a contest for hegemony that features *bourgeois* economics and corporate-dominated politics *contra* environmental studies and planetary needs. Sovereignty resides in righteous indignation and universal values that purportedly transcend government to represent the popular will via civil society (Graeber 2002; McGregor 2015; Yates 2015).

There is some evidence that disruptive climate protests can produce support for more moderate environmental groups—that radical direct action encourages the wider public to think about mainstream but related causes and organizations (Ostarek et al. 2024) via the 'radical flank effect' (Haines 1984). Mild-mannered environmental organizations look appealing by contrast with their radical flank, even when they share certain policies (Simpson et al. 2022; Dasch et al. 2024; Kountouis and Williams 2023; Brehm and Gruhl 2024). In addition, newer forms of activism side-step party-political alignments (de Moor et al. 2021).

But while groups like Extinction Rebellion have succeeded in raising awareness, their efforts at mobilizing the citizenry have not (Kenward and Brick 2024). Prior to its dissolution, Just Stop Oil had alienated even those who opposed its craven criminalization by the state ("New Survey" 2023).

That propels a more critical view of activism, arguing that non-violent direct action has failed. Instead, rationality must be appealed to in discussions of climate change. Why? The silent majority doesn't like vanguardism, marketing outspends graffiti, protests preach to a light-skinned, middle-class eco-choir, serious media coverage is slender, emotion will ultimately fail, and crucial decisions are more likely to be taken in golf carts than streets. Many people experience activists as tree-hugging hippies, dangerous eco-terrorists, anti-growth evangelists, economic ignoramuses, middle-class lay-abouts, vapid vegans, or romantic dreamers. Such otherness may be at the core of some environmentalists' identities (behaving eccentrically in public, grabbing attention, securing column inches) and restrict their ability to communicate across a range of constituencies, leaving them stuck in a self-fulfilling, vanguardist politics (Bashir et al. 2013; M. S. 2019). As the *Economist* puts it, 'reactions to people scaling buildings and unveiling banners range from apathy to mild annoyance. Those tactics seem to belong to another era' ("Childish" 2014).

That critique maintains there is a lack of compelling evidence to prove the efficacy of direct environmental action, and little even to support the idea that progressives are more positive in their responses than the right (Gulliver et al. 2022; Croco et al. 2023). In the USA, only leftists and Democrats are moved by direct action on the environment; Republicans show no interest (Bugden 2020; Feinberg et al. 2020; Tyson and Kennedy 2023), in keeping with right-wing politicians and think tanks ignoring science or demeaning it (Furnas et al. 2025). Justice will seemingly not be served by taking to the streets or explaining formal knowledge to Republicans.

Is everything hopeless, in the light of ambiguous evidence about activism, a divided left, the right contesting the politics of spectacle and the realities of science, plus a seemingly intractable climate crisis? A century ago, the Nobel laureate Romain Rolland called for 'une alliance intime … du pessimisme de l'intelligence qui perce toute illusion, et de l'optimisme de la volonté' [an intimate alliance … of pessimism of the intelligence, which penetrates every illusion, and optimism of the will] (1920). That remarkable blend of realism and hope spoke to Gramsci in a Fascist jail (1978: 175 n.75) and has been a rallying cry for his many followers since (Antonini 2019).

Resistance works by juggling Rolland's paradox of pessimism and optimism. And it's noteworthy that more and more climate scientists are joining environmental movements, which many see as a professional duty inexorably linked to scholarship (Finnerty et al. 2024). The key is constructing a fruitful mobility between public policy and people power—and to go with the findings of huge samples across the globe that indicate massive popular support for urgent action to save the climate, albeit without recognizing the popularity of that very view (Andre et al. 2024).

The Medley

In the sports world, there is less virulent denial of our climate crisis than in many other industries—but more co-optation. Five formations, sometimes interlocking, engage in environmental sports activism. They form a complex medley:

A) Corporate, college, and bureaucratic actors urge one another, athletes, the *bourgeois* media, and the public to support carbon events, claiming they are green (Formation A—executives and their chorines).

B) Big non-government organizations argue for direct action they devise and control (Formation B—NGOs).

C) Public-interest collectives and scholars produce research problematizing polluters (Formation C—critics and academics).

D) Grassroots groups favor the environment (Formation D—the public).

E) Participants support environmental priorities (Formation E—athletes).

Formation A: Carbon sports employ sizeable cohorts of supine spokespeople and policy lobbyists in the fight to look "green" and avoid democratic regulation, selling symbolic cover to themselves and their sponsors. They undertake what they call asymmetric action against critics, basing their strategies on regular armies confronting *guerrilla* (Marshall et al. 2012) in order to hide a 'poor environmental performance' via 'positive communication about environmental performance' (Delmas and Cuerel Burbano 2011: 65). They project an image of commitment to nature, drawing attention away from activities and industries that display nothing of the sort (Miller 2018: 2). Jules Boykoff makes an experiential distinction between purposive actions in crowded streets versus executive suites:

> Over the years, I have found in city alter Olympic city that the streets pulsed with creativity and life, promise and potential, as well as a vigorous skepticism for what the Games would bring. Along the way, I have attended events featuring Olympic officials that were full of pomp, opulence, and top-shelf cheese plates, far from the scruffy vim, vitality, and whimsicality of the street. (2020: 10)

Cheese-plate activism has dominated earlier chapters; not this one.

Formation B: Many environmental NGOs are animated by the truth of science while claiming legitimacy from grassroots connections. Organic community ties are asserted, at the same time as small numbers of people orchestrate direct action. The rhetoric may be hortatory (Diaz 2022), but these organizations favor popular protest organized by their hierarchies

(Eden 2004). So multinational bureaucracies like Greenpeace stage spectacles at sports events to disrupt pleasure and attract media attention. Such "Big Green" entities generally target secondary entities rather than bringing events themselves into question or working with people invested in sports. They blend 'wet suits' with 'business suits' (Grant 2004). Their annual budgets, in the tens of millions, may involve currency speculation and property investment, leading to understandable grassroots skepticism (Oltermann 2014). Clickocracy/clickbait is a vital source of finance, even as there is little scholarly evidence that social-media environmental activism reaches beyond chorines (Schäfer 2012).

Formation C: Critical scholars and public intellectuals produce knowledge about the climate criminality of Formation A, informing an environmental movement dedicated to influencing states and sports administrators (McCullough and Kellison 2018). Advocacy groups like the Green Sports Alliance, Sport and Sustainability International, and the Sports Environment Alliance are notable instances (Kay 2019; Martin 2019). There is some overlap with Formation B, when entities such as Greenpeace undertake and promote science versus engaging in hierarchically driven direct action. Formation C's contributions have been noted throughout this volume. They are key citations below.

Formation D: Organic grassroots opposition to greenwashing and despoliation can be found in Fossil Free Football, Football Supporters Europe, Pledgeball, and Game Changer (Kay 2019; Martin 2019). Formation D works toward environmental justice from the ground up, per *Ellis v. Nike U.S., Inc.* (2024), in which a dissatisfied customer sued a sporting-apparel company for greenwashing products, contesting Formation A. The League of Fans *Sports Manifesto* engages today's almost unbridled commodification. Concerned that this tendency diverts attention from sports' capacity for cultural and civic expression and togetherness, it calls for all principal actors to be involved in decision-making—'citizenship through sports activism' (2011). We see similar tendencies among US college sports fans (Casper et al. 2014). Many such folks engage in artivism, a neologism that describes the merger of urban, outsider, and public art, Situationism, graffiti, and direct-action performance (Aladro-Vico et al. 2018).

Formation E: Notable athletes sometimes embark on environmental activism, though this is low on the list of priorities for many acquisitive or ideologized labor aristocrats (Barquero 2023). There is scant, mostly US-based, research on their impact on the wider public, which doesn't appear to show great interest in stars' environmental or human-rights messages, though black Americans can be engaged politically by African-American athletes (O'Neill et al. 2025; Becker 2013; Thrall et al. 2008; Till et al. 2008; Artime and Hershey 2023; Towler et al. 2020). Consider Boykoff's telling testimony, as the most important academic and public intellectual on carbon sports: 'a former professional soccer player who represented the US Olympic soccer team in international competition,' his 'lack of political awareness as an athlete ran parallel to a wider societal credulousness about the positive social power of the Olympics' (2020: 15). But that tendency may be changing, as it did for Boykoff. For example, three-quarters of today's track-and-field athletes report 'a direct negative impact on their health and performance due to climate change.' Eighty-five percent believe their sports have suffered 'adverse consequences' from the crisis and 90% want their associations to work for sustainability ("Three-Quarters" 2023).

The tasks facing Formations B through E are intense. They confront executive dominion over the world's carbon sporting fixtures—and their own complicity as spectators and athletes. Some of that activism occurs outside particular carbon mega-events, driven by general critiques of the sports involved. The examples below stretch across numerous sites, illustrating what has been essayed and achieved in struggles against immense forces.

The Carbon World Cup

If we leave UEFA and FIFA to one side, some less-oligarchic parts of Formation A are beavering away. In 2018, Forest Green Rovers, then in England's League Two but since relegated, became the only pro team certified for carbon neutrality by the UN ("Doing the Right" 2018). Ajax, the Netherlands' most storied club, powers its ground via thousands of solar panels and a wind turbine (Forbes 2025; Johan Cruijff Arena n.d.). The Europa-Park Stadion in Freiburg boasts one of the biggest solar roofs in the world ("Europa-Park Stadion" n.d.), its 6200 solar panels bested only by Galatasaray, which produces 4.2 megawatts from 10,400 panels (Nestler 2022). But many grounds have not shifted to solar sources, because of technical, financial, regulatory, design, and social problems.

Formation B has done little of note. When Greenpeace sought to interrupt the 2014 Champions' League Final with a vanguardist protest against

Gazprom (saying nothing about the unsustainability of the event itself), its ruse was quickly uncovered and managed. Nevertheless, the organization used embarrassingly *macho* corporate language to claim 'total domination in the sphere of delivery' (quoted in Miller 2018).

More usefully, prior to the 2022 Qatar fiasco, Notre Affaire à Tous insisted that:

> Il est donc essentiel que la FIFA mette immédiatement fin à cette campagne de greenwashing et s'abstienne de revendiquer la neutralité carbone de ses événements à l'avenir sans le niveau de justification requis par les normes de déontologie publicitaire [FIFA must put an immediate end to its greenwashing campaign and refrain from claiming carbon neutrality for its events in the future without the justification required by advertising ethics standards]. (2022: 15)

Such Formation C arguments and associated research are crucial, per the Carbon Trust (2013, 2016, n.d.), Fossil Free Football, Game Changer Sponsorship Pledge, Badvertising (2024a, b), Carbon Market Watch (2022), FairSquare (2024), New Weather Institute with Scientists for Global Responsibility (2025a, b), The Shift Project (2025), and Amnesty International (2024).

Many Formation D football fans see their commitments as questions of collective identity, not just ticket purchase, and oppose "their" teams being capitalist enterprises. They seek to transform the sport via popular versus corporate ownership, as trusteeship—emotional physiocracy rather than shareholder trading and leveraged purchases. Some campaign against clientelist oligarchs, racism, and homophobia and for greener norms (Keoghan 2014; Davies 2019; Football Supporters Association n.d.; Mabon 2023; Weston 2022). Football for Future (n.d.) recognizes our impending catastrophe and the European Football for Development Network has a *Football for Climate Justice* project (2023).

Although Forest Green's claim to be 'the world's first vegan football club' (Forest Green n.d.) has been greeted by opposition crowds chanting 'You dirty vegan bastard' (quoted in Gallan 2019), the informal sector of fans can be productive. Eighteenth-century British football played a key part in rebellions against enclosure, when commoners played on pitches a mile long and used advertisements for matches to organize and riot against capitalist enterprises (Neeson 1996: 194, 171). E.P. Thompson insisted that crowds may be animated by economic conditions, sexual urges, blind rages—and ideological commitments:

> When, in 1799, the magistrates attempted to put down Shrove Tuesday football in the streets of Kingston [London], it was "the populace" and "the mob" who assembled and triumphantly defied their orders. The mob may not have been noted for an impeccable consciousness of class; but the rulers of England were in no doubt at all that it was a horizontal sort of beast. (Thompson 1974: 397)

In our own time, fan research rejects either their romantic annunciation as working-class scions or criminalization via moral panics (Armstrong and Young 1997; Armstrong 1998; Giulianotti 1999: 80–82; Schimmel et al. 2007). Consider today's *ultras* model, with its connotations of carnival or danger, depending on where you sit: the spectacular arena conduct of Argentina's *barras bravas*; Che Guevara fans in Livorno, where the nation's Communist Party was founded, singing 'Bella Ciao,' song of the Italian resistance during the War (Pato 2021; Ruta 2021; Gutiérrez Bayardi 2022); anti-fascist, pro-feminist, pro-queer, pro-migrant Sankt Pauli supporters, players, and musicians (Galbiati 2013; Montague 2010; Totten 2013; Viñas and Parra 2020; Braun 2024); the Green Brigade of Glasgow Celtic; and Beşiktaş' *Çarşi*, who oppose both an oppressive state and nuclear power (Erhart 2014). For Rayo Vallecano's *hinchadas antifascistas*, the *Bukaneros* (Gullo 2022; Flynn 2022), 'el clásico del Rayo Vallecano no es el Atlético de Madrid ni el Real Madrid. En Vallecas, el clásico es contra el capital' [For Rayo Vallecano, our classic isn't against Atléti or Madrid. In Vallecas, it's us against capital/the capital] (Gullo 2022).

Their cries for social justice are sometimes articulated to environmental protection and economic democracy, part of the struggle for 'community-owned renewable energy projects, the creation of jobs in the low-carbon economy, and enabling their fans and local communities to have a voice in the decision[-]making process on climate action' (European Football for Development Network 2023).

Argentina 1978, the World Cup's greatest shame until Russia and Qatar, saw intense artivism by the Collectif pour le Boycott de l'Organisation par l'Argentine de la Coupe du monde de Football [Collective for Boycotting the World Cup in Argentina] protesting FIFA's complicity with a fascist regime (Fig. 6.1):

It has proven an inspiration. Following years of informal and semi-formal networking driven by the concern *#copapraquem?* [#Whose World Cup?], the run-up to the 2014 Finals in Brazil saw mass protests under the banner of *Comitês Populares da Copa* [World Cup People's Committees] (Gaffney 2016). By 2013, a chaotic mixture of spontaneity, surveillance, and subterfuge had emerged:

> anarchists, right-wing anti-government organizations, and undercover police mixed with lower- and middle-class protesters, student groups, syndicalists, political parties, and newly formed civil society organizations. All of these groups were often present at the same protests, creating a melange [*sic.*] of messages, tactics, and interpretations of dissent. (Gaffney 2016: 335–36)

This mixture of vanguardism, infiltration, and grassroots was dogged by differences over stopping the Finals versus reforming social policy, but it remains an important example of how diverse groups can be convened to show there is a far-from-unified welcome to the Cup, despite disappointments

Fig. 6.1 (Source: https://imagoteca.cedinci.org/s/imagoteka/item/960)

such as when popular opinion in Vancouver opposed the allocation of public resources to the 2026 Cup, *sans* discernible impact.

Within Formation E, the Fédération Internacionale des Associations de Footballeurs Professionels (FIFPRO) has acknowledged that as 'well as protecting players from the impact of climate change in their professional lives, football also has a social responsibility to acknowledge and reduce its carbon footprint'. Women have been in the vanguard of such activism (Wheeler 2024), notably via a wonderfully titled letter to Infantino: 'Aramco Sponsorship Is a Middle Finger to Women's Football.' One hundred and thirty signatories from twenty-seven nations argued that 'Saudi Aramco is one of the corporations which is most responsible for burning football's future. Grassroots football across the world is being deleteriously affected by extreme heat, drought, fires and floods. While we pay the consequences Saudi Arabia rakes in profits, with FIFA its cheerleader' ("Aramco Sponsorship" 2024).

FIFPRO, 40 leagues, and 130 nations have complained to the European Commission that FIFA engages in "abuse of dominance" (akin to US doctrines of monopoly conduct) by creating a crowded calendar in search of economic growth, *contra* its duty as a governing body to ensure players' welfare (Sullivan 2024; Sanghera 2024; "European Leagues" 2024). Morten Thorsby founded WePlayGreen, which brings fellow-pros and their followers together to protect the planet (n.d.). It is distinctly more radical than UEFA or FIFA (Wågan 2024).

Tessel Middag and David Wheeler contributed to the work of the New Weather Institute with Scientists for Global Responsibility (2025a). Héctor Bellerín promised to pay for 3000 trees to be planted every time his team won in June 2020 (OneTreePlanted n.d.), acknowledging '[i]t's nice to show what we have our cars, our watches—but good to send a more meaningful message than "look how cool we are."' Bellerín is pro-tax, pro-vegan, pro-Palestine. He's even a *Público* reader (Alonso 2024). Away from European glamor and finance, the Marshall Islands' national strip gradually disintegrates via cut-outs, metaphorizing perilous sea levels (Wong 2025; Ishiwatari et al. 2025).

The Carbon Olympics

Activism against the environmental blight of the Games has a lengthy, if discontinuous, history, along with concerns about human rights, war, and the political economy (Jefferson Lenskyj 2008). In the Global North's history of the 1968 Mexico Games, much attention is rightly paid to Formation E African-American protests against racism, most notably Tommie Smith and John Carlos following the 200-meter sprint, supported by white silver-medalist Peter Norman, and some heed to Czechoslovakian gymnast Vera Čáslavská resisting the Soviet Union (Reynolds 2018). Today, downhill skiers insist their sport 'take the lead in the fight against climate change and make our sport climate neutral as soon as possible' and Kiribati weightlifter David Katoatau draws attention to rising seas endangering his homeland (Friedman 2016).

But Formation A's IOC is forever tightening prohibitions on activism by athletes to '[p]reserve the podium, FoP [Field of Play] and official ceremonies from any kind of protests and demonstrations, or any acts perceived as such' (IOC 2021a)—with the complicity of most Olympians. The Charter's notorious Rule 50 goes further: 'No kind of demonstration or political, religious, or racial propaganda is permitted in any Olympic sites, venues, or other areas' (International Olympic Committee 2024). Needless to say, authoritarian governments proscribing free speech and companies commodifying athletes' bodies are welcomed (Jørgensen and Sejer Andersen 2024: 289). Against that mania, Athletes of the World stands for activism in several areas, including the climate (n.d.).

Formation D is most renowned for Tlatelolco, mentioned in Chap. 3. As noted there, Mexican students took a stand—peacefully—against a repressive, corrupt one-party state whose Olympian charades were flagrantly articulated to an incomplete democracy. The international press corps gathered for the Games was crucial to the protest's timing.

The activists began their demonstration at the Universidad Nacional Autónoma de México. Removed by the police immediately prior to the Games, they shifted to Tlatelolco because it represented a purportedly successful blend of Aztec, Spanish, and *mestizaje* [mixed race] cultures, symbolizing history and contemporaneity. La Plaza de las Tres Culturas was an imagined pacific *mélange*, on terrain where the vanquished Aztecs had found some degree of refuge from their Spanish conquerors.

The students were mostly from working-class and peasant families. They opposed attempts 'to beautify the city' for the Games that neglected 'those who needed it most, the poor.' The protest was less *contra* the Olympics per se than what the event signified in terms of policy choices and priorities (González de Bustamente 2016: 102), at a time when even the government acknowledged a quarter of the population lived in extreme poverty (Piñera 1978). Other sources put the level at between 40% and 70% (Damián 2019).

The protestors decided to move on and stage a hunger strike opposite the Olympic pool. Chanting nursery rhymes, they were about to depart when a helicopter, snipers, and tank gunners arrived, coordinating with the shadowy, shady Batallón Olimpia [Olympia Battalion], paramilitary in disguise who had merged with the protestors as *agents provocateurs*. The police, military, and *paras* took green flares as a sign to open fire and bayonet innocents (Fallaci 2020: 501, 504, 507; "The Ghosts" 2008). People never got the chance to head for the pool; they were cut down. Those who could fled toward a church, only to encounter what was becoming 'an execution wall' (Fournier and Herrera 2009: 155).

Perhaps the most famous survivor was the renowned *l'Europeo* war correspondent Oriana Fallaci, who had recently departed Australian and US killing fields in Vietnam. She described Tlatelolco as 'un massacre supérieure à tous les autres massacres que j'avais vis à la guerre' [worse than any massacre I had seen in wartime]; Fallaci had lived through battles in Đắk Tô and Huế and the Tết offensive (1985: 2). Cornered, like so many others, by Mexican "special forces," she was labeled '*Comunista! Agitadora!*,' shot, and finally hospitalized with infected wounds that left her scarred for life (2020: 512). There is an extraordinary interview from her hospital bed ("Tlatelolco—La testimonianza" n.d.).

Toby used to know another survivor quite well, one of the remaining *brigadistas del 68*. Imprisoned for over two years, he spent the time playing sports and getting a political education. On release, he forged academic then parliamentary careers, but was forever caught in the daring, the suffering, the honor of that day—recollected, experienced, and voiced through dipsomania.

Sartre and Bertrand Russell expressed their outrage:

> The Mexican government has behaved with a barbarity comparable only to the massacres carried out by occupying Nazi troops in Europe or by napalming American planes in Vietnam. Throughout the world people have been aroused to passionate anger and alarm. We express our profound solidarity with the heroic Mexican students. We ask people, organizations and nations to boycott the Olympic games. (1968)

That didn't happen. But the victims' and survivors' cry—'No queremos Olimpiadas, queremos revolución' [We don't want the Olympics, we want revolution]—lives on (Aviña 2007). October 2, the anniversary, serves as a reminder of that horror. For many survivors and contemporary activists, the massacre is a marker of two Mexicos—before and after Tlatelolco (Poniatowska 1971)—'shorthand for a number of different signifiers: tragedy, conspiracy, student-heroism, disenchantment, and government cover-ups' (Rojo 2016: 1). Tlatelolco became a symbol—*the* symbol—of a new generation struggling against an intransigent oligarchy, and the oppression that followed. The protestors who survived did not give up the struggle:

> La parte más desgarrada se incorporó a una lucha guerrillera urbana en la que se desangró durante los siguientes cinco años, en una guerra sucia sin cuartel. Un grupo enorme de jóvenes fueron hacia los barrios, a fundar las organizaciones de colonias que durante los siguientes 20 años ofrecerían un modelo de resistencia popular. Otros fuimos a buscar las fábricas, a encontrar las claves de por qué el movimiento estudiantil se había movido en la soledad. Otros trataron de transformar la Universidad, crearon los sindicatos, impusieron la reforma educativa. Otros llegaron al campo. [The angriest among us joined an urban guerrilla struggle that bled into a merciless dirty war over the next five years. A sizeable group entered neighborhoods and founded community organizations over the next twenty years—a model of popular resistance. Others worked in factories, encountering the spirit and conditions that had inspired the student movement. Some sought to transform the university, created unions, and pushed for educational reform. Still others moved to the country.] (Taibo II 2022: 182)

This was Olympic activism of a special kind, designed not to prevent or even disrupt the Games—rather, it was a political-economic critique that ended in disaster, but led to enduring and deep public memory (González de Bustamente 2016), incarnating ongoing struggles against the misallocation of funds toward such events and a repressive state. Six decades later, '1968' signifies 'Tlatelolco' in Mexico—not the Olympics, nor Paris, nor Prague—least of all Berkeley or Grosvenor Square (Castañeda 2010). It forms part of a 'vast memory culture' (Kriza 2019: 82).

Many realize the power of Rosario Castellanos' famous poem 'Memorial de Tlatelolco' [In Memory of Tlatelolco]:

La oscuridad engendra la violencia

y la violencia pide oscuridad

para cuajar el crimen

[Darkness breeds violence

And violence breeds darkness

That thickens the crime]

Castellanos noted the extraordinary media blockade of the story, but avowed that:

Recuerdo, recordemos

hasta que la justicia se siente entre nosotros

[I'll remember, we'll remember

Until justice flourishes for all] (2018)

Her contribution is part of a renowned movement, *la literature de Tlatelolco*, that is renewed with each anniversary of the tragedy (Carpenter 2015). The powerful re-disposal of the 1968 event's memorable iconography, depicted below, punctured its air of pop-art festivity with references to the state's barbarism (Fig. 6.2).

Less dramatically and painfully, Denver residents voted against hosting the 1976 Winter Olympics because of the rapacious ski industry's environmental destructiveness; progressives successfully opposed Amsterdam bidding for Summer 1992; and the people of Hamburg and Boston rejected their city governments' campaigns to gain the 2024 Olympics ("Colorado Rejects" 1972; Ramaswamy 2015; Vaccaro 2015). Calgary, Innsbruck, Oslo, Bern, and Rome have also declined to play host (Valvur 2018).

Fig. 6.2 (Source: https://muac.unam.mx/exposicion/imagenes-y-revuelta-la-grafica-del-68)

Prior to London's 2012 absurdity, demonstrations by various campaigning groups erupted over Dow Chemical being an official sponsor of the Games, because of the Bhopal disaster. Survivors and their supporters organized "Bhopal's Special Olympics" (Attepuram 2012). And per the images below, artivists resisted the post-Games redevelopment/clearance of East London addressed in Chap. 3 (Fig. 6.3):

These photographs illustrate the inherent contradiction between policy martinets and the technology sector versus artistic and working-class survival, a contradiction played out under the obfuscatory shadow of infrastructural Olympic fantasies. Utopic claims quickly became dystopic realities, their hypocrisy rendered ephemerally visible by activism.

Rio 2016's opening ceremony was disrupted when a young indigenous boy said the event appeared to endorse the notion that climate change could be combated by planting a tree. He was echoing protests by hundreds of thousands removed from the view of foreign cameras prior to the event (Aranda 2016; Boykoff and Mascarenhas 2016).

Fig. 6.3 (Source: Toby Miller)

Fig. 6.3 (continued)

In the run-up to the 2020 Tokyo Olympics, transnational activism targeted the Games' partners in the timber trade, which engaged in dispossession of land, illegal logging, and violence against protestors. Formation C's Global Witness, the Environmental Investigation Agency, and the Rainforest Action Network campaigned with local environmental groups to criticize corporate purchasers of these items (Global Witness 2016; Sim 2018; Lester 2019), deriding the New National Stadium as a veritable 'timber temple' (Hilburg 2019). Following the Games, activists protested at the mistreatment of a horse, mentioned in Chap. 3. The pentathlon was

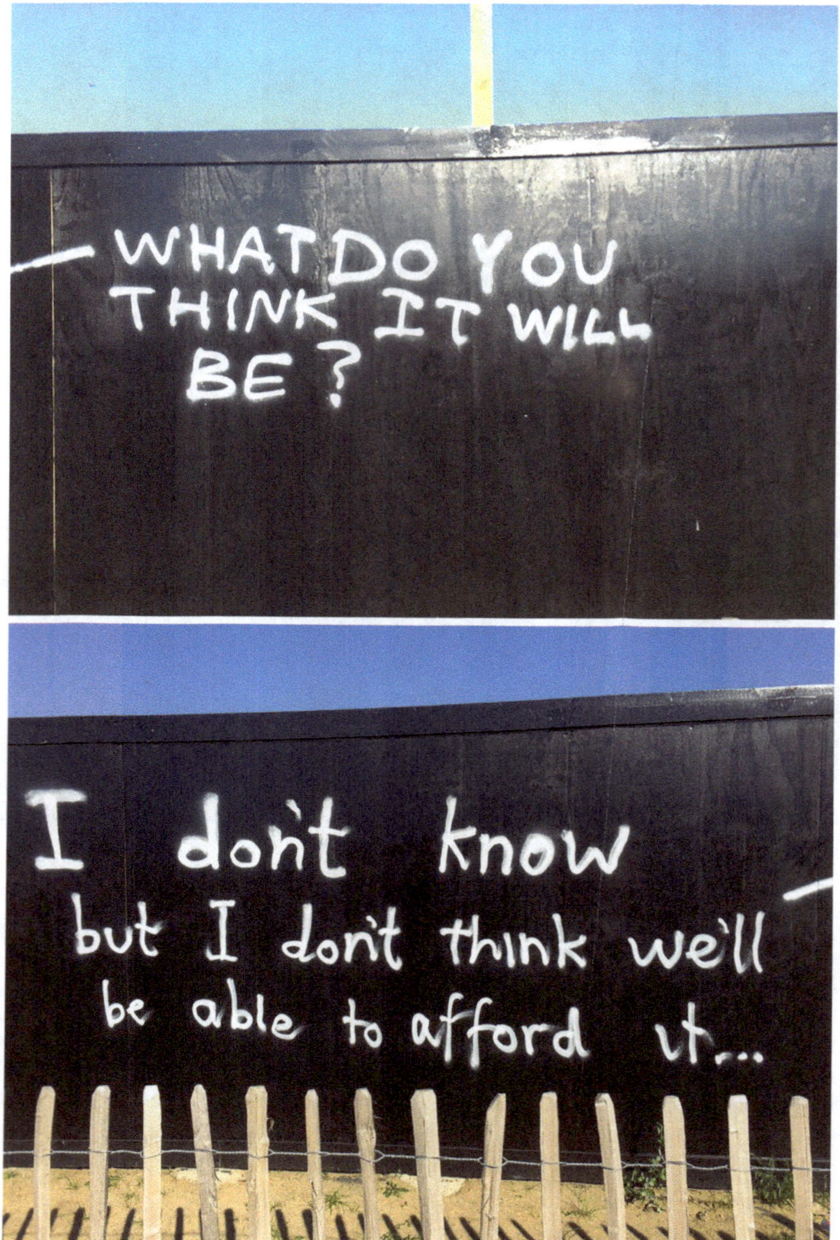

Fig. 6.3 (continued)

altered from the next Olympics, with bike-riding taking the place of equine enslavement (Ropa et al. 2024).

As we saw in that chapter, Paris 2024 was meant to be special. Since *les événements* of 1968 and their subsequent intellectual-political foment, the very word "Paris" has signified new social movements in the Global North. And because the 2015 climate-change accord bears the city's name, it also conjures up ecological (ir)responsibility (United Nations Climate Change n.d.).

There was Formation C resistance in 2024. Supported by *The Lancet* ("The 2024" 2024) and the *British Medical Journal* (Cotter and Mullin 2024), Kick Big Soda Out of Sport focused on the horror done by major sponsor Coca-Cola to public health and the environment (n.d.). Prior to the departure of the Canadian team, Stand.earth hung banners over Lululemon's Toronto headquarters to protest the national team wearing its apparel ("Stand.earth" 2024).

But Formation D was the liveliest. Prior to the Games, undercover activists from People for the Ethical Treatment of Animals (PETA) filmed the appalling torture and slaughter of reptiles that are core features of LVMH's sub-contracted fashion work (Figs. 6.4 and 6.5):

Among many critical French collectives (Kokabi 2024), Saccage [havoc] took its name from Paris 2024's ecological and social destructiveness (Saccage 2021). Dedicated to preserving the Seine-Saint-Denis neighborhood, friendship, mutual aid, and relaxation, Saccage members were demonized by the government as 'Environementalistes d'ultra-gauche' [ultra-left environmentalists] ("JO 2024" 2024; "Une bulle" 2024). The collective Stop violence policières à Saint-Denis [Stop Police Violence in Saint-Denis] denounced the neighborhood's militarization (Gagnepain 2024). When the activists invited journalists on a 'Toxic Tour' to indicate areas adversely affected by clearances, the prefecture interdicted what it deemed a protest march (Guédon 2024). Actual anti-Olympic protests saw Paris police respond heavy-handedly (Saccage 2024), arresting dozens of Climate Extinction members (Muñana 2024) and keeping social and ecological activists on the sidelines wherever possible (Le Foll 2024).

When France's rail and fiber-optic services were disrupted, a bewildered state didn't know whether to blame the Kremlin, local progressives, or a radicalized house marten (Chahuneau 2024; "JO de Paris 2024: une fouine" 2024). Rather wonderfully, the *New York Times* declared the activists 'murky' (Breeden et al. 2024) and *Le Figaro*

Fig. 6.4 (Source: https://www.speciesunite.com/news-stories/olympic-partner-lvmh-slammed-for-animal-abuse)

Fig. 6.5 (Source: https://www.peta.org/media/news-releases/peta-france-takes-over-paris-with-projections-targeting-olympic-parter-lvmh/)

decreed the left's 'l'univers paranoïaque et fantasmé' [paranoid fantasy universe] responsible (Cornevin 2024). Une délégation inattendue [An Unexpected Delegation] issued a *communiqué* explaining the transport intervention in anti-nationalist, anti-capitalist, pro-environmental terms (n.d.). So while the repressive state apparatus quashed some direct action, it was insufficiently nimble or informed to prevent activities that countered the asymmetry of information and power between Parisian publics and propagandists.

NOlympics LA is organizing on the assumption that the 2028 event will follow a typical pattern: 'accelerate policing, evictions, inequality, exploitation, and the erosion of democracy' (n.d.). Such Formation D opponents of the Games now operate transnationally at intellectual and organizational levels (Boykoff 2020). They will not be easily deterred.

The Carbon America's Cup

There is a rich legacy of Formation D grassroots protest and opposition to the America's Cup. Consider València's working-class Poblats Marítims [maritime villages], which date from the thirteenth century. Because they are close to the port, the Poblats have been vulnerable to exploitation, destruction, and clearance by mega-events, from the Cup to Grands Prix. Before and since 2007 and 2010, locals have resisted public and corporate boosters' privatization of space, evictions, and rent increases (Santamarina Campos and Moncusi Ferré 2013). These citizens, left out of the Cup's neoliberal calculus, object to a devastation they must count in social and ecological as well as business terms. State and corporate authoritarianism usually triumph (del Romero Renau and Trudelle 2011). In San Francisco, an environmental impact study prior to the 34th regatta saw ecological organizations and residents of the Bay Area protesting—in vain, as the Board of Supervisors endorsed construction of a Cruise Terminal and Jumbotron (Preza 2012).

After the 2024 finals, the organizers claimed that 'Barcelona welcomed the world with open arms' (Louis Vuitton 37th America's Cup 2024). Think again, please. On the day prior to Toby's visit, thousands of local people had protested against the burgeoning growth of the tourism trade: marching, putting up placards blockading space, and spraying presumed tourists with water pistols, drawing on two decades of anti-tourist action (González-Reverté and Soliguer Guix 2024).

Plataforma No a la Copa Amèrica was the most important movement. It coordinated and promoted the activities of more than 140 collectives and hundreds of individuals opposing the 2024 Cup ("Una plataforma" 2023). The Plataforma's X and Instagram accounts had around 4000 followers each. Some protests and demands were covered by left-wing Catalan media *Directa* and *VilaWeb* ("No a la Copa Amèrica es manifestará" 2024), the Barcelona public-TV channel Betevé (Solans 2024), radical Spanish online outlets such as *Izquierda Diario* (Sabaté 2024), national progressive newspapers in *El Periódico* ("Activistas presentan" 2024) and *La Vanguardia* ("El malestar" 2024) (both headquartered in Barcelona), *20minutos* ("'No a la Copa América' pide" 2024), and *Público* (Camprubí i Pont 2024), and right-wing counterparts *El Confidencial* (Sacristán 2024) and *El Español* (Gil 2024). Deutsche Welle and RTVE produced excellent video reports ("Barcelona's Plans" 2024; *A golpe de evento* 2024) and *Forbes* a useful digest of critical coverage (C Rodriguez 2024). But they were mostly once-only stories. The Plataforma complained of insufficient reportage about the Cup's negative impact and their protests.

Its strength lay elsewhere, in non-violent public action: teach-ins, talks, demonstrations, mockery, murals, and fun. For example, the popular neighborhood Barceloneta saw protests against the impact of the carbon Cup and its tourists-in-tow, not to mention various teams' huge entourages, which had been in situ for over a year. The Plataforma orchestrated folks on Somorrostro beach to recover the coastline from the Cup by making a human chain and sandcastles, holding a *tortilla* contest and water games, and sharing recordings of this pleasurable resistance online. That embedded the movement in a popular locus of leisure; local beachgoers could readily, enjoyably grasp the value of defending the natural environment *contra* the Cup (Sacristán 2024). Such playful protest is a form of empowerment, regardless of whether its policy objectives are met (Bruttomesso 2018).

Protestors chanted 'Copa Amèrica mai més. Enfosem aquest model de ciutat' [No more America's Cup. An end to this model of the city]. Thousands carried banners reading 'Copa América nunca más' [No more America's Cup], 'Vuestra regata nos suda la patata' [We don't give a damn for your regatta], and 'Collboni trilero, devuélvenos el dinero' [Collboni, you trickster, give us back our money], and 'Recuperemos el litoral' [We're reclaiming the coastline]. They whistled at foreigners and insisted 'This is the real Barcelona experience,' 'Tourist go home,' and 'Els vostres velers, ens pugen els lloguers' [Your yachts are raising the rent] (Blanchar 2024; "Miles de personas" 2024). Activists proclaimed, 'No volem cap macroesdeveniment pensat per vendre la ciutat com un producte touristic' [We don't want a mega-event designed to sell the city as a tourism product] (Prieto 2024).

No a la Copa Amèrica's Instagram account featured concerts, interviews, roundtables, protest images, and artivism in various forms, from a parody of the gray functionaries celebrating the carbon event to a constructivist invitation to gather, mixing calls for a discomforting feminism with popular repossession of the beach. The pictures below range between actuality photography, futurist art, and graffiti, in the spirit of *bricolage* (Fig. 6.6).

It goes without saying that none of these images, or anything related, qualified for the Cup's "Photo Galleries" (2024).

The Plataforma also took institutional action. The group asked the Sindicatura de Comptes [Audit Office], Catalunya's external auditor of public accounts and economic management, to investigate the numbers claimed for the event, given the 'fracàs absolut' [absolute failure] to attract television viewers and spectators in situ or to fill fan zones. The activists argued that the real numbers didn't correspond to those provided by the regatta's chorines, among them the Fundació Barcelona Capital Nàutica; TV drew a worldwide audience of 68 million, not close to the billion claimed; and the figure of two and a half million visitors included anyone strolling in the maritime area or simply enjoying the beach (J Rodríguez 2024a).

At a press conference in the Parlament de Catalunya, the Plataforma urged political parties to vote in favor of an audit to study the figures and assess the Cup's real impact. Reactionary forces negated this attempt ("'No a la Copa America' pide" 2024; Rosa and Prieto 2024). The Plataforma also filed suit with the Oficina Antifrau [Anti-Fraud Office] to investigate the event's contracts and use of public money. The activists had been unable to view secret deals between the administration and the organizers and were concerned about potential irregularities in the expenditure of millions of euros of public money. The Office initiated an internal process (Prieto 2024; J Rodríguez 2024b).

The group's position was this: 'Nos dicen que la Copa América aportará riqueza, ocupación y prestigio a nuestra ciudad. Nunca ha sido así … solo sirven para justificar grandes proyectos urbanos impulsados por lobbies turísticos y de la construcción' [We were told the America's Cup would bring riches, employment, and prestige to our city. None of that has happened … it has merely justified huge urban projects lobbied for by the tourism and construction industries] (quoted in Sacristán 2024). The Plataforma opposed vast public expenditure on a carbon event, 'teniendo en cuenta todos los problemas sociales, climáticos y ambientales que ya sufrimos' [bearing in mind all the social, climatic, and environmental problems that plague us] (quoted in Vicente 2023) (Fig. 6.7).

Fig. 6.6 (Source: https://www.instagram.com/nocopaamerica/)

Fig. 6.6 (continued)

Whether or not the malpractices surrounding the Cup came to light and produced prosecutions, the protest movement had been instrumental in making them visible.

Formula One Carbon

Motorsport is the most commodified, élite, and easily policed mega-event. For instance, that lumbering denizen of Formation B, Greenpeace, has deployed its massive resources against Formula One to disrupt Grands Prix via people dressing up in bright colors to mount things they don't own. This played poorly with the significant proportions of Formula One followers fixated on different *macho* heroics and the technological sublime (Cooper 2013; Acaf et al. 2025) (Fig. 6.8).

Fig. 6.6 (continued)

Fig. 6.7 (Source: https://x.com/nocopaamerica?lang=es)

Perhaps more effectively, several Formation E drivers publicize problems within their sport. Retired champion Sebastian Vettel has been an environmental activist since his driving days ("F1 Champion" 2022; Vettel n.d.) and is critical of 'a fossil fuel party that happens nearly every week' (quoted in Y Khan 2025). In 2019, Hamilton acknowledged that 'our carbon footprint is higher than the average homeowner who lives in one city' (quoted in Benson 2019), spoke in favor of veganism, and briefly posted the following on Instagram (Fig. 6.9):

Hamilton's rival Fernando Alonso shot back with '[w]e all know the lifestyle that Lewis has, and that Formula 1 drivers take 200 planes a year. You can't then say: "Don't eat meat"' (quoted in Benson 2019). Some colleagues supported Hamilton's environmentalism, acknowledging their own responsibilities in our climate crisis; others were more skeptical (Richards 2021a, b; Ferrisi 2020). Alonso subsequently adopted a greener position (Portillo 2022). In 2023, Formula One oligarchs sought to silence their stars speaking out on climate change, *inter alia* (Benson 2023). But some would not be muzzled. One can only hope they form alliances with workers among race sponsors who also refuse business as usual. From another grid, former North American Stock Car Auto Racing driver Leilani Münter, who had once identified as 'a vegan hippie chick

Fig. 6.8 (Source: https://archivo-es.greenpeace.org/espana/es/GPmagazine/GPM07/Gracias-a-ti/index.html)

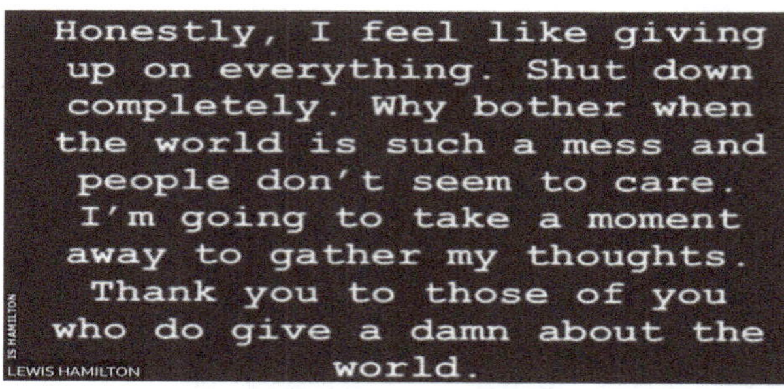

Fig. 6.9 (quoted in Benson 2019)

with a race car' (n.d.), announced that her 'relationship with the combustion engine is over' (quoted in Lockwood 2022).

Despite Rio de Janeiro securing a contract with Formula One, plans to destroy a native forest for a new track were abandoned in 2021, following widespread criticism (Sturm 2023). Australia's Save Albert Park protestors gathered evidence countering Grand Prix boosterism (Campbell 2013), producing kits on the legal, economic, environmental, and traffic implications of the Melbourne event for birdlife, waterways, trees, public access, noise, trash, and carbon emissions (Miller 2018). But special legislation saw Victoria proclaim that '[t]he Grand Prix Act has been principally designed to ensure that the over 400,000 spectators who attend this event won't have their fun spoiled by political protest groups' (quoted in Lowes 2004: 80). The Federación Regional de Asociaciones Vecinales de Madrid [Regional Federation of Madrid Neighborhood Associations] and Más Madrid, the city's left-wing opposition, complained that plans for the 2026 Grand Prix offered inadequate documentation, consultation, and responses to concerns about residents' rights, noise, atmospheric pollution, health, mobility, removal of trees, and imperilment of birds. Their intensely detailed documents were models of public-policy argumentation (Federación 2025; Más Madrid 2025).

Hundreds of Formation D Catalans opposed the environmental devastation of the 2024 Grand Prix, reminding Collboni he was not the chief of a fun park and the 'ciutat no està en venda' [the city is not for sale] (Palmer 2024). Barcelona residents also protested Formula One's promotional roadshow because of its air and noise pollution and use of public funds for private advantage, gaining support from Catalunya's Síndic de Greuges [Ombudsman], which found the event 'difficult to justify' in the light of climate change.

ForaFórmulaFum's [BeyondFormulaFumes] poster below spoke out against commodifying the city while presenting a young person protected from fumes and noise. The Fora convened 200 activist groups, who marched chanting 'Stop contaminació' [End Pollution], 'Fora pijos' [Rich Go Home], and 'Prou autopista urbana' [Enough Urban Freeways], inter alia (Fig. 6.10).

Conclusion

The *faux* activism of Formation A will continue. It must be matched and overwhelmed by progressive voices and forces. The record of Formations B to E is inconsistent, but sometimes impressive, from the sacrifices made by working-class students in Mexico City to citizens opposing mega-city bids. Attending to forces from below, above, and alongside activists will continue to matter, along with the differences and commonalities of approaches soaked in reason versus spectacle and artivism articulated to scientific and popular knowledge.

Fig. 6.10 (Source: https://www.facebook.com/photo/?fbid=1000040325024003&set=a.237180527976657)

Bibliography

NB—Links from US Science Have Been Removed By Its Government

"'No a la Copa América' pide a la Sindicatura de Comptes investigar las cifras del evento deportivo en Barcelona." (2024, October 2). *20minutos* https://www.20minutos.es/noticia/5639660/0/no-a-la-copa-america-pide-a-la-sindicatura-de-comptes-investigar-las-cifras-del-evento-deportivo-en-barcelona/.

"Activistas presentan una denuncia en Antifrau para que investigue las cuentas de la Copa América." (2024, October 10). *El Periódico* https://www.elperiodico.com/es/barcelona/20241010/no-a-la-copa-america-denuncia-antifrau-investigue-cuentas-109152400.

"Act Now." (n.d.). Extinction Rebellion https://extinctionrebellion.uk/act-now/.

"Aramco Sponsorship is a Middle Finger to Women's Football." (2024). Athletes of the World https://img1.wsimg.com/blobby/go/558b37ef-276d-4111-81a8-d0fea75f10fd/Letter%20to%20FIFA_%20Aramco%20Sponsorship%20(3).pdf

"Barcelona's Plans to Manage Tourism Angers [*sic.*] Residents." (2024, July 5). *DW* https://www.dw.com/en/barcelonas-plans-to-manage-tourism-angers-residents/video-69539560.

"Childish Arguments." (2014, October 17). *Economist* http://www.economist.com/blogs/democracyinamerica/2014/10/greenpeace-lego-and-shell.

"Colorado Rejects Olympics." (1972, November 8). *Ludington Daily News*: 5.

"Doing the Right Thing. Forest Green Rovers are World's First UN-Certified Carbon Neutral Club." (2018, August 1). *Inside World Football* http://www.insideworldfootball.com/2018/08/01/right-thing-forest-green-rovers-worlds-first-un-certified-carbon-neutral-club/.

"El malestar por los problemas de acceso a la vivienda se hace oír en varias ciudades españolas." (2024, October 13). *La Vanguardia* https://www.lavanguardia.com/vida/20241013/10018135/dos-mil-personas-reclaman-barcelona-vuelva-acoger-copa-america.html.

"Europa-Park Stadion." (n.d.). *Football Ground Guide.* https://footballground-guide.com/leagues/europe-1/germany/sc-freiburg-stadium-europa-park-stadium.html.

"European Leagues, FIFPRO Europe to File Joint Complaint to European Commission Against FIFA Regarding International Match Calendar." (2024, July 23). FIFPRO https://www.fifpro.org/en/who-we-are/what-we-do/foundations-of-work/european-leagues-and-fifpro-europe-to-file-joint-

complaint-to-european-commission-against-fifa-regarding-international-match-calendar.
"JO 2024. Une bulle de sécurité de 100 policiers et gendarmes autour de la flamme olympique." (2024, January 22). *Ouest-France* https://www.ouest-france.fr/jeux-olympiques/jo-2024-une-bulle-de-securite-de-100-policiers-et-gendarmes-autour-de-la-flamme-olympique-c4c66a90-b934-11ee-879f-a9ffd2809e51#:~:text=Risque%20de%20%C2%AB%20perturbation%20%C2%BB%20du%20parcours,Les%20Soul%C3%A8vements%20de%20la%20Terre%20%C2%BB.
"JO de Paris 2024: une fouine était à l'origine des soupçons de sabotage dans la fan zone du château de Vincennes." (2024, August 8). *franceinfo* https://www.francetvinfo.fr/les-jeux-olympiques/paris-2024/paris-2024-une-fouine-etait-a-l-origine-des-soupcons-de-sabotages-dans-la-fan-zone-du-chateau-de-vincennes6706254.html.
"Miles de personas protestan en Barcelona contra la celebración de la Copa América de vela." (2024, October 13). *Público* https://www.publico.es/sociedad/miles-personas-protestan-barcelona-celebracion-copa-america-vela.html#md=modulo-portada-bloque:2col-t4;mm=mobile-medium.
"New Survey Reveals British Public Generally Think Disruptive, Non-Violent Protestors Should not be Imprisoned." (2023, August 1). University of Bristol https://www.bristol.ac.uk/news/2023/july/public-opinion-on-climate-change-and-protesters.html.
"No a la Copa Amèrica es manifestarà per denunciar la falta de transparència i malbaratament de diners públics." (2024, October 10). *VilaWeb* https://www.vilaweb.cat/noticies/no-copa-america-manifestara-falta-transparencia-malbaratament-diners-publics/.
"Photo Galleries." (2024). https://www.americascup.com/photo.
"The 2024 Olympic Games: End the Health Hypocrisy." (2024, August 10). *The Lancet* https://www.thelancet.com/action/showPdf?pii=S0140-6736%2824%2901641-6.
"The Ghosts of Mexico 1968." (2008, April 24). *Economist* https://www.economist.com/international/2008/04/24/the-ghosts-of-mexico-1968.
"Three-Quarters of Athletes Directly Impacted by Climate Change, World Athletics Survey Finds." (2023, November 29). World Athletics https://worldathletics.org/athletics-better-world/news/climate-change-impact-world-athletics-survey.
"Tlatelolco, la testimonianza di Oriana Fallaci." (n.d.). *YouTube* https://www.youtube.com/watch?v=yW8q5i7mdgo.
"Una Plataforma pide cancelar la Copa del América en Barcelona por "elitista"." (2023, October 10). *iusport* https://iusport.com/art/127377/una-plataforma-pide-cancelar-la-copa-del-america-en-barcelona-por-elitista.
Acaf, Yara, Natalie Brown-Devlin, and Lucy Atkinson. (2025). "Do Fans Have to Choose Between Formula One and Sustainability? A Case Study of Spectators

Balancing Environmental Beliefs and Fandom." *International Journal of Sport Communication* 18, no. 1: 108–18.

A golpe de evento. (2024, November 11). *El Escarabajo Verde* RTVE https://www.rtve.es/play/videos/el-escarabajo-verde/golpe-evento/16313122/.

Aladro-Vico, Eva, Dimitrina Jivkova-Semova, and Olga Bailey. (2018). "Artivismo: Un nuevo lenguaje educativo para la acción social transformadora." *Comunicar* 57: 9–18.

Alonso, Virginia P. (2024, March 27). "Héctor Bellerín: "En el fútbol hay mucho miedo a cambiar y eso provoca esta hipermasculinización tóxica"." *Público* https://www.publico.es/deportes/hector-bellerin-futbol-hay-mucho-miedo-cambiar-provoca-hipermasculinizacion-toxica.html.

Amnesty International. (2024, November 11). *High Stakes Bids: Dangerously Flawed Human Rights Strategies for the 2030 and 2034 FIFA World Cups* https://www.amnesty.org/es/documents/ior10/8712/2024/en/.

Andre, Peter, Teodora Boneva, Felix Chopra, and Armin Falk. (2024). "Globally Representative Evidence on the Actual and Perceived Support for Climate Action." *Nature Climate Change* 14: 253–59.

Antonini, Francesca. (2019). "Pessimism of the Intellect, Optimism of the Will: Gramsci's Political Thought in the Last Miscellaneous Notebooks." *Rethinking Marxism* 31, no. 1: 42–57.

Aranda, Germán. (2016, August 6). "Una ceremonia sostenible entre la selva de atletas." *El Mundo* http://www.elmundo.es/deportes/2016/08/06/57a5397eca4741a6118b45f3.html.

Armstrong, Gary and Malcolm Young. (1997). "Legislators and Interpreters: The Law and 'Football Hooligans'." *Entering the Field: New Perspectives on World Football.* Ed. Gary Armstrong and Richard Giulianotti. Oxford: Berg. 175–91.

Armstrong, Gary. (1998). *Football Hooligans: Knowing the Score.* Oxford: Berg.

Artime, Michael Robert and Megan Hershey. (2023). "The Reel Politics of International Crisis: Benedict Cumberbatch's Appeal for Syrian Refugees." *Political Studies Review* 21, no. 2: 357–75.

Athletes of the World. (n.d.). https://athletesoftheworld.org/.

Attepuram, Sajith. (2012, September 22). "'Olympic Games' & Manifestation of Sports Under Capitalism." *Socialism.* http://www.socialism.in/index.php/olympics-games-manifestation-of-sports-under-capitalism/.

Aviña, Rafael. (2007). *¡No queremos olimpiadas! ¡Queremos revolución! 1968: Una crónica desde la infancia y todas las películas sobre el 2 de octubre.* Mexico City: Palabra de Clio.

Barquero, Marco D. (2023). "Athletes and Environmental Activism: How Much do Athletes Contribute to Protecting the Planet?" *European Journal of Humanities and Social Sciences* 3, no. 3: 11–14.

Bashir, Nadia Y., Penelope Lockwood, Alison L. Chasteen, Daniel Nadolny, and Indra Noyes. (2013). "The Ironic Impact of Activists: Negative Stereotypes

Reduce Social Change Influence." *European Journal of Social Psychology* 43, no. 7: 614–26.

Becker, Amy B. (2013). "Star Power? Advocacy, Receptivity, and Viewpoints on Celebrity Involvement in Issue Politics." *Atlantic Journal of Communication* 21, no. 10: 1–16.

Benson, Andrew. (2019, October 24). "Mexican Grand Prix: Lewis Hamilton Defends Environmental Social Media Posts." *BBC* https://www.bbc.com/sport/formula1/50176037.

Benson, Andrew. (2023, February 7). "Formula 1 President Stefano Domenicali Says the Sport Will 'Never Put a Gag on Anyone'." *BBC* https://www.bbc.co.uk/sport/formula1/64563358.

Blanchar, Clara. (2024, October 13). "Miles de personas se manifiestan en Barcelona contra la Copa del América." *El País* https://elpais.com/espana/catalunya/2024-10-13/miles-de-personas-se-manifiestan-en-barcelona-contra-la-copa-del-america.html.

Blee, Kathleen M. and Kimberly A. Creasap. (2010). "Conservative and Right-Wing Movements." *Annual Review of Sociology* 36: 269–86.

Bobel, Chris. (2007). "'I'm Not an Activist, Though I've Done a Lot of it': Doing Activism, Being Activist and the 'Perfect Standard' in a Contemporary Movement." *Social Movement Studies: Journal of Social, Cultural and Political Protest* 6, no. 2: 147–59.

Boykoff, Jules and Gilmar Mascarenhas. (2016). "The Olympics, Sustainability, and Greenwashing: The Rio 2016 Summer Games." *Capitalism Nature Socialism* 27, no. 2: 1–11.

Boykoff, Jules. (2020). *NOlympians: Inside the Fight Against Capitalist Mega-Sports in Los Angeles, Tokyo and Beyond*. Halifax: Fernwood Publishing.

Braun, Stewart, ed. Cristina Burack. (2024, August 23). "St. Pauli Injects Subculture Into the Bundesliga." *DW* https://www.dw.com/en/st-pauli-injects-subculture-into-the-bundesliga/a-70012361.

Breeden, Aurelien and Catherine Porter with Melissa Eddy. (2024, August 7). "After French Rail Sabotage, Some See Signs of a Murky 'Ultraleft'." *New York Times* https://www.nytimes.com/2024/08/07/world/europe/france-olympics-sabotage-anarchists.html.

Brehm, Johannes and Henri Gruhl. (2024). "Increase in Concerns About Climate Change Following Climate Strikes and Civil Disobedience in Germany." *Nature Communications* 15. https://doi.org/10.1038/s41467-024-46477-4.

Bruttomesso, Elisa. (2018). "Making Sense of the Square: Facing the Touristification of Public Space Through Playful Protest in Barcelona." *Tourist Studies* 18, no. 4: 467–85.

Bugden, Dylan. (2020). "Does Climate Protest Work? Partisanship, Protest, and Sentiment Pools." *Socius: Sociological Research for a Dynamic World* 6: 1–13.

Caiani, Manuela. (2024). "Visual Analysis and the Contentious Politics of the Radical Right." *Sociology Compass* 18, no. 9 https://doi.org/10.1111/soc4.13267.

Campbell, Roderick. (2013). *Blowout! A Cost Benefit Analysis of the Australian Grand Prix.* Economists at Large http://www.ecolarge.com/wp-content/uploads/2013/03/Blowout-A-cost-benefit-analysis-of-the-Australian-Grand-Prix-Ecolarge-Final.pdf.

Camprubí i Pont, Aleix. (2024, October 23). "¿Por qué se va de Barcelona la Copa América? Del discurso triunfalista a las cifras hinchadas y el malestar vecinal." *Público* https://www.publico.es/sociedad/barcelona-copa-america-discurso-triunfalista-cifras-hinchadas-malestar-vecinal.html.

Carbon Market Watch. (2022, May). *Poor Tackling: Yellow Card for 2022 FIFA World Cup's Carbon Neutrality Claim* https://carbonmarketwatch.org/wp-content/uploads/2022/05/Poor-tackling-Yellow-card-for-2022-FIFA-1.pdf.

Carbon Trust. (2013, August 9). "Carbon Trust Unlaces the 'Carbon Bootprint' of Watching Football." https://www.carbontrust.com/news-and-insights/news/carbon-trust-unlaces-the-carbon-bootprint-of-watching-football.

Carbon Trust. (2016). *The "Carbon Bootprint" of Euro 2016: Which Nation's Fans Have the Lowest Carbon Footprint When Watching Games* https://www.carbontrust.com/news/2016/06/the-carbon-bootprint-of-euro-2016-which-nations-fans-have-the-lowest-carbon-footprint-when-watching-games/.

Carbon Trust. (n.d.). *Carbon Bootprint of the FA Community Shield* https://www.carbontrust.com/media/360767/carbon-bootprint-infographic.pdf.

Carpenter, Victoria. (2015). "'You Want the Truth? You Can't Handle the Truth': Poetic Representations of the 1968 Tlatelolco Massacre." *Journal of Iberian and Latin American Research* 21, no. 1: 35–49.

Carter, April. (2010). *Direct Action and Liberal Democracy.* London: Routledge.

Casper, Jonathan M., Michael E. Pfahl, and Brian McCullough. (2014). "Intercollegiate Sport and the Environment: Examining Fan Engagement Based on Athletics Department Sustainability Efforts." *Journal of Issues in Intercollegiate Athletics* 7: 65–91.

Castañeda, Luis. (2010). "Beyond Tlatelolco: Design, Media, and Politics at Mexico '68." *Grey Room* 40: 100–26.

Castellanos, Rosario. (2018). "Memorial de Tlatelolco." *Tercera Via* https://terceravia.mx/2018/10/memorial-de-tlatelolco-poema-de-rosario-castellanos/.

Chahuneau, Louis. (2024, July 29). "Ultragauche, manipulation étrangère … ce que l'on sait de l'enquête sur les actes de sabotage." *France 24* https://www.france24.com/fr/france/20240729-ultragauche-manipulation-%C3%A9trang%C3%A8re-ce-que-l-on-sait-de-l-enqu%C3%AAte-sur-les-actes-de-sabotage.

Chenoweth, Erica. (2021). *Civil Resistance: What Everyone Needs to Know®.* New York: Oxford University Press.

Cooper, Adam. (2013, August 25). "Greenpeace Launches Attack on Shell Oil at Belgian Grand Prix." *Autoweek* https://www.tirebusiness.com/article/20130826/NEWS/130829930/greenpeace-launches-attack-on-shell-at-belgian-grand-prix.

Cornevin, Christophe. (2024, August 1). "Les saboteurs de l'ultragauche galvanisés par les Jeux olympiques." *Le Figaro* https://www.lefigaro.fr/actualite-france/les-saboteurs-de-l-ultragauche-galvanises-par-les-jeux-olympiques-20240801.

Cotter, Trish and Sandra Mullin. (2024). "The Olympic Game's Up: It's Time for the IOC to Stop Promoting Sugary Drinks." *British Medical Journal* https://press.psprings.co.uk/gh/august/bmjgh016586.pdf.

Croco, Sarah E., Kathleen Gallagher Cunningham, and Taylor Vincent. (2023). "Protests and Persuasion: Partisanships [*sic.*] Effect on Evaluating Nonviolent Tactics in the United States." *Journal of Peace Research* 60, no. 1: 26–41.

Damián, Araceli. (2019). "Pobreza y desigualdad en México. La construcción ideológica y fáctica de ciudadanías diversas y desiguales." *El Trimestre Económico* 86, no. 3: 623–66.

Dasch, Sophia T., Millicent Bellm, Eric Shuman, and Martijn van Zomeren. (2024). "The Radical Flank: Curse or Blessing of a Social Movement?" *Global Environmental Psychology* 2 https://doi.org/10.5964/gep.11121.

Davies, Gareth Hugh. (2019, February 5). "Which Football Club Tops the Green League?" http://www.garethhuwdavies.com/environment/environmentblog/which-football-club-top-the-green-league/.

de Coninck, Heleen. (2021, November 18). "Heleen de Coninck on the Value of Activists to Spur Policymakers." *Economist* https://www.economist.com/by-invitation/2021/11/18/heleen-de-coninck-on-the-value-of-activists-to-spur-policymakers.

Delmas, Magali A. and Vanessa Cuerel Burbano. (2011). "The Drivers of Greenwashing." *California Management Review* 54, no. 1: 64–87.

del Romero Renau, Luis and Catherine Trudelle. (2011). "Mega Events and Urban Conflicts in Valencia, Spain: Contesting the New Urban Modernity." *Urban Studies Research* 1 https://doi.org/10.1155/2011/587523.

de Moor, Joost, Michiel De Vydt, Katrin Uba, and Mattias Wahlström. (2021). "New Kids on the Block: Taking Stock of the Recent Cycle of Climate Activism." *Social Movement Studies* 20, no. 5: 619–25.

Diaz, Silvia. (2022, June 15). "The Art of Manipulation: How Oil Giants are Greenwashing with Cultural Sponsorship." *Greenpeace* https://www.greenpeace.org/international/story/54285/the-art-of-manipulation-how-oil-giants-are-greenwashing-with-cultural-sponsorship/.

Eden, Sally. (2004). "Greenpeace." *New Political Economy* 9, no. 4: 595–610.

Ellis v. Nike U.S., Inc. (2024, March 28). 4:23-cv-00632-MTS.

Erhart, Itir. (2014). "United in Protest: From 'Living and Dying with Our Colours' to 'Let All the Colours of the World Unite'." *International Journal of the History of Sport* 31, no. 14: 1724–38.

European Football for Development Network. (2023). *Research Report for Climate Justice* https://footballforclimatejustice.eu/resource/research-reportfootball-forclimate-justice/.

FairSquare. (2024, October). *Substitute: The Case for the External Reform of FIFA*. https://fairsq.org/wp-content/uploads/2024/10/FIFA_Substitute_Report_v2_Single-Pages.pdf.

Falkenberg, Max, Alessandro Galeazzi, Maddalena Torricelli, Niccolò Di Marco, Francesca Larosa, Madalina Sas, Amin Mekacher, Warren Pearce, Fabiana Zollo, Walter Quattrociocchi, and Andrea Baroncheli. (2022). "Growing Polarization Around Climate Change on Social Media." *Nature Climate Change* 12: 1114–121.

Fallaci, Oriana. (1985). *Nada y así sea*. Barcelona: Editorial Noguer.

Fallaci, Oriana. (2020). *La vie, la guerre et puis rien*. Trans. Jacqueline Remillet. Paris: Les Belles Lettres.

Federación Regional de Asociaciones Vecinales de Madrid. (2025, January 8). "Alegaciones de la FRAVM a la solicitud de licencia del Gran Premio de Fórmula 1 de Madrid." https://aavvmadrid.org/?s=gran+premio&et_pb_searchform_submit=et_search_proccess&et_pb_include_posts=yes&et_pb_include_pages=yes.

Feinberg, Matthew, Robb Willer, and Chloe Kovacheff. (2020). "The Activist's Dilemma: Extreme Protest Actions Reduce Popular Support for Social Movements." *Journal of Personality and Social Psychology: Interpersonal Relations and Group Processes* 119, no. 5: 1086–111.

Ferrisi, Michaël. (2020, July 4). "La Formule 1 peut-elle vraiment devenir plus écologique?" *Ecolosport* https://ecolosport.fr/blog/2020/07/04/la-formule-1-peut-elle-vraiment-devenir-plus-ecologique/.

Fians, Guilherme. (2022, March 18). "Prefigurative Politics." *Open Encyclopedia of Anthropology*. Ed. Felix Stein https://www.anthroencyclopedia.com/entry/prefigurative-politics.

Finnerty, Samuel, Jared Piazza, and Mark Levine. (2024). "Scientists' Identities Shape Engagement with Environmental Activism." *Communications Earth & Environment* 5. https://doi.org/10.1038/s43247-024-01412-9.

Flynn, Henry. (2022, March 31). "Spain's Socialist Team Rayo Vallecano is Ready to Make a Heated Return to La Liga." *Forbes* https://www.forbes.com/sites/henryflynn/2021/07/16/spains-socialist-team-rayo-vallecano-awaits-a-heated-top-flight-return/.

Football for Future. (n.d.). https://footballforfuture.org/.

Football Supporters Association. (n.d.). https://thefsa.org.

Forbes, Don Nico. (2025, January 24). "Soccer Tackles Its Carbon Problem." *Wall Street Journal* https://www.wsj.com/articles/soccer-tackles-its-carbon-problem-efff5bcb?utm_source=substack&utm_medium=email.
Forest Green. (n.d.). "The World's First Vegan Football Club." https://www.fgr.co.uk/our-ethos/100-vegan.
Formula 1. (2022). "More Efficient, Less Fuel, and Carbon Net Zero—7 Things You Need to Know about the 2026 F1 Engine Regulations." https://www.formula1.com/en/latest/article.more-efficient-less-fuel-and-carbon-net-zero-7-things-you-need-to-know-about.ZhtzvU3cPCv8QO7jtFxQR.html.
Fossil Free Football, Game Changer Sponsorship Pledge, and Badvertising. (2024a, June). *EURO2024: Green Initiatives Overshadowed by Promotion of Big Polluters* https://www.fossilfreefootball.org/wp-content/uploads/2024/06/EURO24-Sustainability-Briefing.pdf.
Fossil Free Football, Game Changer Sponsorship Pledge, and Badvertising. (2024b, October 7). *UEFA's €500m Qatar Airways Deal Pollutes Like a Coal-Fired Power Plant* https://www.fossilfreefootball.org/2024/10/07/publication-uefas-e500m-qatar-airways-deal-pollutes-like-a-coal-fired-power-plant/.
Fournier, Patricia and Jorge Martínez Herrera. (2009). ""Mexico, 1968": Among Olympic Fanfares, Government Repression and Genocide." *Memories from Darkness: Archaeology of Repression and Resistance in Latin America.* Ed. Pedro Funari, Andres Zarankin, and Melissa Salerno. New York: Springer. 145–74.
Freelon, Deen, Alice Marwick, and Daniel Kreiss. (2020). "False Equivalencies: Online Activism from Left to Right" *Science* 369, no. 6508: 1197–201.
Friedman, Uri. (2016, August 17). "The Saddest Olympic Celebration." *The Atlantic* https://www.theatlantic.com/international/archive/2016/08/david-katoatau-olympics-kiribati/496175/.
Furnas, Alexander C., Timothy M. Lapira, and Dashun Wang. (2025). "Partisan Disparities in the Use of Science in Policy." *Science* 388, no. 6745: 362–67.
Gaffney, Christopher. (2016). "An Anatomy of Resistance: The Popular Committees of the FIFA World Cup in Brazil." *Sport, Protest and Globalisation: Stopping Play.* Ed. Jon Dart and Steven Wagg. London: Palgrave Macmillan. 335–63.
Gagnepain, Névil. (2024, August 3). "À Saint-Denis, une mobilisation citoyenne s'organise face aux violences policières." *Mediapart* https://www.mediapart.fr/journal/france/030824/saint-denis-une-mobilisation-citoyenne-s-organise-face-aux-violences-policieres.
Galbiati, Camilla. (2013, December 16). "St. Pauli F. C. de Hamburgo: Una hinchada que ondea la bandera pirata." *El Libertario* http://periodicoellibertario.blogspot.se/2013/12/st-pauli-fc-de-hamburgo-una-hinchada.html.
Gallan, Daniel. (2019, December 11). "Mocked by Some, but 'Green' Soccer Club Wants to Save the World." *CNN* https://edition.cnn.com/2019/12/11/football/forest-green-rovers-spt-intl/index.html.

Giacomini, Terran and Terisa Turner. (2015). "The 2014 People's Climate March and Flood Wall Street Civil Disobedience: Making the Transition to a Post-Fossil Capitalist, Commoning Civilization." *Capitalism Nature Socialism* 26, no. 2: 27–45.

Gil, Òscar. (2024, September 9). "Convocan una manifestación contra la Copa América en Barcelona." *El Español* https://metropoliabierta.elespanol.com/el-pulso-de-la-ciudad/20240909/convocan-una-manifestacion-contra-la-copa-america-en-barcelona/884661590_0.html.

Giulianotti, Richard. (1999). *Football: A Sociology of the Global Game.* Cambridge: Polity.

Global Witness. (2016). *Wilful Ignorance: How Japan's Voluntary Approach is Failing to Stop the Trade in Illegal Timber* https://www.globalwitness.org/en/reports/wilful-ignorance.

González de Bustamente, Celeste. (2016). "*Memorias del '68*: Media, Massacre, and the Construction of Collective Memories." *Sport, Protest and Globalisation: Stopping Play.* Ed. Jon Dart and Steven Wagg. London: Palgrave Macmillan. 101–30.

González-Reverté, Francesc and Anna Soliguer Guix. (2024). "The Social Construction of Anti-Tourism Protest in Tourist Cities: A Case Study of Barcelona." *International Journal of Tourism Cities* 10, no. 3: 842–59.

Goodwin, Jeff and James M. Jasper. (2015). "Editors' Introduction." *The Social Movements Reader: Cases and Concepts*, 3rd ed. Ed. Jeff Goodwin and James M. Jasper. Malden: Wiley Blackwell. 3–7.

Graeber, David. (2002). "The New Anarchists." *New Left Review* 13 https://newleftreview.org/II/13/david-graeber-the-new-anarchists.

Gramsci, Antonio (1978). *Selections From the Prison Notebooks of Antonio Gramsci.* Trans. and ed. Quentin Hoare and Geoffrey Nowell-Smith. New York: International Publishers.

Grant, Wyn. (2004). "Pressure Politics: The Changing World of Pressure Groups." *Parliamentary Affairs* 57, no. 2: 408–19.

Guédon, Claire. (2024, August 8). "Une interpellation et 17 verbalisations après le « Toxic Tour » avorté des militants anti-JO à Saint-Denis." *Le Parisien* https://www.leparisien.fr/jo-paris-2024/une-interpellation-et-17-verbalisations-apres-le-toxic-tour-avorte-des-militants-anti-jo-a-saint-denis-08-08-2024-ZE7RECPH5BAXFKHXI3JIWPWTOU.php.

Gulliver, Robyn E., Cassandra Star, Kelly S. Fielding, and Winnifred R. Louis. (2022). "A Systematic Review of the Outcomes of Sustained Environmental Collective Action." *Environmental Science and Policy* 133: 180–92.

Gullo, Emiliano. (2022, August 1). "Fútbol y clase obrera: la ideología no se mancha." *Jacobin* https://jacobinlat.com/2022/08/futbol-y-clase-obrera-la-ideologia-no-se-mancha/.

Gutiérrez Bayardi, Alfredo. (2022, November 15). "Tribunas que piensan: las hinchadas con ideologías más marcadas." *Para Ganar on SI* https://www.si.com/deportes/paraganar/futbol/tribunas-que-piensan-las-hinchadas-con-mas-ideologia.

Haines, Herbert H. (1984). "Black Radicalization and the Funding of Civil Rights: 1957–1970." *Social Problems* 32, no. 1: 31–43.

Hilburg, Jonathan. (2019, January 16). "Kengo Kuma is Crafting a Timber Temple to Sports for the 2020 Olympics." *The Architect's Paper* https://www.archpaper.com/2019/01/kengo-kuma-2020-tokyo-olympic-national-stadium/#gallery-0-slide-0.

International Olympic Committee. (2021a, April 21). "IOC Athletes' Commission's Recommendations on Rule 50 and Athlete Expression." https://olympics.com/athlete365/articles/officialstatement/ioc-athletes-commissions-recommendations-on-rule-50-and-athlete-expression.

International Olympic Committee. (2024, July 23). *Olympic Charter* https://stillmed.olympics.com/media/Document%20Library/OlympicOrg/General/EN-Olympic-Charter.pdf.

Ishiwatari, Mikio, Miko Maekawa, and Ryo Fujikura. (2025). "Climate Change Adaptation Strategies and Economic Challenges in Atoll Nations: A Case Study of the Marshall Islands." *Journal of Disaster Research* 20, no. 1: 53–61.

Jacopin, Tanguy. (2006–07). "America's Cup: Value Creation and Stakeholder Management." *Finance & The Common Good/Bien Commun* 26: 64–70.

Jefferson Lenskyj, Helen. (2008). *Olympic Industry Resistance: Challenging Olympic Power and Propaganda*. Albany: State University of New York Press.

Johan Cruijff Arena. (n.d.). https://www.johancruijffarena.nl/en/making-an-impact-togehther/duurzaamheid/.

Jørgensen, Lars and Jens Sejer Andersen. (2024). *When the Silence of Sport Was Broken: Sports Politics in the Times of Play the Game*. Play the Game https://www.playthegame.org/media/adxj333s/when-the-silence-of-sport-was-broken-sports-politics-in-the-times-of-play-the-game-1-mindre.pdf.

Kay, Stanley. (2019, April 22). "Winter is Going: How Climate Change is Imperiling Outdoor Sporting Heritage." *Sports Illustrated* https://www.si.com/nhl/2019/04/22/climate-change-canada-winter-sports-hockey-backyard-rinks.

Kenward, Ben and Cameron Brick. (2024). "Large-Scale Disruptive Activism Strengthened Environmental Attitudes in the United Kingdom." *Global Environmental Psychology* 2 https://doi.org/10.5964/gep.11079.

Keoghan, Jim. (2014). *Punk Football: The Rise of Fan Ownership in English Football*. Worthing: Pitch Publishing.

Khan, Yusuf. (2025, March 12). "Can Formula One Ever Be Sustainable?" *Wall Street Journal* https://www.wsj.com/articles/can-formula-one-ever-be-sustainable-0d63d80c.

Kick Big Soda Out of Sport. (n.d.). https://www.kickbigsodaout.org/.
Kokabi, Alexandre-Reza. (2024, July 23). "Les JO 2024 à Paris, une catastrophe écologique en Seine-Saint-Denis." *Reporterre* https://reporterre.net/Les-JO-2024-a-Paris-une-catastrophe-ecologique-en-Seine-Saint-Denis.
Kountoruis, Yiannis and Eleri Williams. (2023). "Do Protests Influence Environmental Attitudes? Evidence from Extinction Rebellion." *Environmental Research Communications.* https://doi.org/10.1088/2515-7620/ac9aeb.
Krispenz, Anne and Alex Bertrams. (2024). "Understanding Involvement in Environmental Activism: Relationships to Pathological Narcissistic Grandiosity, Virtue Signaling, Dominance, and Sensation Seeking." *Current Psychology* 43: 30011–026.
Kriza, Elisa. (2019). "Anti-Communism, Communism, and Anti-Interventionism in Narratives Surrounding the Student Massacre on Tlatelolco Square (Mexico, 1968)." *Bulletin of Latin American Research* 38, no. 1: 82–96.
Le Foll, Clément. (2024, August 8). "Militants écologistes ou pour la justice sociale, ils ont été écartés des JO." *Mediapart* https://www.mediapart.fr/journal/france/080824/militants-ecologistes-ou-pour-la-justice-sociale-ils-ont-ete-ecartes-des-jo.
Lester, Libby. (2019). *Global Trade and Mediatised Environmental Protest: The View from Here.* Cham: Palgrave Macmillan.
Lockwood, David. (2022, July 31). "Leilani Münter: Former Racing Driver Wins BBC's Evergreen Athlete Award." *BC* https://www.bbc.com/sport/motorsport/62263116.
Louis Vuitton 37th America's Cup. (2024, October 22). "Moltes Gràcies Barcelona—The Jewel in the Sun." https://www.americascup.com/news/3759_MOLTES-GRACIES-BARCELONA-THE-JEWEL-IN-THE-SUN.
Lovato, Roberto. (2004, June 28). "Fear of a Brown Planet." *The Nation*: 17–21.
Lowes, Mark. (2004). "Neoliberal Power Politics and the Controversial Siting of the Australian Grand Prix Motorsport Event in a Public Park." *Loisir et Société/Society and Leisure* 27, no. 1: 69–88.
Mabon, Leslie. (2023). "Football and Climate Change: What Do We Know, and What is Needed for an Evidence-Informed Response?" *Climate Policy* 23, no. 3: 314–28.
Mandard, Stéphane. (2022a, April 21). "Le Pen Vows 'National Environmental Policies'." *Le Monde* https://www.lemonde.fr/en/2022-presidential-election/article/2022/04/17/le-pen-challenges-macron-s-punitive-ecology-with-her-national-ecology_5980790_16.html.
Marshall, Alasdair, Richard Telofski, Udechukwu Ojiako, and Max Chipulu. (2012). "An Examination of "Irregular Competition" Between Corporations and NGOs." *Voluntas* 23, no. 2: 371–91.

Martin, Lisa. (2019, February 6). "Climate Change Set to Disrupt Australia's Summer Sports Calendar." *Guardian* https://www.theguardian.com/environment/2019/feb/06/climate-change-set-to-disrupt-australias-summer-sports-calendar.

Más Madrid. (2025). "Alegaciones a la aprobación inicial del "Plan especial de definición de redes en el Ape 21.10 "Parque Ferial Juan Carlos I y su etorno" situada en La Avenida del Partenón, 5. Distrito de Barajas." https://masmadrid.org/mas-madrid-denuncia-que-el-plan-urbanistico-de-almeida-para-la-formula-1-en-ifema-evita-medir-el-impacto-ambiental-del-proyecto/.

McCarthy, Lauren. (2021, December 2). "Why is Direct Action Essential for Climate Justice?" *Global Citizen* https://www.globalcitizen.org/en/content/direct-action-for-the-climate-crisis/.

McCullough, Bryan P. and Timothy B. Kellison. (2018). "An Introduction to Environmental Sustainability and Sport." *Routledge Handbook of Sport and the Environment*. Ed. Bryan P. McCullough and Timothy B. Kellison. Abingdon: Routledge. 3–10.

McGregor, Callum. (2015). "Direct Climate Action as Public Pedagogy: The Cultural Politics of the Camp for Climate Action." *Environmental Politics* 24, no. 3: 343–62.

Miller, Toby. (2018). *Greenwashing Sport*. London: Routledge.

Montague, James. (2010, August 20). "Punks, Prostitutes and St. Pauli: Inside Soccer's Coolest Club." *CNN* https://edition.cnn.com/2010/SPORT/football/08/18/football.st.pauli.punks/index.html.

M. S. (2019, November 14). "Youth Movements are the Fool's Gold of Politics." *Economist* https://www.economist.com/open-future/2019/11/14/youth-movements-are-the-fools-gold-of-politics.

Muñana, Gustavo. (2024, July 27). "Extinction Rebellion Activists Arrested in Paris for Alleged Sabotage." *Inside the Games* https://www.insidethegames.biz/articles/1147127/extinction-rebellion-activists-arrested.

Münter, Leilani. (n.d.). https://www.leilani.green/.

Neeson, J. M. (1996). *Commoners: Common Right, Enclosure and Social Change in England, 1700–1829*. Cambridge: Cambridge University Press.

Nestler, Stefan. (2022, August 17). "How Football is Responding to the Energy Crisis." *Deutsche Welle* https://www.dw.com/en/how-the-bundesliga-is-responding-to-the-energy-crisis/a-62837284.

New Weather Institute with Scientists for Global Responsibility. (2025a, February 3). *Dirty Tackle: The Growing Carbon Footprint of Football* https://www.newweather.org/wp-content/uploads/2025/01/Dirty_tackle_The_growing_carbon_footprint_of_football.pdf.

New Weather Institute with Scientists for Global Responsibility. (2025b, July). *FIFA's Climate Blind Spot: The Men's World Cup in a Warming World* https://

www.newweather.org/wp-content/uploads/2025/07/FIFAs_climate_blind_spot.pdf.

NOlympics LA. (n.d.). https://nolympicsla.com/.

Notre Affaire à Tous. (2022, November 2). *Plainte devant le Jury de Déontologie Publicitaire* https://climatecasechart.com/wp-content/uploads/non-us-case-documents/2022/20221102_18433_complaint.pdf.

Oltermann, Philip. (2014, June 16). "Greenpeace Loses £3m in Currency Speculation." *Guardian* https://www.theguardian.com/environment/2014/jun/16/greenpeace-loses-3m-pounds-currency-speculation.

O'Neill, Francis, Geoff Dickson, Tim Ströbel, and Ashleigh-Jane Thompson. (2025). "Elite Athlete Activism, Advocacy, and Protest: A Scoping Review." *European Sport Management Quarterly* 25, no. 1: 145–73.

OneTreePlanted. (n.d.). https://bellerintrees.raisely.com/.

Ostarek, Markus, Brent Simpson, Cathy Rogers, and James Ozden. (2024). "Radical Climate Protests Linked to Increases in Public Support for Moderate Organizations." *Nature Sustainability* 7: 1626–632.

Palmer, Jordi. (2024, June 19). "L'altra cara de la Fórmula 1: mig miler de barcelonins es manifesten contra l'exhibició." *El Nacional.cat* https://www.elnacional.cat/ca/barcelona/altra-cara-formula-1-mig-miler-barcelonins-es-manifesten-contra-exhibicio_1238529_102.html.

Pato, Ignacio. (2021, September 1). "In Livorno, Italy's Most Famous Left-Wing Football Club is Fighting to Survive." *Jacobin* https://jacobin.com/2021/09/livorno-football-club-italy-communism.

Piñera, Sebastián. (1978). *Evolución de la pobreza en México: (Períodos 1950–1963; 1963–1968).* Proyecto Interinstitucional de Pobreza Crítica en América Latina https://repositorio.cepal.org/server/api/core/bitstreams/dc713c40-d4ed-45c8-99f9-4c52e10f7008/content.

Poniatowska, Elena. (1971). *La noche de Tlatelolco: Testimonios de historia oral.* Mexico City: Ediciones Era.

Portillo, Miguel. (2022, March 4). "Alonso: 'Estoy al 100% y hay que ser rápidos en la pista y en la fábrica'." *Forbes España* https://forbes.es/lifestyle/141598/alonso-estoy-al-100-y-hay-que-ser-rapidos-en-la-pista-y-en-la-fabrica%ef%bf%bc/.

Preza, Ana. (2012, February 29). "America's Cup Stirs Up Controversy in San Francisco." *Golden Gate Xpress* https://goldengatexpress.org/19459/uncategorized/americas-cup/.

Prieto, Alberto. (2024, October 10). "No a la Copa Amèrica porta la inversió pública en la regata a Antifrau." *MónEconomia* https://elmon.cat/moneconomia/empreses/no-a-la-copa-america-porta-la-inversio-publica-en-la-regata-a-antifrau-79610/.

Ramaswamy, Chitra. (2015, November 30). "Hosting the Olympics: The Competition Noone Wants to Win." *Guardian* https://www.theguardian.com/sport/shortcuts/2015/nov/30/hosting-olympics-hamburg-drop-out-2024-games.

Reynolds, Tom. (2018, October 20). "Vera Caslavska and the Forgotten Story of her 1968 Olympics Protest." *BBC* https://www.bbc.com/sport/olympics/45900544.

Richards, Giles. (2021a, March 31). "Nico Rosberg: 'To Do Good You Need to Get Out There, You Can't Sit in a Cave'." *Guardian* https://www.theguardian.com/sport/2021/mar/31/nico-rosberg-to-do-good-you-need-to-get-out-there-you-cant-sit-in-a-cave.

Richards, Giles. (2021b, November 26). "Climate Emergency Accelerates F1's Efforts to Clean Up Its Image." *Guardian* https://www.theguardian.com/sport/2021/nov/26/climate-emergency-accelerates-f1-efforts-to-clean-up-image.

Rodriguez, Cecilia. (2024, August 23). "Tensions and Hostility Against America's Cup Sailing Competition 'For The Rich' in Barcelona." *Forbes* https://www.forbes.com/sites/ceciliarodriguez/2024/08/23/tensions-and-hostility-against-americas-cup-sailing-competition-for-the-rich-in-barcelona/.

Rodríguez, Jesús. (2024a, September 9). "La Copa Amèrica falseja les dades d'audiència i de visitant." *Directa* https://directa.cat/la-copa-america-falseja-les-dades-daudiencia-i-de-visitants/.

Rodríguez, Jesús. (2024b, October 10). "Un centenar d'entitats presenten una denúncia contra la Copa Amèrica a l'Oficina Antifrau." *Directa* https://directa.cat/un-centenar-dentitats-presenten-una-denuncia-contra-la-copa-america-a-loficina-antifrau/.

Rojo, Juan J. (2016). *Revisiting the Mexican Student Movement of 1968: Shifting Perspectives in Literature and Culture Since Tlatelolco*. New York: Palgrave Macmillan.

Roland, Romain. (1920, March 19). "Un livre de Raymond Lefebvre: *Le Sacrifice d'Abraham*." *l'Humanité* https://gallica.bnf.fr/ark:/12148/bpt6k299613c/f1.item.

Ropa, Anastasija, Ludmila Malahova, and Monta Jakovleva. (2024). "Animal Ethics or Sport Ethics? A Historical Approach to Professional Ethics in Equestrian Sports." *Cheiron: The International Journal of Equine and Equestrian History* 4, no. 1: 95–114.

Rosa, Anton and Alberto Prieto. (2024, November 21). "El Parlament bloqueja l'auditoria dels comptes de la Copa Amèrica." *Tót Barcelona* https://www.totbarcelona.cat/economia/parlament-no-investiga-copa-america-567042/.

Ruta, Alessandro. (2021). "Livorno, la ciudad «más roja», en el centenario del PCI." *Naiz* https://www.naiz.eus/eu/hemeroteca/7k/editions/7k_2021-09-26-07-00/hemeroteca_articles/livorno-la-ciudad-mas-roja-en-el-centenario-del-pci.

Sabaté, Arsen. (2024, October 11). "Una ceremonia deslumbrante para ocultar el negocio capitalista y depredador de la Copa América de vela." *Izquierda Diario* https://www.izquierdadiario.es/Una-ceremonia-deslumbrante-para-ocultar-el-negocio-capitalista-y-depredador-de-la-Copa-America-de.

Saccage. (2021, January). *Un recensement effectué par des Associations et des Collectifs Citoyens* https://saccage2024.noblogs.org/files/2021/03/saccage-2024-argumentaire.pdf.

Saccage. (2024, July 29). "Les Jeux des arrestations arbitraires." https://saccage2024.noblogs.org/archives/3164.

Sacristán, Claudia. (2024, July 14). ""No puedo ir ni a la playa de mi ciudad": los barceloneses, contra de la Copa América de Vela." *El Confidencial* https://www.elconfidencial.com/espana/cataluna/2024-07-14/no-puedo-ir-playa-ciudad-barceloneses-copa-america_3919767/.

Sanghera, Mandeep. (2024, October 14). "Legal Complaint Filed Against FIFA's 'Abuse of Dominance'." *BBC* https://www.bbc.com/sport/football/articles/c981203e61qo.

Santamarina Campos, Beatriz and Albert Moncusí Ferré. (2013). "De huertas y barracas a galaxias faraónicas. Percepciones sociales sobre la mutación de la ciudad de Valencia." *Papers: Revista de sociología* 98, no. 2: 365–91.

Sartre, Jean-Paul and Bertrand Russell. (1968, November 1). "Jean-Paul Sartre and Bertrand Russell Speak-out on Mexico and its Olympics." *Los Angeles Free Press*: 2.

Schäfer, Mike S. (2012). "Online Communication on Climate Change and Climate Politics: A Literature Review." *Wiley Interdisciplinary Reviews: Climate Change* 3, no. 6: 527–43.

Schimmel, Kimberly S., C. Lee Harrington, and Denise D. Bielby. (2007). "Keep Your Fans to Yourself: The Disjuncture between Sport Studies' and Pop Culture Studies' Perspectives on Fandom." *Sport in Society* 10, no. 4: 580–600.

Sexton, Sarah, Nicholas Hildyard, and Larry Lohmann. (2005, April 7). "We're a Small Island: The Greening of Intolerance." *The Corner House* https://www.thecornerhouse.org.uk/sites/thecornerhouse.org.uk/files/SmallIsland.pdf.

Shenker, Jack. (2023, March 6). "The Existential Question for Climate Activists: Have Disruption Tactics Stopped Working?" *Guardian* https://www.theguardian.com/commentisfree/2023/mar/06/british-eco-activists-disruption-extinction-rebellion.

Sim, Walter. (2018, November 12). "Environment Group Accuses Tokyo Olympics Organisers of Using Illegally Sourced Wood." *Straits Times* https://www.straitstimes.com/asia/east-asia/environment-group-accuses-tokyo-olympics-organisers-of-using-illegally-sourced-wood.

Simpson, Brent, Robb Willer, and Matthew Feinberg. (2022). "Radical Flanks of Social Movements Can Increase Support for Moderate Factions." *PNAS Nexus* 1: 1–11.

Solans, Maria. (2024, October 23). "La plataforma No a la Copa Amèrica que la prova marxi de Barcelona: "Bon vent i barca nova"." *Betevé* https://beteve.cat/societat/plataforma-no-a-la-copa-america-celebra-prova-marxi-barcelona/.

STAND.earth. (2024, September). *Prime Polluter.* STAND.earth, Pacific Environment, Clean Mobility Collective, and Ship It Zero https://stand.earth/wp-content/uploads/2024/09/PRIME_POLLUTER-09.2024.pdf.

Sturm, Damion. (2023). "Processes of Greenwashing, Virtue Signaling and Sportwashing in Contemporary Formula One: Formula Façade?" *The Future of Motorsports: Business, Politics and Society.* Ed. Hans Erik Næss and Simon Chadwick. Abingdon: Routledge. 167–82.

Sullivan, Sean P. (2024). "Market Definition." *Research Handbook on Abuse of Dominance and Monopolization.* Ed. Pinar Akman, Or Brook, and Konstantinos Stylianou. Cheltenham: Edward Elgar. 21–25.

Taibo II, Paco Ignacio. (2022). *68.* N.p.: Titivillus.

Taylor, Charles. (2004). *Modern Social Imaginaries.* Durham: Duke University Press.

The League of Fans. (2011). *Sports Manifesto* http://leagueoffans.org/2011/06/29/the-sportsfans-manifesto.

The Shift Project. (2025, February). *Décarbonons le sport: Un premier applicatif au football et au rugby* https://theshiftproject.org/wp-content/uploads/2025/02/Rapport-final-Decarbonons-le-Sport-combine.pdf.

Thompson, E. P. (1974). "Patrician Society, Plebeian Culture." *Journal of Social History* 7, no. 4: 382–405.

Thrall, A. Trevor, Jaime Lollio-Fahkreddine, Jon Berent, Lana Donnelly, Wes Herrin, Zachary Paquette, Rebecca Wenglinski, and Amy Wyatt. (2008). "Star Power: Celebrity Advocacy and the Evolution of the Public Sphere." *International Journal of Press/Politics* 13, no. 4: 362–85.

Till, Brian D., Sarah M. Stanley, and Randi Priluck. (2008). "Classical Conditioning and Celebrity Endorsers: An Examination of Belongingness and Resistance to Extinction." *Psychology & Marketing* 25, no. 2: 179–96.

Totten, Mick. (2013). "Sport Activism and Political Praxis Within the FC Sankt Pauli Fan Subculture." *Play Left Wing* https://www.playleftwing.org/wp-content/uploads/2013/10/Fan-Culture-and-Sport-Activism1.pdf.

Towler, Christopher C., Nyron N. Crawford, and Robert A. Bennett III. (2020). "Shut Up and Play: Black Athletes, Protest Politics, and Black Political Action." *Perspectives on Politics* 18, no. 1: 111–27.

Tyson, Alec and Brian Kennedy. (2023, October 25). *Climate Activism.* Pew Research Center https://www.pewresearch.org/science/2023/10/25/climate-activism/.

Une delegation inattendue. (n.d.). *Revendication du sabotage de lignes TGV quelques heures avant la cérémonie d'ouverture des Jeux Olympiques de Paris 2024* https://reporterre.net/IMG/pdf/lettrerevendicationsabotage.pdf.

United Nations Climate Change. (n.d.). *The Paris Agreement* https://unfccc.int/process-and-meetings/the-paris-agreement.

Vaccaro, Adam. (2015, April 2). "They Just Don't Want the Olympics." *Boston.com* http://www.boston.com/news/local-news/2015/04/02/they-just-dont-want-the-olympics.

Valvur, Andy. (2018, November 14). "6 Cities That Rejected the Olympics." *DW* https://www.dw.com/en/6-cities-that-rejected-the-olympics/a-46289852.

Venzon, Andrea and Colombe Cahen-Salvador. (2019, November 1). "Party Politics is Dying, So Make Citizen Movements the New Unifier." *Economist* https://www.economist.com/open-future/2019/11/01/party-politics-is-dying-so-make-citizen-movements-the-new-unifier.

Vettel Sebastian. (n.d.). https://sebastianvettel.de/en/start/.

Vicente, Sandra. (2023, October 10). "65 entidades se unen contra la Copa América de Barcelona: "Es la síntesis del expolio contra la ciudad"." *El Diario* https://www.eldiario.es/catalunya/65-entidades-unen-copa-america-barcelona-sintesis-expolio-ciudad_1_10585988.html.

Viñas, Carles and Natxo Parra. (2020). *St. Pauli: Another Football is Possible.* Trans. Luke Stobart. London: Pluto Press.

Wågan, Frida Austmo. (2024). "Regenerating the Pitch: Environmental Discourses in Football." *Annals of Leisure Research* 28, no. 1: 73–94.

WePlayGreen. (n.d.). https://weplaygreen.com/.

Weston, Barney. (2022, May 6). "Football Must Do More to Tackle Climate Change: This is How Clubs and Fans Can Help." *Guardian* https://www.theguardian.com/football/blog/2022/may/06/football-must-do-more-to-tackle-climate-change-this-is-how-clubs-and-fans-can-help.

Wheeler, David. (2024, December 8). "Female Footballers Have Shown Us How—Let's Build a Sport Free of Fossil Fuel Deals." *Guardian* https://www.theguardian.com/football/2024/dec/08/female-footballers-have-shown-us-how-lets-build-a-sport-free-of-fossil-fuel-deals.

Wong, Gabriella. (2025, February 12). "Marshall Islands Football Jersey "Disappears" to Visualise Threat of Rising Seas." *Dezeen* https://www.dezeen.com/2025/02/12/marshall-islands-soccer-federation-playerlayer-kit/.

Yates, Luke. (2015). "Rethinking Prefiguration: Alternatives, Micropolitics and Goals in Social Movements." *Social Movement Studies: Journal of Social, Cultural and Political Protest* 14, no. 1: 1–21.

Zacher, Hannes. (2024). "The Dark Side of Environmental Activism." *Personality and Individual Differences* 219 https://doi.org/10.1016/j.paid.2023.112506.

CHAPTER 7

Conclusion

What can be done to conserve sporting pleasures? Carbon mega-events *may* 'bring out the best in athletes.' They can *always* be relied upon to 'bring out the worst in host cities and countries' (Boykoff 2020: 11). For sports are simultaneously victims and perpetrators of climate disaster. At material and ideological levels, they can 'be pressed into the service of any end whatsoever' (Baudrillard 1993: 77)—empty signs waiting to be overloaded with signification but that are ecologically material, all too material. Can they offer solutions to our crises (Gelius et al. 2024)?

There is a duality here: the 'most exciting' sporting tasks 'involve an element of drudgery ... hours of muscle-ache-inducing practice' alongside the joy of success and adulation (Huws 2019: 16) and a carnival-like reversal of norms that may equally incarnate privilege. For Philippe Ariès, caresses exchanged by footballers when their team scores are historically typical, per medieval festivals or today's *mardi gras*, when taboos are broken, 'temporarily compromised in public space but without assumptions re sexuality either feared or asserted' (1987: 75). Anne McClintock interprets things differently:

> Privileged groups can, on occasion, display their privilege precisely by the extravagant display of their right to ambiguity. When the English football star, Paul Gascoigne, returned triumphant from the World Cup, he paraded

T. Miller, J. P. Carañana, *Global Sports Go Green—Or Do They?*,
Global Culture and Sport Series,
https://doi.org/10.1007/978-3-032-14207-8_7

> the streets in plastic women's breasts, as if his excess of heterosexual prowess on the football field licensed his privileged display of gender ambiguity. (1995: 68)

For Gabriel Kuhn, football is 'an increasingly commercialized game of professionals at top, managed by corrupt administrators and greedy corporations, and a grassroots game at the bottom that provides millions of people with joy, confidence, and a sense of community' (2019: 9). Galeano thought similarly:

> Professional soccer does everything to castrate that energy of happiness, but it survives in spite of all the spites. And maybe that's why soccer never stops being astonishing. … The more the technocrats program it down to the smaller detail, the more the powerful manipulate it, soccer continues to be the art of the unforeseeable. When you least expect it the impossible occurs. (2013: 442)

Despite *Modern Times*, sports can offer alternatives to themselves, contra their capitalistic metaphorization. Consider the untrammeled, barely controlled ecstasy of catching a wave or seeing someone else do so. That utopia was celebrated by Adrienne Rich because it 'makes the body shoot … [i]nto its pure and irresistible curve' (1997), akin to Byatt viewing a footballer in close-up as a 'distorted mask of focused breathing energy, like a classical sculpture of a wind god' (2008).

William Blake's poem 'The Ecchoing Green' rhapsodizes the natural rhythm of a day:

> Till the little ones weary
> No more can be merry
> The sun does descend
> And our sports have an end (1789)

This is per Bacon claiming that god and monarchy endorsed the 'innocent and kindly sport of children playing at hide-and-seek, and vouchsafed of his kindness and goodness to admit the human spirit for his playfellow at that game' (1863: 9), or the moment in *The Sound of Music* (Robert Wise 1965) when youngsters are unleashed from their father's strict domestic discipline to frolic outdoors, enjoying ballgames, running, bike-riding, and rowing without any reason beyond emotional and physical self-expression.

As John Dewey wrote, '[l]abor, through its structure and order, lends play its pattern and plot; play then returns the loan with interest to work, in giving it a sense of beginning, sequence and climax' (1958: 157). This is per Trotsky imagining a time when 'Man [*sic.*] at last will begin to harmonize himself in earnest. He will make it his business to achieve beauty by giving the movement of his own limbs the utmost precision, purposefulness and economy in his work, his walk and his play' (1957: 368); what Elizabeth Wilson explained as a 'search for meaning, not through ordered reflection,' but 'kinetic cultural experience' (2022: 6), per Fanon avowing that sports could 'elevate the people, broaden the mind of the people, fill it, differentiate it, humanize it' (2007: 156).

This is no neoliberal or Social Darwinist struggle of all against all; even the most 'competitive activity' is founded 'on a cooperative basis' (Marqusee 2014). And sports need not be about instrumentalization or victory; participants can develop capacities in 'reciprocal self-realisation' with others (Eagleton 2024a), experiencing both the imposition of authority ***and*** the joy of autonomy, via exploitation of their labor process mingling with paradoxically autotelic pleasures.

Specific to environmentalism, philosophical and artistic expressions of human alienation have proliferated since enclosure, colonialism, and industrialization, occasioned by dissociation from nature (Eichberg 2018). In that critique resides hope. Part of the struggle against climate change and greenwashing involves fundamentally rethinking the material world to privilege our fellow-inhabitants of the planet, sentient and otherwise, using a wisdom that eschews the instrumental reason that takes growth as its watchword.

Spinoza maintained that 'the more things the mind knows, the better it understands both its own powers and the order of Nature. … and the better it understand the order of Nature, the more easily it can restrain itself from useless pursuits' (2002: 79). *Contra* anthropocentrism, Hume recognized our fellow-animals' special qualities: they 'learn many things from experience,' developing 'knowledge of the nature of fire, water, earth, stones, heights, depths, etc.' (1955: 112–13) through inference, via what he called 'the reason of animals' (1739). Marcuse realized that:

> the demands of ever more intense exploitation come into conflict with nature itself, since nature is the source and locus of the life-instincts which struggle against the instincts of aggression and destruction. And the demands of exploitation progressively reduce and exhaust resources: the more capitalist productivity increases, the more destructive it becomes. This is one sign of the internal contradictions of capitalism. … [Nature] is a dimension *beyond* labor, a symbol of beauty, of tranquility, of a nonrepressive order. … the very negation of the market society. (1972)

Carbon sports' governing bodies are painful, pitiful, but powerful hangovers from the good-old-boy networks/fiefdoms that ran sports prior to their comprehensive commercialization. The newer hegemons represent the ultimate in corruption, commodification, and governmentalization, with residual participation by greybeards at the highest echelons, in those heady spaces where corrupt young technocrats and old élites mingle (Tomlinson 2014, 2017; Conn 2017; Fruh et al. 2022; Jennings 2016). In denouncing FIFA and the IOC, Marina Hyde typified them as performing an 'eternal contest to be the worst-run governing body … the hardest-fought rivalry in world sport' (2025).

The reformist plans of these oligarchs, exemplified in awarding environmental ratings for stadia (Bullock 2017) or the United Nations Climate Change Conference's *Sports for Climate Action Framework* (2018), tinker at the margins of environmental problems, touting sustainability strategies as prerequisites for growth and development while obscuring real causes and solutions (Gammelsæter and Loland 2023). Carbon sports talk a good game ecologically, even as they engage in despoliation (Millington et al. 2022).

The fantasy of creating a green economy through new technology that forges ongoing links between the economy and the environment increasingly looks like that—a fantasy. Prelates of technology and their chorines in sports studies, journalism, and the state apparatus imagine growth and environmentalism modifying one another in a sustainable post-political cornucopia. That hasn't happened (Johnson and Ali 2018; Kim and Chung 2018). Automobiles, for example, may have become more efficient and less polluting, in part thanks to Formula One, but their overall use has increased, to terrible effect (Dengler 2021)—1.19 million annual deaths from accidents (World Health Organization 2023) and 4.6 metric tons of CO_2 emitted per car, per year (United States Environmental Protection Agency 2023).

We need to think decades ahead when pondering the costs, benefits, causes, and effects of how environmental lifeworlds are altered by sports, both directly and indirectly. Proper consideration of this long shadow is in stark contrast to the duration of carbon mega-events, even those that necessitate years of planning, construction, and competition (Ahmed and Pretorius 2010). Reduced environmental emissions, increased biodiversity, and proper water care are essential.

One might ask why the growth of sports is even an issue. Can't they simply emerge and exist organically? The answer is sonic, onomatopoeic. The most important sound in football, for example, is no longer a ball battering a net, but the Ka-Ching of cash registers (Perkins 2000). Plutocracy frequently prevails over science and democracy, as capitalist businesses elude critical expressions of the public interest by insisting they know what it is and how to serve it (Barrett et al. 2019). That accords with the fashion for "governance," a euphemistic term deployed by politicians and bureaucrats to block bodies of law that could protect the environment from sports and their "governing" mandarins (Baram 2009; Warren 2020; Geeraert 2025). Avoidance of democracy has become a core principle of mega-events. It can be no surprise that a court order was needed for FIFA to introduce hydration breaks during the 2014 Finals (Earls and EE News 2019).

There is great 'tension between the monetary form of values and the principles of natural reproduction' (Koch and Buch-Hansen 2020: 377). That tension finds powerful expression in our own time, when the dominant discourse of economics assumes prosperity depends on "improving" the past, the present, the world. It holds that there are—there can—there must be no limits to growth. Such knowledge is radically separated from nature, which is something to be overcome/admired, i.e., depleted yet fetishized (McMichael 2011). Exemplars of true sustainability are deemed backward-looking points on a path toward economic reality:

> Current generations of forest-dwellers, small farmers, pastoralists and fisherfolk still practising low-carbon lifestyles (and performing 'environmental services') are thus viewed as 'future generations' only insofar as they are expected to out-migrate from their purported impoverished lives within nature, and join the 'planet of slums.' (McMichael 2011: 170)

Instead of such myth-making, we should take as our lodestone a memorable Bobby Kennedy speech querying a definition of Gross National Product that:

> counts air pollution and cigarette advertising, and ambulances to clear our highways of carnage. It counts special locks for our doors and the jails for the people who break them. It counts the destruction of the redwood and the loss of our natural wonder in chaotic sprawl. It counts napalm and counts nuclear warheads and armored cars for the police to fight the riots in our cities. It counts Whitman's rifle and Speck's knife, and the television programs which glorify violence in order to sell toys to our children. Yet the gross national product does not allow for the health of our children, the quality of their education or the joy of their play. It does not include the beauty of our poetry or the strength of our marriages, the intelligence of our public debate or the integrity of our public officials. It measures neither our wit nor our courage, neither our wisdom nor our learning, neither our compassion nor our devotion to our country, it measures everything in short, except that which makes life worthwhile. (1968)

Taking up that challenge, we might consider models of degrowth—an ugly neologism for a complex, almost shocking, but valuable tendency. It means wealthy nations abandoning GDP as a measure of the quality of life, diminishing production to meet existing rather than manufactured needs, and focusing on well-being and stable economies. That necessitates rerouting Global North expenditure from marketing, fast fashion, industrial meat, carbon sports, and fossil-fuel travel to universal free education and health care, guaranteed minimum incomes, and rail and sea transportation (Hickel et al. 2022).

Degrowth requires opposing 'any political system that sees nature only through the lens of demands for unlimited economic growth' (Light 1998: 345, 348). It threatens monetary systems that favor financial rather than productive concerns and rely on exploiting minerals, plants, animals, and labor, where the "failure" to grow economies typically leads to pronouncements of recession, with disproportionately negative outcomes for poor as opposed to rich people in the Global North and South. So this is a tough political project, in both democratic oligarchies, such as Mexico, and Leninist ones, for example, China. But there is 'abundant evidence of the equal or greater productivity of diverse small-farming systems, which in turn restore soil and water health, regenerate natural carbon cycles, sustain biodiversity and stabilize rural populations,' versus the corporate pollution that dominates food provision (McMichael 2011: 170).

Even the *Economist* is prepared to countenance that degrowth has merit (Dobbs 2024). The concept favors alternative visions of life from the dominant, per African *ubuntu* [solidarity], feminist economics, and Latin American *buen vivir* [the good life], reclassifying national accounts to exclude destructive industries and value care (Dengler 2021). And it is a necessity. John Stuart Mill, a darling of neoclassicists, recognized 150 years ago that 'the increase of wealth is not boundless' and there must be a debate over 'to what goal?' (2004: 188). He walked in the shadow of Thomas Malthus, who understood that 'the power of population is indefinitely greater than the power in the earth to produce subsistence' (1798: 4).

Raymond Williams juxtaposed ideology and science:

> Particular estimates of absolute shortages of currently necessary resources, of absolutely unacceptable levels of pollution and environmental damage, of imbalances of population and food which would bring widespread famine, do not have to be accepted in their precise (and in practice always revisable) terms to persuade any reasonable person that the orthodox abstraction of indefinitely expanded production—its version of 'growth'—has to be considered again from the beginning. The current ideological implication that these are simplistic or idealistic or simply 'anti-industrial' arguments could not be further from the truth. What is actually being faced is the full physical and material situation, by contrast with the received mental habits of the most recent versions of human livelihood. (2015: 169)

In that context, we need 'relocalization, work reduction, job-sharing, valuing informed social reproductive activities, redistributing resources, restoring waste and nutrient cycles, urban agriculture, slow cities' (McMichael 2011: 170) with economic forces 'subordinated to the mandates of the Earth' (Acosta and Cajas-Guijarro 2020: 300). One of the prerequisites to achieving those goals is letting people know the extent and the damage of greenwashing and putting a stop to it.

Colombia's then-Ministra de Ambiente y Desarrollo Sostenible [Minister of Environment and Sustainable Development], Susana Muhamad, told the twenty-eighth Conference of the Parties that 'La transición para dejar los combustibles fósiles, debe complementarse con un Tratado que entregue reglas claras de financiamiento para los países, pero además respetando los bosques y las comunidades. ¡Nuestra lucha por la justicia climática y equidad, sigue!' [The transition from fossil fuels must be complemented by a treaty that has clear finance norms and respects forests and communities. Our fight for climate justice and equity continues!] (2023). Muhamad had worked for Shell, until she realized it was 'greenwashing. … There were people trying to push change but the regime of fossil fuels was too strong.' Something 'more systematic' was needed (quoted in Greenfield 2024).

That means facing up to the facts, not the distortions, of climate change and human pollution; facing down mendacious and gullible naysayers; insisting on realistic science education; denouncing supine media coverage that bows before the false idol of capital; and calling for revitalized laws, regulations, and enforcement by governments in keeping with international standards, all in dialog with scientists and everyday people.

The UN explains that 'greenwashing promotes false solutions to the climate crisis that distract from and delay concrete and credible action' (n.d.). Both *Nature* ("Climate Talks" 2019) and *Al Jazeera* (Casey 2019) insist that enforceable legal protection is required to deal with environmental horror. The High-Level Expert Group on the Net-Zero Emissions Commitments of Non-State Entities argues:

> If greenwash premised upon low-quality net zero pledges is not addressed, it will undermine the efforts of genuine leaders, creating both confusion, cynicism and a failure to deliver urgent climate action. Which is why, ultimately, regulations will be required to establish a level playing field. (2022: 12)

That must be our future. "Environmental sustainability and governance?" Not so much. The almost casual references across this book to misconduct by companies that sponsor carbon sports show how abject these firms are.

From the post-industrial core to the manufacturing periphery, universal ecological standards are essential, in addition to contractual deals between multinationals and their hosts. Research on deliberation among the disempowered provides further evidence that decentralized, participatory governance can play a vital role in policy-making by involving community members, experts, and élites alike (Karpowitz et al. 2009: 584).

Again and again, systematic polluters, whose very existence endangers the planet, associate themselves with putatively green leisure, seeking to make it part of their institutional being. Such greenwashing has established a structural and indexical homology between sports and their sponsors, simultaneously exploiting and distorting the brief, fragile impact of star athletes to obscure the lengthy, powerful impact of environmental despoliation. So in addition to attending to their own folly and destructiveness, carbon sports must cut off the supply line to legitimacy. As we have seen, the extractive industries are key sponsors. The record shows thirty-six such companies responsible for almost half the world's CO_2 emissions in 2023 (CarbonMajors 2025).

National boundaries and interests are brought into question by the border-crossing impact of environmental despoliation (Dean 2001). There are several established ways of regulating the trans-territorial challenges multinational corporations pose: 'soft law [protocols of international organizations], hard law [national legislation], codes of conduct [transnational norms] and voluntary self-regulation' (Baram 2009). Private and public sporting policies need to be predicated on transnational environmental justice, on holding polluters accountable for their ecological crimes, be they kangaroo-boot manufacturers or football oligarchs (Lynch et al. 2013; Mullenbach and Baker 2020; Rigolon et al. 2022; Schlosberg 2007).

Along the lines of the World Intellectual Property Organization (n.d.), there must be a World Environmental Organization, with both national accreditation and the credible use of expertise in the face of untethered capital interests. Several sporting agencies and bodies, such as World Athletics (n.d.) and the World Anti-Doping Agency (n.d.), already transcend physical and legislative frontiers. We need a global environmental regulator of sports, in partnership with fans, unions, scientists, governments, and international associations. A Court of Arbitration for Sports & the Environment should match but transcend the Court of Arbitration for Sport (n.d.) in its democratic structure and operation.

There are now over 5000 climate-change laws across the globe. Their number is increasing: people want capital constrained. The European Commission has initiated numerous policies to hold greenwashing in check, which have been partially effective (European Securities and Markets Authority 2024; RepRisk 2024). In 2023, the European Parliament and Council of the European Union introduced a *Corporate Sustainability Reporting Directive* that established reporting rules (2022) and the Commission's *Green Claims Directive* looks to protect 'consumers from greenwashing' (European Parliamentary Research Service 2024). An International Sustainability Standards Board was founded in 2021 to guide firms away from greenwashing (International Financial Reporting Standards 2024). The Australian Securities & Investments Commission levies millions of dollars in fines on greenwashers each year and takes them to court (2024). Britain's Financial Conduct Authority has anti-greenwashing regulations (2024). Caring for workers and the environment are excluded from such initiatives. But something is clearly afoot.

In the USA, needless to say, the Securities and Exchange Commission quietly ended attempts to deal with the matter, abandoning a Climate and ESG Enforcement Task Force after facing reactionary lawfare ("SEC" 2024). For while the EU is concerned about 'the impact of climate change on business, and of business on the climate,' the USA only cares about the first of these problems. It is beset by Republican politicians and judges seeking to reward polluters and punish so-called ethical investors (Bradley 2024).

We need new relationships between popular pleasure and knowledge, scientific reality, and democratic regulation. The good news is that media and scholarly coverage has inflicted significant damage on many greenwashers (Santos et al. 2024). *Josimar*, *Directa*, and a few investigative journalists are unveiling the truth to sports fans, per Solène Alifat and Adelina Paris (2024), Roberto Fuentes Vivar (2019), Jean-Luc Ferré (2020), Sergi López-Egea (2019), and Zirin, who described an overheated Australian Open tennis event as 'the tip of a melting iceberg' (2020).

Even Spain's sports evangelists at *Marca* admit '[e]l deporte contribuye al cambio climático y, a la vez, sufre ya consecuencias' [sports contribute to climate change as well as suffering its consequences] (Carreño 2022) and *Sports Illustrated* ponders "How Phony Environmentalism Came to Sports" via merely 'performative actions' in defense of the climate ("What on Earth" 2022). The *Economist* has its moments, as when a discussion of the 2024 US Open vividly, if implicitly, connected contestants' health to a fetish for growth:

> Players struggled and vomited; in the breaks they packed bags of ice around their necks and heads, or stuck hoses blowing cold air down their shirts in an effort to cool down. Sweltering in the steeply sloping cheaper seats was the largest crowd ever: in all, some 1m spectators attended the competition, fulfilling a goal the organisers had set in 2019. ("Sport is Getting" 2024)

All too often, mainstream sports scholars have been attendants at 'self-avowed service [gas] stations for the capitalist economy' (Eagleton

2024b). Instead, they must make climate change a centerpiece of their teaching and research so that textbooks and syllabi highlight these issues alongside established norms. Contributing to such debates implies new kinds of collaboration and auto-didacticism within sports studies, reading beyond orthodox epistemological fancies, newspaper puff pieces, governmental niceties, and conventional industry research. Inter alia, we must study the science of climate change, lifecycle assessment, and the physics, chemistry, and epidemiology associated with the production, use, and disposal of sporting technologies, from stadium materials to media coverage.

Academics and reporters need to depart their identities as salaried fans (Reef 2018), seeing themselves instead as 'repositories of the mental capital' that belongs to working people (Hall 1978: 12–13)—recalling that without the invisible labor of *Jude the Obscure*'s 'manual toilers … the hard readers could not read, nor the high thinkers live' (Hardy 2002: 109–10).

We must heed the words that greeted 1960 readers of the first issue of *New Left Review*. They were promised a magazine:

> relevant to the imaginative resistances of people who have to live within capitalism—the growing points of social discontent, the projections of deeply-felt needs. Our experience of life today is so extraordinarily fragmented. The task of socialism is to meet people where they are, where they are touched, bitten, moved, frustrated, nauseated—to develop discontent and, at the same time, to give the socialist movement some direct sense of the times and ways in which we live. (Hall 1960: 1)

Popular culture frequently legitimizes socioeconomic-political arrangements in the public mind, per carbon sports, but it is a site of struggle as well as domination. Climate crises, capitalist growth, and mega-events provide moments when contradictions become evident and open to positive transformation. The current conjuncture may be 'dirty, compromised, commercialized, [and] overridden,' but it is not 'uncontested space' (Hall and Back 2009: 681).

Majority and marginal social groups are not simply spectatorial dupes or victims of corporate and governmental carbon power. They are potential and actual producers of new values and languages, with the ability to challenge élites, stimulate policies, and call for the democratic control of capital. Anne Schwenkenbecher argues that 'citizens of states which have the power to achieve an efficient climate regime' should comply 'with the moral duties they have as inhabitants of high emission countries,' not least due to the political power that courses across as well as through democracies (2014: 183). The same applies to the hegemons, athletes, and followers of carbon sports.

The jaw-droppingly dull 2024 closing ceremony in Paris constructed a supposedly dystopian world in which the Olympics were no more; they had to be recreated. Why? Consider Angelique Chrisafis' wise words: 'the very act of holding a planetary sporting event … has to be completely reconsidered if the world is to reach net zero targets in 2050' (2024). Such mega-events are 'an international bacchanal of physical perfection and triumphant will swaddled in human rights abuses and environmental catastrophe' (West 2016).

Two days prior to the conclusion of the Paris jamboree, IOC President Thomas Bach intoned that if 'climate change is continuing in a way that the experts are forecasting … it will be very difficult to organize Olympic Games in summer' (quoted in Livingstone 2024). Within six decades, few world cities will be cool enough to host summer sports safely (Gaind 2016). Whereas the 2004 Games attracted eleven bids, there were two for 2024 (Evans 2018). Australia's third-largest city, Brisbane, a peripheral place in a peripheral place, ended up as the sole bidder to play host in 2032. But potential sites were lining up for the future. Ayuso propelled Madrid toward 2036 (Roces 2024). Her London equivalent, Sadiq Khan, seeks a UK return in 2040 as part of the city alleged status as 'the sporting capital of the world,' because of sports' 'unique power to improve the health and lives of all Londoners and to bring people of different cultures and backgrounds together' (2025: 5). As if incapable of recollection or irony, 2040 would, yet again, be 'the greenest Games ever'; an event that 'inspires the next generation' (quoted in Lawton 2025). All those studies, all those facts, all those interpretations overturned/ignored in favor of routine propaganda.

The lesson of sporting despoliation is clear. Nature's duality—that it is self-generating and self-sustaining, but its survival is contingent on human rhetoric and conduct—makes it vulnerable. The Earth's reaction to our interference will strike back, sooner or later, in mutually assured destruction. Changes to the material world threaten the survival of the planet's most skillful and willful, productive and destructive, inhabitant. Once again, capitalism is perilous; jeopardy for all, wouldn't you say?

Only by rejecting reactionary parthenogenesis in favor of environmental materialism can sports transcend their ignorance and irresponsibility.

Greenwashing and greenwashed sports, sponsors, and states are as unsustainable as they are illegitimate—a ‘realité anti-sociale’ [an anti-social reality] (Saccage 2024). Carbon mega-events must be outlawed—an abolitionist project, not a reformist one.

Will anyone grasp this noxious nettle and bring an end to the slow violence of carbon sports?

Bibliography

NB—Links from US Science Have Been Removed By Its Government

“Britain’s Most Reliable Cars 2024 (Plus the Ones That’ll Let You Down).” (2024,November11). *WhatCar?*https://www.whatcar.com/news/reliability-survey-most-reliable-cars/n26137.

“Climate Talks Must Chart an Equitable Path to Net-Zero Emissions.” (2019, December 5). *Nature* 576: 7.

“SEC Disbands its Climate and ESG Enforcement Task Force.” (2024, September 13). *Crowell*https://www.crowell.com/en/insights/client-alerts/sec-disbands-its-climate-and-esg-enforcement-task-force.

“Sport is Getting Hotter, Harder and Deadlier.” (2024, September 12). *Economist* https://www.economist.com/international/2024/09/12/sport-is-getting-hotter-harder-and-deadlier.

“What on Earth: How Phony Environmentalism Came to Sports.” (2022, November 1). *Sports Illustrated* https://www.si.com/soccer/2022/11/01/sports-greenwashing-daily-cover.

Acosta, Alberto and John Cajas-Guijarro. (2020). “Ghosts, Pluriverse and Hopes. From “Development” to Post-Development.” *The Routledge Handbook to Global Political Economy: Conversations and Inquiries.* Ed. Ernesto Vivares. Abingdon: Routledge. 292–308.

Ahmed, Fathima, and Leon Pretorius. (2010). “Mega-Events and Environmental Impacts: The 2010 FIFA World Cup in South Africa.” *Alternation* 17, no. 2: 274–96.

Alifat, Solène and Adélina Paris. (2024, July 18). “Au parc Georges-Valbon en Seine-Saint-Denis, l’incertitude sur l’héritage environnemental des Jeux olympiques.” *Le Nouvel Obs* https://www.nouvelobs.com/sport/20240716.OBS91185/au-parc-georges-valbon-en-seine-saint-denis-l-incertitude-sur-l-heritage-environnemental-des-jeux-olympiques.html.

Ariès, Philippe. (1987). “Thoughts on the History of Homosexuality.” Trans. Anthony Forster. *Western Sexuality: Practice and Precept in Past and Present Times.* Ed. Philippe Ariès and André Béjin. New York: Basil Blackwell. 62–75.

Bacon, Francis. (1863). *The Works*. Ed. James Spedding, Robert Leslie Ellis, and Douglas Denon Heath. Boston: Taggard and Thompson.

Baram, Michael. (2009). "Globalization and Workplace Hazards in Developing Nations." *Safety Science* 47, no. 6: 756–66.

Barrett, Martin, Kyle S. Bunds, Jonathan M. Casper, and Michael B. Edwards. (2019). "A Descriptive Analysis of Corporate Environmental Responsibility in Major League Professional Sport." *Journal of Applied Sport Management* 11, no. 3: 35–46.

Baudrillard, Jean. (1993). *The Transparency of Evil: Essays on Extreme Phenomena*. Trans. James Benedict. London: Verso.

Blake, William. (1789). "The Ecchoing Green." *Poetry Foundation* https://www.poetryfoundation.org/poems/56592/the-ecchoing-green.

Boykoff, Jules. (2020). *NOlympians: Inside the Fight Against Capitalist Mega-Sports in Los Angeles, Tokyo and Beyond*. Halifax: Fernwood Publishing.

Bradley, Caroline. (2024). "It's Not Easy Being Anti-Greenwashing." *University of Miami Legal Studies Research Paper* https://papers.ssrn.com/sol3/papers.cfm?abstract_id=4784535.

Bullock, Graham. (2017). *Green Grades: Can Information Save the Earth?* Cambridge, Mass.: MIT Press.

Byatt, A. S. (2008, June 29). "Lie Back and Think of Europe." *Observer* https://www.theguardian.com/books/2008/jun/29/sportandleisure.review.

CarbonMajors. (2025, March). *2023 Data Update* https://carbonmajors.org/briefing/The-Carbon-Majors-Database-2023-Update-31397.

Carreño, Fernando M. (2022, December 30). "El cambio climático ya afecta al deporte." *Marca* https://www.marca.com/polideportivo/2022/12/30/63ac75f0ca47410f5f8b4593.html.

Casey, Ruairi. (2019, December 5). "Are Global South Experts Sidelined in Climate Conservations?" *Al Jazeera* https://www.aljazeera.com/indepth/features/global-south-experts-sidelined-climate-conversations-191203132302298.html.

Chrisafis, Angelique. (2024, April 7). "'Greenest Ever Games': How the Paris Olympics Hopes to Inspire a New Era of Global Sporting Events." *Guardian* https://www.theguardian.com/sport/2024/apr/07/paris-olympics-greenest-games-low-carbon.

Conn, David. (2017). *The Fall of the House of FIFA*. New York: Nation Books.

Court, Andrew. (2024, May 14). "'Anti-Sex' Beds Have Arrived at Paris Olympics—After Horny Athletes Admit to Orgies Amid Competition." *New York Post* https://nypost.com/2024/05/14/lifestyle/anti-sex-beds-have-arrived-at-paris-olympics-after-horny-athletes-admit-to-orgies/.

Court of Arbitration for Sport. (n.d.). https://www.tas-cas.org/en/index.html.

Dean, Hartley. (2001). "Green Citizenship." *Social Policy & Administration* 35, no. 5: 490–505.

Dengler, Corinna. (2021). "Degrowth." *The Routledge Handbook of Feminist Economics.* Ed. Günseli Berik and Ebru Kongar. Abingdon: Routledge. 369–77.

Dewey, John. (1958). *Experience and Nature.* New York: Dover.

Dobbs, Rachel. (2024, November 18). "Have Global Emissions Peaked?" *Economist* https://www.economist.com/the-world-ahead/2024/11/18/have-global-emissions-peaked.

Eagleton, Terry. (2024a, April 25). "Where Does Culture Come From?" *London Review of Books* https://www.lrb.co.uk/the-paper/v46/n08/terry-eagleton/where-does-culture-come-from.

Eagleton, Terry. (2024b, October 10). "The Excitement of the Stuff." *London Review of Books* https://www.lrb.co.uk/the-paper/v46/n19/terry-eagleton/the-excitement-of-the-stuff.

Earls, Maya and EE News. (2019, July 2). "As the World Heats Up, Soccer Must Adapt." *Scientific American* https://www.scientificamerican.com/article/as-the-world-heats-up-soccer-must-adapt/.

Eichberg, Henning. (2018). "Play Against Alienation?" *The Philosophy of Play as Life.* Ed. Wendy Russell, Emily Ryall, and Malcolm MacLean. Abingdon: Routledge. 211–26.

European Parliament and Council of the European Union. (2022). *Corporate Sustainability Reporting Directive (EU) 2022/2464 of the European Parliament and of the Council* https://eur-lex.europa.eu/legal-content/EN/TXT/?uri=CELEX:32022L2464.

European Parliamentary Research Service. (2024, October). *'Green Claims' Directive: Protecting Consumers from Greenwashing* https://www.europarl.europa.eu/RegData/etudes/BRIE/2023/753958/EPRS_BRI(2023)753958_EN.pdf.

European Securities and Markets Authority. (2024, June 4). *Final Report on Greenwashing* https://www.esma.europa.eu/sites/default/files/2024-06/ESMA36-287652198-2699_Final_Report_on_Greenwashing.pdf.

Evans, Gareth. (2018, November 18). "Olympic Games: Why Cities Are Snubbing the 'Greatest Show on Earth'." *BBC News* https://www.bbc.co.uk/news/world-46236682.

Fanon, Frantz. (2007). *Los condenados de la tierra.* Trans. Julieta Campos. Rosario: elortiba.

Ferré, Jean-Luc. (2020, January 16). "Les contraintes climatiques obligent le monde du sport à se repenser." *La Croix* https://www.la-croix.com/Sport/contraintes-climatiques-obligent-monde-sport-repenser-2020-01-16-1201072242.

Fruh, Kyle, Alfred Archer, and Jake Wojtowicz. (2022). "Sportswashing: Complicity and Corruption." *Sport, Ethics and Philosophy* 17, no. 1: 101–18.

Gaind, Nisha. (2016). "Most Cities Too Hot to Host 2088 Summer Olympics." *Nature* https://doi.org/10.1038/nature.2016.20503.

Galeano, Eduardo. (2013). *Soccer in Sun and Shadow*, rev. version. Trans. Mark Field. New York: Nation Books.

Gammelsæter, Hallgeir and Sigmund Loland. (2023). "Code Red for Elite Sport: A Critique of Sustainability in Elite Sport and a Tentative Reform Programme." *European Sport Management Quarterly* 23, no. 1: 104–24.

Geeraert, Arnout. (2025). "The Dark Side of Governance Networks: How Private Actors Use Discursive Strategies in Global Sport Governance. *Public Management Review* https://doi.org/10.1080/14719037.2025.2456034.

Gelius, Peter, Sven Messing, Antonina Tcymbal, Leonie Birkholz, and Karim Abu-Omar. (2024). "Physical Activity as a Victim, a Perpetrator, or Part of the Solution to the Climate Crisis?" *Journal of Physical Activity and Health* 21, no. 12: 1220–222.

Greenfield, Patrick. (2024, October 3). "How the 'Frida Kahlo of Environmental Geopolitics' is Lighting a Fire Under Big Oil." *Guardian* https://www.theguardian.com/environment/2024/oct/03/colombia-minister-susana-muhamad-biodiversity-cop16-fossil-fuels.

Hall, Stuart and Les Back. (2009). ""At Home and Not at Home": Stuart Hall in Conversation with Les Back." *Cultural Studies* 23, no. 4: 658–87.

Hall, Stuart. (1960). "Editorial." *New Left Review* 1: 1–3.

Hall, Stuart. (1978). "Marxism and Culture." *Radical History Review* 18: 5–14.

Hardy, Thomas. (2002). *Jude the Obscure*. Ed. Patricia Ingham. Oxford: Oxford University Press.

Hickel, Jason, Giorgos Kallis, Tim Jackson, Daniel W. O'Neill, Juliet B. Schor, Julia K. Steinberger, Peter A. Victor, and Diana Ürge-Vorsatz. (2022, December 15). "Degrowth Can Work—Here's How Science Can Help." *Nature* 612: 400–03.

Hume, David. (1739). *A Treatise of Human Nature* http://michaeljohnsonphilosophy.com/wp-content/uploads/2012/01/5010_Hume_Treatise_Human_Nature.pdf.

Hume, David. (1955). *An Inquiry Concerning Human Understanding with a Supplement: An Abstract of a Treatise of Human Nature*. Ed. Charles W. Hendel. Indianapolis: Bobbs-Merrill.

Huws, Ursula. (2019). *Labour in Contemporary Capitalism: What Next?* London: Palgrave Macmillan.

Hyde, Marina. (2025, March 21). "So It's Victory for Kirsty Coventry in the IOC Elections. Who Says Cards with a Rigged Deck Isn't an Olympic Sport?" *Guardian* https://www.theguardian.com/commentisfree/2025/mar/21/ioc-kirsty-coventry-sebastian-coe-olympic-politics.

International Financial Reporting Standards. (2024, June). *Feedback Statement IFRS® Sustainability Disclosure Standards* https://www.ifrs.org/content/dam/ifrs/project/issb-consultation-on-agenda-priorities/agenda-consultation-feedback-statement-june-2024.pdf.

Jennings, Andrew. (2016). *The Dirty Game: Uncovering the Scandal at FIFA*. N.p.: Century.

Johnson, Jay and Adam Ehsan Ali. (2018). "Ecological Modernization and the 2014 NHL Sustainability Report." *Sociology of Sport Journal* 35, no. 1: 49–57.

Karpowitz, Christopher F., Chad Raphael, and Allen S. Hammond IV. (2009). "Deliberative Democracy and Inequality: Two Cheers for Enclave Deliberation Among the Disempowered." *Politics & Society* 37, no. 4: 576–615.

Kennedy, Robert F. (1968, March 18). "Remarks at the University of Kansas, March 18, 1968." *John F. Kennedy Presidential Library and Museum* https://www.jfklibrary.org/learn/about-jfk/the-kennedy-family/robert-f-kennedy/robert-f-kennedy-speeches/remarks-at-the-university-of-kansas-march-18-1968.

Khan, Sadiq. (2025, April 28). "Foreword." *Major Sports Events—Impact for London 2024 Report* Greater London Authority https://www.london.gov.uk/sites/default/files/2025-04/Major_Sports_Events_Impact_for_London_2024_Report-28April25_0.pdf. 5.

Kim, Kyoung-yim and Heejoon Chung. (2018). "Eco-Modernist Environmental Politics and Counter-Activism Around the 2018 PyeongChang Winter Games." *Sociology of Sport Journal* 35, no. 1: 17–28.

Koch, Max and Hubert Buch-Hansen. (2020). "The IPE of Degrowth and Sustainable Welfare." *The Routledge Handbook to Global Political Economy: Conversations and Inquiries*. Ed. Ernesto Vivares. Abingdon: Routledge. 375–90.

Kuhn, Gabriel. (2019). *Soccer vs. the State: Tackling Football and Radical Politics*, 2nd ed.. Oakland: PM Press.

Lawton, Matt. (2025, April 28). "Sadiq Khan Backs London Bid for 2040 Olympics." *Times* https://www.thetimes.com/article/f67f006f-70ae-405e-bbb3-5a1a0de5238b?shareToken=3a7ca5251ebe72407a61feceb9387c8e.

Light, Andrew. (1998). "Reconsidering Bookchin and Marcuse as Environmental Materialists: Toward an Evolving Social Ecology." *Social Ecology After Bookchin*. Ed. Andrew Light. New York: Guilford Press. 343–84.

Livingstone, Robert. (2024, August 9). ""We are Not Ready to Sell the Olympic Games to the Highest Bidder"—IOC President Thomas Bach on Future Bids." *GamesBids*https://gamesbids.com/eng/summer-olympic-bids/we-are-not-ready-to-sell-the-olympic-games-to-the-highest-bidder-ioc-president-thomas-bach-on-future-bids/.

López-Egea, Sergi. (2019, December 13). "El reto climático en los grandes deportes." *elPeriódico* https://www.elperiodico.com/es/deportes/20191213/deporte-alerta-cambio-climatico-7773643.

Lynch, Michael J., Michael A. Long, Kimberly L. Barrett, and Paul B. Stretesky. (2013). "Is it a Crime to Produce Ecological Disorganization? Why Green Criminology and Political Economy Matter in the Analysis of Global Ecological Harms." *British Journal of Criminology* 53, no. 6: 997–1016.

Malthus, Thomas. (1798). *An Essay on Population, as it Affects the Future Improvement of Society with Remarks on the Speculations of Mr. Godwin, M. Condorcet, and Other Writers*. London: J. Johnson.

Marcuse, Herbert. (1972). "Ecology and Revolution." *Liberation* 16: 10–12.

Marqusee, Mike. (2014, June). "Uneven Playing Fields." *Red Pepper* https://www.mikemarqusee.co.uk/1591-a-level-playing-field-global-sport-in-the-neo-liberal-age/.

McClintock, Anne. (1995). *Imperial Leather: Race, Gender and Sexuality in the Colonial Context*. New York: Routledge.

McMichael, Philip. (2011). "Revaluing Low Carbon Cultures: Learning to Live with the Earth." *Development* 54, no. 2: 169–71.

Mill, John Stuart. (2004). *Principles of Political Economy With Some of Their Applications to Social Philosophy*. Ed. Stephen Nathanson. Indianapolis: Hackett.

Millington, Rob, Audrey R. Giles, Nicolien van Luijk, and Lyndsay M. C. Hayhurst. (2022). "Sport for Sustainability? The Extractives Industry, Sport, and Sustainable Development." *Journal of Sport and Social Issues* 46, no. 3: 293–317.

Muhamad, Susana. (2023, December 13). Speech https://x.com/susanamuhamad/status/1735049874540744800?ref_src=twsrc%5Etfw%7Ctwcamp%5Etweetembed%7Ctwterm%5E1735049874540744800%7Ctwgr%5Ec0e274aa9c8c0603b25025d40ea0c902c7a15c7e%7Ctwcon%5Es1_&ref_url=https%3A%2F%2Fwww.climatechangenews.com%2F2023%2F12%2F13%2Fdubai-deal-ministers-and-observers-react-to-the-uae-consensus%2F.

Mullenbach, Lauren E. and Birgitta L. Baker. (2020). "Environmental Justice, Gentrification, and Leisure: A Systematic Review and Opportunities for the Future." *Leisure Sciences: An Interdisciplinary Journal* 42, nos. 5–6: 430–47.

Perkins, Sean. (2000). "Exploring Future Relationships Between Football Clubs and Local Government." *Soccer & Society* 1, no. 1: 102–13.

Reed, Sada. (2018). ""I'm not a Fan. I'm a Journalist": Measuring American Sports Journalists' Sports Enthusiasm." *Journal of Sports Media* 13, no. 1: 27–47.

RepRisk. (2024, October). *A Turning Tide in Greenwashing? Exploring the First Decline in Six Years* https://www.reprisk.com/research-insights/reports/a-turning-tide-in-greenwashing-exploring-the-first-decline-in-six-years.

Rich, Adrienne. (1997). "The Springboard." *A Whole Other Ball Game: Women's Literature on Women's Sport*. Ed. Joli Sandoz. New York: Noonday Press. 1.

Rigolon, Alessandro, Mariela Fernandez, Brandon Harris, and William Stewart. (2022). "An Ecological Model of Environmental Justice for Recreation." *Leisure Sciences: An Interdisciplinary Journal* 44, no. 6: 655–76.

Roces, Pablo R. (2024, February 2). "Ayuso resucita el Proyecto de los Juegos Olímpicos para Madrid en 2036: "Estamos en ello"." *GranMadrid* https://www.elmundo.es/madrid/2024/02/02/65bcdcb2fdddffa64d8b4587.html.

Saccage. (2024, October). *Les jeux olympiques et paralympiques de Paris 2024: Bilan des saccages. Chapitre 2: Le saccage social* https://saccage2024.noblogs.org/files/2024/10/Bilan-de-Paris-2024-saccage-social.pdf.

Santos, Célia, Arnaldo Coelho, and Alzira Marques. (2024). "A Systematic Literature Review on Greenwashing and its Relationship to Stakeholders: State of Art and Future Research Agenda." *Management Review Quarterly* 74, no. 3: 1397–421.

Schlosberg, David. (2007). *Defining Environmental Justice: Theories, Movements, and Nature.* Oxford: Oxford University Press.

Schwenkenbecher, Anne. (2014). "Is There an Obligation to Reduce One's Individual Carbon Footprint?" *Critical Review of International Social and Political Philosophy* 17, no. 2: 168–88.

Spinoza, Baruch. (2002). *Complete Works.* Ed. Michael L. Morgan. Trans. Samuel Shirley. Indianapolis: Hackett.

Tomlinson, Alan. (2014). *FIFA (Fédération Internationale de Football Association): The Men, the Myths and the Money.* London: Routledge.

Tomlinson, Alan, ed. (2017). *The Olympic Legacy: Social Scientific Explorations.* Abingdon: Routledge.

Trotsky, Leon. (1957). *Literature and Revolution.* Trans. Rose Strunsky. New York: Russell & Russell.

United Nations. (n.d.). *Greenwashing—The Deceptive Tactics Behind Environmental Claims* https://www.un.org/en/climatechange/science/climate-issues/greenwashing.

United Nations Climate Change Conference. (2018). *Sports for Climate Action Framework* https://unfccc.int/sites/default/files/resource/SportsforClimateActionDeclarationandFramework.pdf.

United Nations' High-Level Expert Group on the Net-Zero Emissions Commitments of Non-State Entities. (2022). *Integrity Matters: Net Zero Commitments by Business, Financial Institutions, Cities and Regions* https://www.un.org/sites/un2.un.org/files/high-levelexpertgroupn7b.pdf.

United States Environmental Protection Agency. (2023, June). *Tailpipe Greenhouse Gas Emissions from a Typical Passenger Vehicle* https://nepis.epa.gov/Exe/ZyPDF.cgi?Dockey=P1017FP5.pdf.

Vivar, Robero Fuentes. (2019, September 24). "Deporte y cambio climático." *Milenio* https://www.milenio.com/opinion/roberto-fuentes-vivar/las-otras-competencias/deporte-y-cambio-climatico.

Warren, Gina. (2020). "Mega Sports Events Have Mega Environmental and Social Consequences." *Missouri Law Review* 85, no. 2 https://papers.ssrn.com/sol3/papers.cfm?abstractid=3463119.

West, Lindy. (2016, August 10). "How to Talk about Female Olympians Without Being a Regressive Creep—A Handy Guide." *Guardian* https://www.theguardian.com/commentisfree/2016/aug/09/female-olympians-guide-gaffes-athletes-sports-makeup-shorts-marital-status-lindy-west.

Williams, Raymond. (2015). *A Short Counter-Revolution—Towards 2000 Revisited*, 2nd ed. Ed. Jim McGuigan. London: Sage Publications.

Wilson, Elizabeth. (2022). *Unfolding the Past*. London: Bloomsbury Visual Arts.

World Anti-Doping Agency. (n.d.). https://www.wada-ama.org/.

World Athletics. (n.d.). https://www.worldathletics.org/home.

World Health Organization. (2023, December 13). *Road Traffic Injuries* https://www.who.int/news-room/fact-sheets/detail/road-traffic-injuries.

World Intellectual Property Organization. (n.d.). https://www.wipo.int/portal/en/index.html.

Zirin, Dave. (2020, January 17). "The Australian Open is the Tip of a Melting Iceberg." *The Nation*https://www.thenation.com/article/environment/australian-open-fires-climate-change/.

Bibliography

NB—Links from US Science Have Been Removed By Its Government

"'No a la Copa América' pide a la Sindicatura de Comptes investigar las cifras del evento deportivo en Barcelona." (2024, October 2). *20minutos* https://www.20minutos.es/noticia/5639660/0/no-a-la-copa-america-pide-a-la-sindicatura-de-comptes-investigar-las-cifras-del-evento-deportivo-en-barcelona/.

"10 Great Motor Racing Films." (2023, December 14). BFI https://www.bfi.org.uk/lists/10-great-motor-racing-films.

"1964: Memories of the Tokyo Olympics." (2014). *Japan News* http://tokyo-olympics.the-japan-news.com/#titlePage.

"2025 Budget Notes: FIFA World Cup 2026 Toronto." (2025). City of Toronto https://www.toronto.ca/legdocs/mmis/2025/bu/bgrd/backgroundfile-252534.pdf.

"37th America's Cup Was Amazingly Boring." (2024, December 1). *Scuttlebutt Sailing News* https://www.sailingscuttlebutt.com/2024/12/01/37th-americas-cup-was-amazingly-boring/.

"A First Look at the 2024 National and Federation Kits." (2024, April 12). Nike https://about.nike.com/en/newsroom/collections/2024-national-and-federation-kits.

T. Miller, J. P. Carañana, *Global Sports Go Green—Or Do They?*, Global Culture and Sport Series,
https://doi.org/10.1007/978-3-032-14207-8

"A Global Gold Rush is Changing Sport." (2024, July 25). *Economist* https://www.economist.com/leaders/2024/07/25/a-global-gold-rush-is-changing-sport.

"A Greener Games? Tokyo's Environmental Impact." (2021, July 17). *VOA* https://www.voanews.com/a/east-asia-pacificgreener-games-tokyos-environmental-impact/6208269.html.

"A New Way to Recycle Plastic is Here." (2025, April 16). *Economist* https://www.economist.com/business/2025/04/16/a-new-way-to-recycle-plastic-is-here.

"A Problem at the Heart of Formula One." (2023, December 15). *Economist* https://www.economist.com/culture/2023/12/15/a-problem-at-the-heart-of-formula-one.

"AC-Free Ambition for Paris Olympics Melts Away as Organisers Order 2,500 Cooling Units." (2024, July 2). *France 24* https://www.france24.com/en/live-news/20240702-no-ac-ambition-for-paris-olympic-village-melts-away.

"Activistas presentan una denuncia en Antifrau para que investigue las cuentas de la Copa América." (2024, October 10). *El Periódico* https://www.elperiodico.com/es/barcelona/20241010/no-a-la-copa-america-denuncia-antifrau-investigue-cuentas-109152400.

"Act Now." (n.d.). Extinction Rebellion https://extinctionrebellion.uk/act-now/.

"AIRBNB ATHLETE TRAVEL GRANT 2025." (2025, January 21). https://www.olympics.com/athlete365/offers/airbnb/airbnb-athlete-travel-grant-2025.

"Almost Half the World Tuned in at Home to Watch 2010 FIFA World Cup South Africa™." (2011, July 20). FIFA www.fifa.com/worldcup/archive/southafrica2010/organisation/media/newsid=1473143/index.html.

"America's Cup: Dalton Looks to the Future, and Expansion of 38th Cup. Venue to Move?" (2024, October 23). *Sail-World* https://www.sail-world.com/news/280760/Defenders-want-to-expand-on-AC37-progress.

"America's Cup Line in the Sand." (2021, June 19). *Otago Daily Times* https://www.odt.co.nz/opinion/editorial/america%E2%80%99s-cup-line-sand.

"America's Cup Update." (2024, October 7). *SuperyachtNews* https://www.superyachtnews.com/operations/americas-cup-update-.

"America's Cup: Your Winning Formula." (n.d.). BWA Yachting https://bwayachting.com/superyacht-programmes/americas-cup/.

"America's Gambling Boom Should be Celebrated, not Feared." (2024, December 7). *Economist* https://www.economist.com/leaders/2024/12/05/americas-gambling-boom-should-be-celebrated-not-feared.

"Antitrust: Commission Fines Car Manufacturers €875 Million for Restricting Competition in Emission Cleaning for New Diesel Passenger Cars." (2021,

July 8). European Commission https://ec.europa.eu/commission/presscorner/detail/en/ip_21_3581.

"Aramco Sponsorship is a Middle Finger to Women's Football." (2024). Athletes of the World https://img1.wsimg.com/blobby/go/558b37ef-276d-4111-81a8-d0fea75f10fd/Letter%20to%20FIFA_%20Aramco%20Sponsorship%20(3).pdf

"ASA Ruling on BMW (UK) Ltd." (2024, February 7). Advertising Standards Authority [UK]. Complaint A23-1209400 BMW (UK) Ltd https://www.asa.org.uk/rulings/bmw%2D%2Duk%2D%2Dltd-a23-1209400-bmw%2D%2Duk%2D%2Dltd.html.

"ASA Ruling on Hyundai Motor UK Ltd." (2021, June 9). Advertising Standards Authority [UK]. Complaint A21-1096716 Hyundai Motor UK Ltd https://www.asa.org.uk/rulings/hyundai-motor-uk-ltd-a21-1096716-hyundai-motor-uk-ltd.html.

"ASA Ruling on Hyundai Motor UK Ltd." (2023, June 28). Advertising Standards Authority [UK]. Complaint G22-1147624 Hyundai Motor UK Ltd https://www.asa.org.uk/rulings/hyundai-motor-uk-ltd-g22-1147624-hyundai-motor-uk-ltd.html.

"ASA Ruling on Repsol SA." (2023, October 18). Advertising Standards Authority [UK]. Complaint A23-1206254 Repsol SA https://www.asa.org.uk/rulings/repsol-sa-a23-1206254-repsol-sa.html.

"A Selection of Formula 1 Sponsorship Trends." (2024, July 15). caytoo https://caytoo.co.uk/article/formula-1-sponsorship-trends/.

"Australian Athletes Benefit from Extra Comforts as Olympic Village Criticised." (2024, July 31). *New Daily* https://www.thenewdaily.com.au/sport/olympics-2024/2024/07/31/olympics-australians-village.

"Barcelona *Blue Economy*." (n.d.). Barcelona Activa https://www.barcelonactiva.cat/en/blueeconomy.

"Barcelona: manifestantes piden quitar estatua de Colón." (2020, June 14). *DW* https://www.dw.com/es/barcelona-manifestantes-piden-quitar-estatua-de-col%C3%B3n/a-53805920.

"Barcelona's Plans to Manage Tourism Angers [*sic.*] Residents." (2024, July 5). *DW* https://www.dw.com/en/barcelonas-plans-to-manage-tourism-angers-residents/video-69539560.

"BP Ad: Carbon Footprint." (2003). https://www.youtube.com/watch?v=ywrZPypqSB4.

"Britain's Most Reliable Cars 2024 (Plus the Ones That'll Let You Down)." (2024, November 11). *What Car?* https://www.whatcar.com/news/reliability-survey-most-reliable-cars/n26137.

"Brundage en Defensa de las Olimpiadas." (1968, October 8). *El Informador*: 1, 6.

"BT Fined £17.5m for 999 Call-Handling Failures." (2024, July 22). Ofcom https://www.ofcom.org.uk/phones-and-broadband/telecoms-infrastructure/bt-fined-17.5m-for-999-call-handling-failures.
"BT Fined £2.8m Over Failures Affecting 1.1 Million EE and PlusNet Customers." (2024, May 22). Ofcom https://news.sky.com/story/bt-fined-2-8m-over-failures-affecting-1-1-million-ee-and-plusnet-customers-13141349.
"Budweiser Budvar Pays Charity £414,000 for Recycling Failure." (2024, August 15). Environment Agency https://www.gov.uk/government/news/budweiser-budvar-pays-charity-414000-for-recycling-failure.
"Calculan costo ambiental del Mundial." (2014, June 24). *Revista Mundo Verde* http://www.revistamundoverde.net/noticias-ambientales/calcula-costo-ambiental-del-mundial.
"Can Computing Clean Up its Act?" (2023, August 16). *Economist* https://www.economist.com/science-and-technology/2023/08/16/can-computing-clean-up-its-act.
"Childish Arguments." (2014, October 17). *Economist* http://www.economist.com/blogs/democracyinamerica/2014/10/greenpeace-lego-and-shell.
"China's Lethal Milk Scandal Reverberates a Decade Later." (2019, January 22). *Straits Times* https://www.straitstimes.com/asia/east-asia/chinas-lethal-milk-scandal-reverberates-a-decade-later.
"Climate Change Cannot Be Ignored." (2024). FIFPRO https://fifpro.org/en/supporting-players/health-and-performance/extreme-weather-and-climate-change.
"Climate Talks Must Chart an Equitable Path to Net-Zero Emissions." (2019, December 5). *Nature* 576: 7.
"Climate X Report Projects $800 Million in Climate-Related Losses for Fifa World Cup Venues by 2050." (2024, December 3). Climate X https://www.climate-x.com/articles/industry/climate-x-report-projects-800-million-in-climate-related-losses-for-fifa-world-cup-venues-by-2050.
"Colorado Rejects Olympics." (1972, November 8). *Ludington Daily News*: 5.
"Complaints Decisions." (2024, January). Advertising Standards Authority [Ireland]. C&C Group—Budweiser Reference 41478 https://adstandards.ie/complaint/alcohol-35/.
"Copa América: Barcelona espera 2,5 millones de turistas y un impacto de más de mil millones." (2024, April 2). *Palco23* https://www.palco23.com/competiciones/copa-america-barcelona-espera-25-millones-de-turistas-y-un-impacto-de-mas-de-mil-millones.
"Dalian Wanda Settles Dispute with China Vanke Over Mega Project, Wanda Official Says." (2023, July 28). *Reuters* https://www.reuters.com/markets/asia/dalian-wanda-settles-dispute-with-china-vanke-over-mega-project-

wanda-official-2023-07-28/#:~:text=Dalian%20Wanda%20has%20suffered%20from,China%20due%20to%20payment%20disputes.

"DAZN Secures Exclusive Fifa Club World Cup Rights." (2024, December 4). *BBC* https://www.bbc.com/sport/football/articles/crl387gr18ro

"Descarbonització d'embaracions d'esbarjo." (n.d.). La Fundació Barcelona Capital Nàutica https://www.barcelonacapitalnautica.org/sostenibilitat-i-innovacio/descarbonitzacio-embarcacions/.

"DHL and PostNL Adjust Their Sustainability Claims Following an ACM Action." (2024, October 3). Authority for Consumers and Markets https://www.acm.nl/en/publications/dhl-and-postnl-adjust-their-sustainability-claims-following-acm-action.

"Doing the Right Thing. Forest Green Rovers are World's First UN-Certified Carbon Neutral Club." (2018, August 1). *Inside World Football* http://www.insideworldfootball.com/2018/08/01/right-thing-forest-green-rovers-worlds-first-un-certified-carbon-neutral-club/.

"Dossier L'affaire Squarcini." (2024, November 29). *Mediapart* https://www.mediapart.fr/journal/france/dossier/l-affaire-squarcini.

"Driven to Watch: How a Sports Docuseries Drove U.S. Fans to Formula 1." (2022, May). *Nielsen* https://www.nielsen.com/insights/2022/driven-to-watch-how-a-sports-docuseries-drove-u-s-fans-to-formula-1/.

"Ecológica y sostenible, así es la Villa Olímpica de París 2024." (2024, July 7). *TeleMadrid* https://www.telemadrid.es/deportes/Ecologica-y-sostenible-asi-es-la-Villa-Olimpica-de-Paris-2024-0-2684731547%2D%2D20240703090000.html.

"Electing Olympic and Paralympic Hosts: Targeted Dialogue Explained." (2023, November 27). International Olympic Committee https://olympics.com/ioc/news/electing-olympic-and-paralympic-hosts-targeted-dialogue-explained.

"El Grand Prix de Fórmula 1 en Madrid afectará a 729 árboles y el Ayuntamiento tendrá que suspender los estándares de calidad acústica." (2025, February 6). *Madrid On* https://www.madridon.com/2025/02/06/el-grand-prix-de-formula-1-en-madrid-afectara-a-729-arboles-y-el-ayuntamiento-tendra-que-suspender-los-estandares-de-calidad-acustica/.

"El malestar por los problemas de acceso a la vivienda se hace oír en varias ciudades españolas." (2024, October 13). *La Vanguardia* https://www.lavanguardia.com/vida/20241013/10018135/dos-mil-personas-reclaman-barcelona-vuelva-acoger-copa-america.html.

"El Port de Barcelona impulsa la innovación en el mundo portuario." (2024, August 1). *Público* https://www.publico.es/sociedad/port-barcelona-impulsa-innovacion-mundo-portuario.html#md=modulo-portada-fi.

"El PSOE pide explicaciones sobre los costes del Gran Premio: "Madrid merece la F1, pero sobre todo merece rigor"." (2025, February 8). *infobae* https://

www.infobae.com/espana/agencias/2025/02/05/el-psoe-pide-explicaciones-sobre-los-costes-del-gran-premio-madrid-merece-la-f1-pero-sobre-todo-merece-rigor/.

"Europa-Park Stadion." (n.d.). *Football Ground Guide*. https://footballground-guide.com/leagues/europe-1/germany/sc-freiburg-stadium-europa-park-stadium.html.

"European Leagues, FIFPRO Europe to File Joint Complaint to European Commission Against FIFA Regarding International Match Calendar." (2024, July 23). FIFPRO https://www.fifpro.org/en/who-we-are/what-we-do/foundations-of-work/european-leagues-and-fifpro-europe-to-file-joint-complaint-to-european-commission-against-fifa-regarding-international-match-calendar.

"Expert Slams Climate Greenwashing in F1." (2023, August 23). *GrandPrix* https://www.grandprix.com/news/expert-slams-climate-greenwashing-in-f1.html.

"Fees Paid to Host Formula One Races in 2023, by Circuit." (2024, April 16). Statista Research Department https://www.statista.com/statistics/1360331/formula-1-hosting-fees/.

"Felipe Calderón dio su primer informe como miembro de la FIA, máximo órgano de automovilismo." (2022, March 11). *Infobae* https://www.infobae.com/america/deportes/2022/03/11/felipe-calderon-dio-su-primer-informe-como-miembro-de-la-fia-maximo-organo-de-automovilismo/.

"F1 Champion, Activist Vettel Bids Farewell After Final Race." (2022, November 20). *AP* https://apnews.com/article/sports-auto-racing-formula-one-sebastian-vettel-48d191a936a15cb7803e0892f6593ecf.

"FIA Officially Launches an Application Process for Prospective Formula 1 Teams." (2023, February 2). Fédération Internationale de l'Automobile https://www.fia.com/news/fia-officially-launches-application-process-prospective-formula-1-teams.

"FIFA and UEFA Welcome Helsinki Report on Sport from the European Commission." (1999, December 4). FIFA http://www.fifa.com/news/y=1999/m=12/news=fifa-and-uefa-welcome-helsinki-reportsport-from-the-european-commissi-71582.html.

"FIFA Announces Deal with Saudi Oil Company Aramco." (2024, April 25). *BBC* https://www.bbc.com/sport/football/articles/c51n3ld43yro.

"FIFA President Infantino Defends Saudi World Cup, Trump Relationship." (2025, March 1). *RTL Today* https://today.rtl.lu/sport/international/a/2281240.html.

"FIFA Produces Whitewash of Saudi Arabia World Cup Bid and Ignores Own Report to Provide Compensation for Migrant Workers in Qatar." (2024, November 30). Amnesty International https://www.amnesty.org/en/latest/

news/2024/11/global-fifa-produces-whitewash-of-saudi-arabia-world-cup-bid-and-ignores-own-report-to-provide-compensation-for-migrant-workers-in-qatar/.

"Five Years of Bad Decisions in Spain's Rental Market." (2024, October 29). *Idealista/news* https://www.idealista.com/en/news/property-for-rent-in-spain/2024/10/29/820581-five-years-of-bad-decisions-in-spain-s-rental-market.

"Football for Friendship Raises Environmental Awareness Whilst Breaking Records." (2021, July 23). *Euronews* [advertorial for Gazprom] https://www.euronews.com/business/2021/07/21/football-for-friendship-raises-environmental-awareness-whilst-breaking-records.

"Formula 1 and LVMH Announce Historic 10-Year Global Partnership." (2024, January 10). LVMH https://www.lvmh.com/en/news-lvmh/formula-1-and-lvmh-announce-historic-10-year-global-partnership.

"Formula 1 Partners." (n.d.). Formula 1 https://www.formula1.com/en/information/formula-1-partners.2shz7bIVhGgS3ovP69N3Rs.

"Formula 1's Revenue Growth." (2025, March 2). theracingpilot https://www.threads.net/@theracingpilot/post/DGs9GJ7N7N3/formula-ones-revenue-growth-as-stated-by-liberty-media-f1-formula1.

"*Formula 1: The Impossible Collection.*" (2025). Assouline https://www.assouline.com/products/formula-1-the-impossible-collection?srsltid=AfmBOorZbZ5vx-rDoptmqhBAlnmXcObosmhHEdu35-NhXtNEWBkVBXS4.

"Formule 1 : la série Netflix 'Formula 1 : Pilotes de leur destin' critiquée pour sa promotion du tabac." (2023, March 9). *Le Figaro* https://www.lefigaro.fr/flash-sport/f1-la-serie-netflix-formula-1-pilotes-de-leur-destin-critiquee-pour-sa-promotion-du-tabac-20230308.

"Fremantle Prepares for Invasion: Australians Earmark $960 Million in Improvements for the 1987 America's Cup Race." (1985, May 5). *Los Angeles Times* https://www.latimes.com/archives/la-xpm-1985-05-05-sp-8715-story.html.

"French Energy Regulator (CRE) Fines EDF with the First Sanction for Insider Trading Under REMIT, and EDF Trading Limited (EDFT) for Market Manipulation." (2022, May 10). European Union Agency for the Cooperation of Energy Regulators https://www.acer.europa.eu/news-and-events/news/french-energy-regulator-cre-fines-edf-first-sanction-insider-trading-under-remit-and-edf-trading-limited-edft-market-manipulation.

"GE: Decades of Misdeeds and Wrongdoing." (2001, July–August). *Multinational Monitor* 22, nos. 7 & 8 https://www.multinationalmonitor.org/mm2001/01july-august/julyaug01corp4.html.

"General Electric to Pay $2.25 Million for Violating Federal and State Environmental Laws in Waterford, New York." (2015, December 3).

U.S. Department of Justice https://www.justice.gov/opa/pr/general-electric-pay-225-million-violating-federal-and-state-environmental-laws-waterford-new.

"GE's Vernova Power Business Spinoff Risks Being Greenwash." (n.d.). Market Forces https://marketforces.net/ge/.

"Greenwashing." (n.d.). ClientEarth https://www.clientearth.org/what-we-do/greenwashing/.

"Heineken Heiress Finds New Way to Avoid Dutch Taxes, Sheltering Billions in Tax Havens: NRC." (2024, April 4). *NL Times* https://nltimes.nl/2024/04/04/heineken-heiress-finds-new-way-avoid-dutch-taxes-sheltering-billions-tax-havens-nrc.

"He Outran the Flames and Lost His Mum. Now He Could be Shot." (2025, January 16). *Animals Australia* https://animalsaustralia.org/our-work/shooting-and-hunting/urge-adidas-to-stop-using-bushfire-victims-for-shoes/.

"Honda Unveil New Livery." (2007, February 22). *Eurosport* https://www.eurosport.com/formula-1/season/2007/honda-unveil-new-liverysto1098918/story.shtml.

"How Sports Gambling Became Ubiquitous." (2024, December 5). *Economist* https://www.economist.com/finance-and-economics/2024/12/05/how-sports-gambling-became-ubiquitous.

"How Will Los Angeles Benefit from the Games?" (n.d.). *LA28* https://la28.org/en/faqs/what-benefit-will-the-games-bring-to-los-angeles--.html.

"Hydrogen Cars Risk Derailing Green Credibility of Paris Olympics: Open Letter from Scientists Concerned that Hydrogen Cars are Misaligned with Net-Zero." (2024, July). Centre for Sustainable Road Freight https://www.csrf.ac.uk/wp-content/uploads/2024/07/Open-letter-from-scientists_-Hydrogen-cars_final.pdf.

"Hyundai Plant Caught Again in Child Labor Scandal." (2024, May 31). Freedom United https://www.freedomunited.org/news/hyundai-plant-child-labor-scandal/.

"INEOS Wants to Drill UK Fracking Test Site in Attempt to Show it is Safe." (2022, April 11). *Guardian* https://www.theguardian.com/environment/2022/apr/11/ineos-wants-to-drill-uk-fracking-test-site-in-attempt-to-show-it-is-safe-jim-ratcliffe#:~:text=2%20years%20old-,Ineos%20wants%20to%20drill%20UK%20fracking%20test%20site,to%20show%20it%20is%20safe&text=The%20chemicals%20and%20energy%20company,wrestles%20with%20high%20energy%20prices.

"International Sporting Events are Increasingly Held in Autocracies." (2022, November 17). *Economist* https://www.economist.com/graphic-detail/2022/11/17/international-sporting-events-are-increasingly-held-in-autocracies.

"Is it Recession-Proof?" (2009, February 12). *Economist* https://www.economist.com/international/2009/02/12/is-it-recession-proof.

"JO 2024: Coca et ses 10 millions de boissons en plastique, médaille d'or du green-washing." (2024, June 5). *France Nature Environnement* https://fne.asso.fr/actualites/jo-2024-coca-et-ses-10-millions-de-boissons-en-plastique-medaille-d-or-du-green-washing.

"JO 2024. Une bulle de sécurité de 100 policiers et gendarmes autour de la flamme olympique." (2024, January 22). *Ouest-France* https://www.ouest-france.fr/jeux-olympiques/jo-2024-une-bulle-de-securite-de-100-policiers-et-gendarmes-autour-de-la-flamme-olympique-c4c66a90-b934-11ee-879f-a9ffd2809e51#:~:text=Risque%20de%20%C2%AB%20perturbation%20%C2%BB%20du%20parcours,Les%20Soul%C3%A8vements%20de%20la%20Terre%20%C2%BB.

"JO de Paris 2024: une fouine était à l'origine des soupçons de sabotage dans la fan zone du château de Vincennes." (2024, August 8). *franceinfo* https://www.francetvinfo.fr/les-jeux-olympiques/paris-2024/paris-2024-une-fouine-etait-a-l-origine-des-soupcons-de-sabotages-dans-la-fan-zone-du-chateau-de-vincennes6706254.html.

"JO-2024: Des surfeurs mobilisés pour dénoncer la tour controversée des juges à Tahiti." (2023, December 12). *France 24* https://www.france24.com/fr/info-en-continu/20231217-jo-2024-des-surfeurs-mobilis%C3%A9s-pour-d%C3%A9noncer-la-tour-controvers%C3%A9e-des-juges-%C3%A0-tahiti.

"Jugadores piden a la FIFA que rompa con la petrolera Aramco porque Arabia Saudí oprime a las mujeres y al colectivo LGTBI+." (2024, October 21). *Público* https://www.publico.es/deportes/centenar-futbolistas-piden-fifa-rompa-aramco-arabia-saudi-oprime-mujeres-gais.html#md=modulo-portada-bloque:2col-t4;mm=mobile-medium.

"La Copa América reconecta de nuevo a los barceloneses con el mar." (2024, August 21). *Mundo Deportivo* https://www.mundodeportivo.com/vela/20240821/1002301649/copa-america-reconecta-nuevo-barceloneses-mar.html.

"La Generalitat refuerza la promoción del turismo náutico por su "amplia transversalidad" con otros sectores." (2024, October 23). *ValenciaPlaza* https://valenciaplaza.com/generalitat-refuerza-promocion-turismo-nautico-amplia-transversalidad-otros-sectores.

"La iniciativa del Ayuntamiento de Barcelona 'Viu la Vela', gran legado de la Copa América." (2024, September 25). *Mundo Deportivo* https://www.mundodeportivo.com/vela/20240925/1002321946/iniciativa-ayuntamiento-barcelona-viu-vela-gran-legado-copa-america.html.

"La plataforma No a la Copa Amèrica anuncia les primeres mobilitzacions." (2023, September 6). *Fosbury* https://fosbury.cat/fosbudiary/vela/la-plataforma-no-a-la-copa-america-anuncia-les-primeres-mobilitzacions/.

"La serie de Netflix sobre la Fórmula 1 recibe criticas [*sic.*] por promocionar el tabaco." (2023, March 8). *Life and Style* https://lifeandstyle.expansion.mx/entretenimiento/2023/03/08/la-serie-de-netflix-sobre-la-formula-1-recibe-criticas-por-promocionar-el-tabaco.

"Lawsuit Filed Against EDF for Alleged Failure to Respect French Duty of Vigilance Law Over Wind Project in Mexico." (2024, June 18). Business & Human Rights Resource Centre https://www.business-humanrights.org/en/latest-news/devoir-de-vigilance-edf-assign%C3%A9-en-justice-pour-ses-activit%C3%A9s-au-mexique/.

"Lenovo Settles FTC Charges it Harmed Consumers With Preinstalled Software on its Laptops that Compromised Online Security." (2017, September 5). Federal Trade Commission https://www.ftc.gov/news-events/news/press-releases/2017/09/lenovo-settles-ftc-charges-it-harmed-consumers-preinstalled-software-its-laptops-compromised-online.

"Les Jeux Olympiques de Paris 2024 sont sponsorisés par la maltraitance animale." (2024, July 24). PETA France https://www.petafrance.com/actualites/les-jeux-olympiques-de-paris-2024-sont-sponsorises-par-la-maltraitance-animale/.

"Liberty Media Corporation Reports Fourth Quarter and Year End 2024 Financial Results." (2025, February 27). Liberty Media https://www.libertymedia.com/investors/news-events/press-releases/detail/557/liberty-media-corporation-reports-fourth-quarter-and-year.

"Los japoneses, los mejores turistas. Los peores, los británicos." (2014, August 26). *CapitalMadrid* https://www.capitalmadrid.com/2014/8/26/35025/los-japoneses-los-mejores-turistas-los-peores-los-britanicos.html.

"Lululemon: Canadian Sportswear Brand Accused of Greenwashing Ahead of Olympic Opening Ceremony." (2024, July 25). *Euronews* https://www.euronews.com/green/2024/07/25/greenwashing-lululemon-hit-with-legal-complaint-before-olympic-opening-ceremony.

"LVMH Chief Executive Testifies in Trial of Former French Spy Agency Chief." (2024, November 28). *Guardian* https://www.theguardian.com/world/2024/nov/28/lvmh-chief-executive-testifies-trial-french-spy-agency-arnault-squarcini?CMP=Share_iOSApp_Other.

"LVMH Clinches 10-Year Sponsorship Deal with Formula 1." (2024, October 2). *Reuters* https://www.reuters.com/business/lvmh-strikes-sponsorship-deal-with-formula-1-2024-10-02/.

"LVMH est partenaire premium des jeux olympiques de Paris 2024." (2024). LVMH https://www.lvmh.com/fr/lvmh-et-les-jeux-olympiques-et-paralympiques-de-paris-2024.

"LVMH is Splurging on the Olympics." (2024, July 25). *Economist* https://www.economist.com/business/2024/07/25/lvmh-is-splurging-on-the-olympics.

"Marina Port Vell Barcelona firma un acuerdo con la Copa América para ser la marina preferente de las grandes esloras que visiten Barcelona durante la regata en 2024." (n.d.). Marina Por Vell https://www.marinaportvell.com/es/articulos/marina-port-vell-barcelona-se-convierte-en-la-marina-preferente-de-superyates-en-la-37a-edicion-de-copa-america/.

"Meares: Stop Bickering Over Brisbane's Olympics." (2024, August 10). *ESPN* https://global.espn.com/olympics/story//id/40806298/australian-anna-meares-stop-bickering-brisbane-olympics.

"Métro gratuit, cérémonie d'ouverture, environnement … Les promesses pas forcément tenues des Jeux olympiques de Paris 2024." (2024, July 20). *rtbf actus* https://www.rtbf.be/article/metro-gratuit-ceremonie-d-ouverture-environnement-les-promesses-pas-forcement-tenues-des-jeux-olympiques-de-paris-2024-11406393.

"Mexico '68 Legacy Gets Fresh Impetus on 50th Anniversary." (2018). International Olympic Committee https://olympics.com/ioc/news/mexico-68-legacy-gets-fresh-impetus-on-50th-anniversary.

"Mike Hawthorn Killed." (1959). *Pathé News* https://www.youtube.com/watch?v=SyiJQ5_CbXM.

"Miles de personas protestan en Barcelona contra la celebración de la Copa América de vela." (2024, October 13). *Público* https://www.publico.es/sociedad/miles-personas-protestan-barcelona-celebracion-copa-america-vela.html#md=modulo-portada-bloque:2col-t4;mm=mobile-medium.

"Multimilionaris a tota vela." (2024, September 9). *Directa* https://directa.cat/multimilionaris-a-tota-vela/.

"Naples to Host America's Cup 2027 in Strategic Soft-Power Move." (2025, May 15). *Decode39* https://decode39.com/10787/naples-to-host-americas-cup-2027-in-strategic-soft-power-move/.

"Nearly Half of All Professional Football Stadiums Put at Risk by Climate Change." (2023, October 9). Zurich UK https://www.zurich.co.uk/media-centre/football-and-climate-change.

"New Survey Reveals British Public Generally Think Disruptive, Non-Violent Protestors Should not be Imprisoned." (2023, August 1). University of Bristol https://www.bristol.ac.uk/news/2023/july/public-opinion-on-climate-change-and-protesters.html.

"Nigeria: Shell Must be Held Fully Accountable for Human Rights Harms Before Being Allowed to Sell its Niger Delta Business." (2024, May 14). Amnesty International https://www.amnesty.org/en/latest/news/2024/05/nigeria-shell-must-be-held-fully-accountable-for-human-rights-harms-before-being-allowed-to-sell-its-niger-delta-business/.

"No a la Copa Amèrica es manifestarà per denunciar la falta de transparència i malbaratament de diners públics." (2024, October 10). *Vila Web* https://www.vilaweb.cat/noticies/no-copa-america-manifestara-falta-transparencia-malbaratament-diners-publics/.

"Olympic Games to Become "Climate Positive" From 2030." (2020, March 4). International Olympic Committee https://www.anocolympic.org/olympic-movement/olympic-games-to-become-climate-positive-from-2030/.

"Olympic Glory Tends to Go Hand-in-Hand with Economic Clout." (2021, August 8). *Economist* https://www.economist.com/graphic-detail/2021/08/08/olympic-glory-tends-to-go-hand-in-hand-with-economic-clout.

"Olympic Highlights 15/11/2024." (2024, November 15). International Olympic Committee https://olympics.com/ioc/news/olympic-highlights-15-11-2024.

"Olympics Equipment Heads for "Second Life" Post-Games." (2024, August 12). *Inside the Games* https://www.insidethegames.biz/articles/1147766/olympics-equipment-heads-second-life.

"Olympics President: Paris 2024 on Track to Reach 'More than Half the World's Population'." (2024, August 7). *IBC365* https://www.ibc.org/news/olympics-president-paris-2024-on-track-to-reach-more-than-half-the-worlds-population/11205.article.

"Olympics Under Pressure to Dump Coca-Cola Sponsorship After Nearly 100 Years Over Health and Environmental Concerns." (2024, August 7). *Fortune* https://fortune.com/europe/2024/08/07/olympics-under-pressure-to-dump-coca-cola-sponsorship-after-nearly-100-years-over-health-environmental-concerns-paris-nike/.

"One of Every Four People in Barcelona is Now a Foreign National." (2024, June 11). *Barcelona International Welcome* https://www.barcelona.cat/internationalwelcome/en/news/one-of-every-four-people-in-barcelona-is-now-a-foreign-national-1408362.

"Our Sustainability." (n.d.). Lululemon https://www.eu.lululemon.com/en-lu/c/community/about-us/our-sustainability.

"Pandemic Olympics Endured Heat, and Now a Typhoon's En Route." (2021, July 26). *NBC* https://www.nbcnews.com/news/world/pandemic-olympics-endured-heat-now-typhoon-s-en-route-n1274976.

"Paris 2024 Carbon Contribution Projects." (2024, June 28). Paris 2024 https://press.paris2024.org/assets/paris-2024-carbon-contribution-projects-2eda-7578a.html.

"Paris 2024: les forces de l'ordre, l'attraction inattendue des Jeux." (2024, August 8). *franceinfo* https://www.francetvinfo.fr/les-jeux-olympiques/paris-2024-les-forces-de-l-ordre-l-attraction-inattendue-des-jeux6714135.html.

"Paris 2024 Presents the Latest Advance in its Low Carbon Strategy." (2024, June 27). Paris 2024 https://press.paris2024.org/news/press-release-paris-2024-presents-the-latest-advance-in-its-low-carbon-strategy-175c-7578a.html.

"Paris's Stunning Vision for the Olympics Wins a Gold Medal." (2024, August 10). *Economist* https://www.economist.com/europe/2024/08/10/pariss-stunning-vision-for-the-olympics-wins-a-gold-medal.

"Partners." (n.d.). LVMH https://www.americascup.com/partners.

"Petronor y Repsol se convierten en los partners de energía del Athletic." (2024, April 19). Athletic Club de Bilbao https://www.athletic-club.eus/noticias/2024/04/19/petronor-y-repsol-se-convierten-en-los-partners-de-energia-del-athletic/.

"Photo Galleries." (2024). https://www.americascup.com/photo.

"Political Will to Fight Climate Change is Fading, Warns UN Chief." (2019, May 12). *BBC* https://www.bbc.com/news/av/world-asia-48244315/political-will-to-fight-climate-change-is-fading-warns-un-chief.

"Prada Blackface Scandal Results in Major Settlement." (2020, February 5). Center for Constitutional Rights https://ccrjustice.org/home/press-center/press-releases/prada-blackface-scandal-results-major-settlement.

"Premier League Agrees Record £6.7bn Domestic TV Rights Deal." (2023, December 4). *BBC* https://www.bbc.com/sport/football/67619756.

"Premier League Clubs' Flight Habits Revealed." (2025, May 28). Campaign for Better Transport https://bettertransport.org.uk/media/premier-league-clubs-flight-habits-revealed/.

"Prime Minister Sets Out Blueprint to Turbocharge AI." (2025, January 13). UK Government https://www.gov.uk/government/news/prime-minister-sets-out-blueprint-to-turbocharge-ai.

"Procès "Squarcini" : la responsabilité du fondateur de LVMH doit être clairement établie quant à l'espionnage de la rédaction de *Fakir*." (2024, November 27). Reporters sans Frontières https://rsf.org/fr/proc%C3%A8s-squarcini-la-responsabilit%C3%A9-du-fondateur-de-lvmh-doit-%C3%AAtre-clairement-%C3%A9tablie-quant-%C3%A0-l

"PSG Coach Sets Off Storm With Quip About Flying." (2022, September 6). https://www.youtube.com/watch?v=gRlhBqofTdM.

"PSG Face Backlash for Mocking Eco-Friendly Travel Plea." (2022, September 6). *France 24* https://www.france24.com/en/france/20220906-psg-face-backlash-for-mocking-eco-friendly-travel-plea.

"Qatar's World Cup Will Emit More CO2 Than Any Recent Sporting Event." (2022, December 1). *Economist* https://www.economist.com/graphic-detail/2022/12/01/qatars-world-cup-will-emit-more-co2-than-any-recent-sporting-event.

"¿Qué ha sido de las estatuas de Cristóbal Colón en América Latina y Estados Unidos?" (2022, September 7). *CNN Español* https://cnnespanol.cnn.com/2022/09/07/que-ha-sido-de-las-estatuas-de-cristobal-colon-en-america-latina#:~:text=La%20estatua%20del%20navegante%20ubicada,de%20la%20independencia%2C%20fue%20destruida.

"Racing Legends—Stirling Moss." (2012). https://www.youtube.com/watch?v=UtRWhPQVAY8&t=1466s.

"Report Questions Environmental Impact of Madrid Grand Prix." (2025, February 7). *GrandPrix247* https://www.grandprix247.com/2025/02/07/report-questions-environmental-impact-of-madrid-grand-prix/.

"Repsol patrocinará a la selección peruana tras haber causado el mayor desastre ecológico de su historia." (2023, September 15). *El Salto* https://www.elsaltodiario.com/greenwashing/repsol-patrocinara-seleccion-peruana-haber-causado-mayor-desastre-ecologico-historia.

"Saudi Arabia is Spending a Fortune on Sport." (2023, August 10). *Economist* https://www.economist.com/briefing/2023/08/10/saudi-arabia-is-spending-a-fortune-on-sport.

"Saudi Arabia/FIFA: Organizations Demand a Credible Human Rights Assessment for 2034 World Cup Bid." (2024, October 28). Amnesty International https://www.amnesty.org/en/latest/news/2024/10/saudi-arabia-fifa-organizations-demand-a-credible-human-rights-assessment-for-2034-world-cup-bid/.

"Saudi Crown Prince Interview with Fox News." (2023). https://www.youtube.com/watch?v=w0NxI44yBDM.

"SEC Disbands its Climate and ESG Enforcement Task Force." (2024, September 13). *Crowell* https://www.crowell.com/en/insights/client-alerts/sec-disbands-its-climate-and-esg-enforcement-task-force.

"Session is Done: FIFA World Cup 2026 Gets $21m." (2024, March 11). The Chamber https://www.seattlechamber.com/news/2024/03/11/policy-advocacy/session-is-done-fifa-world-cup-2026-gets-21m/.

"Seven Incredible Private Jets Owned by F1 Drivers: Lewis Hamilton, Max Verstappen and More." (2023, June 15). *PlanetF1* https://www.planetf1.com/features/f1-private-jets-owned-lewis-hamilton-max-verstappen.

"Shoes." (n.d.). Prada https://www.prada.com/eu/en/mens/shoes/americas-cup/c/10297EU.

"Sport is Getting Hotter, Harder and Deadlier." (2024, September 12). *Economist* https://www.economist.com/international/2024/09/12/sport-is-getting-hotter-harder-and-deadlier.

"Sport Takes Prominent Place at World Urban Forum." (2024, November 11). International Olympic Committee https://olympics.com/ioc/news/sport-takes-prominent-place-at-world-urban-forum.

"Sports Mega-Events: Social Scientific Analyses of a Global Phenomenon." (2006). *Sociological Review* 54, no. 2: 1–187.

"Stand.earth Hangs Banner Over Toronto Home of Team Canada, Raising Concerns Over Integrity of Athletes' Olympic Apparel." (2024, July 17). Stand.earth https://stand.earth/press-releases/stand-earth-hangs-banner-over-toronto-home-of-team-canada-raising-concerns-over-integrity-of-athletes-olympic-apparel/.

"Study on Greenwashing in Online Marketing Funded by CCCS Finds Use of Vague Environmental Claims and Confusing Technical Jargon." (2023, November 16). ASEAN https://asean-competition.org/read-cases-study-on-greenwashing-in-online-marketing-funded-by-cccs-finds-use-of-vague-environmental-claims-and-confusing-technical-jargon.

"Suffering from the Bhopal Disaster in India Continues, 40 Years on." (2024, December 1). *Economist* https://www.economist.com/asia/2024/12/01/suffering-from-the-bhopal-disaster-in-india-continues-40-years-on.

"Swiss Commission for Fairness Upholds Complaints Against FIFA." (2023, June 7). Commission Suisse pour la Loyauté https://www.faire-werbung.ch/de/swiss-commission-for-fairness-upholds-complaints-against-fifa/.

"Szilard Petition." (1945). *Atomic Heritage Foundation* https://ahf.nuclearmuseum.org/ahf/key-documents/szilard-petition/.

"The 2024 Olympic Games: End the Health Hypocrisy." (2024, August 10). *The Lancet* https://www.thelancet.com/action/showPdf?pii=S0140-6736%2824%2901641-6.

"The America's Cup is Becoming Formula One on the Water." (2017, June 17). *Economist* https://www.economist.com/game-theory/2017/06/17/the-americas-cup-is-becoming-formula-one-on-the-water.

"The Benefits of the America's Cup." (2024). *Port de Barcelona* https://www.portdebarcelona.cat/en/port-vell/americas-cup-barcelona-2024/benefits-americas-cup.

"The Changing Climate of Sustainability Has Reached a Critical Moment." (2022, November 30). Nielsen https://nielseniq.com/global/en/insights/analysis/2022/the-changing-climate-of-sustainability-has-reached-a-critical-moment/.

"The Data-Centre Investment Spree." (2025, February 5). *Economist* https://www.economist.com/business/2025/02/05/the-data-centre-investment-spree-shows-no-signs-of-stopping.

"The Evolution of TV: A Brief History of TV Technology in Japan." (2010). Sec. 12: "Color TV Broadcasts and the Tokyo Olympic Games." http://www.nhk.or.jp/strl/aboutstrl/evolution-of-tv-en/p12.html.

"The Fundamental Contradiction of ESG is Being Laid Bare." (2022, September 29). *Economist* https://www.economist.com/leaders/2022/09/29/the-fundamental-contradiction-of-esg-is-being-laid-bare.

"The Gentlemen Drivers." (2022, February 13). Formula One https://www.formula1.com/en/video/top-10-formula-1s-gentleman-drivers.1724488121066642766.

"The Ghosts of Mexico 1968." (2008, April 24). *Economist* https://www.economist.com/international/2008/04/24/the-ghosts-of-mexico-1968.

"The Legacy of the London Olympics." (2022, July 27). *Economist* https://www.economist.com/britain/2022/07/27/the-legacy-of-the-london-olympics.

"The Remarkable Rise of "Greenhushing"." (2025, July 29). *Economist* https://www.economist.com/business/2025/07/29/the-remarkable-rise-of-greenhushing.

"The Timekeeper's Watch." (n.d.). Omega https://www.omegawatches.com/watches/seamaster/diver-300-m/america-s-cup-edition/product.

"The Ultimate List of Formula One Romance Novels." (2024, August 3). *Ana's Attic* https://anasattic.com/formula-one-romance-novels/.

"Three-Quarters of Athletes Directly Impacted by Climate Change, World Athletics Survey Finds." (2023, November 29). World Athletics https://worldathletics.org/athletics-better-world/news/climate-change-impact-world-athletics-survey.

"Time for LIV, Golf's Saudi-Funded Rebel Tour, to Drift into Obscurity." (2024, October 4). *Economist* https://www.economist.com/culture/2024/10/04/time-for-liv-golfs-saudi-funded-rebel-tour-to-drift-into-obscurity.

"Tlatelolco, la testimonianza di Oriana Fallaci." (n.d.). *YouTube* https://www.youtube.com/watch?v=yW8q5i7mdgo.

"Tlatelolco, según los periódicos del 3 de octubre de 1968." (2015, October1). *Expansión* https://expansion.mx/nacional/2015/10/01/tlatelolco-segun-los-periodicos-del-3-de-octubre-de-1968.

"Tokyo 2020, 100 Days to Go—For the Planet and the People." (2021, April 13). https://olympics.com/ioc/news/tokyo-2020-100-days-to-go-for-the-planet-and-the-people.

"Tokyo Olympics: German Pentathlon Coach Thrown Out." (2021, August 7). *BBC* https://www.bbc.com/sport/olympics/58127366.

"Total Revenue of the Formula One Group From 2017 to 2022." (2024, April 10). Statista Research Department https://www.statista.com/statistics/1137226/formula-one-revenue/.

"Trojan Horsepower: Formula 1 and the Saudi Oil Lobby." (2023, December 7). *SourceMaterial* https://www.source-material.org/saudi-aramco-formula-1-fossil-fuel-lobby/.

"Una Plataforma pide cancelar la Copa del América en Barcelona por "elitista"." (2023, October 10). *iusport* https://iusport.com/art/127377/una-plataforma-pide-cancelar-la-copa-del-america-en-barcelona-por-elitista.

"UN ENCUENTRO CON LA VELOCIDAD Y LA EMOCIÓN." (n.d.). IFEMA https://www.ifema.es/f1.

"What on Earth: How Phony Environmentalism Came to Sports." (2022, November 1). *Sports Illustrated* https://www.si.com/soccer/2022/11/01/sports-greenwashing-daily-cover.

"When Playboys Ruled the World: Barry Sheene and James Hunt." (2011). https://www.youtube.com/watch?v=-4UXfyxgZsk.

"When Sex Was Safe & Racing Was Dangerous." (2015). https://www.youtube.com/watch?v=GGP5FEIMza4.

"Why Louis Vuitton is Struggling but Hermès is Not." (2024, December 16). *Economist* https://www.economist.com/business/2024/12/16/why-louis-vuitton-is-struggling-but-hermes-is-not.

"Why the World is Teeming with So Many New Sports Leagues." (2024, August 23). *Economist* https://www.economist.com/culture/2024/08/23/why-the-world-is-teeming-with-so-many-new-sports-leagues.

Abbas Ahmadi, Ali. (2025, May 15). "LA 2028 Olympics Plans Air Taxi Service for Spectators." *BBC* https://www.bbc.com/news/articles/cvgnp103zr9o.

Abel, Jessamyn R. (2015). *The International Minimum: Creativity and Contradiction in Japan's Global Engagement, 1933–1964.* Honolulu: University of Hawaii Press.

Abel, Jessamyn R. (2022). *Dream Super-Express: A Cultural History of the World's First Bullet Train.* Stanford: Stanford University Press.

Abusin, Sana, Noora Lari, Salma Khaled, and Noor Al Emadi. (2020). "Effective Policies to Mitigate Food Waste in Qatar." *African Journal of Agricultural Research* 15, no. 3: 343–50.

Abzouz, Yunnes and Khedidja Zerouali. (2024, September 18). "Liberté de la presse : LVMH voit rouge et met Mediapart sur liste noire." *Mediapart* https://www.mediapart.fr/journal/economie-et-social/180924/liberte-de-la-presse-lvmh-voit-rouge-et-met-mediapart-sur-liste-noire.

Acaf, Yara, Natalie Brown-Devlin, and Lucy Atkinson. (2025). "Do Fans Have to Choose Between Formula One and Sustainability? A Case Study of Spectators Balancing Environmental Beliefs and Fandom." *International Journal of Sport Communication* 18, no. 1: 108–18.

Accenture. (2022). *Accelerating Global Companies Toward Net Zero by 2050* https://www.accenture.com/content/dam/accenture/final/capabilities/strategy-and-consulting/strategy/document/Accenture-Net-Zero-By-2050-Global-Report-2022.pdf#zoom=40.

Acosta, Alberto and John Cajas-Guijarro. (2020). "Ghosts, Pluriverse and Hopes. From "Development" to Post-Development." *The Routledge Handbook to Global Political Economy: Conversations and Inquiries.* Ed. Ernesto Vivares. Abingdon: Routledge. 292–308.

Action Speaks Louder. (2024, April 18). *'Be Planet'? Toxic Pollution and Fossil Fuel Reliance in Lululemon's Supply Chain* https://speakslouder.org/report/be-planet-toxic-pollution-and-fossil-fuel-reliance-in-lululemons-supply-chain/.

Adorno, Theodor and Max Horkheimer. (1979). *Dialectic of the Enlightenment.* Trans. John Cumming. London: Verso.

Adorno, Theodor W. (1945). "A Social Critique of Radio Music." *Kenyon Review* 7, no. 2: 208–17.

Adorno, Theodor W. (1975) "Culture Industry Reconsidered." Trans. Anson G. Rabinbach. *New German Critique* 7: 12–19.

African Union Sports Council. (n.d.). “Introduction to Sport for Development.” https://sportscouncil.au.int/en/introduction-sport-development.

Agini, Samuel. (2025, March 15). “Formula 1 Owner Bets on Brad Pitt’s Pulling Power to Shift into US Mainstream.” *Financial Times* https://www.ft.com/content/61b0156f-0025-4f2b-8037-e7477c2e92c1?utm_source=semafor.

A golpe de evento. (2024, November 11). *El Escarabajo Verde* RTVE https://www.rtve.es/play/videos/el-escarabajo-verde/golpe-evento/16313122/.

Aguad, Felipe and Charles Thraves. (2024). “Optimizing Pit Stop Strategies in Formula 1 with Dynamic Programming and Game Theory.” *European Journal of Operational Research* 319, no. 3: 908–19.

Aguar, Ximo. (2024, October 29). “Déjà vu 2007: Mazón redobla la presión contra el PSPV y el Gobierno por la Copa América.” *ValenciaPlaza* https://valenciaplaza.com/deja-vu-2007-mazon-redobla-la-presion-contra-el-pspv-y-el-gobierno-por-la-copa-america.

Ahmed, Fathima, and Leon Pretorius. (2010). “Mega-Events and Environmental Impacts: The 2010 FIFA World Cup in South Africa.” *Alternation* 17, no. 2: 274–96.

Ahmed, Muriad and Michael Pooler. (2020, March 15). “INEOS: Why Jim Ratcliffe is Mixing Petrochemicals and Sports.” *Financial Times* https://www.ft.com/content/31b5e7e0-5c76-11ea-b0ab-339c2307bcd4.

Ajuntament de Barcelona. (2022, March 29). “Barcelona, escollida seu de la 37a. Copa Amèrica de Vela.” https://ajuntament.barcelona.cat/premsa/2022/03/29/barcelona-escollida-seu-de-la-37a-copa-america-de-vela-2/.

Aladro-Vico, Eva, Dimitrina Jivkova-Semova, and Olga Bailey. (2018). “Artivismo: Un nuevo lenguaje educativo para la acción social transformadora.” *Comunicar* 57: 9–18.

Alarcón, Adriana. (2025, June 12). “Mexico’s World Cup 2026: Infrastructure Race Amid Investment Cuts.” *Mexico Business News* https://mexicobusiness.news/infrastructure/news/mexicos-world-cup-2026-infrastructure-race-amid-investment-cuts.

Alcalá, César. (2024, July 15). “1.200 millones y 19.000 puestos de trabajo: los beneficios para Barcelona de acoger la Copa América.” *El Debate* https://www.eldebate.com/espana/cataluna/barcelona/20240715/1200-millones-19000-puestos-trabajo-beneficios-barcelona-acoger-copa-america212766.html.

Albert, Eric. (2024, January 5). “What Legacy for Paris 2024? The Exemplary Conversion of the London 2012 Olympic Park.” *Le Monde* https://www.lemonde.fr/en/sports/article/2024/01/05/what-legacy-for-paris-2024-the-exemplary-conversion-of-the-london-2012-olympic-park_6402549_9.html.

Alcívar, Estíbaliz. (2024, September 25). “GP de España en Madrid sin inversores: Ninguna empresa quiere asumir el “riesgo” de organizar el evento de F1.” *Car and Driver* https://www.caranddriver.com/es/formula-1/a62364002/gp-espana-madrid-ningun-inversor-riesgo-organizar-f1/.

Ali, Mona. (2022, April 27). "Regime Change? The Evolution and Weaponization of the World Dollar." *Phenomenal World* https://www.phenomenalworld.org/analysis/regime-change/.

Alifat, Solène and Adélina Paris. (2024, July 18). "Au parc Georges-Valbon en Seine-Saint-Denis, l'incertitude sur l'héritage environnemental des Jeux olympiques." *Le Nouvel Obs* https://www.nouvelobs.com/sport/20240716.OBS91185/au-parc-georges-valbon-en-seine-saint-denis-l-incertitude-sur-l-heritage-environnemental-des-jeux-olympiques.html.

Aljerian, Khaldoon. (2019). "Death Certificate Errors in One Saudi Arabian Hospital." *Death Studies* 43, no. 5: 311–15.

Allen, James. (2013, November 1). "Hybrid Power Fits with Greener Agenda." *Financial Times*: 2.

Allen, James. (2014, March 14). "F1 Sponsors and Supporters Drive the Shift to Sustainability." *Financial Times* https://www.ft.com/content/ceed5302-a446-11e3-9cb0-00144feab7de.

Alonso, Virginia P. (2024, March 27). "Héctor Bellerín: "En el fútbol hay mucho miedo a cambiar y eso provoca esta hipermasculinización tóxica"." *Público* https://www.publico.es/deportes/hector-bellerin-futbol-hay-mucho-miedo-cambiar-provoca-hipermasculinizacion-toxica.html.

ALQST for Human Rights. (2023, February). *The Dark Side of Neom: Expropriation, Expulsion and Prosecution of the Region's Inhabitants* https://alqst.org/uploads/the-dark-side-of-neom-expropriation-expulsion-and-prosecution-en.pdf.

Althusser, Louis. (1971). *Lenin and Philosophy and Other Essays.* Trans. Ben Brewster. New York: Monthly Review Press.

Amazon Employees for Climate Justice. (2024). *Burns Trust: The Amazon Unsustainability Report* https://static1.squarespace.com/static/65681f099d7c3d48feb86a5f/t/668ebf702516716ca72bbf98/1720631157044/unsustainability-report.pdf.

Ambrose, Jillian. (2025, February 26). "BP Blames 'Misplaced' Faith in Green Transition for its Renewed Focus on Fossil Fuels." *Guardian* https://www.theguardian.com/business/2025/feb/26/bp-oil-and-gas-spending-green-energy-scale-back?CMP=Share_iOSApp_Other.

America's Cup Event Ltd. (n.d.). *36th America's Cup Presented by Prada Auckland 2021: America's Cup Event Limited Final Event Report.* https://www.americascup.com/files/m5301_Americas-Cup-Event-Limited-Final-Event-Report_FA_2.pdf.

Ames, Nick. (2024, October 28). "Saudi Arabia's World Cup Bid Gathers Pace—But at What Cost to Human Life?" *Guardian* https://www.theguardian.com/football/2024/oct/28/saudi-arabia-world-cup-2034-qatar-fifa-neom.

Ames, Nick. (2025, February 14). "Ineos and Ratcliffe's Sporting Empire Risks Atrophy with Horizons Narrowing." *Guardian* https://www.theguardian.com/football/2025/feb/14/ineos-jim-ratcliffe-manchester-united-sporting-empire-ben-ainslie?CMP=Share_iOSApp_Other.

Amnesty International. (2024, November 11). *High Stakes Bids: Dangerously Flawed Human Rights Strategies for the 2030 and 2034 FIFA World Cups* https://www.amnesty.org/es/documents/ior10/8712/2024/en/.

Amnistía Internacional. (2025, June 11). *Copa Mundial: Cuando falta un año para su celebración, aumentan los ataques a los derechos* https://www.amnesty.org/es/latest/news/2025/06/world-cup-a-year-out-growing-attacks-on-rights/.

Ampere. (2025). *A Year of Sport 2024: Full Stream Ahead in 2025.* https://19957038.fs1.hubspotusercontent-na1.net/hubfs/19957038/2024%20in%20Sports-1.pdf?utm_medium=email&_hsenc=p2ANqtz-_dXEmIwhUZXNR8DZTyC-GUGWDN8MjPoO5EcDUzk62EM2P1PCObUjAH9Y8GOaZ2ESRWnL-vWSERAUBeKBETg6joUvYAj0w&_hsmi=307835695&utm_content=307835695&utm_source=hs_automation.

Anani, Zak. (2024, December 9). "The True Size of the Global Sports Industry." Global Sports Insights https://gis.sport/news/the-true-size-of-the-global-sports-industry/.

Anderson, Benedict. (2006). *Imagined Communities: Reflections on the Origin and Spread of Nationalism*, rev. ed. London: Verso.

Anderson, Christopher. (2005). "Little Green Men." *Flow* 2, no. 8 http://flowtv.org/2005/07/advertising-ge-ecoimagination.

Anderson, Perry. (2004, June 3). "Stand-Off in Taiwan." *London Review of Books*: 12–17.

Anderson, Perry. (2011). *The New Old World.* London: Verso.

Andre, Peter, Teodora Boneva, Felix Chopra, and Armin Falk. (2024). "Globally Representative Evidence on the Actual and Perceived Support for Climate Action." *Nature Climate Change* 14: 253–59.

An Ecomodernist Manifesto. (n.d.) https://www.ecomodernism.org/manifesto-english.

Anenberg, Susan, Joshua Miller, Daven Henze, and Ray Minares. (2019). *A Global Snapshot of the Air Pollution-Related Health Impacts of Transportation Sector Emissions in 2010 and 2015.* International Council on Clean Transportation/Climate & Clean Air Coalition https://theicct.org/sites/

default/files/publications/Global_health_impacts_transport_emissions_2010-2015_20190226.pdf.

Ang, Ulysses. (2011, January 19). "Race on Sunday, Sell on Monday." *The Philippine Star* http://www.philstar.com/motoring/649035/race-sunday-sell-monday.

Antich, José. (2025, February 7). "Problemas para la Fórmula 1 en Madrid." (2025, February 7). *El Nacional* https://www.elnacional.cat/es/editorial/problemas-formula-1-madrid-jose-antich_1360055_102.html.

Antonini, Francesca. (2019). "Pessimism of the Intellect, Optimism of the Will: Gramsci's Political Thought in the Last Miscellaneous Notebooks." *Rethinking Marxism* 31, no. 1: 42–57.

Arafat, Yasir, Muhammad Rizwan Azhar, and Waqas Uzair. (2024, June 24). "Formula One is Moving Towards Hybrid Engines and Renewable Fuel. Major Environmental Progress or Just 'Greenwashing'? *The Conversation* https://theconversation.com/formula-one-is-moving-towards-hybrid-engines-and-renewable-fuel-major-environmental-progress-or-just-greenwashing-232375#:~:text=According%20to%20a%20report%20from,the%20cars%20on%20the%20track.

Aranda, Germán. (2016, August 6). "Una ceremonia sostenible entre la selva de atletas." *El Mundo* http://www.elmundo.es/deportes/2016/08/06/57a5397eca4741a6118b45f3.html.

Araújo, Maria Cristina B. and Monica F. Costa. (2019). "A Critical Review of the Issue of Cigarette Butt Pollution in Coastal Environments." *Environmental Research* 172: 137–49.

Arbena, Joseph L. (1991). "Sport, Development, and Mexican Nationalism, 1920–1970." *Journal of Sport History* 18, no. 3: 350–64.

Arias-Maldonado, Manuel. (2007). "An Imaginary Solution? The Green Defence of Deliberative Democracy." *Environmental Values* 16, no. 2: 233–52.

Ariès, Philippe. (1987). "Thoughts on the History of Homosexuality." Trans. Anthony Forster. *Western Sexuality: Practice and Precept in Past and Present Times.* Ed. Philippe Ariès and André Béjin. New York: Basil Blackwell. 62–75.

Aristotle. (1998). *Politics.* Trans. C. D. C. Reeve. Indianapolis: Hackett Publishing.

Armstrong, Chris. (2022). *A Blue New Deal: Why We Need a New Politics for the Ocean.* New Haven: Yale University Press.

Armstrong, Gary and Malcolm Young. (1997). "Legislators and Interpreters: The Law and 'Football Hooligans'." *Entering the Field: New Perspectives on World Football.* Ed. Gary Armstrong and Richard Giulianotti. Oxford: Berg. 175–91.

Armstrong, Gary. (1998). *Football Hooligans: Knowing the Score.* Oxford: Berg.

Arnold, Matthew. (1910). *Essays in Criticism: Third Series.* Boston: The Ball Publishing Co.

Arnold, Matthew. (2003). *Culture and Anarchy.* Project Gutenberg http://www.gutenberg.org/dirs/etext03/7cltn10.txt.

Arnold, Richard. (2021). "Nationalism and Sport: A Review of the Field." *Nationalities Papers* 49, no. 1: 2–11.

Arsenal. (n.d.). https://www.arsenal.com/news/sustainability.

Artime, Michael Robert and Megan Hershey. (2023). "The Reel Politics of International Crisis: Benedict Cumberbatch's Appeal for Syrian Refugees." *Political Studies Review* 21, no. 2: 357–75.

Arufe, Fernando. (2025, April 20). "Valencia tiene infraestructuras, experiencia y empresas preparadas para recuperar la America's Cup." *Economía Digital* https://www.economiadigital.es/valencia/economia/valencia-infraestructuras-empresas-recuperar-americas-cup.html.

Ashoo, Sam. (2021, December 3). "Climate Crisis: How Can Football Make a Difference?" *Euronews* https://www.euronews.com/2021/10/28/the-climate-crisis-why-football-can-no-longer-hide.

Asimov, Isaac. (1964, August 16). "Visit to the World's Fair of 2014." *New York Times* http://www.nytimes.com/books/97/03/23/lifetimes/asi-v-fair.html.

Association of Southeast Asian Nations. (n.d.). "Overview." https://asean.org/our-communities/asean-socio-cultural-community/sports/.

Assunta, Mary and Phil Chamberlain. (2024). "Ensuring a Safe, Tobacco Free Future for the Young: Protecting Children from Tobacco Industry Interference." *Tobacco Control* 33, no. 3: 281–82.

Athletes of the World. (n.d.). https://athletesoftheworld.org/.

Atkinson, Anthony B. (2015). *Inequality: What Can Be Done?* Cambridge, Mass: Harvard University Press.

Attali, Jacques. (2001). *Bruits: Essais sur l'économie politique de la musique*, 2nd ed.. Paris: Librairie Arthème Fayard/Presses Universitaires de France.

Attepuram, Sajith. (2012, September 22). "'Olympic Games' & Manifestation of Sports Under Capitalism." *Socialism*. http://www.socialism.in/index.php/olympics-games-manifestation-of-sports-under-capitalism/.

Aubenas, Florence. (2024, August 3). "JO 2024: « C'est au-delà de la beauté, de la magie », dans les rues de Paris, un enthousiasme débordant." *Le Monde* https://www.lemonde.fr/societe/article/2024/08/03/jo-2024-c-est-au-dela-de-la-beaute-de-la-magie-dans-les-rues-de-paris-un-enthousiasme-debordant62661403224.html.

Auclair, Philippe, Andy Brown, Jack Kerr, Sam Kunti, and Steve Menary. (2024a, August 12). "White Elephants and Green Lies." *Josimar* https://josimarfootball.com/2024/08/12/white-elephants-and-green-lies/.

Auclair, Philippe, Andy Brown, Jack Kerr, Sam Kunti, and Steve Menary. (2024b, October 21). "Child's Play." *Josimar* https://josimarfootball.com/2024/10/21/childs-play/.

Auclair, Philippe. (2021, August 12). "The Trillion-Dollar Gambling Game." *Josimar* https://josimarfootball.com/2021/08/12/the-trillion-dollar-gambling-game/.

Auclair, Philippe. (2022, January 31). "Gambling with Lives." *Josimar* https://josimarfootball.com/2022/01/31/gambling-with-lives/.

Auclair, Philippe. (2023, June 23). "In the Name of Charity." *Josimar* https://josimarfootball.com/2023/06/23/in-the-name-of-charity/.

Auclair, Philippe. (2024, June 27). "Uefa's Lofty Environmental Ambitions and the Elephant in the Room." *Guardian* https://www.theguardian.com/football/article/2024/jun/27/uefa-lofty-environmental-ambitions-euro-2024-elephant-in-the-room.

Auclair, Philippe. (2025, July 29). "Plus ça change." *Josimar* https://josimarfootball.com/2025/07/29/plus-ca-change/.

Aurelius, Marcus. (1862). *Meditations.* Trans. George Long. https://lexundria.com/m_aur_med/1/lg.

Austin, Daniel. (2025, April 18). "From Beer to Betting—How Have Football Shirt Sponsors Changed?" *BBC* https://www.bbc.com/sport/football/articles/cge1lp0734jo.

Australian Communications and Media Authority. (2021, December 13). *Investigation Report no. BI-607* https://www.acma.gov.au/publications/2022-02/report/investigation-report-bi-607.

Australian Securities & Investments Commission. (2024, August). *ASIC's Interventions on Greenwashing Misconduct: 2023–2024* Report 791 https://download.asic.gov.au/media/lbygvudn/rep791-published-23-august-2024.pdf.

Aviña, Rafael. (2007). *¡No queremos olimpiadas! ¡Queremos revolución! 1968: Una crónica desde la infancia y todas las películas sobre el 2 de octubre.* Mexico City: Palabra de Clio.

Ayuso, Glòria. (2024, September 25). "Barcelona celebra la gran fiesta de la vela con una marea de jóvenes navegantes." *El Periódico* https://www.elperiodico.com/es/blau/copa-america/20240925/fiesta-viu-la-vela-barcelona-deportes-acuaticos-108562226.

Bachelor, Lois and Clea Skopoleti. (2024, November 26). "Selfies and Surf Simulators: The Young Cruisers Driving Boom in Sea Holidays." *Guardian* https://www.theguardian.com/environment/2024/nov/26/selfies-and-surf-simulators-the-young-cruisers-driving-boom-in-sea-holidays.

Bacon, Francis. (1863). *The Works.* Ed. James Spedding, Robert Leslie Ellis, and Douglas Denon Heath. Boston: Taggard and Thompson.

Badiou, Alain. (2007). *The Century.* Trans. Alberto Toscano. Cambridge: Polity.

Bandiera, David, Sebastien Racinais, Frédéric Garrandes, Paolo Emilio Adami, Stéphane Bermon, Yannis P. Pitsiladis, and Antonio Tessitore. (2024). "Heat-Related Risk at Paris 2024: A Proposal for Classification and Review of International Federations Policies." *British Journal of Sports Medicine* 58: 860–69.

Banking on Climate Chaos. (2024). *Fossil Fuel Finance Report 2024* https://www.bankingonclimatechaos.org/wp-content/uploads/2024/07/BOCC2024vF3.pdf.

Baram, Michael. (2009). "Globalization and Workplace Hazards in Developing Nations." *Safety Science* 47, no. 6: 756–66.

Baraniuk, Chris. (2023, September 30). "Lego is a Company Haunted by its Own Plastic." *Wired* https://www.wired.com/story/lego-haunted-by-its-own-plastic/.

Barber, Daniel A., Lee Stickells, Daniel J. Ryan, Maren Koehler, Andrew Leach, Philip Goad, Deborah van der Plaat, Cathy Keys, Farhan Karim, and William M. Taylor. (2018). "Architecture, Environment, History: Questions and Consequences." *Architectural History Review* 22, no. 2: 249–86.

Barcelona FC. (n.d.). https://www.fcbarcelona.cat/ca/club/sostenibilitat.

Barcelona Capital Nàutica. (2024, March). *Por la vela y por el mar: Memoria de actividades 2023* https://www.barcelonacapitalnautica.org/wp-content/uploads/2024/03/FBCNMemoria2023-CAST.pdf.

Barker, Alexander B., Magdalena Opazo Breton, Rachael L. Murray, Bruce Grant-Braham, and John Britton. (2019). "Exposure to 'Smokescreen' Marketing During the 2018 Formula 1 Championship." *Tobacco Control* 28: e154–55.

Barquero, Marco D. (2023). "Athletes and Environmental Activism: How Much do Athletes Contribute to Protecting the Planet?" *European Journal of Humanities and Social Sciences* 3, no. 3: 11–14.

Barrass, Tony. (2015, June 9). "WA's Cup Ran Over with Champagne." *West Australian* https://thewest.com.au/news/australia/was-cup-ran-over-with-champagne-ng-ya-118643.

Barrett, Martin, Kyle S. Bunds, Jonathan M. Casper, and Michael B. Edwards. (2019). "A Descriptive Analysis of Corporate Environmental Responsibility in Major League Professional Sport." *Journal of Applied Sport Management* 11, no. 3: 35–46.

Barry, John. (2006). "Resistance is Fertile: From Environmental to Sustainability Citizenship." *Environmental Citizenship*. Ed. Andre Dobson and Derek Bell. Cambridge, Mass.: MIT Press. 21–48.

Barthes, Roland. (1997). "Of Sport and Men." Trans. Scott MacKenzie. *Canadian Journal of Film Studies/Revue canadienne d'études cinématographiques* 6, no. 2: 75–83.

Bashir, Nadia Y., Penelope Lockwood, Alison L. Chasteen, Daniel Nadolny, and Indra Noyes. (2013). "The Ironic Impact of Activists: Negative Stereotypes Reduce Social Change Influence." *European Journal of Social Psychology* 43, no. 7: 614–26.

Battaglio, Stephen. (2022, June 14). "Why Apple TV Has Become the Exclusive Streaming Home for Major League Soccer." *Los Angeles Times* https://www.

latimes.com/entertainment-arts/business/story/2022-06-14/apple-tv-will-become-the-exclusive-streaming-home-for-major-league-soccer.

Battan, Carrie. (2022, March 11). "How 'Drive to Survive' Remade Formula 1." *New Yorker* https://www.newyorker.com/culture/culture-desk/how-drive-to-survive-remade-formula-1.

Baudrillard, Jean. (1983). *In the Shadow of the Silent Majorities … Or the End of the Social and Other Essays.* Trans. Paul Foss, Paul Patton, and John Johnston. New York: Semiotext(e).

Baudrillard, Jean. (1993). *The Transparency of Evil: Essays on Extreme Phenomena.* Trans. James Benedict. London: Verso.

Bauman, Adrian E., Masamitsu Kamada, Rodrigo S. Reis, Ricardo P. Troiano, Ding Ding, Karen Milton, Niamh Murphy, and Pedro C. Hallal. (2021). "An Evidence-Based Assessment of the Impact of the Olympic Games on Population Levels of Physical Activity." *The Lancet* 393, no. 10298: pp. 456–64.

Bautista-González, Manuel A. (2023, October 30). "The Environmental Footprint of Payment Cards." *CashEssentials* https://cashessentials.org/the-environmental-footprint-of-payment-cards/.

Baxter, Kevin with Tracy Wilkinson. (2025, January 29). "Visa Approval Crisis Threatens to Cost 2026 World Cup and L.A. Olympics Millions." *Los Angeles Times* https://www.latimes.com/sports/story/2025-01-29/sports-visas-delay-olympics-world-cup.

Beachcomber Resorts & Hotels. (2024, September). *Environmental & Social Charter: Our 52 Commitments* https://www.beachcomber-hotels.com/en/environmental-responsibility.

Beck, Ulrich. (2009). *World at Risk.* Trans. Ciaran Cronin. Cambridge: Polity.

Becker, Amy B. (2013). "Star Power? Advocacy, Receptivity, and Viewpoints on Celebrity Involvement in Issue Politics." *Atlantic Journal of Communication* 21, no. 10: 1–16.

Beckhart Coppinger, Sarah. (2021). ""Loco Tlatelolco:" The Conflicting Past of this Historic Neighborhood, 1960–1964." *The Latin Americanist* 65, no. 4: 431–55.

Bell, Amy with Samuel Agini. (2024, December 3). "America's Cup Fights to Boost Audience While Preserving Heritage." *Financial Times* https://www.ft.com/content/e6848cc8-0c10-429e-86ec-2ae4bd4e4773.

Bell, Sebastien. (2022, February 20). "How Can an F1 Team Spend $450 Million in a Year and Still Make a Profit?" *CARSCOOPS* https://www.carscoops.com/2022/02/how-does-an-f1-team-spend-450-million-in-a-year-and-still-make-a-profit/.

Benítez-Cano, Daniela, Patricia González-Marín, Anna Gómez-Gutiérrez, Marc Marí-Dell'Olmo, and Laura Oliveras. (2024). "Association of Drought Conditions and Heavy Rainfalls with the Quality of Drinking Water in Barcelona

(2010–2022)." *Journal of Exposure Science & Environmental Epidemiology* 34: 175–83.

Benjamin, Walter. (2003). "On the Concept of History." *Walter Benjamin: Selected Writings Volume 4, 1938–1940.* Ed. Howard Eiland and Michael W. Jennings. Cambridge, Mass.: Harvard University Press. 389–400.

Bennett, Nathan James, Jessica Blythe, Carole Sandrine White, and Cecilia Campero. (2021). "Blue Growth and Blue Justice: Ten Risks and Solutions for the Ocean Economy." *Marine Policy* 125 https://www.sciencedirect.com/science/article/abs/pii/S0308597X20310381.

Benson, Andrew. (2019, October 24). "Mexican Grand Prix: Lewis Hamilton Defends Environmental Social Media Posts." *BBC* https://www.bbc.com/sport/formula1/50176037.

Benson, Andrew. (2022, January 21). "Formula 1 Bans Military Air Displays at Grands Prix." *BBC* https://www.bbc.com/sport/formula1/60082961.

Benson, Andrew. (2023, February 7). "Formula 1 President Stefano Domenicali Says the Sport Will 'Never Put a Gag on Anyone'." *BBC* https://www.bbc.co.uk/sport/formula1/64563358.

Benton, Ted, ed. (1996). *The Greening of Marxism.* New York: Guilford.

Berik, Günseli and Ebru Kongar, eds. (2021). *The Routledge Handbook of Feminist Economics.* Abingdon: Routledge.

Berman, Marshall. (2017). *Modernism in the Streets: A Life and Times in Essays.* Ed. David Marcus and Shellie Sclan. London: Verso.

Bernard, Paquito, Guillaume Chevance, Celia Kingsbury, Aurélie Baillot, Ahmed-Jérôme Romain, Virginie Molinier, Tegwen Gadais, and Kelsey N. Dancause. (2021). "Climate Change, Physical Activity and Sport: A Systematic Review." *Sports Medicine* 51, no. 5: 1041–059.

Bernard, Raphaël. (2023, August 25). "Seine-Saint-Denis: contre les Jeux Olympiques, Saccage 2024 fait son "Toxic Tour"." *Citoyens* https://94.citoyens.com/2023/seine-saint-denis-contre-les-jeux-olympiques-saccage-2024-fait-son-toxic-tour,25-08-2023.html.

Bernstock, Penny, Sue Brownill, Juliet Davis, Clare Melhuish, Anna Minton and Saffron Woodcraft, eds. (2022). *State of the Legacy: Reviewing a Decade of Writing on the 'Regeneration' Promises of London 2012* https://www.ucl.ac.uk/urban-lab/sites/urban_lab/files/state_of_the_legacy.pdf.

Bertrand, Aliénor. (2019). "A Rupture Between Human Beings and Earth: A Philosophical Critical Approach to Coviability." *Coviability of Social and Ecological Systems: Reconnecting Mankind to the Biosphere in an Era of Global Change. Vol. 1: The Foundations of a New Paradigm.* Ed. Olivier Barrière, Mohammed Behnassi, Gilbert David, Vincent Douzal, Mireille Fargette, Thérèse Libourel, Maud Loireau, Laurence Pascal, Catherine Prost, Voyner Ravena-Cañete, Frédérique Seyler, and Serge Morand. Cham: Springer. 269–84.

Betts, Kellyn. (2006). "PBDEs and PCBs in Computers, Cars, and Homes." *Environmental Science & Technology* 40, no. 24: 7452.

Biel, Jonas, Vincent Reinke, Tobias Finger, and Arne Niemann. (2024). "No Longer Sidelined? Football Fandom, Belonging, and the Boundaries of Europe." *Journal of Contemporary European Studies* 32, no. 4: 1343–365.

Bilton, Dean. (2024, August 12). "The Paris Olympics, a Timely Celebration of Humanity, Taught Us About Joy, Hope and Each Other." *ABC* https://www.abc.net.au/news/2024-08-12/what-the-paris-olympics-taught-us-about-joy-hope-and-each-other/104211818.

Bioregional and WWF. (2012). *Towards a One Planet Olympics Revisited* http://assets.wwf.org.uk/downloads/towardsaoneplanetolympicsrevisited.pdf.

Birley, Derek. (1995). *Playing the Game: Sport and British Society, 1910–45.* Manchester: Manchester University Press.

Black, David. (2020). "Sailing." *Routledge Handbook of Global Sport.* Ed. John Nauright and Sara Zipp. Abingdon: Routledge. 312–19.

Black, Richard. (2010, June 30). "Formula One Embarks on Carbon-Cutting Drive." *BBC* http://www.bbc.com/news/10456984.

Blake, Heidi and Jonathan Calvert. (2016). *The Ugly Game: The Qatari Plot to Buy the World Cup.* New York: Simon & Schuster.

Blake, Heidi. (2022, December 1). "Qatar's Tarnished World Cup." *New Yorker* https://www.newyorker.com/culture/cultural-comment/qatars-tarnished-world-cup.

Blake, William. (1789). "The Ecchoing Green." *Poetry Foundation* https://www.poetryfoundation.org/poems/56592/the-ecchoing-green.

Blanchar, Clara. (2024, October 13). "Miles de personas se manifiestan en Barcelona contra la Copa del América." *El País* https://elpais.com/espana/catalunya/2024-10-13/miles-de-personas-se-manifiestan-en-barcelona-contra-la-copa-del-america.html.

Blee, Kathleen M. and Kimberly A. Creasap. (2010). "Conservative and Right-Wing Movements." *Annual Review of Sociology* 36: 269–86.

Blin, Simon. (2024, December 2). "Bernard Arnault élu à l'Académie: le patron qui veut être partout, tout le temps, éternellement." *Libération* https://www.liberation.fr/idees-et-debats/opinions/bernard-arnault-elu-a-lacademie-le-patron-qui-veut-etre-partout-tout-le-temps-eternellement-20241202_LI6M6QQJD5AR7DXN22JQDUSLN4/.

Blitz, Roger. (2013, March 15). "New Season, New Faces—The Drama Continues." *Financial Times*: 1, 3.

Bloch, Ernst. (1995). *The Principle of Hope Volume Two.* Ed. Neville Plaice, Stephen Plaice, and Paul Knight. Cambridge, Mass.: MIT Press.

Blum, Alan. (1991). "The Marlboro Grand Prix—Circumvention of the Television Ban on Tobacco Advertising." *New England Journal of Medicine* 324: 913–17.

Boardman, Anthony E. and Shaun P. Hargreaves-Heap. (1999). "Network Externalities and Government Restrictions on Satellite Broadcasting of Key Sporting Events." *Journal of Cultural Economics* 23, no. 3: 167–81.

Bobel, Chris. (2007). "'I'm Not an Activist, Though I've Done a Lot of it': Doing Activism, Being Activist and the 'Perfect Standard' in a Contemporary Movement." *Social Movement Studies: Journal of Social, Cultural and Political Protest* 6, no. 2: 147–59.

Bodie, Matthew T. (2011). "NASCAR Green: The Problem of Sustainability in Corporations and Corporate Law." *Wake Forest Law Review* 46, no. 3: 491–522.

Bogaards, Rodney. (2022, March 29). *Brisbane Olympic Games 2032: Will the Olympics Deliver a Net Benefit for the Australian Community?* Parliamentary Library Research Paper https://www.aph.gov.au/-/media/05_About_Parliament/54_Parliamentary_Depts/544_Parliamentary_Library/Research_Papers/2021-22/BrisbaneOlympicGames2032NetBenefit-epub.epub?la=en&hash=CA50DAC37173B3329EB354307454202BBEE4907A.

Bogado, Ignacio. (2025, February 2). "Madrid evaluá reubicar residentes y ajustes del trazado para suavizar el impacto del GP de F1." *GP* https://grandepremio.com/es/f1/madrid-evalua-reubicar-residentes-ajustes-trazado-f1-2026/.

Boganda, Cyprien. (2024, August 9). "Saint-Denis à l'heure de Paris 2024: une fête sous surveillance." *l'Humanité* https://www.humanite.fr/societe/jeux-olympiques-paris-2024/saint-denis-a-lheure-de-paris-2024-une-fete-sous-surveillance.

Bohrer, Ashley J. (2019). *Marxism and Intersectionality: Race, Gender, Class and Sexuality Under Contemporary Capitalism.* Bielefeld: [transcript].

Bonet, Enric. (2024, August 11). "La cara B de una fiesta olímpica "inclusiva" y "popular"." *El Salto* https://www.elsaltodiario.com/juegos-olimpicos/cara-b-una-fiesta-olimpica-inclusiva-popular.

Bonet Torra, Enric. (2024, August 16). "Francia afronta la resaca económica de los Juegos Olímpicos de París." *Público* https://www.publico.es/internacional/francia-afronta-resaca-economica-juegos-olimpicos-paris.html#md=modulo-portada-bloque:2col-t4;mm=mobile-medium.

Bono, Ferran. (2021, August 9). "The Shanty Town on Valencia's Abandoned Formula 1 Circuit." *El País* https://english.elpais.com/society/2021-08-09/the-shanty-town-on-valencias-abandoned-formula-1-circuit.html.

Boren, Zach and Emma Howard. (2024, November 20). "Companies Behind Campaign to 'End Plastic Waste' Produced 1,000 Times More Plastic Than it Cleaned Up." *Unearthed* https://unearthed.greenpeace.org/2024/11/20/alliance-to-end-plastic-waste-oil-chemical-exxon-shell-total/.

Borges, Jorge Luis with Richard Burgin, Daniel Bourne, Stephen Cape, Charles Silver, and Gloria López Lecube. (2013). *The Last Interview and Other Conversations.* Trans. Kit Maude. New York: Melville House.

Borja, Jordi with Majda Drnda, Mariela Iglesias, Mirela Fiori, and Zaida Muxí. (2003). *La ciudad conquistada*. Madrid: Alianza Editorial.

Bouchama, Abderrezak, Bisher Abuyassin, Cynthia Lehe, Orlando Laitano, Ollie Jay, Francis G. O'Connor, and Lisa R. Leon. (2022). "Classic and Exertional Heatstroke." *Nature Reviews: Disease Primers* 8, no. 8. https://doi.org/10.1038/s41572-021-00334-6.

Bougault, Valerie, Christopher Carlsten, Paolo Emilio Adami, Nicola Sewry, Wolfgang Schobersberger, Torbjørn Soligard, Lars Engebretsen, Richard Budgett, Martin Schwellnus, and Ken Fitch. (2025). "Air Quality, Respiratory Health and Performance in Athletes: A Summary of the IOC Consensus Subgroup Narrative Review on 'Acute Respiratory Illness in Athletes." *British Journal of Sports Medicine* 59, no. 7: 480–90.

Bourdieu, Pierre. (1978). "Sport and Social Class." *Social Science Information* 17, no. 6: 819–40.

Bourdieu, Pierre. (1999). "The State, Economics and Sport." Trans. Hugh Dauncey and Geoff Hare. *France and the 1998 World Cup: The National Impact of a World Sporting Event*. Ed. Hugh Dauncey and Geoff Hare. London: Frank Cass. 15–21.

Boutry, Timothée. (2024, November 28). "Procès Squarcini : « François Ruffin est bien meilleur sur le plan cinématographique que politique », réagit Bernard Arnault." *Les Parisiens* https://www.leparisien.fr/faits-divers/proces-squarcini-francois-ruffin-est-bien-meilleur-sur-le-plan-cinematographique-que-politique-reagit-bernard-arnault-28-11-2024-3KNLKOAWKBFALPYFAB6ZBX4F5E.php.

Bowen, Howard R. (2013). *Social Responsibilities of the Businessman*. Iowa City: University of Iowa Press.

Boxall-Legge, Jake. (2020, June 30). "How Mercedes Can Address the Lack of Minority Representation." *Motorsport* https://www.motorsport.com/f1/news/how-mercedes-can-address-the-lack-of-minority-representation-in-f1/4816804/.

Boykoff, Jules and Christopher Gaffney. (2020). "The Tokyo 2020 Games and the End of Olympic History." *Capitalism Nature Socialism* 31, no. 2: 1–19.

Boykoff, Jules and Gilmar Mascarenhas. (2016). "The Olympics, Sustainability, and Greenwashing: The Rio 2016 Summer Games." *Capitalism Nature Socialism* 27, no. 2: 1–11.

Boykoff, Jules and Dave Zirin. (2024, August 9). "The Paris Olympics Has Launched a War on the Poor." *The Nation* https://www.thenation.com/article/world/paris-olympics-homeless

Boykoff, Jules and Dave Zirin. (2025, January 15). "The Olympics Aren't the Way to Rebuild Los Angeles." *The Nation* https://www.thenation.com/article/society/los-angeles-olympics-fires/.

Boykoff, Jules. (2016). *Power Games: A Political History of the Olympics.* London: Verso.

Boykoff, Jules. (2020). *NOlympians: Inside the Fight Against Capitalist Mega-Sports in Los Angeles, Tokyo and Beyond.* Halifax: Fernwood Publishing.

Boykoff, Jules. (2021). “Celebration Capitalism, Sports Mega-Events, and the Tokyo 2020 Olympics.” *Journal of Sport and Gender Studies* 19: 50–66.

Boykoff, Jules. (2022, November 23). “The World Cup in Qatar is a Climate Catastrophe.” *Scientific American* https://www.scientificamerican.com/article/the-world-cup-in-qatar-is-a-climate-catastrophe/.

Boykoff, Jules. (2024a). *What Are the Olympics For?* Bristol: Bristol University Press.

Boykoff, Jules. (2024b, July 14). “Jeux olympiques: le pari manqué de l’écologie?” *Pour la Science* 124 https://www.pourlascience.fr/sd/environnement/jeux-olympiques-le-pari-manque-de-l-ecologie-26597.php.

Boykoff, Jules. (2025a). “The Olympics in Crisis: Challenges, Reforms, and the Path Forward.” *International Sports Law Journal* 25, no. 1: 29–48.

Boykoff, Jules. (2025b). “Greenwash Gold?: The Paris 2024 Olympics.” *Capitalism Nature Socialism* 36, no. 2: 171–78.

Bradbury, John Charles, Dennis Coates, and Brad R. Humphreys. (2023). “Public Policy Toward Professional Sports Stadiums: A Review.” *Journal of Policy Analysis and Management* 43, no. 3: 899–937.

Bradley, Caroline. (2024). “It’s Not Easy Being Anti-Greenwashing.” *University of Miami Legal Studies Research Paper* https://papers.ssrn.com/sol3/papers.cfm?abstract_id=4784535.

Branch, John and Motoko Rich. (2021, July 30). “Tokyo Braces for the Hottest Olympics Ever.” *New York Times* https://www.nytimes.com/2019/10/10/sports/tokyo-braces-for-the-hottest-olympics-ever.html.

Braun, Stewart, ed. Cristina Burack. (2024, August 23). “St. Pauli Injects Subculture Into the Bundesliga.” *DW* https://www.dw.com/en/st-pauli-injects-subculture-into-the-bundesliga/a-70012361.

Break Free From Plastic. (2023). *Brand Audit Report 2023* https://drive.google.com/file/d/1YFyfRv4mviZZXa8b1HdpucDX3WEwJzv/view.

Brecht, Bertolt. (1964). “Emphasis on Sport.” *Brecht on Theatre.* Ed. John Willett. Basingstoke: Macmillan. 6–9.

Brecht, Bertolt. (2018). *The Collected Poems of Bertolt Brecht.* Trans. and Ed. Tom Kuhn and David Constantine with Charlotte Ryland. New York: Liveright Publishing.

Breeden, Aurelien and Catherine Porter with Melissa Eddy. (2024, August 7). “After French Rail Sabotage, Some See Signs of a Murky ‘Ultraleft’.” *New York Times* https://www.nytimes.com/2024/08/07/world/europe/france-olympics-sabotage-anarchists.html.

Brehm, Johannes and Henri Gruhl. (2024). "Increase in Concerns About Climate Change Following Climate Strikes and Civil Disobedience in Germany." *Nature Communications* 15. https://doi.org/10.1038/s41467-024-46477-4.

Breland, Ali. (2021, June 25). "How Amazon Bullies, Manipulates, and Lies to Reporters." *Mother Jones* https://www.motherjones.com/politics/2021/06/amazon-journalists-pr-tactics/.

Brentin, Dario and Laurence Cooley. (2015). "Sport and Nationalism: A Review of the Literature." *Studies on National Movements* 3 http://snm.nise.eu/index.php/studies/article/view/0306s.

Brevini, Benedetta and Justin Lewis, eds. (2018). *Climate Change and the Media*. New York: Peter Lang.

Brigham, Katie. (2022, January 29). "How the Fossil Fuel Industry is Pushing the Plastics on the World." *CNBC* https://www.cnbc.com/2022/01/29/how-the-fossil-fuel-industry-is-pushing-plastics-on-the-world-.html.

Brinsford, James. (2023, January 18). "Formula 1 Carbon Footprint Slammed After 2023 Calendar Release." *Newsweek* https://www.newsweek.com/formula-one-carbon-footprint-slammed-2023-calendar-1774806.

British Screen Forum. (2021, April). *Business Briefing: High-End Television Drama Investment: An Update* https://britishscreenforum.co.uk/wp-content/uploads/2021/04/British-Screen-Forum-High-end-Television-Drama-An-Update.pdf.

Brittle, Cian. (2024a, February 29). "F1 Revenue Soars 25% YoY to US$3.2bn." *SportsPro* https://www.sportspro.com/finance-investment/finances/f1-2023-revenue-accounts-financial-year-liberty-media/.

Brittle, Cian. (2024b, April 8). "McLaren Racing Agrees Multi-Year BAT Sponsorship Extension." *Blackbook Motorsport* https://www.blackbookmotorsport.com/news/mclaren-racing-f1-formula-e-bat-british-american-tobacco-vuse/.

Brittle, Cian. (2025, January 30). "Louis Vuitton Officially Joins F1 as Part of LVMH Deal." *SportsPro* https://www.sportspro.com/news/f1-louis-vuitton-lvmh-sponsorship-january-2025/.

Brocherie, Franck, Olivier Girard, Adèle Mornas, Janne Bouten, and Grégoire P. Millet. (2024). ""Citius, Altius, Fortius" in the Face of Global Warming: Not as Simple as it Seems." *Climate and Atmospheric Science* 7. https://doi.org/10.1038/s41612-024-00774-3.

Bronner, Luc. (2024, August 6). "Anne Hidalgo et les Jeux olympiques: « C'est pas complètement foutu, on peut être ensemble et être heureux ensemble! »." *Le Monde* https://www.lemonde.fr/politique/article/2024/08/06/anne-hidalgo-avec-les-jo-les-gens-se-disent-c-est-pas-completement-foutu-on-peut-etre-ensemble-et-on-peut-etre-heureux-ensemble6269386823448.html?lmdmedium=al&lmdcampaign=envoye-par-appli&lmdcreation=ios&lmdsource=default.

Brown, Katherine. (2016). "Association Between Alcohol Sports Sponsorship and Consumption: A Systematic Review." *Alcohol and Alcoholism* 51, no. 6: 747–55.

Brown, Andy, Philippe Auclair, Jack Kerr, Steve Menary, and Samindra Kunti. (2025, February 6). "Strangelove: How the Premier League Can Enable Illegal Gambling and Money Laundering." *Josimar* https://josimarfootball.com/2025/02/06/strangelove-how-premier-league-learned-to-stop-worrying-and-embrace-illegal-gambling-and-money-laundering/.

Brown, Maury. (2023, March 29). "Inside The Numbers That Show Formula 1's Popularity and Financial Growth." *Forbes* https://www.forbes.com/sites/maurybrown/2023/03/29/inside-the-numbers-that-show-formula-1s-popularity-and-financial-growth/.

Brown, Wendy. (2019). *In the Ruins of Neoliberalism: The Rise of Anti-Democratic Politics in the West.* New York: Columbia University Press.

Bruttomesso, Elisa. (2018). "Making Sense of the Square: Facing the Touristification of Public Space Through Playful Protest in Barcelona." *Tourist Studies* 18, no. 4: 467–85.

Budzier, Alexander and Bent Flyvbjerg. (2024). *The Oxford Olympics Study 2024: Are Cost and Cost Overrun at the Games Coming Down?* Said Business School Working Paper https://arxiv.org/pdf/2406.01714.

Buesa, Cristina. (2024, October 25). "El deporte catalán también pide "reflexionar" sobre la decisión de Barcelona de no repetir la Copa América de vela." *El Periódico* https://www.elperiodico.com/es/blau/copa-america/20241025/ufec-deporte-catalan-pide-reflexionar-decision-no-repetir-copa-america-de-vela-110411271.

Bugden, Dylan. (2020). "Does Climate Protest Work? Partisanship, Protest, and Sentiment Pools." *Socius: Sociological Research for a Dynamic World* 6: 1–13.

Building and Wood Workers' International. (2024). *Migrant Worker Conditions in the Kingdom of Saudi Arabia: An Overview of Worker Abuses Linked with Forced Labour* https://www.diarioconstitucional.cl/wp-content/uploads/2024/06/Vea-texto-investigativo-de-la-denuncia.pdf.

Bulbeck, Chilla. (1998). *Re-Orienting Western Feminisms: Women's Diversity in a Postcolonial World.* Cambridge: Cambridge University Press.

Bull, Andy. (2024a, August 12). "Adieu, Paris: You Were Chic, Spectacular and Showed How to Stage a Handsome Games." *Guardian* https://www.theguardian.com/sport/article/2024/aug/12/paris-olympic-games.

Bull, Andy. (2024b, October 20). "New Zealand Win America's Cup as Britain's Challenge Falls Short." *Guardian* https://www.theguardian.com/sport/2024/oct/19/new-zealand-win-americas-cup-britain.

Bullock, Graham. (2017). *Green Grades: Can Information Save the Earth?* Cambridge, Mass.: MIT Press.

Burgaya, Josep. (2024, October 26). "Copa América en Barcelona: el engaño, solo una vez." *Metropoli* https://metropoliabierta.elespanol.com/opin-

ion/20241026/copa-america-en-barcelona-el-engano-solo-una-vez/896540341_13.html.
Burgen, Stephen. (2024, August 26). "'It's About Money': Doubt Over America's Cup Benefits in Tourist Hotspot Barcelona." *Guardian* https://www.theguardian.com/world/article/2024/aug/26/barcelona-americas-cup-overtourism-super-rich-local-economy.
Burgen, Stephen. (2025, May 20). "Thinking of a Trip to Barcelona This Summer? Beware—Here's What You'll Find." *Guardian* https://www.theguardian.com/commentisfree/2025/may/20/barcelona-summer-tourism-locals-city.
Burke, Peter. (2004). *Languages and Communities in Early Modern Europe: The 2002 Wiles Lectures Given at Queen's University, Belfast*. Cambridge: Cambridge University Press.
Burrows, Tom. (2024, August 2). "The Olympic Games and Sex—Is it a Myth or a 'Free for All'?" *The Athletic* https://www.nytimes.com/athletic/5671602/2024/08/02/olympic-games-paris-2024-sex/.
Bustad, Jacob J. and David L. Andrews. (2023). "The Circus Comes to Town: Formula 1, Globalization, and the Uber-Sport Spectacle." *Lives in the Fast Lane: Essays on the History and Politics of Motor Racing*. Ed. Damion Sturm, Stephen Wagg, and David L. Andrews. Cham: Palgrave Macmillan. 595–617.
Buzzoni, Lorenzo, Marta Portocarrero, and Chris Matthews. (2025, March 20). "Revealed: How Gambling Embedded Itself into European Football." *Investigate Europe* https://www.investigate-europe.eu/posts/revealed-how-gambling-betting-industry-sponsors-european-football.
Byatt, A. S. (2008, June 29). "Lie Back and Think of Europe." *Observer* https://www.theguardian.com/books/2008/jun/29/sportandleisure.review.
Byrne, Thomas H., Benjamin F. Henwood, and Anthony W. Orlando. (2021). "A Rising Tide Drowns Unstable Boats: How Inequality Creates Homelessness." *Annals of the American Academy of Political and Social Sciences* 693, no. 1: 28–45.
Cabot, Cyrielle. (2022, October 28). "Qatar: derrière "la neutralité carbone", la Coupe du monde du "greenwashing"." *France 24* https://www.france24.com/fr/sports/20221028-qatar-derri%C3%A8re-la-neutralit%C3%A9-carbone-la-coupe-du-monde-du-greenwashing.
Cabral, Amílcar. (1979). *Unity and Struggle: Speeches and Writings: Texts Selected by the PAIGC* [Partido Africano da Independência da Guiné e Cabo Verde]. Trans. Michael Wolfers. New York: Monthly Review Press.
Cadoret, NnoMan and Justine Guitton-Boussion. (2021, September 6). "À Aubervilliers l'« écœurement » après la destruction des jardins ouvriers." *Reporterre* https://reporterre.net/A-Aubervilliers-l-ecoeurement-apres-la-destruction-des-jardins-ouvriers.

Caiani, Manuela. (2024). "Visual Analysis and the Contentious Politics of the Radical Right." *Sociology Compass* 18, no. 9 https://doi.org/10.1111/soc4.13267.

Cain, Alison and Michael Callan. (2025). "Principles in Practice? A Policy Review of the IOC's Environmental Sustainability Agenda." *Frontiers in Sports and Active Living* 7 https://doi.org/10.3889/fsport.2025.1511092.

Calma, Justine and Zoe Schiffer. (2021, May 25). "Amazon Workers Demand Company Quit Polluting Near Communities of Color." *The Verge* https://www.theverge.com/2021/5/25/22453516/amazon-workers-pollution-petition-letter-demand.

Campbell, Roderick. (2013). *Blowout! A Cost Benefit Analysis of the Australian Grand Prix.* Economists at Large http://www.ecolarge.com/wp-content/uploads/2013/03/Blowout-A-cost-benefit-analysis-of-the-Australian-Grand-Prix-Ecolarge-Final.pdf.

Camprubí i Pont, Aleix. (2024, October 23). "¿Por qué se va de Barcelona la Copa América? Del discurso triunfalista a las cifras hinchadas y el malestar vecinal." *Público* https://www.publico.es/sociedad/barcelona-copa-america-discurso-triunfalista-cifras-hinchadas-malestar-vecinal.html.

Cañada, Ernest, Carla Izcara, Bastien Montovert, and Mariciana Zorz. (2024). *Proposals for the Design of Public Policies on Popular Tourism* Alba Sud Policy Brief #3 https://www.albasud.org/publ/docs/135.en.pdf.

Canela, Joan. (2022, February 28). "La Copa América no volverá a Valècia tras la negativa de Ayuntamiento y Generalitat de pagar el evento." *Público* https://www.publico.es/deportes/copa-america-copa-america-no-volvera-valencia-negativa-ayuntamiento-generalitat-pagar-evento.html.

Cannesant, Raphael. (2023, December 1). "Coupe du Monde de football: la FIFA se moque-t-elle des enjeux environnementaux?" *Radiofrance* https://www.radiofrance.fr/mouv/coupe-du-monde-de-football-la-fifa-se-moque-t-elle-des-enjeux-environnementaux-spoiler-oui-9717302.

Carbon Literacy Project. (n.d.). *Sport* https://carbonliteracy.com/sport/#:~:text=The%20global%20sports%20industry%20is,the%20'Race%20to%20Zero'.

CarbonMajors. (2025, March). *2023 Data Update* https://carbonmajors.org/briefing/The-Carbon-Majors-Database-2023-Update-31397.

Carbon Market Watch and éclaircies. (2024). *Going for Green: Assessing the Climate Strategy and Communication of the 2024 Paris Olympics* https://carbonmarketwatch.org/wp-content/uploads/2024/04/Going-for-green-Olympics-report.pdf.

Carbon Market Watch. (2022, May). *Poor Tackling: Yellow Card for 2022 FIFA World Cup's Carbon Neutrality Claim* https://carbonmarketwatch.org/wp-content/uploads/2022/05/Poor-tackling-Yellow-card-for-2022-FIFA-1.pdf.

Carbon Trust. (2013, August 9). "Carbon Trust Unlaces the 'Carbon Bootprint' of Watching Football." https://www.carbontrust.com/news-and-insights/news/carbon-trust-unlaces-the-carbon-bootprint-of-watching-football.

Carbon Trust. (2016). *The "Carbon Bootprint" of Euro 2016: Which Nation's Fans Have the Lowest Carbon Footprint When Watching Games* https://www.carbontrust.com/news/2016/06/the-carbon-bootprint-of-euro-2016-which-nations-fans-have-the-lowest-carbon-footprint-when-watching-games/.

Carbon Trust. (n.d.). *Carbon Bootprint of the FA Community Shield* https://www.carbontrust.com/media/360767/carbon-bootprint-infographic.pdf.

Cárdenas Rodríguez, Oscar Javier. (2009). "Poverty Reduction Approaches in Mexico Since 1950: Public Spending for Social Programs and Economic Competitiveness Programs." *Journal of Business Ethics* 88, Suppl. 2: 269–81.

Carlini, Joan, Adele Pavlidis Alana Thomson, and Clare Morrison. (2025). "Delivering on Social Good—Corporate Social Responsibility and Professional Sport: A Systematic Quantitative Literature Review." *Journal of Strategic Marketing* 33, no. 3: 323–36.

Carmichael, Andy. (2022). "Time for Practice; Sport and the Environment." *Managing Sport and Leisure* 27, no. 3: 189–98.

Carnicé, Alba. (2024a, July 21). "Barcelona permite a la Fundación de la Copa América pedir créditos de hasta dos millones de euros." *Metropoli* https://metropoliabierta.elespanol.com/economia/20240721/barcelona-permite-la-fundacion-de-copa-america-pedir-creditos-hasta-dos-millones-euros/8696631060.html.

Carnicé, Alba. (2024b, October 23). "Barcelona también descarta repetir la Copa América en 2026: "Es económicamente insostenible"." *Metropoli* https://metropoliabierta.elespanol.com/informacion-municipal/20241023/barcelona-tambien-descarta-repetir-la-copa-america-en-es-economicamente-insostenible/895660434_0.html.

Carosella, Vitas. (2023, October 6). "FIFA Announces 2030 and 2034 World Cup Plans Amongst Climate Concerns." *Forbes* https://www.forbes.com/sites/vitascarosella/2023/10/06/fifa-announces-2030-and-2034-world-cup-plans-amongst-climate-concerns/.

Carpenter, Victoria. (2015). "'You Want the Truth? You Can't Handle the Truth': Poetic Representations of the 1968 Tlatelolco Massacre." *Journal of Iberian and Latin American Research* 21, no. 1: 35–49.

Carreño, Fernando M. (2022, December 30). "El cambio climático ya afecta al deporte." *Marca* https://www.marca.com/polideportivo/2022/12/30/63ac75f0ca47410f5f8b4593.html.

Carter, April. (2010). *Direct Action and Liberal Democracy.* London: Routledge.

Carter Bill with Richard Sandomir. (2000, September 20). "Sydney 2000: Television. NBC Ratings for Olympics Are Worst Ever." *New York Times* https://www.nytimes.com/2000/09/20/sports/sydney-2000-television-nbc-s-ratings-for-olympics-are-worst-ever.html.

Casey, Ruairi. (2019, December 5). "Are Global South Experts Sidelined in Climate Conservations?" *Al Jazeera* https://www.aljazeera.com/indepth/features/global-south-experts-sidelined-climate-conversations-191203132302298.html.

Casper, Jonathan M., Michael E. Pfahl, and Brian McCullough. (2014). "Intercollegiate Sport and the Environment: Examining Fan Engagement Based on Athletics Department Sustainability Efforts." *Journal of Issues in Intercollegiate Athletics* 7: 65–91.

Castañeda, Luis. (2010). "Beyond Tlatelolco: Design, Media, and Politics at Mexico '68." *Grey Room* 40: 100–26.

Castañeda, Luis. (2014). *Spectacular Mexico: Design, Propaganda, and the 1968 Olympics*. Minneapolis: University of Minnesota Press.

Castellanos, Rosario. (2018). "Memorial de Tlatelolco." *Tercera Via* https://terceravia.mx/2018/10/memorial-de-tlatelolco-poema-de-rosario-castellanos/.

Castelló, Carlos Navarro. (2024, September 17). "Carreras illegals en el abandonado circuito urbano de Fórmula 1 de València." *El Diario* https://www.eldiario.es/comunitat-valenciana/valencia/carreras-ilegales-abandonado-circuito-urbano-formula-1-valencia_1_11656041.html.

Catton, William R., Jr., and Riley E. Dunlap. (1978). "Environmental Sociology: A New Paradigm." *American Sociologist* 13, no. 1: 41–49.

Cave, Andrew and Alex Miller. (2015, July 22). "Passion Drives Sponsorship of Formula One." *Telegraph* http://www.telegraph.co.uk/investing/business-of-sport/formula-one-sponsorship/.

Ceccon, Angelique, Andrew Hargrove, and Jamie Sommer. (2024). "Do the Olympics Impact CO_2 Emissions? A Cross-National Analysis." *Global Transitions* 6: 241–48.

Centre on Housing Rights and Evictions. (2007). *Fair Play for Housing Rights: Mega-Events, Olympic Games and Housing Rights* https://www.ruig-gian.org/ressources/Report%20Fair%20Play%20FINAL%20FINAL%20070531.pdf.

Cesare, Benjamin. (2024, November–December). "The Amazing America's Cup." *WindCheck* https://www.windcheckmagazine.com/article/the-amazing-americas-cup/.

Charfeddine, Lanouar, Afnan Yousef Al-Malk, and Kholoud Al Korbi. (2018). "Is it Possible to Improve Environmental Quality Without Reducing Economic Growth: Evidence from the Qatar Economy." *Renewable and Sustainable Energy Reviews* 82, no. 1: 25–39.

Chahuneau, Louis. (2024, July 29). "Ultragauche, manipulation étrangère ... ce que l'on sait de l'enquête sur les actes de sabotage." *France 24* https://www.france24.com/fr/france/20240729-ultragauche-manipulation-%C3%A9trang%C3%A8re-ce-que-l-on-sait-de-l-enqu%C3%AAte-sur-les-actes-de-sabotage.

Chandler, Jenna. (2018, July 12). "LA 'Sterilized' its Streets for the '84 Olympics—How Will it Treat the Homeless in 2028?" *Curbed* https://la.curbed.com/2018/7/12/17454676/los-angeles-olympics-homeless-police-militarization-security.

Chandler, Raymond. (1949). *The Little Sister*. New York: Houghton Mifflin.

Chartier, Roger. (1989). "Texts, Printings, Readings." *The New Cultural History*. Ed. Lynn Hunt. Berkeley: University of California Press. 154–75.

Chartier, Roger. (2005). "Crossing Borders in Early Modern Europe: Sociology of Texts and Literature." Trans. Maurice Elton. *Book History* 8: 37–50.

Chaumier, Morgan. (2023, November 30). "En Seine-Saint-Denis, « la promesse républicaine n'est pas tenue »." *l'Humanité* https://www.humanite.fr/politique/education/en-seine-saint-denis-la-promesse-republicaine-nest-pas-tenue.

Chebil, Mehdi. (2024, August 4). "Olympic Fervour in Pictures: 'My Friends Who Left Paris Now Regret It'." *France 24* https://www.france24.com/en/sport/20240804-olympic-fervour-in-pictures-my-friends-who-left-paris-now-regret-it.

Chen, Chwen Chwen, Cinzia Colapinto, and Qing Luo. (2012). "The 2008 Beijing Olympics Opening Ceremony: Visual Insights into China's Soft Power." *Visual Studies* 27, no. 2: 188–95.

Chenoweth, Erica. (2021). *Civil Resistance: What Everyone Needs to Know®*. New York: Oxford University Press.

Chisholm, Justin. (2025, May 27). "Italian Prime Minister Giorgia Meloni Welcomes America's Cup in Rome." *Yacht Racing Life* https://www.yachtracinglife.com/italian-prime-minister-giorgia-meloni-welcomes-americas-cup-in-rome/?ref=yacht-racing-life-daily-newsletter.

Chrisafis, Angelique. (2024a, April 7). "'Greenest Ever Games': How the Paris Olympics Hopes to Inspire a New Era of Global Sporting Events." *Guardian* https://www.theguardian.com/sport/2024/apr/07/paris-olympics-greenest-games-low-carbon.

Chrisafis, Angelique. (2024b, August 7). "Greenwash Games? French Public Opinion Points Finger at Coca-Cola Over Olympics Plastic Waste." *Guardian* https://www.theguardian.com/sport/article/2024/aug/07/greenwash-games-french-public-points-finger-at-coca-cola-over-olympics-plastic-waste?CMP=ShareiOSAppOther.

Churchwell, Sarah. (2025, April 4). "How Gatsby Foretold Trump's America." *Financial Times* https://www.ft.com/content/dd6f1a32-398e-47f2-85a3-41120c952554?emailId=6e35221e-171a-48ce-9e8d-bdede0f9e5f0&segmentId=9264b0f7-e7ac-8f9b-044f-c10729049333.

Cicero. (2008). *Selected Letters*. Trans. P. G. Walsh. Oxford: Oxford University Press.

Clarey, Christopher. (1998, December 22). "Bribery Allegations Suddenly Raise International Olympic Committee's Visibility: Just Who Are the Members of the Arcane IOC?" *New York Times* http://www.nytimes.com/1998/12/22/sports/bribery-allegations-suddenly-raise-international-olympic-committees.html.

Clark, Aaron and Jack Wittels. (2023, July 31). "The Invisible Climate Impact of a Cruise Ship." *Bloomberg* https://www.bloomberg.com/news/articles/2023-07-31/cruise-ship-methane-leak-shows-hidden-climate-impact.

ClientEarth. (n.d.). *Greenwashing Files* https://www.clientearth.org/projects/the-greenwashing-files/.

ClientEarth. (2024, September 17). "Environmental Groups are Once Again Taking INEOS to Court." https://www.clientearth.org/latest/press-office/press/environmental-groups-are-once-again-taking-ineos-to-court/.

Climate Action Tracker. (2024, December 3). "Saudi Arabia." https://climateactiontracker.org/countries/saudi-arabia/policies-action/.

Coch, Matt. (2022, February 23). "FI Teams Battle Development-Cost Cap Equation." *Speed Cafe* https://speedcafe.com/f1-teams-battle-development-cost-cap-equation/.

Cohen, Roger and Jeré Longman. (1999, February 7). "Olympic Chief's Expansion Goals Left Little Zeal to Pursue Abuses." *New York Times*: 1, 26.

Coleman, Madeline. (2023, February 25). "'Drive to Survive' Season Five: Our Top Five Takeaways." *The Athletic* https://theathletic.com/4251298/2023/02/25/drive-to-survive-season-5-takeaways-f1/.

Coleman, Madeline. (2024, May 9). "How F1's Miami GP's Food Surplus Feeds the City and Fights Climate Change." *The Athletic* https://www.nytimes.com/athletic/5482029/2024/05/09/f1-miami-gp-food-rescue/.

Coll, Bernat. (2024, July 11). "Copa del América de vela: Los espías toman la costa de Barcelona." *El País* https://elpais.com/espana/cataluna/2024-07-11/copa-del-america-de-vela-los-espias-toman-la-costa-de-barcelona.html.

Collier, Eric and Anthony Hernandez. (2022, July 23). "The Nike Saga: Winning at All Costs." *Le Monde* https://www.lemonde.fr/en/summer-reads/article/2022/07/23/the-nike-saga-winning-at-all-costs_5991163_183.html.

Collins, Andrea, Calvin Jones, and Max Munday. (2009). "Assessing the Environmental Impacts of Mega Sporting Events: Two Options?" *Tourism Management* 30, no. 6: 828–37.

Collins, Andrea and Andrew Flynn. (2008). "Measuring the Environmental Sustainability of a Major Sporting Event: A Case Study of the FA Cup Final." *Tourism Economics* 14, no. 4: 751–68.

Collins, Marybeth. (2025, July 14). "World Cup Projects in Mexico Threaten Ecosystems. *Environment + Energy Leader* https://www.environmentenergyleader.com/stories/world-cup-projects-in-mexico-threaten-ecosystems,84079?.

Collins, Sandra. (2011). "East Asian Olympic Desires: Identity on the Global Stage in the 1964 Tokyo, 1988 Seoul and 2008 Beijing Games." *International Journal of the History of Sport* 28, no. 16: 2240–260.

Commission of the European Communities. (2007, July 11). *White Paper on Sport* https://eur-lex.europa.eu/legal-content/EN/TXT/?uri=celex%3A52007DC0391.

Committee for Economic Development. (n.d.). *Climate, Energy & Nature* https://www.conference-board.org/us/committee-economic-development/climate-energy-environment.

Committee on Women, Children, Family Relations and Gender Equality. (2022, January 26). *Executive Summary Committee Report Senate Resolution No. 131* https://legacy.senate.gov.ph/lis/committee_rpt.aspx?congress=18&q=542.

Commonwealth of Virginia, Joint Legislative Audit and Review Commission. (2024). *Data Centers in Virginia: Report to the Governor and the General Assembly of Virginia Commission Draft* https://jlarc.virginia.gov/pdfs/reports/Rpt598-2.pdf.

Congostrina, Alfonso L. (2024, July 16). "Barcelona argumenta que no puede informar sobre los ingresos por el desfile de Louis Vuitton por cuestiones de "confidencialidad"." *El País* https://elpais.com/espana/catalunya/2024-07-16/barcelona-se-escuda-en-el-secreto-empresarial-para-no-revelar-cuanto-cobro-de-louis-vuitton-por-el-desfile-en-el-park-guell.html.

Congostrina, Alfonso L. (2024b, September 30). "Dos de cada tres cruceros que atracan en Barcelona permanecen menos de doce horas en la ciudad." *El País* https://elpais.com/espana/catalunya/2024-09-30/dos-de-cada-tres-cruceros-que-atracan-en-barcelona-permanecen-menos-de-doce-horas-en-la-ciudad.html.

Conn, David. (2017). *The Fall of the House of FIFA*. New York: Nation Books.

Connell, Raewyn. (2020). *Southern Theory: The Global Dynamics of Knowledge in Social Science*. Abingdon: Routledge.

Connell, R. W. (1984). "Class Formation on a World Scale." *Review (Fernand Braudel Center)* 7, no. 3: 407–40.

Constitución de Ecuador. (2008). http://www.asambleanacional.gob.ec/documentos/constituciondebolsillo.pdf.

Cooley, Charles Horton. (1910). *Social Organization: A Study of the Larger Mind*. New York: Charles Scribner's Sons.

Cooper, Adam. (2013, August 25). "Greenpeace Launches Attack on Shell Oil at Belgian Grand Prix." *Autoweek* https://www.tirebusiness.com/article/20130826/NEWS/130829930/greenpeace-launches-attack-on-shell-at-belgian-grand-prix.

Coopersmith, Jonathan. (2015). *Faxed: The Rise and Fall of the Fax Machine*. Baltimore: Johns Hopkins University Press.

Corlett, Eva. (2025, January 30). "Taranki Mounga: New Zealand Mountain Granted Same Legal Rights as a Person." *Guardian* https://www.theguardian.com/world/2025/jan/30/taranaki-mounga-new-zealands-second-highest-mountain-granted-same-legal-rights-as-a-person.

Cornelissen, Scarlett, Bob Urmilla, and Kamilla Swart. (2011). "Towards Redefining the Concept of Legacy in Relation to Sport Mega-Events: Insights from the 2010 FIFA World Cup." *Development Southern Africa* 28, no. 3: 307–18.

Cornevin, Christophe. (2024, August 1). "Les saboteurs de l'ultragauche galvanisés par les Jeux olympiques." *Le Figaro* https://www.lefigaro.fr/actualite-france/les-saboteurs-de-l-ultragauche-galvanises-par-les-jeux-olympiques-20240801.

Correia, Mickaël. (2018). *Une histoire populaire du football.* Paris: Éditions La Découverte.

Correia, Mickaël. (2022). *Criminels climatiques.* Paris: Éditions La Découverte.

Cortes, Anthony. (2024, August 12). "À gauche, quelle suite donner aux Jeux Olympiques?" *l'Humanité* https://www.humanite.fr/politique/jeux-olympiques-paris-2024/a-gauche-quelle-suite-donner-aux-jeux-olympiques.

Cotter, Trish and Sandra Mullin. (2024). "The Olympic Game's Up: It's Time for the IOC to Stop Promoting Sugary Drinks." *British Medical Journal* https://press.psprings.co.uk/gh/august/bmjgh016586.pdf.

Coulibaly, Thierry Yerema, Mihoko Tegawa Wakamatsu, and Shunsuke Managi. (2022). "The Use of Geographically Weighted Regression to Improve Information from Satellite Night Light Data in Evaluating the Economic Effects of the 2010 FIFA World Cup." *Area Development and Policy* 7, no. 4: 463–81.

Court, Andrew. (2024, May 14). "'Anti-Sex' Beds Have Arrived at Paris Olympics—After Horny Athletes Admit to Orgies Amid Competition." *New York Post* https://nypost.com/2024/05/14/lifestyle/anti-sex-beds-have-arrived-at-paris-olympics-after-horny-athletes-admit-to-orgies/.

Court of Arbitration for Sport. (n.d.). https://www.tas-cas.org/en/index.html.

Cowger, Win, Kathryn A. Willis, Sybil Bullock, Katie Conlon, Jorge Emmanuel, Lisa M. Erdle, Marcus Eriksen, Trisia A. Farrelly, Britta Denise Hardesty, Kristinna Kerge, Natalie Li, Yedan Li, Adam Liebman, Neil Tangri, Martin Thiel, Patricia Villarrubia-Gómez, Tony R. Walker, and Mengjiao Wang. (2024). "Global Producer Responsibility for Plastic Pollution." *Science Advances* 10, no. 17 https://doi.org/10.1126/sciadv.adj8275.

Crabb, Lauren A. H. (2018). "Debating the Success of Carbon-Offsetting Projects at Sports Mega-Events. A Case from the 2014 FIFA World Cup." *Journal of Sustainable Forestry* 3, no. 2: 178–96.

Crafton, Adam. (2024, April 30). "NBC to Continue Push for Premier League Games in U.S., Says Leading Exec." *The Athletic* https://www.nytimes.com/athletic/5460003/2024/04/30/premier-league-games-united-states-nbc/.

Craig, Christopher and Ismail Karabas. (2024). "Forecasting Thermal Stress for Sports Tourists at the 2026 FIFA World Cup." *International Journal of Biometeorology* 68: 2731–742.

Cray, Charlie. (2001, July/August). "Toxics on the Hudson: The Saga of GE, PCBs and the Hudson River." *Multinational Monitor* 22, nos. 7/8 https://www.multinationalmonitor.org/mm2001/072001.

Croci, Osvaldo and Julian Ammirante. (1999). "Soccer in the Age of Globalization." *Peace Review* 11, no. 4: 499–504.

Croco, Sarah E., Kathleen Gallagher Cunningham, and Taylor Vincent. (2023). "Protests and Persuasion: Partisanships [*sic.*] Effect on Evaluating Nonviolent Tactics in the United States." *Journal of Peace Research* 60, no. 1: 26–41.

Cronin, Ben and Martin Ross. (2024, March 5). "EXCLUSIVE: FIFA Withdraws Wanda Sponsorship Rights Over Missed Payments." *SportBusiness* https://www.sportbusiness.com/news/exclusive-fifa-withdraws-wanda-sponsorship-rights-over-missed-payments/.

Cronin, Ben. (2024, September 16). "Qatar Airways Ties Down UCL in 6-Year, €500m Deal." *SportBusiness* https://www.sportbusiness.com/news/exclusive-qatar-airways-ties-down-ucl-in-6-year-e500m-deal/.

Crook, Clive. (2005, January 22). "The Good Company." *Economist*: Survey 3–4.

Cruise Lines International. (2024, May). *State of the Cruise Industry Report* https://cruising.org/-/media/clia-media/research/2024/2024-state-of-the-cruise-industry-report_updated-050824_web.ashx.

Cummins, Steven, Charlotte Clark, Daniel Lewis, Neil Smith, Claire Thompson, Melanie Smuk, Stephen Stansfeld, Stephanie Taylor, Amanda Fahy, Trish Greenhalgh, and Sandra Eldridge. (2018). "The Effects of the London 2012 Olympics and Related Urban Regeneration on Physical and Mental Health: The ORiEL Mixed-Methods Evaluation of a Natural Experiment." *Public Health Research* 6, no. 12 https://www.ncbi.nlm.nih.gov/books/NBK532750/pdf/Bookshelf_NBK532750.pdf.

Cupples, Julie and Ramón Grosfoguel, eds. (2018). *Unsettling Eurocentrism in the Westernized University*. New York: Routledge.

Curry, Patrick. (2006). *Ecological Ethics: An Introduction*. Cambridge: Polity Press.

Cusce Nobre, Eduardo Alberto. (2017). "Preface." *Sports Mega-Events and Urban Legacies: The 2014 FIFA World Cup, Brazil*. Ed. Eduardo Alberto Cusce Nobre. Cham: Palgrave. v-vi.

Daddi, Tiberio, Francesco Rizzi, Gaia Pretner, Niccolò Todaro, Eleonora Annunziata, Marco Frey, and Fabio Iraldo. (2022). "Environmental Management of Sport Events: A Focus on European Professional Football." *Sport, Business and Management: An International Journal* 12, no. 2: 208–32.

Dai, Xiaodong, and Guo-Ming Chen. (2014). "The Construction of National Image in the Media and the Management of Intercultural Conflicts." *The Handbook of Media and Mass Communication Theory*. Ed. Robert S. Fortner and P. Mark Fackler. Malden: Wiley-Blackwell. 708–24.

daMatta, Roberto. (1991). *Carnivals, Rogues, and Heroes: An Interpretation of the Brazilian Dilemma*. Trans. John Dury. Notre Dame: University of Notre Dame Press.

Damián, Araceli. (2019). "Pobreza y desigualdad en México. La construcción ideológica y fáctica de ciudadanías diversas y desiguales." *El Trimestre Económico* 86, no. 3: 623–66.

Dang, Sheila and Mimosa Spencer. (2024, August 12). "Paris Sets New Trend as the Product Placement Olympics." *Reuters* https://www.reuters.com/sports/olympics/paris-sets-new-trend-product-placement-olympics-2024-08-10/.

Danjon, Alexis. (2023, June 7). "La FIFA a bien fait du greenwashing pour la Coupe du monde 2022." *L'Équipe* https://www.lequipe.fr/Football/Actualites/La-fifa-a-bien-fait-du-greenwashing-pour-la-coupe-du-monde-2022/1401281.

Das, Leah. (2022, April 12). "Greenwash: What It is and How Not to Fall for It." Greenpeace https://www.greenpeace.org.uk/news/what-is-greenwashing/.

Das, Shanti. (2024, November 2). "'It's Created an Internal Shitstorm': Turmoil at UK Law Firm Accused of 'Whitewashing' Saudi World Cup Report." *Guardian* https://www.theguardian.com/world/2024/nov/02/top-uk-law-firm-accused-of-whitewashing-saudi-human-rights-report-for-world-cup-bid.

Dasch, Sophia T., Millicent Bellm, Eric Shuman, and Martijn van Zomeren. (2024). "The Radical Flank: Curse or Blessing of a Social Movement?" *Global Environmental Psychology* 2 https://doi.org/10.5964/gep.11121.

Dauphin, Sandy. (2022, November 8). "L'empreinte carbone du Mondial au Qatar serait près de deux fois supérieure à celle annoncée par la FIFA." *FranceInter* https://www.radiofrance.fr/franceinter/l-empreinte-carbone-du-mondial-au-qatar-serait-pres-de-deux-fois-superieure-a-celle-annoncee-par-la-fifa-8733673.

Davies, Gareth Hugh. (2019, February 5). "Which Football Club Tops the Green League?" http://www.garethhuwdavies.com/environment/environment-blog/which-football-club-top-the-green-league/.

Davis, Juliet and Penny Bernstock. (2023). "From Inclusive Legacy Promises to Exclusive Realities: Planning, Design, and Displacement in Post-Olympic East London." *Urbanizing Suburbia Hyper-Gentrification, the Financialization of Housing, and the Remaking of the Outer European City*. Ed. Tahl Kaminer, Leonard Ma, and Helen Runting. Berlin: jovis Verlag. 212–34.

Davis, Juliet. (2016). "The Making and Remaking of Hackney Wick, 1870–2014: From Urban Edgeland to Olympic Fringe." *Planning Perspectives* 31, no. 3: 425–57.

Davis, Mike. (2006). *Planet of Slums*. London: Verso.

Davis, Rob, Agnel Philip, Alex Mierjeski, and Matthew Kish with Ryanne Mena and Jeff Frankl. (2024, August 13). "After Nike Leaders Promised Climate Action, Their Corporate Jets Kept Flying—and Polluting." *ProPublica* https://www.propublica.org/article/nike-corporate-jet-travel-carbon-emissions.

Davis, Tricia Holly. (2009, July 26). "A Mixed Record So Far on Environmental Issues." *Financial Times* https://www.ft.com/content/31503024-77eb-11de-9713-00144feabdc0.

Davis, Whitney. (2017). "Visuality and Vision: Questions for a Post-Culturalist Art History." *Estetika: The Central European Journal of Aesthetics* 54, no. 2: 238–57.

Day, Thomas, Silke Mooldijk, Frederic Hans, Sybrig Smit, Eduardo Posada, Reena Skribbe, Santiago Woollands, Harry Fearnehough, Takeshi Kuramochi, Carsten Warnecke, Aki Kachi, and Niklas Höhne with Carbon Market Watch. (2023, February). *Corporate Climate Responsibility Monitor 2023: Assessing the Transparency and Integrity of Companies' Emission Reduction and Net-Zero Targets*. Carbon Market Watch and the New Climate Institute. https://carbonmarketwatch.org/wp-content/uploads/2023/02/CorporateClimateResponsibilityMonitor2023.pdf.

Dean, Hartley. (2001). "Green Citizenship." *Social Policy & Administration* 35, no. 5: 490–505.

de Anda Alanis, Enrique. (2021). "The 19th Olympic Games Celebrated in Mexico in 1968 and Their Architecture." *Das moderne Erbe der Olympischen Spiele: Historische Sportstätten zwischen Konservierung und Konversion/The Modern Heritage of the Olympic Games: Historic Sports Sites between Conservation and Conversion*. Ed. Sigrid Brandt, Jörg Haspel, Ralph Paschke, and John Ziesemer. Berlin: ICOMOS. 66–72.

Death, Carl. (2011). "Greening the 2010 FIFA World Cup: Environmental Sustainability and the Mega-Event in South Africa." *Journal of Environmental Policy & Planning* 13, no. 2: 99–117.

d'Eaubonne, Françoise. (1978). *Écologie/féminisme: Révolution ou mutation?* Paris: Éditions ATP.

Debord, Guy. (2004). *The Society of the Spectacle*. Trans. Ken Knabb. London: Rebel Press.

de Coninck, Heleen. (2021, November 18). "Heleen de Coninck on the Value of Activists to Spur Policymakers." *Economist* https://www.economist.com/by-invitation/2021/11/18/heleen-de-coninck-on-the-value-of-activists-to-spur-policymakers.

de Coubertin, Pierre. (1913). *Essais de psychologie sportive* https://fr.wikisource.org/wiki/Essais_de_psychologie_sportive/Texte_entier.

De Griff, Jan. (2024, November 13). "Spy Scandal and Slowing Sales: Challenges Mount for LVMH." *Atelier de Griff* https://atelierdegriff.com/2024/11/13/spy-scandal-and-slowing-sales-challenges-mount-for-lvmh/.

de la Cruz, José Luis. (2022, December 20). "Mundial Catar 2022: la insostenibilidad ambiental de los macro eventos deportivos." *Público* https://blogs.publico.es/otrasmiradas/67095/mundial-catar-2022-la-insostenibilidad-ambiental-de-los-macro-eventos-deportivos/.

Delaney, Tim. (2015). "The Functionalist Perspective on Sport." *Routledge Handbook of the Sociology of Sport.* Ed. Richard Giulianotti. London: Routledge. 18–28.

Delaplace, Marie and Martin Müller. (2024, June 18). "Can We Really Measure the Ecological Footprint of the Olympics?" *Polytechnique Insights: A Review by the Institut Polytechnique de Paris* https://www.polytechnique-insights.com/en/columns/planet/can-we-really-measure-the-ecological-footprint-of-the-olympics/.

Delmas, Aurélie. (2017, September 14). "Merci de l'avoir posée—Paris 2024: Jeux verts ou greenwashing?" *Libération* https://www.liberation.fr/france/2017/09/14/paris-2024-jeux-verts-ou-greenwashing1596104/.

Delmas, Magali A. and Vanessa Cuerel Burbano. (2011). "The Drivers of Greenwashing." *California Management Review* 54, no. 1: 64–87.

De Loera-Brust, Antonio. (2024, August 11). "The World Cup is Coming to America—And Heading for Disaster." *Foreign Policy* https://foreignpolicy.com/2024/08/11/world-cup-united-states-2026-germany-qatar-russia-olympics/.

Deloitte. (2025, January 23). *Deloitte Football Money League 2025* https://www.deloitte.com/uk/en/services/financial-advisory/analysis/deloitte-football-money-league.html?utm_source=substack&utm_medium=email.

del Romero Renau, Luis and Catherine Trudelle. (2011). "Mega Events and Urban Conflicts in Valencia, Spain: Contesting the New Urban Modernity." *Urban Studies Research* 1 https://doi.org/10.1155/2011/587523.

de Menezes, Jack. (2015, November 28). "Arsenal Criticized for Taking 'Ridiculous, Ludicrous and Farcical' 14-Minute Plane Flight to Norwich." *Independent* https://www.independent.co.uk/sport/football/premier-league/arsenal-criticised-for-taking-ridiculous-14-minute-plane-flight-to-norwich-a6752486.html.

de Moor, Joost, Michiel De Vydt, Katrin Uba, and Mattias Wahlström. (2021). "New Kids on the Block: Taking Stock of the Recent Cycle of Climate Activism." *Social Movement Studies* 20, no. 5: 619–25.

Demosthenes. (2015). *The Complete Works of Demosthenes.* Hastings: Delphi Classics.

Dengler, Corinna. (2021). "Degrowth." *The Routledge Handbook of Feminist Economics*. Ed. Günseli Berik and Ebru Kongar. Abingdon: Routledge. 369–77.

Department for Culture, Media and Sport and Department for Digital, Culture, Media & Sport. (2008). *Our Promise for 2012: How the UK Will Benefit from the Olympic Games and Paralympic Games* https://www.gov.uk/government/publications/our-promise-for-2012-how-the-uk-will-benefit-from-the-olympic-games-and-paralympic-games.

Depledge, Joanna, Kari De Pryck, and J. Timmons Roberts. (2023, November). *Decades of Systematic Obstructionism: Saudi Arabia's Role in Slowing Progress in UN Climate Negotiations*. Climate Social Science Network https://cssn.org/wp-content/uploads/2023/11/Decades-of-Systematic-Obstructionism-CSSN-Issue-Paper3.pdf.

de Senneville, Valérie. (2024, November 28). "Procès Squarcini : François Ruffin « est bien meilleur cinéaste que politique », estime Bernard Arnault." *Les Echos* https://www.lesechos.fr/politique-societe/societe/proces-squarcini-francois-ruffin-est-bien-meilleur-cineaste-que-politique-estime-bernard-arnault-2134679.

de Villepin, Paul with Pauline de Saint Remy. (2023, August 25). "World's Richest Man lobbied Macron to Ditch Prime Minister." *Politico* https://www.politico.eu/article/emmanuel-macron-bernard-arnault-elisabeth-borne-gerald-darmanin-meet-the-worlds-richest-man-favorite-politician/.

Dewey, John. (1958). *Experience and Nature*. New York: Dover.

Dewhirst, Timothy, Wonkyong Beth Lee, and Lauren Czaplicki. (2023). "Philip Morris International's Formula 1 Sponsorship-Linked Marketing: Transformation from Marlboro to Mission Winnow." *Nicotine and Tobacco Research* 25: 1838–846.

Dewhirst, Timothy and A. Hunter. (2002). "Tobacco Sponsorship of Formula One and CART Auto Racing: Tobacco Brand Exposure and Enhanced Symbolic Imagery Through Co-Sponsors' Third Party Advertising." *Tobacco Control* 11: 146–50.

Dewhirst, Tim. (2023). "The Geopolitics of Money Versus Morals: Location, Location, Location of the Formula 1 Race Calendar." *The Geopolitical Economy of Sport: Power, Politics, Money, and the State*. Ed. Simon Chadwick, Paul Widdop, and Michael M. Goldman. Abingdon: Routledge. 235–41.

Diaz, Silvia. (2022, June 15). "The Art of Manipulation: How Oil Giants are Greenwashing with Cultural Sponsorship." *Greenpeace* https://www.greenpeace.org/international/story/54285/the-art-of-manipulation-how-oil-giants-are-greenwashing-with-cultural-sponsorship/.

Díaz Ordaz, Gustavo. (1968, September 1). *Informe*. Legislatura XLVII Año II Período Ordinario Fecha 19680901 Número de Diario 3. https://cronica.diputados.gob.mx/DDebates/47/2do/Ord/19680901-I.html.

Díaz-Pont, Joana. (2023). "Visibilising the Climate in Social Policies in Barcelona: Connections in the Urban Context." *Critical Social Policy* 43, no. 1: 119–39.

Dijkstra, Constance and Valentin Simon. (2023). *The Return of the Cruise: How Luxury Cruises Are Polluting Europe's Cities.* Transport & Environment https://www.transportenvironment.org/uploads/files/2023-Cruise-ship-study_2024-05-07-143832_lapq.pdf.

Ding, Jaimie and Associated Press. (2024, November 1). "Coca-Cola and Pepsi, Two of the World's Top Plastic Polluters, Just Got Slapped with a Lawsuit Alleging Exaggerated Recycling Claims and Downplayed Health Effects." *Fortune* https://fortune.com/2024/11/01/coca-cola-pepsi-plastic-pollution-lawsuit-los-angeles-county/.

Dingle, Gregory. (2009). "Sustaining the Race: A Review of Literature Pertaining to the Environmental Sustainability of Motorsport." *International Journal of Sports Marketing and Sponsorship* 11, no. 1: 75–91.

Dmowski, Seweryn. (2015). "Football Sites of Memory in the Eastern Bloc 1945–1991." *European Football and Collective Memory.* Ed. Wolfram Pyta and Nils Havermann. Basingstoke: Palgrave Macmillan. 171–84.

Dobbs, Rachel. (2024, November 18). "Have Global Emissions Peaked?" *Economist* https://www.economist.com/the-world-ahead/2024/11/18/have-global-emissions-peaked.

Doe v. Cisco Systems, Inc. (2024). Human Rights Law Foundation https://hrlf.net/doe-vs-cisco-systems/.

Domingo Rodríguez, Diego. (2025, January 9). "El Gran Premio de Fórmula 1 en Madrid: entusiasmo, críticas y debate sobre su impacto." *El Diario de Madrid* https://www.eldiariodemadrid.es/articulo/sociedad/gran-premio-formula-1-madrid-entusiasmo-criticas-debate-impacto/20250109162501087476.html.

Dos Passos, John. (1937). *U.S.A.: I. The 42nd Parallel. II. Nineteen Nineteen. III. The Big Money.* New York: Modern Library.

Dosumu, Adekunle, Ian Colbeck, and Rachel Elizabeth Bragg. (2017). "Greenhouse Gas Emissions as a Result of Spectators Travelling to Football in England." *Nature* 7: 6986.

Douglas, Susan J. (1987). *Inventing American Broadcasting 1899–1922.* Baltimore: The Johns Hopkins University Press.

Doustaly, Cécile and Geneviève Zembri-Mary. (2024). "Is Urban Planning Returning to the Past in Search of a Sustainable Future? Exploring the Six Paris and London Olympic Games." *Planning Perspectives* 39, no. 3: 675–700.

Dowling, Matthew and Spencer Harris. (2022, July 26). "London 2012 Olympics: How it Boosted Medal Winning but Failed to Inspire a Generation." *The Conversation* https://theconversation.com/london-2012-olympics-how-it-boosted-medal-winning-but-failed-to-inspire-a-generation-187383.

Dowling, Matthew. (2025). "London 2012: Did it Inspire a Generation?" *Events and Society: Bridging Theory and Practice*. Ed. Mike Duignan. Abingdon: Routledge. 64–72.

Doyle, Julie. (2011). *Mediating Climate Change*. Abingdon: Routledge.

Du Bois, W. E. B. (2022). *W. E. B. Du Bois: International Thought*. Ed. Adom Getachew and Jennifer Pitts. Cambridge: Cambridge University Press.

Dunne, Daisy and Patrick Greenfield. (2024, October 15). *COP16: More Than 85% of Countries Miss UN Deadline to Submit Nature Pledges* https://www.carbonbrief.org/cop16-countries-miss-un-deadline-to-submit-nature-pledges/.

Durkheim, Émile. (1961). "The Solidarity of Occupational Groups." *Theories of Society: Foundations of Modern Sociological Theory. Volume I*. Ed. Talcott Parsons, Edward Shils, Kaspar D. Naegle, and Jesse R. Pitts. New York: Free Press of Glencoe. 356–63.

Duval, Antoine and Ben Van Rompuy, eds. (2016). *The Legacy of Bosman: Revisiting the Relationship Between EU Law and Sport*. The Hague: Asser/Cham: Springer.

Eagleton, Terry. (1990). *The Ideology of the Aesthetic*. Oxford: Blackwell.

Eagleton, Terry. (2007). *The Meaning of Life: A Very Short Introduction*. Oxford: Oxford University Press.

Eagleton, Terry. (2010, June 15). "Football: A Dear Friend to Capitalism." *Guardian* https://www.theguardian.com/commentisfree/2010/jun/15/football-socialism-crack-cocaine-people.

Eagleton, Terry. (2024a, April 25). "Where Does Culture Come From?" *London Review of Books* https://www.lrb.co.uk/the-paper/v46/n08/terry-eagleton/where-does-culture-come-from.

Eagleton, Terry. (2024b, October 10). "The Excitement of the Stuff." *London Review of Books* https://www.lrb.co.uk/the-paper/v46/n19/terry-eagleton/the-excitement-of-the-stuff.

Earls, Maya and EE News. (2019, July 2). "As the World Heats Up, Soccer Must Adapt." *Scientific American* https://www.scientificamerican.com/article/as-the-world-heats-up-soccer-must-adapt/.

Echikson, William. (1998, February 9). "The Olympics: Governance: Making the Games Run on Time." *Business Week*: 66–68.

Eckerman, Ingrid. (2005). *The Bhopal Saga: Causes and Consequences of the World's Largest Industrial Disaster*. Hyderabad: Universities Press.

Eco, Umberto. (1981). *The Role of the Reader: Explorations in the Semiotics of Texts*. London: Hutchinson.

Eco, Umberto. (1987). *Travels in Hyperreality*. Trans. William Weaver. London: Picador.

Eden, Sally. (2004). "Greenpeace." *New Political Economy* 9, no. 4: 595–610.

Eichberg, Henning. (1986). "The Enclosure of the Body—On the Historical Relativity of 'Health', 'Nature' and the Environment of Sport." *Journal of Contemporary History* 21, no. 1: 99–121.

Eichberg, Henning. (2018). "Play Against Alienation?" *The Philosophy of Play as Life*. Ed. Wendy Russell, Emily Ryall, and Malcolm MacLean. Abingdon: Routledge. 211–26.

Eisenstein, Sergei. (1987). *Nonindifferent Nature*. Trans. Herbert Marshall and Roberta Reader. New York: Cambridge University Press.

Elder, Bryce. (2025, April 8). "Yacht Envy: A Visit to England's Half-a-Billion Boatyard." *Financial Times* https://www.ft.com/content/aba8786e-7d9c-4270-9e56-61945995e62b?emailId=bb331970-754c-44a4-9404-832459231fda&segmentId=9264b0f7-e7ac-8f9b-044f-c10729049333.

Elías, Alex. (2021). *Mexico City's Olympic Games: Citizenship and Nation Building, 1963–1968*. Cham: Palgrave.

Elias, Norbert and Eric Dunning. (1986). *Quest for Excitement: Sport and Leisure in the Civilizing Process*. Oxford: Basil Blackwell.

Elias, Norbert. (1978). "On Transformations of Aggressiveness." *Theory and Society* 5, no. 2: 229–42.

Elliott, Larry. (2014, March 23). "How F1 and Champagne Might Help Us to Solve Global Warming." *Guardian* http://www.theguardian.com/business/2014/mar/23/solve-global-warming-pension-champagne-formula-1.

Ellis v. Nike U.S., Inc. (2024, March 28). 4:23-cv-00632-MTS.

Ellul, Jacques. (1964). *The Technological Society*. Trans. John Wilkinson. New York: Vintage Books.

Elsborg, Stanis and Karim Zidan with Nathalie Zahle and Lasse Kjær Nielsen. (2024). *Saudi Arabia's Grip on World Sport*. Play the Game https://www.playthegame.org/media/rxflxlpt/saudi-arabias-grip-on-world-sport.pdf.

Elsborg, Stanis. (2023). "The Power Players Behind Saudi Arabia's Sports Strategy." *Play the Game* https://www.playthegame.org/news/the-power-players-behind-saudi-arabias-sports-strategy/.

Emerson, Ralph Waldo. (2000). *The Essential Writings of Ralph Waldo Emerson*. Ed. Brooks Atkinson. New York: The Modern Library.

Engar, Aadil and Jacques du Toit. (2023). "Cape Town's 2010 FIFA World Cup Stadium Location and its Spatial and Environmental Justice Implications." *Sport Stadiums and Environmental Justice*. Ed. Timothy Kellison. London: Routledge. 158–71.

Engels, Friedrich. (2010). *The Origin of the Family, Private Property and the State*. London: Penguin.

England, Phil. (2012, April 25). "#London2012: An Olympian Exercise in Corporate Greenwashing." *Ceasefire* https://ceasefiremagazine.co.uk/london2012-olympian-exercise-corporate-greenwashing/.

Environmental Defenders Office. (2023, March 2). "Complaint About Misleading Advertising by Toyota Motor Corporation Australia Ltd." https://climatecasechart.com/wp-content/uploads/non-us-case-documents/2023/20230302_19330_complaint.pdf.

EPRI. (2024, May 28). *Powering Intelligence: Analyzing Artificial Intelligence and Data Center Energy Consumption* https://www.epri.com/research/products/000000003002028905.

Erhart, Itir. (2014). "United in Protest: From 'Living and Dying with Our Colours' to 'Let All the Colours of the World Unite'." *International Journal of the History of Sport* 31, no. 14: 1724–38.

Escobar, Arturo. (2020). *Pluriversal Politics*. Trans. David Frye. Durham: Duke University Press.

Estirado, Laura. (2024, August 29). "Felipe VI inaugural la Copa América 2024: "Barcelona revive el espíritu de los JJOO del 92"." *El Periódico* https://www.elperiodico.com/es/estilo-y-famosos/20240829/rey-felipe-bienvenida-copa-america-vela-museu-maritim-salvador-illa-jaume-collboni-107491405.

Estrada, Daniela. (2010, June 3). "World Cup 2010: Climate Change Fouls and Goals." *Guardian* www.guardian.co.uk/environment/2010/jun/03/climate-change-world-cup.

European Banking Authority, European Insurance and Occupational Pensions Authority, and European Securities and Markets Authority. (2022). *ESAs Call for Evidence on Better Understanding Greenwashing* https://www.esma.europa.eu/sites/default/files/library/esas_call_for_evidence_on_greenwashing.pdf

European Football for Development Network. (2023). *Research Report for Climate Justice* https://footballforclimatejustice.eu/resource/research-reportfootball-forclimate-justice/.

European Parliament and Council of the European Union. (2022). *Corporate Sustainability Reporting Directive (EU) 2022/2464 of the European Parliament and of the Council* https://eur-lex.europa.eu/legal-content/EN/TXT/?uri=CELEX:32022L2464.

European Parliamentary Research Service. (2024, October). *'Green Claims' Directive: Protecting Consumers from Greenwashing* https://www.europarl.europa.eu/RegData/etudes/BRIE/2023/753958/EPRS_BRI(2023)753958_EN.pdf.

European Securities and Markets Authority. (2024, June 4). *Final Report on Greenwashing* https://www.esma.europa.eu/sites/default/files/2024-06/ESMA36-287652198-2699_Final_Report_on_Greenwashing.pdf.

Evans, Gareth. (2018, November 18). "Olympic Games: Why Cities Are Snubbing the 'Greatest Show on Earth'." *BBC News* https://www.bbc.co.uk/news/world-46236682.

Fair, Leslie. (2022, April 8). "$5.5 Million Total FTC Settlements with Kohl's and Walmart Challenge "Bamboo" and Eco Claims, Shed Light on Penalty Offense Enforcement." Federal Trade Commission https://www.ftc.gov/business-guidance/blog/2022/04/55-million-total-ftc-settlements-kohls-and-walmart-challenge-bamboo-and-eco-claims-shed-light.

Fairley, Sheranne, B. David Tyler, Pamm Kellett, and Kari D'Elia. (2011). "The Formula One Australian Grand Prix: Exploring the Triple Bottom Line." *Sport Management Review* 14, no. 2: 141–52.

FairSquare. (2024, October). *Substitute: The Case for the External Reform of FIFA.* https://fairsq.org/wp-content/uploads/2024/10/FIFA_Substitute_Report_v2_Single-Pages.pdf.

FairSquare. (2025a, May 14). *Underlying Causes: Unexplained Migrant Worker Deaths in Saudi Arabia* https://fairsq.org/wp-content/uploads/2025/05/Fairsquare_Underlying-Issues-Report-May-2025.pdf.

FairSquare. (2025b, May 27). "Joint Statement on FIFA Misgovernance" https://fairsq.org/wp-content/uploads/2025/05/FIFA-Reforms-Statement-27-May-2025.pdf.

Falkenberg, Max, Alessandro Galeazzi, Maddalena Torricelli, Niccolò Di Marco, Francesca Larosa, Madalina Sas, Amin Mekacher, Warren Pearce, Fabiana Zollo, Walter Quattrociocchi, and Andrea Baroncheli. (2022). "Growing Polarization Around Climate Change on Social Media." *Nature Climate Change* 12: 1114–121.

Fallaci, Oriana. (1985). *Nada y así sea.* Barcelona: Editorial Noguer.

Fallaci, Oriana. (2020). *La vie, la guerre et puis rien.* Trans. Jacqueline Remillet. Paris: Les Belles Lettres.

Fals-Borda, Orlando. (1988). *Knowledge and People's Power: Lessons with Peasants in Nicaragua, Mexico and Colombia.* New Delhi: Indian Social Institute; New York: New Horizons Press.

Fanon, Frantz. (2007). *Los condenados de la tierra.* Trans. Julieta Campos. Rosario: elortiba.

Fanon, Frantz. (2018). *Alienation and Freedom.* Trans. Steven Corcoran. Ed. Jean Khalfa and Robert J. C. Young. London: Bloomsbury.

Farrell, Justin. (2019). "The Growth of Climate Change Misinformation in US philanthropy: Evidence from Natural Language Processing." *Environmental Research Letters* 14, no. 3: 1–10.

Faye, Olivier. (2023, September 28). "Les interrogations de Tracfin sur une transaction entre des sociétés de l'oligarque russe Nikolaï Sarkisov et de Bernard Arnault." *Le Monde* https://www.lemonde.fr/societe/article/2023/09/28/les-interrogations-de-tracfin-sur-une-transaction-entre-des-societes-de-l-oligarque-russe-nikolai-sarkisov-et-de-bernard-arnault_6191354_3224.html.

Fédération Internationale des Associations de Footballeurs Professionnels. (2024). *Excessive Workload Demands: Player Performance, Recovery and Health* https://fifpro.org/media/ad4lyibk/en_pwm_men-s-annual-report-2024.pdf.

Fédération Internationale de l'Automobile. (n.d.). *FIA Environmental Strategy 2020–2030* https://www.fia.com/sites/default/files/fiaenvironmentalstrategyv4web.pdf.

Fédération Internationale de l'Automobile. (2023). "FIA Purpose Driven." https://purposedriven.fia.com/page/frontpage.

Fédération internationale de football association (FIFA) contre BZ, en présence de : Union royale belge des sociétés de football association ASBL (URBSFA), Sporting du Pays de Charleroi SA, Fédération internationale des footballeurs professionnels, Fédération internationale des footballeurs professionnels – Division Europe, Union nationale des footballeurs professionnels (UNFP). (2024, October 4). C-650/22 ECLI:EU:C:2024:824.

Federación Regional de Asociaciones Vecinales de Madrid. (2025, January 8). "Alegaciones de la FRAVM a la solicitud de licencia del Gran Premio de Fórmula 1 de Madrid." https://aavvmadrid.org/?s=gran+premio&et_pb_searchform_submit=et_search_proccess&et_pb_include_posts=yes&et_pb_include_pages=yes.

Feinberg, Matthew, Robb Willer, and Chloe Kovacheff. (2020). "The Activist's Dilemma: Extreme Protest Actions Reduce Popular Support for Social Movements." *Journal of Personality and Social Psychology: Interpersonal Relations and Group Processes* 119, no. 5: 1086–111.

Ferrando, Ramón. (2024, October 24). "Samaranch: "Es una buena idea invertir en la Copa del América en València"." *El Periódico* https://www.elperiodico.com/es/blau/copa-america/20241024/samaranch-buena-idea-invertir-copa-america-vela-110311523.

Ferré, Jean-Luc. (2020, January 16). "Les contraintes climatiques obligent le monde du sport à se repenser." *La Croix* https://www.la-croix.com/Sport/contraintes-climatiques-obligent-monde-sport-repenser-2020-01-16-1201072242.

Ferrisi, Michaël. (2020, July 4). "La Formule 1 peut-elle vraiment devenir plus écologique?" *Ecolosport* https://ecolosport.fr/blog/2020/07/04/la-formule-1-peut-elle-vraiment-devenir-plus-ecologique/.

Fett, Matthias. (2020). "The Game Has Changed—A Systematic Approach to Classify FIFA World Cups." *International Journal of Sport Policy and Politics* 12, no. 3: 455–70.

Fians, Guilherme. (2022, March 18). "Prefigurative Politics." *Open Encyclopedia of Anthropology*. Ed. Felix Stein https://www.anthroencyclopedia.com/entry/prefigurative-politics.

FIFA Sub-Committee on Human Rights & Social Responsibility. (2023, December 22). *Report on Remedy and Legacy for Workers in the Context of the FIFA World*

Cup Qatar 2022 https://digitalhub.fifa.com/m/774a7297f704cebb/original/FIFA-Sub-Committee-on-Human-Rights-Social-Responsibility-Report.pdf.

FIFA. (n.d.). *Football Unites the World: Strategic Objectives for the Global Game: 2023–2027* https://digitalhub.fifa.com/m/38577b5ce4c3cc23/original/FIFA-Strategic-Objectives-for-the-Global-Game_2023_2027_EN.pdf.

FIFA. (2019). *2018 FIFA World Cup Russia™ Sustainability Report* https://img.fifa.com/image/upload/ya7pgcyslxpzlqmjkykg.pdf.

FIFA. (2024). *FIFA World Cup 2034™ Bid Evaluation Report* https://digitalhub.fifa.com/m/51d8a13714827f1d/original/FIFA-World-Cup-2034-Bid-Evaluation-Report.pdf.

FIFA. (2025). "Partners." *Inside FIFA* https://inside.fifa.com/tournament-organisation/partners.

Fifka, Matthias and Johannes Jaeger. (2020). "CSR in Professional European Football: An Integrative Framework." *Soccer & Society* 21, no. 1: 61–78.

FilmLA. (2019a). *2018 Television Report* https://filmla.com/wp-content/uploads/2019/02/2018_TV_Report_web.pdf.

FilmLA. (2019b). *2019 Television Report* https://filmla.com/wp-content/uploads/2019/10/2019_TV_Report-WEB.pdf.

Financial Conduct Authority. (2024, April 25). "FCA Confirms Anti-Greenwashing Guidance and Proposes Extending Sustainability Framework." https://www.fca.org.uk/news/press-releases/fca-confirms-anti-greenwashing-guidance-and-proposes-extending-sustainability-framework.

Finnerty, Samuel, Jared Piazza, and Mark Levine. (2024). "Scientists' Identities Shape Engagement with Environmental Activism." *Communications Earth & Environment* 5. https://doi.org/10.1038/s43247-024-01412-9.

Fisher, Daniel. (2005, August 15). "GE Turns Green." *Forbes*: 80–85.

Fisk, Milton. (1980). "Materialism and Dialectic." *Critique: Journal of Socialist Theory* 12, no. 1: 97–116.

Fisné-Koch, Audrey. (2024, August 9). "Les Jeux olympiques de Paris ont-ils été exemplaires sur le plan social?" *Alternatives Economiques* https://www.alternatives-economiques.fr/jeux-olympiques-de-paris-ont-ete-exemplaires-plan-social/00112042?utmsource=emailing&utmmedium=email&utmcontent=09082024&utmcampaign=hebdo.

Fitzgerald, F. Scott. (1925). *The Great Gatsby*. New York: Scribner's.

Flaming, Daniel, Gary L. Blasi, Anthony W. Orlando, and Sam Tsemberis. (2024, September 4). *Excelling for the 2028 Olympics: Restoring, Not Displacing, LA's Unsheltered Residents*. Economic Roundtable https://economicrt.org/publication/excelling-for-the-2028-olympics/.

Flynn, Henry. (2022a, March 31). "Spain's Socialist Team Rayo Vallecano is Ready to Make a Heated Return to La Liga." *Forbes* https://www.forbes.com/

sites/henryflynn/2021/07/16/spains-socialist-team-rayo-vallecano-awaits-a-heated-top-flight-return/.
Flynn, Henry. (2022b, November 28). "Qatar Has Spent Well Over $200 Billion On A Flawed World Cup." *Forbes* https://www.forbes.com/sites/henryflynn/2022/11/27/qatar-has-spent-well-over-200-billion-on-a-flawed-world-cup/.
Flynn, Jack. (2023, November 14). "20+ Shocking Greenwashing Statistics [2023]." *Zippia* https://www.zippia.com/advice/greenwashing-statistics/.
Földesi, Gyöngyi Szabó. (2014). "The Impact of the Global Economic Crisis on Sport." *Physical Culture and Sport* 63, no. 1: 22–30.
Food and Water Europe. (2017, November). *INEOS' Chequered Environmental Track Record in Europe* https://www.foodandwatereurope.org/wp-content/uploads/2017/11/FoodandWaterEuropeReportIneosChequeredRecordinEuropeNov2017.pdf.
Football for Friendship. (n.d.). https://footballforfriendship.com/.
Football for Future. (n.d.). https://footballforfuture.org/.
Football Supporters Association. (n.d.). https://thefsa.org.
Forbes, Don Nico. (2025, January 24). "Soccer Tackles Its Carbon Problem." *Wall Street Journal* https://www.wsj.com/articles/soccer-tackles-its-carbon-problem-efff5bcb?utm_source=substack&utm_medium=email.
Forero-Ortiz, Edwar, Eduardo Martínez-Gomariz, and Robert Monjo. (2020). "Climate Change Implications for Water Availability: A Case Study of Barcelona City." *Sustainability* 12, no. 5 https://doi.org/10.3390/su12051779.
Forest Green. (n.d.). "The World's First Vegan Football Club." https://www.fgr.co.uk/our-ethos/100-vegan.
Formula 1. (2019). "Formula 1 Sustainability Strategy." https://corp.formula1.com/sustainability/.
Formula 1. (2022). "More Efficient, Less Fuel, and Carbon Net Zero—7 Things You Need to Know about the 2026 F1 Engine Regulations." https://www.formula1.com/en/latest/article.more-efficient-less-fuel-and-carbon-net-zero-7-things-you-need-to-know-about.ZhtzvU3cPCv8QO7jtFxQR.html.
Forrest, Christopher B., Lauren J. Koenigsberg, Francis Eddy Harvey, Mitchell G. Maltenfort, and Neal Halfon. (2025). "Trends in US Children's Mortality, Chronic Conditions, Obesity, Functional Status, and Symptoms." *Journal of the American Medical Association* https://doi.org/10.1001/jama.2025.9855.
Fossil Free Football, Game Changer Sponsorship Pledge, and Badvertising. (2024a, June). *EURO2024: Green Initiatives Overshadowed by Promotion of Big Polluters* https://www.fossilfreefootball.org/wp-content/uploads/2024/06/EURO24-Sustainability-Briefing.pdf.
Fossil Free Football, Game Changer Sponsorship Pledge, and Badvertising. (2024b, October 7). *UEFA's €500m Qatar Airways Deal Pollutes Like a Coal-Fired Power Plant* https://www.fossilfreefootball.org/2024/10/07/

publication-uefas-e500m-qatar-airways-deal-pollutes-like-a-coal-fired-power-plant/.

Foster, Aurelia. (2022, July 20). "London Olympics: Residents 'Betrayed' Over Housing Promise." *BBC* https://www.bbc.co.uk/news/uk-england-london-62138346.

Foster, John Bellamy. (1999). "Marx's Theory of Metabolic Rift: Classical Foundations for Environmental Sociology." *American Journal of Sociology* 105, no. 2: 366–405.

Foth, Marcus. (2025, March 30). "Brisbane 2032 is no Longer Legally Bound to be 'Climate Positive'. Will it Still Leave a Green Legacy?" *The Conversation* https://theconversation.com/brisbane-2032-is-no-longer-legally-bound-to-be-climate-positive-will-it-still-leave-a-green-legacy-246672.

Foucault, Michel. (1986). *The Use of Pleasure: The History of Sexuality Volume Two.* Trans. Robert Hurley. New York: Vintage.

Foucault, Michel. (1988). *The Care of the Self: Volume 3 of The History of Sexuality.* Trans. Robert Hurley. New York: Vintage.

Foucault, Michel. (2008). *The Birth of Biopolitics: Lectures at the Collège de France, 1978–79.* Trans. Graham Burchell. Ed. Michel Senellart. Houndmills: Palgrave Macmillan.

Foucault, Michel. (2021). *Confessions of the Flesh: The History of Sexuality Volume 4.* Ed. Frédéric Gros. Trans. Robert Hurley. New York: Pantheon.

Foucault, Michel. (2024). *The Japan Lectures: A Transnational Critical Encounter.* London: Routledge.

Fournier, Patricia and Jorge Martínez Herrera. (2009). ""Mexico, 1968": Among Olympic Fanfares, Government Repression and Genocide." *Memories from Darkness: Archaeology of Repression and Resistance in Latin America.* Ed. Pedro Funari, Andres Zarankin, and Melissa Salerno. New York: Springer. 145–74.

Fox, George. (1654). *The Vials of the Wrath of God Poured Forth Upon the Seat of the Man of Sin, and Upon All Professors of the World, Who Denieth the Light of Christ Which He Hath Enlightned Every One Withal, and Walk Contrary to it, with it They are Condemned : And a Vvarning from the Lord to All who are Walking Headlong to Destruction in the Lusts of the Flesh, and Deceits of the World, That They May Repent and Turn to the Lord, Lest the Overflowing Scourge Sweep Them All Into the Pit.* Early English Books Online 2 https://quod.lib.umich.edu/e/eebo2/A40314.0001.001?rgn=main;view=fulltext.

Foxhall, Lin. (1998). "Pandora Unbound: A Feminist Critique of Foucault's *History of Sexuality.*" *Rethinking Sexuality: Foucault and Classical Antiquity.* Ed. David H. Larmour, Paul Allen Miller, and Charles Platter. Princeton: Princeton University Press. 122–38.

Foxman, Simone and Adveith Nair. (2022, November 18). "What Qatar Built for the Most Expensive World Cup Ever." *Bloomberg* https://www.bloomberg.

com/graphics/2022-what-qatar-built-for-the-world-cup/?leadSource=uverify%20wall&embedded-checkout=true.
Foxtons. (2025). https://www.foxtons.co.uk/living-in/hackney-wick.
Foye, Lauren. (2024, July 9). "Racing Towards Net Zero: How F1 is Going Green in Response to Criticism Around its Impact on the Environment." *Zero Carbon Academy* https://www.zerocarbonacademy.com/posts/racing-towards-net-zero-how-f1-is-going-green-in-response-to-criticism-around-its-impact-on-the-environment.
FRA European Union Agency for Fundamental Rights. (2024, March 7). *Enforcing Consumer Rights to Combat Greenwashing* https://fra.europa.eu/en/publication/2024/enforcing-consumer-rights-combat-greenwashing?page=2#read-online.
franceinfo. (2024, March 25). *Paris 2024: 37% des Français attendent les Jeux olympiques "avec impatience", alors que 45% suivent l'actualité sportive régulièrement* https://www.francetvinfo.fr/les-jeux-olympiques/paris-2024/sondage-paris-2024-37-des-francais-attendent-les-jeux-olympiques-avec-impatience-alors-que-45-suivent-l-actualite-sportive-regulierement6444229.html.
Francis, Annes Elsa, Matthew Webb, Cheryl Desha, Sharyn Rundle-Thiele, and Savindi Caldera. (2023). "Environmental Sustainability in Stadium Design and Construction: A Systematic Literature Review." *Sustainability* 15, no. 8 https://doi.org/10.3390/su15086896.
Freedom House. (2024). *Freedom in the World 2024: Saudi Arabia* https://freedomhouse.org/country/saudi-arabia/freedom-world/2024.
Freelon, Deen, Alice Marwick, and Daniel Kreiss. (2020). "False Equivalencies: Online Activism from Left to Right" *Science* 369, no. 6508: 1197–201.
Fresh Info. (2021). *36th America's Cup Impact Evaluation: Final Report* https://www.majorevents.govt.nz/dmsdocument/15674-36th-americas-cup-impact-evaluation-report.
Friedman, Uri. (2016, August 17). "The Saddest Olympic Celebration." *The Atlantic* https://www.theatlantic.com/international/archive/2016/08/david-katoatau-olympics-kiribati/496175/.
Frieiro, Leonardo. (2022, November 15). "El fútbol, una mercancía global: Una entrevista con Pablo Alabarces." *Jacobin* https://jacobinlat.com/2022/11/el-futbol-una-mercancia-global/.
Friends of the Earth. (2024a, January). *From False Promises to Real Solutions: How to Navigate Greenwashing by Corporations and Governments* https://friendsoftheearth.eu/wp-content/uploads/2024/02/FRI-23-False-Promises_WEB-corrected.pdf.
Friends of the Earth. (2024b, November). *2024 Cruise Ship Report Card* https://foe.org/wp-content/uploads/2024/11/CruiseShipReportCard_2024_final.pdf.
Frow, John. (2019). *On Interpretive Conflict*. Chicago: University of Chicago Press.

Fruh, Kyle, Alfred Archer, and Jake Wojtowicz. (2022). “Sportswashing: Complicity and Corruption.” *Sport, Ethics and Philosophy* 17, no. 1: 101–18.

Fu, Rachel. (2024, February 14). “Revving Up Tourism: Formula One and Other Big Events Look Set to Drive Growth in the Hospitality Industry.” *The Conversation* https://theconversation.com/revving-up-tourism-formula-one-and-other-big-events-look-set-to-drive-growth-in-the-hospitality-industry-221706.

Furnas, Alexander C., Timothy M. Lapira, and Dashun Wang. (2025). “Partisan Disparities in the Use of Science in Policy.” *Science* 388, no. 6745: 362–67.

Gable, Margaux. (2024, August 8). “A Paris, le nouveau visage Olympique de la Porte de la Chapelle accusé d’être un cache-misère.” *Libération* https://www.liberation.fr/societe/ville/a-paris-le-nouveau-visage-olympique-de-la-porte-de-la-chapelle-accuse-detre-un-cache-misere-202408087TC6A3PFHJG7ZKF6IZTNS4HCME/.

GAe Engineering. (n.d). *Juventus Stadium* https://www.gae-engineering.com/project/juventus-stadium/.

Gaffney, Christopher. (2016). “An Anatomy of Resistance: The Popular Committees of the FIFA World Cup in Brazil.” *Sport, Protest and Globalisation: Stopping Play*. Ed. Jon Dart and Steven Wagg. London: Palgrave Macmillan. 335–63.

Gagnepain, Névil. (2024a, August 3). “À Saint-Denis, une mobilisation citoyenne s’organise face aux violences policières.” *Mediapart* https://www.mediapart.fr/journal/france/030824/saint-denis-une-mobilisation-citoyenne-s-organise-face-aux-violences-policieres.

Gagnepain, Névil. (2024b, August 5). “Prostitution: les femmes du bois de Vincennes contraintes de déserter.” *Mediapart* https://www.mediapart.fr/journal/france/050824/prostitution-les-femmes-du-bois-de-vincennes-contraintes-de-deserter.

Gaind, Nisha. (2016). “Most Cities Too Hot to Host 2088 Summer Olympics.” *Nature* https://doi.org/10.1038/nature.2016.20503.

Galbiati, Camilla. (2013, December 16). “St. Pauli F. C. de Hamburgo: Una hinchada que ondea la bandera pirata.” *El Libertario* http://periodicoellibertario.blogspot.se/2013/12/st-pauli-fc-de-hamburgo-una-hinchada.html.

Galeano, Eduardo. (1995). *El fútbol a sol y sombra y otros escritos*. Ciudad de México: Siglo XXI Editores.

Galeano, Eduardo. (2013). *Soccer in Sun and Shadow*, rev. version. Trans. Mark Field. New York: Nation Books.

Galeano, Eduardo. (2016, September 16). “Fútbol a sol y a sombra.” *Don Patadon* https://www.don-patadon.com/2015/02/futbol-sol-y-sombra-de-eduardo-galeano.html.

Gallan, Daniel. (2019, December 11). "Mocked by Some, but 'Green' Soccer Club Wants to Save the World." *CNN* https://edition.cnn.com/2019/12/11/football/forest-green-rovers-spt-intl/index.html.

Gamage, Prasanna, Lauren V. Fortington, and Carline F. Finch. (2020). "Epidemiology of Exertional Heat Illnesses in Organised Sports: A Systematic Review." *Journal of Science and Medicine in Sport* 23, no. 8: P701–09.

Gambling Commission. (2025, May 16). "TGP Europe Leaves GB Market Following Commission Investigation." https://www.gamblingcommission.gov.uk/news/article/tgp-europe-leaves-gb-market-following-commission-investigation.

Game Changer, Fossil Free Football, and Badvertising. (2024, February). *Promoting the Most Polluting Form of Travel? Why UEFA Should Drop Airline Sponsors* https://www.fossilfreefootball.org/2024/02/13/this-valentines-day-football-needs-to-break-up-with-airlines/.

Gammelsæter, Hallgeir and Sigmund Loland. (2023). "Code Red for Elite Sport: A Critique of Sustainability in Elite Sport and a Tentative Reform Programme." *European Sport Management Quarterly* 23, no. 1: 104–24.

Gandhi, M. K. (1943). *Conquest of Self (Being Gleanings from His Writings and Speeches.* Comp. R. K. Prabhu and U. R. Rao. Bombay [Mumbai]: Thacker & Co.

Ganguly, Bani Bandana, Shouvik Mandal, Nalok Banerjee, Nitin N. Kadam, and Rita Abbi. (2018). "Effect of Age at Exposure on Chromosome Abnormalities in MIC-Exposed Bhopal Population Detected 30 Years Post-Disaster." *Mutation Research* 809: 32–50.

García Márquez, Gabriel. (1996). *Noticia de un secuestro.* Mexico City: Editorial Diana.

Garot, Blandine. (2024, August 7). "JO Paris 2024: découvrez notre pódium du greenwashing." *novethic* https://www.novethic.fr/environnement/climat/jo-paris-2024-decouvrez-notre-podium-du-greenwashing.

Garrahan, Matthew. (2025, March 2). "Football's Man in the Middle: The Premier League CEO on Keeping the Peace." *Financial Times* https://www.ft.com/content/ade619c9-7191-4fe2-8ae9-e14992ccc4b9.

Gasparetto, Thadeu, Marina Orlova, and Anton Vernikovskiy. (2024). "Same, Same but Different: Analyzing Uncertainty of Outcome in Formula One Races." *Managing Sport and Leisure* 29, no. 4: 651–65.

Gavin, Mike. (2024, May 7). "How the Paris Olympics is Going Green to Help Protect the Environment." *NBC Boston* https://www.nbcboston.com/paris-2024-summer-olympics/paris-olympics-2024-carbon-footprint/3345708/.

Gaynor, Jennifer L. (2013). "Ages of Sail, Ocean Basins, and Southeast Asia." *Journal of World History* 24, no. 2: 309–33.

Geeraert, Arnout. (2025). "The Dark Side of Governance Networks: How Private Actors Use Discursive Strategies in Global Sport Governance. *Public Management Review* https://doi.org/10.1080/14719037.2025.2456034.

Gelius, Peter, Sven Messing, Antonina Tcymbal, Leonie Birkholz, and Karim Abu-Omar. (2024). "Physical Activity as a Victim, a Perpetrator, or Part of the Solution to the Climate Crisis?" *Journal of Physical Activity and Health* 21, no. 12: 1220–222.

Gelles, David. (2015, October 17). "Social Responsibility That Rubs Right Off." *New York Times* https://www.nytimes.com/2015/10/18/business/energy-environment/social-responsibility-that-rubs-right-off.html.

Gellner, Ernest. (1988). *Plough, Sword and Book: The Structure of Human History.* Chicago: University of Chicago Press.

Gelmini, Stefano. (2021, June 29). "We're Living in a Golden Age of Greenwash." Greenpeace https://www.greenpeace.org.uk/news/golden-age-of-greenwash/.

Genovese, Jack. (2024, October 11). "Why Do US Sports Media Rights Keep Climbing?" *SportsPro* https://www.sportspromedia.com/insights/opinions/sports-tv-rights-values-usa-nba-nfl-cfp-nhl-ampere-analysis-data/.

Germano, Sara, Anna Nicolaou, and Hannah Murphy. (2022, December 22). "Google Agrees NFL Streaming Deal as Big Tech Chases Sports Rights." *Irish Times* https://www.irishtimes.com/business/2022/12/22/google-agrees-nfl-streaming-deal-as-big-tech-chases-sports-rights/.

Germano, Sara, Josh Noble, and Adrienne Klasa. (2024, August 1). "The Official Medal Selfie: Paris Olympics Steps Up Product Placement." *Financial Times* https://www.ft.com/content/4db45132-13b9-4304-8cb3-514ea921a318?FTCamp=engage/CAPI/website/Channel_muckrack//B2B.

Gezici, Ferhan and Serra Er. (2014). "What Has Been Left After Hosting the Formula 1 Grand Prix in Istanbul?" *Cities* 41, Part A: 44–53.

Giacomini, Terran and Terisa Turner. (2015). "The 2014 People's Climate March and Flood Wall Street Civil Disobedience: Making the Transition to a Post-Fossil Capitalist, Commoning Civilization." *Capitalism Nature Socialism* 26, no. 2: 27–45.

Gibbs, Carole, Meredith L. Gore, Edmund F. McGarrell, and Louie Rivers III. (2010). "Introducing Conservation Criminology: Towards Interdisciplinary Scholarship on Environmental Crimes and Risks." *British Journal of Criminology* 50, no. 1: 124–44.

Gibson-Graham, J. K. (2006a). *The End of Capitalism (As We Knew It): A Feminist Critique of Political Economy.* Minneapolis: University of Minnesota Press.

Gibson-Graham, J. K. (2006b). *A Postcapitalist Politics.* Minneapolis: University of Minnesota Press.

Gil, Òscar. (2024, September 9). "Convocan una manifestación contra la Copa América en Barcelona." *El Español* https://metropoliabierta.elespanol.com/el-pulso-de-la-ciudad/20240909/convocan-una-manifestacion-contra-la-copa-america-en-barcelona/884661590_0.html.

Gilbert, Olivier. (2023). "The Olympic Games: Social and Environmental Innovation." *Field Actions Science Reports* 25: 20–23.

Ginsky, Kaya. (2024, August 21). "LVMH Puts Mark on Olympics as Luxury Brands Embrace Sports." *CNBC* https://www.cnbc.com/2024/08/11/lvmh-puts-mark-on-olympics-as-luxury-brands-embrace-sports.html.

Giulianotti, Richard. (1999). *Football: A Sociology of the Global Game*. Cambridge: Polity.

Gladwell, Richard. (2025, May 22). "America's Cup: The Blitz of Statements Answered with Protocol Release." *Sail-World* https://www.sail-world.com/news/286349/Americas-Cup-The-Blitz-of-Statements.

Glanville, Brian. (2018, November 8). ""Infantino is as Rotten as Blatter"." *World Soccer* https://www.worldsoccer.com/world-soccer-latest/infantino-rotten-blatter-407740.

Glasson, Ben and Brett Hutchins. (2024). ""Carbon Partners" and Collaborative Greenwashing: The Sustainability Partnership Between Dow Chemical and the Olympic Games." *Journal of Sport & Social Issues* 48, no. 5: 271–89.

Glasson, Ben. (2024). "Reality Offsets: Climate Meets Capitalism at the Olympic Games." *European Journal of Cultural Studies* 28, no. 1: 137–55.

Glendenning, Barry. (2016, October 9). "The Lewis Hamilton Enigma is Quite Why He Thinks He's a Mysterious Riddle." *Guardian* https://www.theguardian.com/sport/blog/2016/oct/09/lewis-hamilton-national-treasure-paranoid-f1.

Global Solo Challenge. (2022, May 7). "Why Do People Become so Passionate About Ocean Sailing?" *Sail-World* https://www.sail-world.com/news/248757/What-is-the-passion-for-ocean-sailing.

Global Web Index. (2019a). *Sports Media Landscape* https://www.gwi.com/reports/sports-media-landscape.

Global Web Index. (2019b). *Sports Around the World* https://iccopr.com/wp-content/uploads/2019/03/Sports-Around-the-World-report.pdf.

Global Witness. (2016). *Wilful Ignorance: How Japan's Voluntary Approach is Failing to Stop the Trade in Illegal Timber* https://www.globalwitness.org/en/reports/wilful-ignorance.

GMM. (2016, June 16). "Critics Lash out at F1 Over New Heineken Sponsorship." *Autoweek* https://www.autoweek.com/racing/formula-1/a1848126/critics-lash-out-f1-over-new-major-beer-sponsor-deal/.

Godson, Matilda. (2024, April 10). "The Uncomfortable Truth About Kangaroo Leather." *Sokito* https://www.sokito.com/live-coverage/the-uncomfortable-truth-about-kangaroo-leather/.

Goffman, Erving. (1986). *Stigma: Notes on Management of Spoiled Identity*. New York: Touchstone.

Gogishvili, David and Martin Müller. (2025). "Mega-Events and the Climate: What's the Game?" *Current Issues in Sport Science* 10, no. 2 https://doi.org/10.36950/2025.2ciss074.

Goldblatt, David. (2016). *The Games: A Global History of the Olympics.* New York: WW Norton.

Goldman, Michael and Rachel A. Schurman. (2000). "Closing the "Great Divide": New Social Theory on Society and Nature." *Annual Review of Sociology* 26: 563–84.

González de Bustamente, Celeste. (2016). "*Memorias del '68*: Media, Massacre, and the Construction of Collective Memories." *Sport, Protest and Globalisation: Stopping Play.* Ed. Jon Dart and Steven Wagg. London: Palgrave Macmillan. 101–30.

González-Reverté, Francesc and Anna Soliguer Guix. (2024). "The Social Construction of Anti-Tourism Protest in Tourist Cities: A Case Study of Barcelona." *International Journal of Tourism Cities* 10, no. 3: 842–59.

Goodman, Al. (2025, March 22). "Barcelona Finally Turned on its Crowds of Tourists. Now the City Faces a Major Problem." *CNN* https://edition.cnn.com/travel/barcelona-reckons-with-overtourism-summer-2025?cid=external-feeds_iluminar_flipboard.

Goodwin, Bill. (2019, August 1). "Cisco Pays $8.6m After Whistleblower Discloses Security Flaws in Video Surveillance System." *Computer Weekly* https://www.computerweekly.com/news/252467624/Cisco-pays-83m-after-whistleblower-discloses-security-flaws-in-video-surveillance-system.

Goodwin, Jeff and James M. Jasper. (2015). "Editors' Introduction." *The Social Movements Reader: Cases and Concepts*, 3rd ed. Ed. Jeff Goodwin and James M. Jasper. Malden: Wiley Blackwell. 3–7.

Google. (2023). *Environmental Report 2023* https://www.gstatic.com/gumdrop/sustainability/google-2023-environmental-report.pdf.

Gordon, Josh. (2022, December 21). "Grand Prix Costs Victoria Half a Billion Dollars Over 10 Years." *The Age* https://www.theage.com.au/national/victoria/grand-prix-costs-victoria-half-a-billion-dollars-over-10-years-20221221-p5c831.html.

Gorky, Maxim. (1952). *Collected Works in Ten Volumes. Volume VIII: The Artamonovs.* Trans. Helen Altschuler. Moscow: Progress Publishers.

Gössling, Stefan, Andreas Humpe, and Jorge Cardoso Leitão. (2024). "Private Aviation is Making a Growing Contribution to Climate Change." *Communications Earth & Environment* 5. https://doi.org/10.1038/s43247-024-01775-z.

Gough, Christina. (2023, March 10). "Sports on U.S. TV—Statistics & Facts." *Statista* https://www.statista.com/topics/2113/sports-on-tv/#topicOverview.

Graeber, David. (2002). "The New Anarchists." *New Left Review* 13 https://newleftreview.org/II/13/david-graeber-the-new-anarchists.

Graham, Caroline. (2013, May 27). "To You, James Hunt Was an F1 Playboy Who Bedded 5000 Women. To Me, He Was Dad—Who Doted on Me ... and His 300 Budgies." *Mail on Sunday* https://www.dailymail.co.uk/femail/article-2331043/James-Hunt-F1-playboy-bedded-5-000-women-To-Tom-Hunt-Dad%2D%2Ddoted%2D%2D300-budgies.html.

Graham, Christopher J. (2025). "Alcohol Normalisation in Europe: The Role of Alcohol Sports Sponsorship." *The Lancet Regional Health—Europe* 48 doi:https://doi.org/10.1016/j.lanepe.2024.101156.

Gramsci, Antonio (1978). *Selections From the Prison Notebooks of Antonio Gramsci.* Trans. and ed. Quentin Hoare and Geoffrey Nowell-Smith. New York: International Publishers.

Gramsci, Antonio. (1994). *Pre-Prison Writings.* Ed. Richard Bellamy. Trans. Virginia Cox. Cambridge: Cambridge University Press.

Grant, Thomas L, Jr. (2014). "Green Monsters: Examining the Environmental Impact of Sports Stadiums." *Environmental Law Journal* 25, no. 1: 149–76.

Grant, Wyn. (2004). "Pressure Politics: The Changing World of Pressure Groups." *Parliamentary Affairs* 57, no. 2: 408–19.

Gray, James. (2023, October 6). "Fifa 'Prioritising Profit Over People' with 2030 World Cup Plans to Fly Teams 5,000 Miles." *The i Paper* https://inews.co.uk/sport/football/fifa-prioritising-profit-over-people-world-cup-2667248?srsltid=AfmBOoosMSupyfcOTXv3EWkaz6JV9AfLvzBoPy0fsxLogDS878sIAzUC.

Greater London Authority. (2025, April 28). *Major Sports Events—Impact for London 2024 Report* https://www.london.gov.uk/sites/default/files/2025-04/Major_Sports_Events_Impact_for_London_2024_Report-28April25_0.pdf.

Green, Molly. (2024, July 16). "Energy Demand Saw Rapid 1300MW Pick Up During Euro 2024 Football Final Halftime." *Current* +- https://www.current-news.co.uk/energy-demand-saw-rapid-1300mw-pick-up-during-euro-2024-football-final-halftime/.

Greene, Shayna. (2024, August 16). "General Electric Loses Bid for Quick Win in Contamination Suit." *Bloomberg Law* https://news.bloomberglaw.com/environment-and-energy/general-electric-loses-bid-for-quick-win-in-contamination-suit.

Greenfield, Patrick. (2024, October 3). "How the 'Frida Kahlo of Environmental Geopolitics' is Lighting a Fire Under Big Oil." *Guardian* https://www.theguardian.com/environment/2024/oct/03/colombia-minister-susana-muhamad-biodiversity-cop16-fossil-fuels.

Greenhalgh, Susan. (2024). *Soda Science: Making the World Safe for Coca-Cola.* Chicago: University of Chicago Press.

Grez, Matias. (2020, December 11). "Some Motorsport Teams Remain Addicted to Tobacco Company Sponsorship Deals, Despite Tobacco Causing 8 Million Deaths Each Year." *CNN* https://edition.cnn.com/2020/12/11/motorsport/formula-one-tobacco-sponsorship-deals-spt-itl/index.html.

Griggs, Lee. (1963, October 28). "A Very Dry Run in Tokyo." *Sports Illustrated* https://web.archive.org/web/20140222142617/http://sportsillustrated.cnn.com/vault/article/magazine/MAG1075316/index.htm.

Groves, Richard H. (1995). *Green Imperialism: Colonial Expansion, Tropical Island Edens and the Origins of Environmentalism 1600–1860.* Cambridge: Cambridge University Press.

Gruneau, Richard and John Horne. (2016). "Mega-Events and Globalization: A Critical Introduction." *Mega-Events and Globalization: Capital and Spectacle in a Changing World Order.* Ed. Richard Gruneau and John Horne. London: Routledge. 1–28.

Guattari, Pierre-Félix. (1996). *The Guattari Reader.* Ed. Gary Genosko. Oxford: Blackwell.

Guédon, Claire. (2024, August 8). "Une interpellation et 17 verbalisations après le « Toxic Tour » avorté des militants anti-JO à Saint-Denis." *Le Parisien* https://www.leparisien.fr/jo-paris-2024/une-interpellation-et-17-verbalisations-apres-le-toxic-tour-avorte-des-militants-anti-jo-a-saint-denis-08-08-2024-ZE7RECPH5BAXFKHXI3JIWPWTOU.php.

Guitton-Boussion, Justine. (2024, August 7). "LVMH, Decathlon : ces sponsors des JO qui maltraitent les animaux." *Reporterre* https://reporterre.net/LVMH-Decathlon-ces-sponsors-des-JO-qui-maltraitent-les-animaux.

Gulliver, Robyn E., Cassandra Star, Kelly S. Fielding, and Winnifred R. Louis. (2022). "A Systematic Review of the Outcomes of Sustained Environmental Collective Action." *Environmental Science and Policy* 133: 180–92.

Gullo, Emiliano. (2022, August 1). "Fútbol y clase obrera: la ideología no se mancha." *Jacobin* https://jacobinlat.com/2022/08/futbol-y-clase-obrera-la-ideologia-no-se-mancha/.

Gutiérrez Bayardi, Alfredo. (2022, November 15). "Tribunas que piensan: las hinchadas con ideologías más marcadas." *Para Ganar on SI* https://www.si.com/deportes/paraganar/futbol/tribunas-que-piensan-las-hinchadas-con-mas-ideologia.

Gutiérrez Cañas, Marta. (2025, February 13). "Adiós a la Ley del Ruido para el Gran Premio de F1 en Madrid." *La Razón* https://www.larazon.es/madrid/adios-ley-ruido-gran-premio-formula-uno-madrid-p7m_2025021367add7e5417ec20001f86058.html.

Habtemariam, Dawit. (2024a, May 7). "World Cup Tickets Will Go to Corporations, Not Fans." *Skift* https://skift.com/2024/05/07/world-cup-tickets-will-go-to-corporations-not-fans/.

Habtemariam, Dawit. (2024b, August 16). "Las Vegas Tourism Agency to Spend $1.7 Million on Formula 1 Tickets." *Skift* https://skift.com/2024/08/16/las-vegas-tourism-agency-to-spend-1-7-million-on-formula-1-tickets/.

Haines, Herbert H. (1984). "Black Radicalization and the Funding of Civil Rights: 1957–1970." *Social Problems* 32, no. 1: 31–43.

Hallisat, Ismaël and Caurentin Courtois. (2024, September 19). "Les images du déchaînement de violences subi par un homme dans un commissariat parisien." *Libération* https://www.liberation.fr/societe/police-justice/les-images-revoltantes-des-violences-subies-par-un-homme-dans-un-commissariat-parisien-20240919_Q2XURAOL7BDCTFWJH6LBF46RN4/.

Hall, M. Ann. (1996). *Feminism and Sporting Bodies: Essays on Theory and Practice.* Urbana: Human Kinetics.

Hall, Stuart and Les Back. (2009). ""At Home and Not at Home": Stuart Hall in Conversation with Les Back." *Cultural Studies* 23, no. 4: 658–87.

Hall, Stuart. (1960). "Editorial." *New Left Review* 1: 1–3.

Hall, Stuart. (1978). "Marxism and Culture." *Radical History Review* 18: 5–14.

Hall, Stuart. (1998). "The Windrush Issue: *Postscript.*" *Soundings* 10: 188–92.

Hall, Stuart. (2011). "The Neo-Liberal Revolution." *Cultural Studies* 25, no. 6: 705–28.

Hall, Stuart. (2016). *Cultural Studies 1983: A Theoretical History.* Ed. Jennifer Daryl Slack and Lawrence Grossberg. Durham: Duke University Press.

Halliday, Fred. (2009). "International Relations in a Post-Hegemonic Age." *International Affairs* 85, no. 1: 50.

Hanley, Steve. (2019, April 30). "The Gigantic Carbon Footprint of Formula One." *Clean Technica* https://cleantechnica.com/2019/04/30/the-gigantic-carbon-footprint-of-formula-one/.

Hardin, Garrett. (1968). "The Tragedy of the Commons." *Science* 162, no. 3859: 1243–248.

Hardy, Thomas. (2002). *Jude the Obscure.* Ed. Patricia Ingham. Oxford: Oxford University Press.

Hare, David. (2015). *The Blue Touch Paper: A Memoir.* New York: WW Norton.

Harraghy, Daniel. (2024, August 16). "How the Premier League's Global Popularity is Driving its Revenue Growth." *SportsPro* https://www.sportspromedia.com/insights/opinions/premier-league-tv-rights-revenue-global-popularity-sponsorship-data-ampere/.

Harris Interactive. (2024, July). *Réaction des Français à la cérémonie d'ouverture des Jeux Olympiques de Paris 2024* https://harris-interactive.fr/wp-content/uploads/sites/6/2024/07/Rapport-Toluna-Harris-La-perception-de-la-ceremonie-douverture-Jeux-Olympiques-de-Paris-2024-par-les-Francais-Paris-2024.pdf.

Harris, Marvin. (1979). *Cultural Materialism: The Struggle for a Science of Culture.* New York: Random House.

Hart, Matt. (2020). *Win at All Costs: Inside Nike Running and its Culture of Deception*. New York: Dey St.

Harte, Tim. (2009). *Fast Forward: The Aesthetics and Ideology of Speed in Russian Avant-Garde Culture, 1910–1930*. Madison: University of Wisconsin Press.

Harte, Tim. (2020). *Faster, Higher, Stronger, Comrades! Sport, Art, & Ideology in Late Russian and Early Soviet Culture*. Madison: University of Wisconsin Press.

Hartle, Sam. (2025, January 29). "Missouri Gov. Mike Kehoe's Budget Outlines Additional $40m for 2026 World Cup." *KSHB* https://www.kshb.com/sports/world-cup/missouri-gov-mike-kehoes-budget-outlines-additional-40m-for-2026-world-cup.

Harney, Stefano. (2006). *Nationalism and Identity: Culture and the Imagination in a Caribbean Diaspora*. Jamaica: University of the West Indies Press.

Harvey, David. (2020a). *The Anti-Capitalist Chronicles*. Ed. Jordan T. Camp and Chris Caruso. London: Pluto Press.

Harvey, Fiona. (2020b, September 10). "Toxic Pesticides Banned for EU Use Exported from UK." *Guardian* https://www.theguardian.com/environment/2020/sep/10/toxic-pesticides-banned-for-eu-use-exported-from-uk.

Hegel, Georg Wilhelm Friedrich. (1988). *Lectures on the Philosophy of World History. Introduction: Reason in History*. Trans. H. B. Nisbet. Cambridge: Cambridge University Press.

Heller, Agnes. (1979). *On Instincts*. Trans. Mario Fenyö. Assen: Van Gorcum.

Heller, Agnes. (1984). *Everyday Life*. Trans. G. L. Campbell. London: Routledge & Kegan Paul.

Hemphill, Thomas A. (2004). "Corporate Citizenship: The Case for a New Corporate Governance Model." *Business and Society Review* 109, no. 3: 339–61.

Hendlin, Yogi Hale and Stella A. Bialous. (2020). "The Environmental Externalities of Tobacco Manufacturing: A Review of Tobacco Industry Reporting." *Ambio: A Journal of Environment and Society* 49, no. 1: 17–34.

Henley, Jon. (2024, July 12). "'It Might Actually be Cool': Wary Parisians Seduced by Prospect of Olympics." *Guardian* https://www.theguardian.com/sport/article/2024/jul/12/it-might-actually-be-cool-wary-parisians-seduced-by-prospect-of-olympics?CMP=ShareiOSAppOther.

Henry, Nick, Tim Angus, Mark Jenkins, and Chris Aylett. (2007). *Motorsport Going Global: The Challenges Facing the World's Motorsport Industry*. Cham: Palgrave Macmillan.

Herder, Johann Gottfried von. (2002). *Philosophical Writings*. Trans. and Ed. Michael N. Forster. Cambridge: Cambridge University Press.

Heynes, George. (2022, December 14). "Avoiding a 'Messi' Situation: Managing Demand Pick-Ups During the FIFA World Cup 2022." *Current* +- https://www.current-news.co.uk/avoiding-a-messi-situation-managing-demand-pick-ups-during-the-fifa-world-cup-2022/.

Hibberd, James. (2021, April 16). "Amazon's 'The Lord of the Rings' to Cost $465M for Just One Season." *Hollywood Reporter* https://www.hollywoodreporter.com/tv/tv-news/amazons-lord-of-the-rings-cost-465-million-one-season-4167791/.

Hickel, Jason, Giorgos Kallis, Tim Jackson, Daniel W. O'Neill, Juliet B. Schor, Julia K. Steinberger, Peter A. Victor, and Diana Ürge-Vorsatz. (2022, December 15). "Degrowth Can Work—Here's How Science Can Help." *Nature* 612: 400–03.

Hilburg, Jonathan. (2019, January 16). "Kengo Kuma is Crafting a Timber Temple to Sports for the 2020 Olympics." *The Architect's Paper* https://www.archpaper.com/2019/01/kengo-kuma-2020-tokyo-olympic-national-stadium/#gallery-0-slide-0.

Hobsbawm, E. J. (1989). *The Age of Empire 1875–1914.* New York: Vintage Books.

Hobsbawm, Eric and Terence Ranger, eds. (2002). *La invención de la tradición.* Trans. Omar Rodríguez. Barcelona: Editorial Crítica.

Hodgson, Camilla with Tom Wilson. (2023, June 7). "UK Watchdog Bans Shell, Repsol and Petronas Greenwashing Ads." *Financial Times* https://www.ft.com/content/fec9e504-7006-495a-964a-ba2ded4df5ad.

Hogan, Michael. (2015, January 11). "Suzi Perry on F1 Technology, Her Love of Apple and Useful Emojis." *Guardian* http://www.theguardian.com/technology/2015/jan/11/suzi-perry-tv-presenter-motor-sport-favourite-technology.

Hoggart, Richard. (1965). "The World We Offer." *Health Education Journal* 23: 34–40.

Hollander, Karsten, Milan Klöwer, Andy Richardson, Laurent Navarro, Sébastien Racinais, Volker Scheer, Andrew Murray, Pedro Branco, Toomas Timpka, Astrid Junge, and Pascal Edouard. (2021). "Apparent Temperature and Heat-Related Illnesses During International Athletic Championships: A Prospective Cohort Study." *Scandinavian Journal of Medicine & Science in Sport* 31: 2092–102.

Homer. (1967). *The Odyssey*. Trans. Albert Cook. New York: WW Norton.

Hometree. (2022). "How Much Will it Cost to Power the World Cup This Winter?" https://www.hometree.co.uk/articles/how-much-will-it-cost-to-power-the-world-cup-this-winter/?referrer=611041&awc=1701916780217206262d94f1cc551a7a7ddc83d0101c338&utmsource=AWIN&utmmedium=affiliate&utmcampaign=Sub+Networks&utmcontent=78888-Skimlinks.

Hooper, Simon. (2012, August 12). "Olympics 2012: A Spectacular Triumph for London." *CNN* https://edition.cnn.com/2012/08/12/opinion/hooper-london-triumph/index.html.

Horkheimer, Max. (1973). *Critical Theory: Selected Essays.* Trans. Matthew J. O'Connell *et al.* New York: Continuum.

Horkheimer, Max. (2012). *Critique of Instrumental Reason.* Trans. Matthew J. O'Connell *et al.* London: Verso.

Horky, Thomas and Jörg-Uwe Nieland, eds. (2013). *International Sports Press Survey 2011 (Sport & Communication Volume 5).* Hamburg: Horky Sport & Communication.

Hornbeck, Kenton. (2023, November 1). "DHL Faces New Labor Controversy: Unfair Practices Allegations Escalate." *LINK nky* https://linknky.com/business/2023/11/01/dhl-faces-new-labor-controversy-unfair-practices-allegations-escalate/.

Horton, Helena. (2024, December 3). "Coca-Cola Accused of Quietly Dropping its 25% Reusable Packaging Target." *Guardian* https://www.theguardian.com/environment/2024/dec/03/coca-cola-accused-dropping-reusable-packaging-target.

Houghton, Frank, Sharon Houghton, Diane O'Doherty, Derek McInerney, and Bruce Duncan. (2018). "'Greenwashing' Tobacco Products Through Ecological and Social/Equity Labelling: A Potential Threat to Tobacco Control." *Tobacco Prevention and Cessation* 4: 37.

Houlihan, Barrie. (1997). *Sport, Policy and Politics: A Comparative Analysis.* London: Routledge.

House of Lords Select Committee on a National Plan for Sport and Recreation. (2021, December). *A National Plan for Sport, Health and Wellbeing* https://committees.parliament.uk/publications/8094/documents/84813/default/.

Housman, A. E. (1919). *A Shropshire Lad* https://www.gutenberg.org/ebooks/5720.

Hsu, Tiffany. (2021, August 5). "NBC Tries to Salvage a Difficult Olympics." *New York Times* https://www.nytimes.com/2021/08/05/business/media/nbc-olympics-tv-ratings.html.

Hua, Jing, Martin Lundqvist, Shanti Naidu, Mikael T. Ekvall, and Tommy Cedervall. (2024). "Environmental Risks of Breakdown Nanoplastics from Synthetic Football Fields." *Environmental Pollution* 347 doi:https://doi.org/10.1016/j.envpol.2024.123652.

Hughes, Matt. (2025, August 4). "Fifa Facing Multibillion-Pound Compensation Claim from Players." *Guardian* https://www.theguardian.com/football/2025/aug/04/fifa-multibillion-pound-compensation-claim-former-players-transfer-regulations?CMP=Share_iOSApp_Other.

Huijuan, Cao, Hidemichi Fujii, and Shunsuke Managi. (2013). *Environmental Impact of the 2008 Beijing Olympic Games.* Economics Discussion Paper 2013-30, Kiel Institute for the World Economy https://www.economics-ejournal.org/economics/discussionpapers/2013-30.

Human Rights Watch. (2024, August). *"You Have to Move!" The Cruel and Ineffective Criminalization of Unhoused People in Los Angeles* https://www.hrw.org/report/2024/08/14/you-have-move/cruel-and-ineffective-criminalization-unhoused-people-los-angeles#:~:text='%20The%20cruel%20and%20ineffective%20criminalization,governmental%20failures%20to%20prioritize%20eviction.

Hume, David. (1739). *A Treatise of Human Nature* http://michaeljohnsonphilosophy.com/wp-content/uploads/2012/01/5010_Hume_Treatise_Human_Nature.pdf.

Hume, David. (1955). *An Inquiry Concerning Human Understanding with a Supplement: An Abstract of a Treatise of Human Nature*. Ed. Charles W. Hendel. Indianapolis: Bobbs-Merrill.

Hume, David. (2007). *Dialogues Concerning Natural Religion and Other Writings*. Ed. Dorothy Coleman. Cambridge: Cambridge University Press.

Humphreys, David. (2009). "Environmental and Ecological Citizenship in Civil Society." *The International Spectator: Italian Journal of International Affairs* 44, no. 1: 171–83.

Hunt, Jeremy. (2010, June 28). "Sports Keynote Speech" https://www.gov.uk/government/speeches/sports-keynote-speech.

Hurst, Alexander. (2024, August 12). "France Specialises in Glum Self-Doubt. Has This Joyful Olympics Finally Changed That?" *Guardian* https://www.theguardian.com/commentisfree/article/2024/aug/12/france-olympics-paris-city-cool?CMP=ShareiOSAppOther.

Hussain, Umer, Adam Ali, MacIntosh Ross, and Shakiba Moghadam. (2024, May 8). "Paris 2024 Olympics: How the Games Are Being Used to Marginalize the Most Vulnerable." *The Conversation* https://theconversation.com/paris-2024-olympics-how-the-games-are-being-used-to-marginalize-the-most-vulnerable-228855.

Hutchins, Brett, Bo Li, and David Rowe. (2019). "Over-the-Top Sport: Live Streaming Services, Changing Coverage Rights Markets, and the Growth of Media Sport Portals." *Media, Culture & Society* 41, no. 7: 975–94.

Huws, Ursula. (2019). *Labour in Contemporary Capitalism: What Next?* London: Palgrave Macmillan.

Hyde, Marina. (2025, March 21). "So It's Victory for Kirsty Coventry in the IOC Elections. Who Says Cards with a Rigged Deck Isn't an Olympic Sport?" *Guardian* https://www.theguardian.com/commentisfree/2025/mar/21/ioc-kirsty-coventry-sebastian-coe-olympic-politics.

Hyde-Clarke, Nathalie, Rune Ottosen, and Toby Miller. (2014). "Nation-Building and the FIFA World Cup, South Africa 2010." *African Football, Identity Politics and Global Media Narratives: The Legacy of the FIFA 2010 World Cup*. Ed. Chari Tenda and Mhiripiri Nhamo Anthony. London: Palgrave Macmillan. 15–28.

Infantino, Gianni. (2024). https://www.instagram.com/gianni_infantino/p/CzErr6Uo1QJ/?hl=en.

Ingle, Sean. (2022, November 19). "'I Feel Gay, Disabled … Like a Woman Too:': Infantino Makes Bizarre Attack on Critics." *Guardian* https://www.theguardian.com/football/2022/nov/19/fifa-gianni-infantino-world-cup-qatar.

Institut national de la statistique et des études économiques. (2023, March 30). *Immigrés et descendants d'immigrés: Édition 2023* https://www.insee.fr/fr/statistiques/6793282?sommaire=6793391#:~:text=Les%20immigr%C3%A9s%20r%C3%A9sident%20fr%C3%A9quemment%20dans,Mayotte)%20(figure%201).

Instituto Nacional de Estadística. (2025, February 3). *Estadística de Movimientos Turísticos en Fronteras* https://www.ine.es/dyngs/Prensa/FRONTUR1224.htm.

Inteligência Esportiva Research Institute and Play the Game. (2023). *Research on the Sports Legacy of Brazilian Stadiums from the 2014 FIFA World Cup* https://www.playthegame.org/media/z3bkh5fh/research-on-the-sports-legacy-of-brazilian-stadiums-from-the-2014-fifa-world-cupupdated-10.pdf.

Intergovernmental Panel on Climate Change. (2022). *Climate Change 2022: Mitigation of Climate Change* https://www.ipcc.ch/report/ar6/wg3/downloads/report/IPCC_AR6_WGIII_FullReport.pdf.

International Civil Aviation Organization. (2016). *ICAO Carbon Emissions Calculator* https://www.icao.int/environmental-protection/CarbonOffset/Pages/default.aspx.

International Energy Agency. (2014). *More Data, Less Energy: Making Network Standby More Efficient in Billions of Connected Devices* https://www.iea.org/publications/freepublications/publication/MoreDataLessEnergy.pdf.

International Energy Agency. (2018). *The Future of Petrochemicals: Towards More Sustainable Plastics and Fertilisers* https://iea.blob.core.windows.net/assets/bee4ef3a-8876-4566-98cf-7a130c013805/TheFutureofPetrochemicals.pdf.

International Energy Agency. (2024). *World Energy Outlook 2024* https://iea.blob.core.windows.net/assets/47a9a222-78e4-4c43-9bab-977b4ad5326b/WorldEnergyOutlook2024.pdf.

International Financial Reporting Standards. (2024, June). *Feedback Statement IFRS® Sustainability Disclosure Standards* https://www.ifrs.org/content/dam/ifrs/project/issb-consultation-on-agenda-priorities/agenda-consultation-feedback-statement-june-2024.pdf.

International Olympic Committee. (n.d.-a). *Becoming an Olympic Games Host* https://olympics.com/ioc/becoming-an-olympic-games-host.

International Olympic Committee. (n.d.-b). "XII Olympic Congress Paris 1994" https://www.olympics.com/ioc/paris-1994-olympic-congress.

International Olympic Committee. (2018). *Carbon Footprint Methodology for the Olympic Games* https://library.olympic.org/Default/doc/SYRACUSE/184686/carbon-footprint-methodology-for-the-olympic-games-international-olympic-committee?lg=fr-FR.

International Olympic Committee. (2021a, April 21). "IOC Athletes' Commission's Recommendations on Rule 50 and Athlete Expression." https://olympics.com/athlete365/articles/officialstatement/ioc-athletes-commissions-recommendations-on-rule-50-and-athlete-expression.

International Olympic Committee. (2021b). *IOC Marketing Report Tokyo 2020* https://stillmed.olympics.com/media/Documents/International-Olympic-Committee/IOC-Marketing-And-Broadcasting/IOC-Marketing-report-Tokyo-2020.pdf.

International Olympic Committee. (2022a). *Annual Report 2021: Faster, Higher, Stronger—Together* https://stillmed.olympics.com/media/Documents/International-Olympic-Committee/Annual-report/IOC-Annual-Report-2021.pdf.

International Olympic Committee. (2022b, with addendum). *Olympic Host Contract: Operational Requirements* https://stillmed.olympics.com/media/Documents/Olympic-Games/Brisbane-2032/Host-Contract/OHC-Operational-Requirements2022-addendum.pdf.

International Olympic Committee. (2024a, July 23). *Olympic Charter* https://stillmed.olympics.com/media/Document%20Library/OlympicOrg/General/EN-Olympic-Charter.pdf.

International Olympic Committee. (2024b). *Paris 2024: More Sustainable Games* https://olympics.com/ioc/paris-2024-sustainable-games.

International Olympic Committee. (2024c). *Olympic Games Paris 2024* https://www.olympics.com/en/olympic-games/paris-2024.

International Olympic Committee. (2024d, July 22). *Olympic Truce* https://www.olympics.com/ioc/olympic-truce.

International Olympic Committee. (2024e, July). *Olympic Host Contract* https://stillmed.olympics.com/media/Documents/Olympic-Games/Salt-Lake-City-Utah-2034/Olympic-Host-Contract-Principles-2034.pdf.

International Olympic Committee. (2025). *IOC Members* https://www.olympics.com/ioc/members.

International Organization of Securities Commissions. (2023, December). *Supervisory Practices to Address Greenwashing: Final Report* https://www.iosco.org/library/pubdocs/pdf/IOSCOPD750.pdf.

Ionescu, Diana. (2024, December 27). "Mexico City Raises Mobility, Public Works Budget Ahead of World Cup." *Planetizen* https://www.planetizen.com/news/2024/12/133448-mexico-city-raises-mobility-public-works-budget-ahead-world-cup.

Ireland, Robin, Magdalena Muc, Christopher Bunn, and Emma Boyland. (2024). "Marketing of Unhealthy Brands During the 2018 Fédération Internationale de Football Association (FIFA) World Cup UK Broadcasts—A Frequency Analysis." *Journal of Strategic Marketing* 32, no. 6: 762–77.

Irigaray, Luce, ed. (2004). *Luce Irigaray: Key Writings*. New York: Continuum.
Irimia, Silvian. (2022, March 11). "Best Sponsors in Formula One History." *Autoevolution* https://www.autoevolution.com/news/best-sponsors-in-formula-one-history-183650.html.
Ishiwatari, Mikio, Miko Maekawa, and Ryo Fujikura. (2025). "Climate Change Adaptation Strategies and Economic Challenges in Atoll Nations: A Case Study of the Marshall Islands." *Journal of Disaster Research* 20, no. 1: 53–61.
Iskander, Natasha. (2021). *Does Skill Make Us Human? Migrant Workers in 21st-Century Qatar and Beyond*. Princeton: Princeton University Press.
Isocrates. (1929). *Isocrates in Three Volumes. II*. Trans. George Norlin. London: William Heinemann/New York: GP Putnam's Sons.
Isocrates. (1945). *Isocrates in Three Volumes. III*. Trans. Larue Van Hook. Cambridge, Mass.: Harvard University Press/London: William Heinemann.
Ito, Eiji, James Higham, and Joseph M. Cheer. (2022). "Carbon Emission Reduction and the Tokyo 2020 Olympics." *Annals of Tourism Research Empirical Insights* 3, no. 2 https://doi.org/10.1016/j.annale.2022.100056.
Ivanov, Denis and Gaygysyz Ashyrov. (2024). "Wealth Inequality and Mega Events." *European Journal of Political Economy* 85 https://doi.org/10.1016/j.ejpoleco.2024.102615.
Jabkhiro, Juliette and Mimosa Spencer. (2025, March 7). "France's Former Spy Chief Found Guilty of Using Public Resources to Aid LVMH." *Reuters* https://www.reuters.com/world/europe/frances-former-spy-chief-found-guilty-using-public-resources-aid-lvmh-2025-03-07/.
Jacobs, Jane. (1961). *The Death and Life of American Cities*. New York: Vintage.
Jacopin, Tanguy. (2006–07). "America's Cup: Value Creation and Stakeholder Management." *Finance & The Common Good/Bien Commun* 26: 64–70.
Jakar, Gidon and Christina Philippou. (2025). "Legacy and Urban Regeneration: Long-Term Local Impact of the London 2012 Olympic Games." *European Planning Studies* 33, no. 2: 183–203.
James I and Charles I. (1914). "The King's Majesty's Declaration to His Subjects Concerning Lawful Sports to be Used." *Documents Illustrative of English Church History*. Ed. Henry Gee and William John Hardy. New York: Macmillan. 528–32.
Janetsky, Megan. (2024, July 26). "Migrants and Homeless People Are Cleared Out of Paris During the Olympics." *AP* https://apnews.com/article/olympics-2024-paris-migrant-camp-3ef2a08d8da1085148ed409dcb44d6f6.
Jefferson Lenskyj, Helen. (2008). *Olympic Industry Resistance: Challenging Olympic Power and Propaganda*. Albany: State University of New York Press.
Jennings, Andrew and Clare Sambrook. (2000). *The Great Olympic Swindle: When the World Wanted its Games Back*. New York: Simon & Schuster.
Jennings, Andrew. (1996). *The New Lords of the Rings: Olympic Corruption and How to Buy Gold Medals*. London: Pocket Books.

Jennings, Andrew. (2016). *The Dirty Game: Uncovering the Scandal at FIFA*. N.p.: Century.

Jensen, Jonathan A., Joe B. Cobbs, Alex Mazer, and B. David Tyler. (2024). "Analyzing Brand Strategy on an International Scale: The Sponsorship Performance Cycle in Formula One Racing." *Journal of International Marketing* 32, no. 3: 23–42.

Jevons, William Stanley. (1866). *The Coal Question: An Inquiry Concerning the Progress of the Nation, and the Probable Exhaustion of Our Coal-Mines*, 2nd ed.. London: Macmillan.

Jhally, Sut. (2016). "Stuart Hall: The Last Interview." *Cultural Studies* 30, no. 2: 332–45.

Jiménez-García, Mercedes, José Ruiz-Chico, Antonio Rafael Peña-Sánchez, and José Antonio López-Sánchez. (2020). "A Bibliometric Analysis of Sports Tourism and Sustainability (2002–2019)." *Sustainability* 12, no. 7: 2840–858.

Jing, Gao. (2024, August 12). "Triumphs, Challenges and Unity in Paris: A Glorious Farewell to the 2024 Olympics." *NBC* https://www.nbcnews.com/sports/olympics/paris-summer-olympics-conclusion-rcna164908.

Johan Cruijff Arena. (n.d.). https://www.johancruijffarena.nl/en/making-an-impact-togehther/duurzaamheid/.

John, Alistair and Steve J. Jackson. (2011). "Call Me Loyal: Globalization, Corporate Nationalism and the America's Cup." *International Review for the Sociology of Sport* 46, no. 4: 399–417.

Johnson, Jay and Adam Ehsan Ali. (2018). "Ecological Modernization and the 2014 NHL Sustainability Report." *Sociology of Sport Journal* 35, no. 1: 49–57.

Johnson, Lamar. (2024, September 4). "Coca-Cola Must Face Greenwashing Lawsuit: D.C. Appeals Court." *ESG Dive* https://www.esgdive.com/news/d-c-appeals-court-rules-coca-cola-must-face-greenwashing-lawsuit/726071/.

Jolly, Jasper. (2023, September 29). "Transactions Involving Bernard Arnault Investigated Over Suspected Money Laundering." *Guardian* https://www.theguardian.com/business/2023/sep/29/bernard-arnault-lvmh-nikolai-sarkisov.

Jolly, Patricia. (2024, September 9). "Du chaos à la «parenthèse enchantée» : le 26 juillet, la journée qui a tout changé aux Jeux de Paris 2024." *Le Monde* https://www.lemonde.fr/sport/article/2024/09/08/jeux-de-paris-le-26-juillet-2024-la-parenthese-devint-enchantee_6306999_3242.html.

Jones, Simon. (2015, August 11). "America's Cup Dispute Set for Court." *Royal Gazette* https://www.royalgazette.com/event/news/article/20150810/americas-cup-dispute-set-for-court/.

Jørgensen, Lars and Jens Sejer Andersen. (2024). *When the Silence of Sport Was Broken: Sports Politics in the Times of Play the Game*. Play the Game https://

www.playthegame.org/media/adxj333s/when-the-silence-of-sport-was-broken-sports-politics-in-the-times-of-play-the-game-1-mindre.pdf.

Jorgenson, Andrew K. and Brett Clark. (2016). "The Temporal Stability and Developmental Differences in the Environmental Impacts of Militarism: The Treadmill of Destruction and Consumption-Based Carbon Emissions." *Sustainability Science* 26, no. 11: 505–14.

Joseph, Janelle. (2017). *Sport in the Black Atlantic: Cricket, Canada and the Caribbean Diaspora*. Manchester: Manchester University Press.

Joseph, May. (1995). "Diaspora, New Hybrid Identities, and the Performance of Citizenship." *Women and Performance* 7, no. 2/8, no. 1: 3–13.

Joshi, Saumya. (2024, July 12). "Paris Olympics 2024: Why Are Scientists Urging the Olympics to Ditch Its Official Car?" *Oneindia* https://www.oneindia.com/sports/paris-olympics-2024-why-are-scientists-urging-the-olympics-to-ditch-its-official-car-011-3876999.html.

Jukes, Peter. (2012, August 13). "Take a Bow, London: The Olympics Were a Triumph." *The Daily Beast* https://www.thedailybeast.com/take-a-bow-london-the-olympics-were-a-triumph/.

Justice for Players. (n.d.). https://justiceforplayers.com/.

Juvenal. (1998). *The Sixteen Satires*, 3rd ed. Trans. and Ed. Peter Green. Harmondsworth: Penguin.

Kafka, Peter. (2021, March 18). "Amazon's $10 Billion NFL Deal is Huge—And a Sign That Sports is Staying on TV for a While Longer." *Vox* https://www.vox.com/recode/22338794/amazon-nfl-10-billion-tv-analysis.

Kang, Jasnoor. (2020, January 31). "Lifecycle of a Soccer Ball." *StoryMaps* https://storymaps.arcgis.com/stories/ef946b742331425fb773a2d06f6b44f0.

Kant, Immanuel. (2006). *Anthropology from a Pragmatic Point of View*. Trans. and Ed. Robert B. Louden. Cambridge: Cambridge University Press.

Karali, Nihan, Nina Khanna, and Nihar Shah. (2024). *Climate Impact of Primary Plastic Production*. Energy Analysis & Environmental Impacts Division, Berkeley Lab https://live-lbl-eta-publications.pantheonsite.io/sites/default/files/2024-09/climateandplasticreportlbnl-2001585.pdf.

Karaman, M. Ihsan. (2019). "Smoking as an Environmental Health Problem." *International Journal of Human and Health Sciences* 3, no. 3: 123–26.

Karamichas, John. (2013). *The Olympic Games and the Environment*. New York: Palgrave Macmillan.

Karpowitz, Christopher F., Chad Raphael, and Allen S. Hammond IV. (2009). "Deliberative Democracy and Inequality: Two Cheers for Enclave Deliberation Among the Disempowered." *Politics & Society* 37, no. 4: 576–615.

Kasztelan, Marta and Sam Quashie-Idun. (2022, August 8). "Revealed: Garment Waste from Nike, Clarks and Other Leading Brands Burned to Fuel Toxic Kilns in Cambodia." *Unearthed* https://unearthed.greenpeace.org/2022/08/08/garment-waste-nike-clarks-cambodia-bonded-workers-toxic/.

Kaufman, Mark. (2020). "The Carbon Footprint Sham: A 'Successful, Deceptive' PR Campaign." *Mashable* https://mashable.com/feature/carbon-footprint-pr-campaign-sham.

Kay, Stanley. (2019, April 22). "Winter is Going: How Climate Change is Imperiling Outdoor Sporting Heritage." *Sports Illustrated* https://www.si.com/nhl/2019/04/22/climate-change-canada-winter-sports-hockey-backyard-rinks.

Kazim, Hasnain. (2010, March 16). "The Football Stitchers of Sialkot." *Spiegel International* https://www.spiegel.de/international/world/globalization-in-pakistan-the-football-stitchers-of-sialkot-a-683873.html.

Keegan, Brendan. (2024, November 1). "Racing's Role in Shaping Modern Masculinity: Redefining Strength and Vulnerability." *Rolling Stone* https://www.rollingstone.com/culture-council/articles/racings-shaping-modern-masculinity-redefining-strength-vulnerability-1235148485/.

Keith, Felix. (2024, February 21). "F1 Sponsor in Spotlight for 'Misleading' Fans with Sustainable Fuels Amid Climate Crisis." *Daily Mirror* https://www.mirror.co.uk/sport/formula-1/f1-aramco-greenwashing-sustainable-fuels-32173647.

Kennedy, John F. (1960–61, December 26–January 7). "The Soft American." *Sports Illustrated*: 15–17.

Kennedy, Lindsey and Nathan Southern. (2024, August 16). "How Premier League Sponsors Run Human Trafficking Syndicates." *Seasons of Crime* https://www.seasonsofcrime.com/p/how-premier-league-sponsors-run-human.

Kennedy, Robert F. (1968, March 18). "Remarks at the University of Kansas, March 18, 1968." *John F. Kennedy Presidential Library and Museum* https://www.jfklibrary.org/learn/about-jfk/the-kennedy-family/robert-f-kennedy/robert-f-kennedy-speeches/remarks-at-the-university-of-kansas-march-18-1968.

Kennedy, Niamh. (2023, August 25). "Heineken Sells Russian Business for $1 as it Completes Exit." *CNN* https://edition.cnn.com/2023/08/25/business/heineken-exits-russia/index.html.

Kenward, Ben and Cameron Brick. (2024). "Large-Scale Disruptive Activism Strengthened Environmental Attitudes in the United Kingdom." *Global Environmental Psychology* 2 https://doi.org/10.5964/gep.11079.

Keoghan, Jim. (2014). *Punk Football: The Rise of Fan Ownership in English Football.* Worthing: Pitch Publishing.

Kew, Matt. (2023, December 2). "Hamilton: F1 Diversity Not Moved "Anywhere Near as Much as it Should Have"." *Motorsport* https://www.motorsport.com/

f1/news/hamilton-f1-diversity-not-moved-anywhere-near-as-much-as-it-should-have-/10554313/.

Khan, Sadiq. (2025a, April 28). "Foreword." *Major Sports Events—Impact for London 2024 Report* Greater London Authority https://www.london.gov.uk/sites/default/files/2025-04/Major_Sports_Events_Impact_for_London_2024_Report-28April25_0.pdf. 5.

Khan, Yusuf. (2025b, March 12). "Can Formula One Ever Be Sustainable?" *Wall Street Journal* https://www.wsj.com/articles/can-formula-one-ever-be-sustainable-0d63d80c.

Kick Big Soda Out of Sport. (n.d.). https://www.kickbigsodaout.org/.

Kidd, Patrick. (2017, July 26). "America's Cup: The Rising Cost of Sailing's Ultimate Prize." *Boat International* https://www.boatinternational.com/yachts/americas-cup/americas-cup-the-rising-cost-of-sailings-ultimate-prize%2D%2D31985.

Kim, Kyoung-yim and Heejoon Chung. (2018). "Eco-Modernist Environmental Politics and Counter-Activism Around the 2018 PyeongChang Winter Games." *Sociology of Sport Journal* 35, no. 1: 17–28.

King, Ed. (2013, March 11). "Formula One: The Petrol Heads Driving the Green Economy." *Responding to Climate Change* https://www.rtcc.org/2013/03/11/formula-one-the-petrolheads-driving-the-green-economy/.

Kingdom of Saudi Arabia. (2023). *Saudi Arabia: A Story of Transformation: Vision 2030* https://www.vision2030.gov.sa/media/jvllaxpl/story-of-transformation-2023.pdf.

Kipling, Rudyard. (1902). "The Islanders." https://www.kiplingsociety.co.uk/poem/poems_islanders.htm.

Klasa, Adrienne with Sara Germano. (2024, October 2). "LVMH Strikes Sponsorship Deal with Formula 1." *Financial Times* https://www.ft.com/content/5bee32c7-3305-474e-a85b-93ff7a5e333f.

Klein, Paul. (2012, December 28). "Three Ways to Secure Your Social License to Operate in 2013." *Forbes* https://www.forbes.com/sites/csr/2012/12/28/three-ways-to-secure-your-social-license-to-operate-in-2013/.

KnowTheChain. (2023). *Apparel & Footwear Benchmark Company Scorecard 2023* https://knowthechain.org/wp-content/uploads/2023_KTC-AF-Scorecard-LVMH-1.pdf.

Koch, Max and Hubert Buch-Hansen. (2020). "The IPE of Degrowth and Sustainable Welfare." *The Routledge Handbook to Global Political Economy: Conversations and Inquiries.* Ed. Ernesto Vivares. Abingdon: Routledge. 375–90.

Kokabi, Alexandre-Reza. (2024, July 23). "Les JO 2024 à Paris, une catastrophe écologique en Seine-Saint-Denis." *Reporterre* https://reporterre.net/Les-JO-2024-a-Paris-une-catastrophe-ecologique-en-Seine-Saint-Denis.

Konyn, Carol. (2024, April 2). "The Carbon Footprint of Football: Unveiling the Dark Side of the Game." *Earth* https://earth.org/the-carbon-footprint-of-football/.

Kostov, Nick and Joshua Robinson. (2024, October 2). "LVMH Took Over the Paris Olympics. Now it's Snagged Formula One." *Wall Street Journal* https://www.wsj.com/business/retail/lvmh-took-over-the-paris-olympics-now-its-snagged-formula-one-b570c0f4.

Kountoruis, Yiannis and Eleri Williams. (2023). "Do Protests Influence Environmental Attitudes? Evidence from Extinction Rebellion." *Environmental Research Communications*. https://doi.org/10.1088/2515-7620/ac9aeb.

Kraus, Hans and Ruth P. Hirschland. (1953). "Muscular Fitness and Health." *Journal of the American Association for Health, Physical Education, and Recreation* 24, no. 10: 17–19.

Krauze, Enrique. (1994, June 28). "México en un balón." *El País* https://elpais.com/diario/1994/06/28/opinion/772754409850215.html.

Krispenz, Anne and Alex Bertrams. (2024). "Understanding Involvement in Environmental Activism: Relationships to Pathological Narcissistic Grandiosity, Virtue Signaling, Dominance, and Sensation Seeking." *Current Psychology* 43: 30011–026.

Kristanadjaja, Gurvan. (2022, December 5). "Sans-papiers sur les chantiers de Paris 2024: «Les JO ne pourraient pas se faire sans nous»." *Libération* https://www.liberation.fr/societe/sans-papiers-sur-les-chantiers-les-jeux-olympiques-ne-pourraient-pas-se-faire-sans-nous-20221205 3IOA3NFTMZBADLDM35V7OVI3TI/.

Kristeva, Julia. (2024). *Revolution in Poetic Language*. Trans. Margaret Waller. New York: Columbia University Press.

Kriza, Elisa. (2019). "Anti-Communism, Communism, and Anti-Interventionism in Narratives Surrounding the Student Massacre on Tlatelolco Square (Mexico, 1968)." *Bulletin of Latin American Research* 38, no. 1: 82–96.

Kuhn, Gabriel. (2019). *Soccer vs. the State: Tackling Football and Radical Politics*, 2nd ed.. Oakland: PM Press.

Kunti, Sam and Andreas Selliaas. (2025, January 6). "The Zone of Interest." *Josimar* https://josimarfootball.com/2025/01/06/the-zone-of-interest/.

Kunti, Sam, Steve Menary, Andy Brown, Philippe Auclair, and Jack Kerr. (2024, September 12). "Fly Another Day." *Josimar* https://josimarfootball.com/2024/09/12/fly-another-day/.

L, Jennifer. (2024, February 9). "Beyond Touchdowns and Trophies: Unveiling the Carbon Footprint of Superbowl LVIII." *Carbon Credits* https://carboncredits.com/beyond-touchdowns-and-trophies-unveiling-the-carbon-footprint-of-superbowl-lviii/.

Lambert, Stéphane. (2024, August 15). "JO de Paris 2024: pouvez-vous vibrer sans nous tyranniser?" *Libération* https://www.liberation.fr/idees-et-debats/

tribunes/jo-de-paris-2024-pouvez-vous-vibrer-sans-nous-tyranniser-20240815GIL4Q6QHNNGBXNHRAZXEHOKZ7A/.

Lamborelle, Aymone and Beatriz Rios. (2019, September 4). *Sustainability in Sports? What's Football's Carbon Footprint?* Euractiv https://www.euractiv.com/section/health-consumers/infographic/whats-footballs-carbon-footprint/.

Landrigan, Philip J., Sarah Dunlop, Marina Treskova, Hervé Raps, Christos Symeonides, Jane Muncke, Margaret Spring, John Stegeman, Bethanie Carney Almroth, Thomas C. Chiles, Maureen Cropper, Megan Deeney, Lizzie Fuller, Roland Geyer, Rachel Karasik, Tiza Mafira, Alexander Mangwiro, Denise Margaret Matias, Yannick Mulders, Yongjoon Park, Costas A. Velis, Roel Vermeulen, Martin Wagner, Zhanyung Wang, Ella M. Whitman, Tracey J. Woodruff, and Joacim Rocklöv. (2025). *The Lancet Countdown on Health and Plastics.* https://doi.org/10.1016/S0140-6736(25)01447-3.

Lasch, Christopher. (1991). *The Culture of Narcissism: American Life in an Age of Diminishing Expectations.* New York: WW Norton.

Lassalle, Frederic. (2014). "America's Cup 2007: Sailing, Glitter and Power." *Sport, Business and Management: An International Journal* 4, no. 1: 38–51.

Latham-Coyle, Harry. (2024, August 6). "Adam Peaty Claims Athletes Found Worms in Food at Olympic Village: 'It's Just Not Good Enough'." *Independent* https://www.independent.co.uk/sport/olympics/adam-peaty-olympic-village-food-worms-paris-2024-b2591821.html.

Latour, Bruno, Pablo Jensen, Tommaso Venturini, Sébastian Grauwin, and Dominique Boullier. (2012). "The Whole is Always Smaller Than its Parts: A Digital Test of Gabriel Tarde's Monads." *British Journal of Sociology* 63, no. 4: 590–615.

Latta, P. Alex. (2007). "Locating Democratic Politics in Ecological Citizenship." *Environmental Politics* 16, no. 3: 377–93.

Lawrence, Andrew. (2021, December 17). "'Big Egos, Power Struggles, Stunning Betrayals': How Netflix's *Drive to Survive* Turned Americans into F1 Fans." *Guardian* https://www.theguardian.com/media/2021/dec/17/netflixs-drive-to-survive-americans-f1-fans.

Lawrence, Geoffrey and David Rowe. (1986). "Towards a Sociology of Australian Sport." *Australian Sport. Essays in the Sociology of Australian Sport.* Ed. Geoffrey Lawrence and David Rowe. Sydney: Hale & Iremonger. 13–45.

Lawton, Matt. (2025, April 28). "Sadiq Khan Backs London Bid for 2040 Olympics." *Times* https://www.thetimes.com/article/f67f006f-70ae-405e-bbb3-5a1a0de5238b?shareToken=3a7ca5251ebe72407a61feceb9387c8e.

Laye, Sébastien and Matthieu Verry. (2023, November 16). "JO 2024: «Pourquoi la mairie de Paris risque de provoquer une catastrophe écologique»." *Le Figaro* https://www.lefigaro.fr/vox/societe/jo-2024-pourquoi-la-mairie-de-paris-risque-de-provoquer-une-catastrophe-ecologique-20231116.

Lazard, Violette. (2024, November 28). "« J'ai fait plus pour la France que François Ruffin ! », au procès Squarcini, le témoignage du milliardaire Bernard Arnault." *Le Nouvel Obs* https://www.nouvelobs.com/justice/20241128.OBS97111/j-ai-fait-plus-pour-la-france-que-francois-ruffin-au-proces-squarcini-le-temoignage-du-milliardaire-bernard-arnault.html.

Lazarsfeld, Paul F. (1941). "Remarks on Administrative and Critical Communications Research." *Studies in Philosophy and Social Science* 9, no. 1: 2–16.

Lecardane, Renzo. (2008). "Le front de mer de Valencia et l'héritage de l'America's Cup 2007." *Méditerranée* 111: 109–14.

Lee, Kelley, Natalia Carrillo Botero, and Thomas Novotny. (2016). "'Manage and Mitigate Punitive Regulatory Measures, Enhance the Corporate Image, Influence Public Policy': Industry Efforts to Shape Understanding of Tobacco-Attributable Deforestation." *Globalization and Health* 12. https://doi.org/10.1186/s12992-016-0192-6.

Lee White, Lauren. (2022, August 19). "The 2028 L.A. Olympics Are Already Creating a Housing Disaster." *The New Republic* https://newrepublic.com/article/167453/2028-los-angeles-olympics-housing.

Lefebvre, Sylvain and Romain Roult. (2011). "Formula One's New Urban Economies." *Cities* 28, no. 4: 330–39.

Le Foll, Clément. (2024a, July 30). "En catimini, Saint-Denis achète un logiciel de surveillance algorithmique." *Mediapart* https://www.mediapart.fr/journal/france/300724/en-catimini-saint-denis-achete-un-logiciel-de-surveillance-algorithmique.

Le Foll, Clément. (2024b, August 8). "Militants écologistes ou pour la justice sociale, ils ont été écartés des JO." *Mediapart* https://www.mediapart.fr/journal/france/080824/militants-ecologistes-ou-pour-la-justice-sociale-ils-ont-ete-ecartes-des-jo.

Lehne, Johanna and Felix Preston. (2018). *Making Concrete Change: Innovation in Low-Carbon Cement and Concrete.* Chatham House https://www.chathamhouse.org/sites/default/files/publications/2018-06-13-making-concrete--change-cement-lehne-preston-final.pdf.

Lei, Tze-Huan and Faming Wang. (2021). "Looking Ahead of 2021 Tokyo Summer Olympic Games: How Does Humid Heat Affect Endurance Performance? Insight into Physiological Mechanism and Heat-Related Illness Prevention Strategies." *Journal of Thermal Biology* 99 https://www.sciencedirect.com/science/article/abs/pii/S0306456521001431.

Leighton, Neil. (2005, June 3). "Old Shin Sport Alive and Kicking." *BBC* https://news.bbc.co.uk/2/hi/uk_news/england/gloucestershire/4605157.stm.

Lennon, John. (1971, January 21). "Interview with Tariq Ali and Robin Blackburn." *Red Mole* https://www.beatlesinterviews.org/db1971.0121.beatles.html.

Le Revers de la Medalle. (n.d.). https://lereversdelamedaille.fr/.

Lester, Libby. (2019). *Global Trade and Mediatised Environmental Protest: The View from Here.* Cham: Palgrave Macmillan.

Le Vaillant, Luc and Service Médias. (2012, September 10). "Le «Riche con !» provoque du buzz et une plainte." *Libération* https://www.liberation.fr/medias/2012/09/10/le-riche-con-provoque-du-buzz-et-une-plainte_845366/.

Levermore, Roger. (2010). "CSR for Development Through Sport: Examining its Potential and Limitations." *Third World Quarterly* 31, no. 2: 223–41.

Levintova, Hannah. (2012, August 11). "BP's Olympic Ads Seek to Erase Oil Spill from Memory." *Mother Jones* https://www.motherjones.com/politics/2012/08/bps-olympic-ads-are-helping-deepwater-horizon-go-away/.

Lewandowsky, Stephan, Naomi Oreskes, James S. Risbey, Ben R. Newell, and Michael Smithson. (2015). "Seepage: Climate Change Denial and its Effect on the Scientific Community." *Global Environmental Change* 33: 1–13.

Liang, Annabelle. (2024, June 4). "Toyota Raided as Safety Testing Scandal Grows." *BBC* https://www.bbc.com/news/articles/c1wwj1p2wdyo.

Lidón, Inma and Héctor Atienza. (2024, October 22). "Los 90 días que tiene Valencia para traer la Copa América de nuevo "a casa"." *El Mundo* https://www.elmundo.es/comunidad-valenciana/2024/10/22/6717b56be4d4d8e5028b4591.html.

Lies, Elaine. (2021, July 1). "'Sad, Lonely Feeling': Tokyo Man Evicted Twice, 50 Years Apart, for Olympic Construction." *Reuters* https://www.reuters.com/world/asia-pacific/sad-lonely-feeling-tokyo-man-evicted-twice-50-years-apart-olympic-construction-2021-07-01/.

Light, Andrew. (1998). "Reconsidering Bookchin and Marcuse as Environmental Materialists: Toward an Evolving Social Ecology." *Social Ecology After Bookchin.* Ed. Andrew Light. New York: Guilford Press. 343–84.

Lim, Vanessa. (2022, February 10). "Race to Reduce Singapore's F1's Carbon Footprint a Good Start, but Experts are Mixed on its Impact." *Channel News Asia* https://www.channelnewsasia.com/singapore/f1-singapore-formula-one-carbon-footprint-environment-2487536.

Lindgaard, Jade. (2024). *Une ville face à la violence olympique.* Quimperlé: Éditions Divergences.

Lind Gustavussen, Mathilde. (2023, April 9). "The Gentrification Scheme Behind Los Angeles's "No-Build" Olympics." *Jacobin* https://jacobin.com/2023/04/los-angeles-olympics-2028-development-politics-gentrification.

Lindner-Cendrowska, Katarzyna, Kamil Leziak, Peter Bröde, Dusan Fiala, and Marek Konefał. (2024). "Prospective Heat Stress Risk Assessment for Professional Soccer Players in the Context of the 2026 FIFA World Cup." *Scientific Reports* 14. https://doi.org/10.1038/s41598-024-77540-1.

Lindwall, Courtney. (2023, February 9). "What is Greenwashing?" NRDC https://www.nrdc.org/stories/what-greenwashing.

Lins Ribeiro, Gustavo and Arturo Escobar, eds. (2006). *World Anthropologies: Disciplinary Transformations Within Systems of Power*. Oxford: Berg.

Livingstone, Robert. (2024, August 9). ""We are Not Ready to Sell the Olympic Games to the Highest Bidder"—IOC President Thomas Bach on Future Bids." *GamesBids* https://gamesbids.com/eng/summer-olympic-bids/we-are-not-ready-to-sell-the-olympic-games-to-the-highest-bidder-ioc-president-thomas-bach-on-future-bids/.

Lo, Jo and Marlene Jacobsen. (2021, August 6). "'Hottest Games Ever': At the Tokyo Olympics, Elite Sport Met the Climate Crisis." *Climate Home News* https://www.climatechangenews.com/2021/08/06/hottest-games-ever-tokyo-olympics-elite-sport-met-climate-crisis/.

Lockwood, David and Matt Hickson. (2023, October 24). "What Could More European Football Mean for Planet?" *BBC* https://www.bbc.com/sport/football/67159156.

Lockwood, David. (2022, July 31). "Leilani Münter: Former Racing Driver Wins BBC's Evergreen Athlete Award." *BC* https://www.bbc.com/sport/motorsport/62263116.

Lockwood, David. (2023a, March 23). "Premier League Domestic Flights: BBC Sport Research Shows 81 Flights from 100 Games." *BBC* https://www.bbc.co.uk/sport/football/65017565.

Lockwood, David. (2023b, September 15). "World Cup 2026: Could Climate Crisis Impact the Men's Tournament?" *BBC* https://www.bbc.com/sport/football/66211285.

Lockwood, David. (2023c, November 2). "Saudi Arabia: New Research on Extent of Country's Investment in Sport." *BBC* https://www.bbc.com/sport/67295892.

Lockwood, David. (2024, July 28). "Which Premier League Clubs Fly the Most in Pre-Season?" *BBC* https://www.bbc.com/sport/football/articles/cm52013g52jo.

Logan, Joan. (2024, March 2). "Melbourne's F1 Grand Prix a Financial Car Crash for Victoria." *Pearls and Irritations* https://johnmenadue.com/melbournes-f1-grand-prix-a-financial-car-crash-for-victoria/.

Logsdon, Jeanne M. (2004). "Global Business Citizenship: Applications to Environmental Issues." *Business and Society Review* 109, no. 1: 67–87.

London Assembly Plenary. (2022, October 6). *Transcript of Agenda Item 3—Question and Answer Session: Legacy of the 2012 London Olympic Games and Paralympic Games* https://gla.moderngov.co.uk/documents/s102250/Minutes%20-%20Appendix%201%20-%20Transcript%20of%20Agenda%20Item%203.pdf.

López-Castro, Fernanda. (2025, January 10). "Gobierno de Sheinbaum remodelará AICM y AIFA rumbo al Mundial 2026: estos serán los cambios." *infobae* https://www.infobae.com/mexico/2025/01/10/gobierno-de-sheinbaum-remodelara-aicm-y-aifa-rumbo-al-mundial-2026-estos-seran-los-cambios/.

López-Egea, Sergi. (2019, December 13). "El reto climático en los grandes deportes." *elPeriódico* https://www.elperiodico.com/es/deportes/20191213/deporte-alerta-cambio-climatico-7773643.

López, Estela. (2022, March 29). "Barcelona capta la Copa América de Vela 2024 con 70 millones." *El Economista* https://www.eleconomista.es/catalunya/noticias/11692411/03/22/Barcelona-capta-la-Copa-America-de-Vela-2024-con-70-millones.html.

Louis Vuitton 37th America's Cup. (2024a). *Press Kit* https://www.americascup.com/press-kit.

Louis Vuitton 37th America's Cup. (2024b, October 22). "Moltes Gràcies Barcelona—The Jewel in the Sun." https://www.americascup.com/news/3759_MOLTES-GRACIES-BARCELONA-THE-JEWEL-IN-THE-SUN.

Lovato, Roberto. (2004, June 28). "Fear of a Brown Planet." *The Nation*: 17–21.

Lowes, Mark. (2004). "Neoliberal Power Politics and the Controversial Siting of the Australian Grand Prix Motorsport Event in a Public Park." *Loisir et Société/Society and Leisure* 27, no. 1: 69–88.

Lowes, Mark. (2018). "Toward a Conceptual Understanding of Formula One Motorsport and Local Cosmopolitanism Discourse in Urban Placemarketing Strategies." *Communication & Sport* 6, no. 2: 203–18.

Loy, John W. and Douglas Booth. (2000). "Functionalism, Sport and Society." *Handbook of Sports Studies*. Ed. Jay Coakley and Eric Dunning. London: Sage Publications. 8–27.

Lukács, Georg. (1972). *History and Class Consciousness: Studies in Marxist Dialectics*. Trans. Rodney Livingstone. Cambridge, Mass.: MIT Press.

Lusetich, Robert. (1999, January 11). "Games Elite Faces Purge Over Bribes." *Australian*: 1.

Luxemburg, Rosa. (1970). *Rosa Luxemburg Speaks*. Ed. Mary-Alice Waters. New York: Pathfinder Press.

Luxemburg, Rosa. (2004). *The Rosa Luxemburg Reader*. Ed. Peter Hudis and Kevin B. Anderson. New York: Monthly Review Press.

Luxemburg, Rosa. (2014). *The Complete Works of Rosa Luxemburg. Volume I: Economic Writings*. London: Verso.

Lyjak, Adam. (2023, January 10). "The Finances Behind the 2022 World Cup." *MichiganJournalofEconomics*https://sites.lsa.umich.edu/mje/2023/01/10/the-finances-behind-the-2022-world-cup/.

Lynch, Declan. (2021, August 1). "Our Ruling Class is at Ease with Itself and Just Sailing By." *Irish Independent* https://www.independent.ie/opinion/comment/our-ruling-class-is-at-ease-with-itself-and-just-sailing-by/40707762.html.

Lynch, Michael J., Michael A. Long, Kimberly L. Barrett, and Paul B. Stretesky. (2013). "Is it a Crime to Produce Ecological Disorganization? Why Green Criminology and Political Economy Matter in the Analysis of Global Ecological Harms." *British Journal of Criminology* 53, no. 6: 997–1016.

Lynd, Robert S. and Helen Merrell Lynd. (1956). *Middletown: A Study in Modern American Culture*. New York: Harcourt, Brace and Company.

Lynd, Robert S. and Helen Merrell Lynd. (1965). *Middletown in Transition: A Study in Cultural Conflicts*. New York: Harcourt Brace Jovanovich.

Mabon, Leslie. (2023). "Football and Climate Change: What Do We Know, and What is Needed for an Evidence-Informed Response?" *Climate Policy* 23, no. 3: 314–28.

Macbeth, Jim, John Selwood, and Sarah Veitch. (2012). "Paradigm Shift or a Drop in the Ocean? The America's Cup Impact on Fremantle." *Tourism Geographies: An International Journal of Tourism Space, Place and Environment* 14, no. 1: 162–82.

Macdonald, Ted. (2021, April 13). "In 2020, Broadcast TV Networks Mentioned Climate Solutions in Less Than 30% of Overall Climate Change Coverage." *Media Matters* https://www.mediamatters.org/broadcast-networks/2020-broadcast-tv-networks-mentioned-climate-solutions-less-30-overall-climate.

MacGregor, Sherilyn. (2006). *Beyond Mothering Earth: Ecological Citizenship and the Politics of Care*. Vancouver: University of British Columbia Press.

MacInnes, Paul. (2024, November 30). "Fifa Ignores Own Report Into Qatar World Cup Over Workers' Compensation." *Guardian* https://www.theguardian.com/football/2024/nov/30/fifa-ignores-own-report-into-qatar-world-cup-over-workers-compensation?CMP=Share_iOSApp_Other.

MacInnes, Paul. (2025, March 11). "Fifa Accused on Blocking Labour Inspection at 2026 World Cup Stadium." *Guardian* https://www.theguardian.com/football/2025/mar/11/fifa-accused-blocking-labour-inspection-azteca-stadium-2026-world-cup?CMP=Share_iOSApp_Other.

Maguire, Joseph. (1993). "Bodies, Sportscultures and Societies: A Critical Review of Some Theories in the Sociology of the Body." *International Review for the Sociology of Sport* 28, no. 1: 33–52.

Maguire, Kieran and Christina Philippou. (2024, October 17). "Why America is Buying Up English Football—And What it Means for the Future of 'Soccer'." *The Conversation* https://theconversation.com/why-america-is-buying-up-english-football-and-what-it-means-for-the-future-of-soccer-240695.

Mahiri, Jabari and Derek Van Rheenen. (2010). *Out of Bounds: When Scholarship Athletes Become Academic Scholars.* New York: Peter Lang.

Mair, Judith and Mike Duignan. (2025). "How Does Hosting the Olympics Benefit Local Communities?" *Events and Society: Bridging Theory and Practice.* Ed. Mike Duignan. Abingdon: Routledge. 107–14.

Malhado, Acácia Cristina Mendes and Rainer Rothfuss. (2013). "Transporting 2014 FIFA World Cup to Sustainability: Exploring Residents' and Tourists' Attitudes and Behaviours." *Journal of Policy Research in Tourism, Leisure and Events* 5, no. 3: 252–69.

Malik, Nasrine. (2025, July 22). "The Go-Between: How Qatar Became the Global Capital of Diplomacy." *Guardian* https://www.theguardian.com/world/2025/jul/22/how-qatar-became-the-global-capital-of-diplomacy.

Malthus, Thomas. (1798). *An Essay on Population, as it Affects the Future Improvement of Society with Remarks on the Speculations of Mr. Godwin, M. Condorcet, and Other Writers.* London: J. Johnson.

Manchester United. (n.d.). https://csr.manutd.com/.

Mandard, Stéphane. (2022a, April 21). "Le Pen Vows 'National Environmental Policies'." *Le Monde* https://www.lemonde.fr/en/2022-presidential-election/article/2022/04/17/le-pen-challenges-macron-s-punitive-ecology-with-her-national-ecology_5980790_16.html.

Mandard, Stéphane. (2022b, November 18). "Coupe du monde 2022: Le mirage de la compensation carbone pour atteindre la «neutralité» promise par la FIFA." *Le Monde* https://www.lemonde.fr/football/article/2022/11/18/coupe-du-monde-2022-le-mirage-de-la-compensation-carbone-pour-atteindre-la-neutralite-promise-par-la-fifa_6150404_1616938.html.

Mangan, J. A. (1981). *Athleticism in the Victorian and Edwardian Public School: The Emergence and Consolidation of an Educational Ideology.* Cambridge: Cambridge University Press.

Mao, Zedong. (1917). "A Study of Physical Education." https://www.commonprogram.science/documents/a%20study%20of%20physical%20education.pdf.

Marcus, George. (1995). "Ethnography in/of the World System: The Emergence of Multi-Sited Ethnography." *Annual Review of Anthropology* 24: 95–117.

Marcuse, Herbert. (1972). "Ecology and Revolution." *Liberation* 16: 10–12.

Marcuse, Herbert. (2009). *Negations: Essays in Critical Theory.* Trans. Jeremy J. Shapiro. London: MayFlyBooks.

Marinetti, F. T. (2009a). "The Founding and Manifesto of Futurism." *Futurism: An Anthology.* Ed. Lawrence Rainey, Christine Poggi, and Laura Wittman. New Haven: Yale University Press. 49–53.

Marinetti, F. T. (2009b). "The New Religion-Morality of Speed." *Futurism: An Anthology.* Ed. Lawrence Rainey, Christine Poggi, and Laura Wittman. New Haven: Yale University Press. 224–29.

Marire, Juniours. (2022). "Effect of Government Spending on Total Factor Productivity in South Africa." *European Journal of Economics, Law and Social Sciences* 6, no. 1: 435–52.

Marlet, Pierre. (2024, July 23). "Les Jeux Olympiques de Londres 2012 : un feu d'artifice ou un pétard mouillé ?" *RTBF* https://www.rtbf.be/article/les-jo-de-londres-2012-du-feu-d-artifice-de-promesses-pour-les-britanniques-au-petard-mouille-du-brexit-11408173.

Marmot, Michael G. (2004). "Evidence Based Policy or Policy Based Evidence?" *British Medical Journal* 328, no. 7445: 906–07.

Marquis, Christopher, Michael W. Toffel, and Yanhua Zhou. (2016). "Scrutiny, Norms, and Selective Disclosure: A Global Study of Greenwashing." *Organization Science* 27, no. 2: 483–504.

Marqusee, Mike. (2012, August 7). "London 2012: Spare Us the Jingoistic Olympic Hype." *Guardian* https://www.theguardian.com/commentisfree/2012/aug/07/london-2012-olympic-hype.

Marqusee, Mike. (2014, June). "Uneven Playing Fields." *Red Pepper* https://www.mikemarqusee.co.uk/1591-a-level-playing-field-global-sport-in-the-neo-liberal-age/.

Marrero-Guillamón, Isaac. (2011). "London 2012: espacio de excepción." *urbe. Revista Brasileira de Gestão Urbana* 3, no. 2: 179–90.

Marrero-Guillamón, Isaac. (2016). "The Politics and Aesthetics of Assembling: (Un)building the Common in Hackney Wick, London." *Urban Cosmopolitics: Agencements, Assemblies, Atmospheres.* Ed. Anders Blok and Ignacio Farías. Abingdon: Routledge. 125–45.

Marshall, Alasdair, Richard Telofski, Udechukwu Ojiako, and Max Chipulu. (2012). "An Examination of "Irregular Competition" Between Corporations and NGOs." *Voluntas* 23, no. 2: 371–91.

Marshall, T. H. (1950). *Citizenship and Social Class and Other Essays.* Cambridge: Cambridge University Press.

Martin, Alexander. (2013, September 5). "The 1964 Tokyo Olympics: A Turning Point for Japan." *Wall Street Journal* https://blogs.wsj.com/japanrealtime/2013/09/05/the-1964-tokyo-olympics-a-turning-point-for-japan/.

Martin, Daniel, Jason Macarlino, Felix Joensson, and Kate Hislop. (2017, June 15). "How Did the America's Cup Instigate Perth's Green Sand Dunes?" *Foreground* https://www.foreground.com.au/planning-policy/perth-and-americas-cup/.

Martin, Declan, Liss Ralston, and Tom Alves. (2024). *South East Queensland Displacement Monitoring: Baseline Report.* Australian Housing and Urban Research Institute for QShelter https://www.ahuri.edu.au/research/south-east-queensland-displacement-monitoring.

Martin, Lisa. (2019, February 6). "Climate Change Set to Disrupt Australia's Summer Sports Calendar." *Guardian* https://www.theguardian.com/environment/2019/feb/06/climate-change-set-to-disrupt-australias-summer-sports-calendar.

Martínez-Alier, Joan. (2012). "Environmental Justice and Economic Degrowth: An Alliance Between Two Movements." *Capitalism Nature Socialism* 23, no. 1: 51–73.

Martínez, David García. (2024, May 7). "El ABC de la America's Cup: Origen, equipos, formato e impacto económico." *Palco23* https://www.palco23.com/competiciones/americas-cup-la-competicion-que-se-reta-en-los-mares-desde-1851.

Marx, Karl. (1987). *Capital: Vol. 1: A Critical Analysis of Capitalist Production*, 3rd ed. Ed. Friedrich Engels. Trans. Samuel Moore and Edward Aveling. New York: International Publishers.

Marx, Leo. (1964). *The Machine in the Garden: Technology and the Pastoral Ideal in America*. New York: Oxford University Press.

Mas de Xaxàs, Xavier. (2024, August 19). "My Barcelona is Being Destroyed by Mass Tourism—But Kicking Visitors Out Isn't the Answer." *Guardian* https://www.theguardian.com/commentisfree/article/2024/aug/19/barcelona-mass-tourism-visitors-city-industry?CMP=ShareiOSAppOther.

Más Madrid. (2025). "Alegaciones a la aprobación inicial del "Plan especial de definición de redes en el Ape 21.10 "Parque Ferial Juan Carlos I y su etorno" situada en La Avenida del Partenón, 5. Distrito de Barajas." https://masmadrid.org/mas-madrid-denuncia-que-el-plan-urbanistico-de-almeida-para-la-formula-1-en-ifema-evita-medir-el-impacto-ambiental-del-proyecto/.

Matheson, Victor A. (2024). "The Economics of the FIFA World Cup." *Research Handbook on Major Sporting Events*. Ed. Harry A. Solberg, Rasmus K. Storm, and Kamilla Swart. Cheltenham: Edward Elgar. 663–77.

Matos Serodio, Paulo, Martin Mckee, Deborah Cohen, and David Stuckler. (2016). "Coca-Cola—A Model of Transparency in Research Partnerships?" *European Journal of Public Health* 26, no. 1. https://doi.org/10.1093/eurpub/ckw170.021.

Matras, Tomasz. (2024). "'Less Democracy Is Sometimes Better': Sportswashing and Hosting World Championships by Non-Democratic Countries, 1993–2022." *International Journal of the History of Sport* 41, no. 6: 531–48.

Matthews, Jordan J. K. and Elizabeth C. J. Pike. (2016). "'What On Earth Are They Doing in a Racing Car?': Towards an Understanding of Women in Motorsport." *International Journal of the History of Sport* 33, no. 13: 1532–550.

Mavrokefalidis, Dimitris. (2023, August 21). "Nation's Roar for Lionesses Ignites Power Surge." *Energy Live News* https://www.energylivenews.com/2023/08/21/nations-roar-for-lionesses-ignites-enormous-power-surge/.

Maxwell, Richard, Jon Raundalen, and Nina Lager Vestberg, eds. (2015). *Media and the Ecological Crisis.* New York: Routledge.

McBride, James, Noah Berman, and Melissa Manno. (2024, July 20). "The Economics of Hosting the Olympic Games." *Council on Foreign Relations* https://www.cfr.org/backgrounder/economics-hosting-olympic-games.

McCarthy, Lauren. (2021, December 2). "Why is Direct Action Essential for Climate Justice?" *Global Citizen* https://www.globalcitizen.org/en/content/direct-action-for-the-climate-crisis/.

McCarthy, Michael. (2009, December 8). "Football's Carbon Footprint Comes Under Fire." *Independent* https://www.independent.co.uk/environment/climate-change/footballs-carbon-footprint-comes-under-fire-1836035.html.

McCartney, Gerry, Sian Thomas, Hilary Thomson, John Scott, Val Hamilton, Phil Hanlon, David S. Morrison, and Lyndal Bond. (2010). "The Health and Socioeconomic Impacts of Major Multi-Sport Events: A Systematic Review." *British Medical Journal* 340 https://doi.org/10.1136/bmj.c2369.

McClintock, Anne. (1995). *Imperial Leather: Race, Gender and Sexuality in the Colonial Context.* New York: Routledge.

McClintock, Anne. (2010, June 24). "Militarizing the Gulf Oil Crisis." *CounterPunch* https://www.counterpunch.org/2010/06/24/militarizing-the-gulf-oil-crisis/.

McCordick, Jack. (2024, April 23). "Inside the Brutal Business Practices of Amazon—And How It Became "Too Toxic to Touch"." *Vanity Fair* https://www.vanityfair.com/news/story/inside-amazon-business-practices.

McCubbing, Gus and Zoe Samios. (2024, November 28). "Grand Prix Has Cost Victoria Nearly $700m in the Last Decade." *Australian Financial Review* https://www.afr.com/companies/sport/grand-prix-has-cost-victoria-nearly-700m-in-the-last-decade-20241128-p5kua5.

McCullough, Bryan P. and Timothy B. Kellison. (2018). "An Introduction to Environmental Sustainability and Sport." *Routledge Handbook of Sport and the Environment.* Ed. Bryan P. McCullough and Timothy B. Kellison. Abingdon: Routledge. 3–10.

McCurry, Justin and Associated Press. (2023, July 3). "'Tokyo Would Lose its Soul': Anger Over Plans to Redevelop Historic City Park." *Guardian* https://www.theguardian.com/world/2023/jul/01/tokyo-anger-plans-redevelop-historic-city-park-meiji-jingu-gaien?CMP=sharebtntw.

McDonald, Scott. (2020, March 25). "The Reason Why Olympics in 2021 Will Still be Called the 2020 Olympic Games." *Newsweek* https://www.newsweek.com/reason-why-olympics-2021-will-still-called-olympic-2020-games-1494333.

McGowan, Elliot. (2025, January 30). "Brisbane 2032 in "Hot Mess"." *inside the games* https://www.insidethegames.biz/articles/1151589/brisbane-2032-olympic-games-concerns.

McGregor, Callum. (2015). "Direct Climate Action as Public Pedagogy: The Cultural Politics of the Camp for Climate Action." *Environmental Politics* 24, no. 3: 343–62.

McHoul, Alec. (2015). "The Humdrum." *The Routledge Companion to Global Popular Culture*. Ed. Toby Miller. New York: Routledge. 112–18.

McIntyre, Isaac. (2023, October 26). "Lando Norris Sets New Twitch Record After F1 Cancels Aus Grand Prix." *Dexerto* https://www.dexerto.com/entertainment/lando-norris-twitch-record-f1-cancels-aus-grand-prix-1341664/.

McKay, Jim and Debbie Huber. (1992). "Anchoring Media Images of Technology and Sport." *Women's Studies International Forum* 15, no. 2: 205–18.

McKay, Jim. (2015). "Assessing the Sociology of Sport: On Revisiting the Sociological Imagination." *International Review for the Sociology of Sport* 50, nos. 4–5: 547–52.

McKenna, Kate. (2024, November 29). "Second Review into Olympic Infrastructure Launched, with 100 Days to Identify Priorities for 2032." *ABC* https://www.abc.net.au/news/2024-11-29/100-day-review-olympic-and-paralympic-infrastructure-brisbane/104665538.

McKie, Robin, Michael Savage, and Philip Cornwall. (2019, May 11). "As English Fans Get Set to Cross Europe, Anger Rises at Football's Carbon Bootprint." *Guardian* https://www.theguardian.com/environment/2019/may/11/anger-carbon-bootprint-english-football-finals-champions-league-europa-league.

McLaren Racing. (2021, April 14). *Sustainability at McLaren Racing* https://static-cdn.mclaren.com/static/pdf/SustainabilityatMcLarenRacing14April2021.pdf.

McMichael, Philip. (2011). "Revaluing Low Carbon Cultures: Learning to Live with the Earth." *Development* 54, no. 2: 169–71.

McMillan, Stephen. (2024, August 12). "28 Hours Later: An Olympic Zombie in Paris After the Games Are Over." *Guardian* https://www.theguardian.com/sport/article/2024/aug/12/28-hours-later-an-olympic-zombie-in-paris-after-the-games-are-over?CMP=ShareiOSAppOther.

McRae, Donald. (2024, September 27). "Ben Ainslie: 'We Were Having a Tussle and Then I Saw a Guy Pull a Knife Out'." *Guardian* https://www.theguardian.com/sport/2024/sep/27/ben-ainslie-sailing-americas-cup-interview.

Meager, Dani and WA Maritime Museum. (2023, November 5). "The Transformative Impact of the America's Cup on Fremantle." *Perth Now* https://www.perthnow.com.au/lifestyle/culture/the-transformative-impact-of-the-americas-cup-on-fremantle%2D%2Dc-12455806.

Meghji, Ali. (2021). *Decolonizing Sociology*. Cambridge: Polity Press.

Mehta, Amar. (2023, March 25). "Lewis Hamilton: Diversity in Formula 1 About Behind Scenes." *Sky Sports* https://www.skysports.com/f1/

news/12433/12842284/lewis-hamilton-diversity-in-formula-1-about-behind-scenes-i-felt-like-lone-ranger.

Ménage, Gaëlle. (2024, August 5). "Paris donne aux JO un bel environnement." *Forbes* https://www.forbes.fr/societe/issu-du-magazine-paris-donne-aux-jo-un-bel-environnement/.

Mercader, Aleix. (2024, October 24). "Barcelona Global solicita una reunión "urgente" para intentar salvar la Copa América." *El Economista* https://www.eleconomista.es/actualidad/noticias/13050484/10/24/barcelona-global-solicita-una-reunion-urgente-para-intentar-salvar-la-copa-america.html.

Merino, Daniel. (2020, July 23). "The First Undeniable Climate Change Deaths." *Slate* https://slate.com/technology/2020/07/climate-change-deaths-japan-2018-heat-wave.html.

Merton, Robert K. (1968). *Social Theory and Social Structure*. New York: The Free Press.

Microsoft. (2022). *2022 Environmental Sustainability Report* https://query.prod.cms.rt.microsoft.com/cms/api/am/binary/RW15mgm.

Mill, John Stuart. (2004). *Principles of Political Economy With Some of Their Applications to Social Philosophy*. Ed. Stephen Nathanson. Indianapolis: Hackett.

Miller, Sue. (2023, August 9). "Beware the Brutal Side of the Olympic Spirit." *Guardian* https://www.theguardian.com/lifeandstyle/2023/aug/09/beware-the-brutal-side-of-the-olympic-spirit.

Miller, Toby, Brett Hutchins, Libby Lester, and Richard Maxwell. (2023). "Formula One and the Insanity of Car-Based Transportation." *The History and Politics of Motor Racing: Lives in the Fast Lane*. Ed. Damion Sturm, Stephen Wagg, and David L. Andrews. Cham: Palgrave Macmillan. 733–60.

Miller, Toby, Geoffrey Lawrence, Jim McKay and David Rowe. (2001). *Globalization and Sport: Playing the World*. London: Sage Publications.

Miller, Toby. (1998). *Technologies of Truth: Cultural Citizenship and the Popular Media*. Minneapolis: University of Minnesota Press.

Miller, Toby. (2007). *Cultural Citizenship: Cosmopolitanism, Consumerism, and Television in a Neoliberal Age*. Philadelphia: Temple University Press.

Miller, Toby. (2010a). *Television Studies: The Basics*. London: Routledge.

Miller, Toby. (2010b, June 4). "Soccer Conquers the World." *Chronicle of Higher Education*: B6–B9.

Miller, Toby. (2018). *Greenwashing Sport*. London: Routledge.

Miller, Toby. (2019). "La crisis ambiental: El Marxismo contra el antropocentrismo filosófico." *Das Kapital. Marx, actualidad y crítica*. Ed. Paulina Aroch Fugellie, Enrique Gallegos, Miriam Madureira, and Felipe Victoriano. Mexico City: Siglo Veintiuno/Universidad Autónoma Metropolitana, Unidad Cuajimalpa. 200–14.

Milliken, Robert. (1996, October 15). "Sport is Murdoch's 'Battering Ram' for Pay TV." *Independent* https://www.independent.co.uk/sport/sport-is-murdoch-s-battering-ram-for-pay-tv-1358686.html.

Millington, Rob, Audrey R. Giles, Nicolien van Luijk, and Lyndsay M. C. Hayhurst. (2022). "Sport for Sustainability? The Extractives Industry, Sport, and Sustainable Development." *Journal of Sport and Social Issues* 46, no. 3: 293–317.

Mills, C. Wright. (2000). *The Sociological Imagination*. New York: Oxford University Press.

Mills, Mark P. (2013). *The Cloud Begins with Coal: Big Data, Big Networks, Big Infrastructure, and Big Power*. National Mining Association and American Coalition for Clean Coal Electricity https://www.tech-pundit.com/wpcontent/uploads/2013/07/CloudBeginsWithCoal.pdf?c761ac.

Ministerio de Trabajo y Economía Social. (2024, July 24). *Real Decreto 712/2024, de 23 de julio, por el que se desarrollan las bonificaciones de cuotas aplicables a las actividades relacionadas con la organización y celebración de la XXXVII edición de la Copa América en la ciudad de Barcelona* https://www.boe.es/diario_boe/txt.php?id=BOE-A-2024-15206.

Mitchell, Carleton. (1958, May 12). "Defenders of the Cup." *Sports Illustrated* https://vault.si.com/vault/1958/05/12/defenders-of-the-cup.

Mitchell, James. (2007). "Sustainable Soccer: How Green Projects at International Sporting Events Benefit the Fans, the Global Climate, and Local Populations." *Sustainable Development Law & Policy* 7, no. 2: 52, 84.

Mitchell, Kevin. (2024, November 20). "Farewell, Rafael Nadal: It Was My Pleasure to Know One of Sport's Most Gracious Champions." *Guardian* https://www.theguardian.com/sport/blog/2024/nov/20/farewell-rafael-nadal-one-of-sports-most-gracious-champions-tennis.

Mitchell, Scott. (2019, February 11). "Former BAR Team Owner BAT Back into Formula 1 with McLaren Deal." *Autosport* https://www.autosport.com/f1/news/former-bar-team-owner-bat-back-into-formula-1-with-mclaren-deal-5283606/5283606/.

Mohamed Radhi, Nor Ain. (2023, May 26). "MACC Opens Investigation Paper on Corruption Involving Petronas, International Oil and Gas Firm." *New Straits Times Business Times* https://www.nst.com.my/news/crime-courts/2023/05/913656/macc-opens-investigation-paper-corruption-involving-petronas.

Molinari, Letizia. (2024, August 8). "Sin tregua olímpica para los sectores populares." Trans. Florencia Oroz. *Jacobin* https://jacobinlat.com/2024/08/sin-tregua-olimpica-para-los-sectores-populares/.

Molodkin, Andrei. (2022). *Líbero: The Dirtiest Cup* https://thedirtiestcup.com/.

Montague, James. (2010, August 20). "Punks, Prostitutes and St. Pauli: Inside Soccer's Coolest Club." *CNN* https://edition.cnn.com/2010/SPORT/football/08/18/football.st.pauli.punks/index.html.

Montaigne, Michel de. (1966). *The Complete Essays of Montaigne*. Trans. Donald M. Frame. Stanford: Stanford University Press.

Moore, Dan. (2024, May 8). "Taxpayers Are About to Subsidize a Lot More Sports Stadiums." *The Atlantic* https://www.theatlantic.com/ideas/archive/2024/05/sports-stadium-subsidies-taxpayer-funding/678319/.

Moore, Malcolm. (2025, April 21). "Will Going Back to Basics Restore BP's Fortunes?" *Financial Times* https://www.ft.com/content/ab60efc1-4abf-4e92-bd1b-b09046ced3f5?emailId=6520e7be-7310-4050-95aa-88ec4689ffc0&segmentId=9264b0f7-e7ac-8f9b-044f-c10729049333.

More, Thomas. (2012). *Open Utopia*. Ed. Stephen Duncombe. New York: Minor Compositions.

Morenilla, Juan. (2024, August 12). "París ilumina el camino a Los Ángeles." *El País* https://elpais.com/deportes/juegos-olimpicos/2024-08-11/paris-ilumina-el-camino-a-los-angeles.html.

Morgan, Rhys and Yohanes Scarlett. (2021, July). *Accelerating Change: Improving Representation of Black People in UK Motorsport*. The Hamilton Commission and the Royal Academy of Engineering https://static1.squarespace.com/static/5f29736c8982c82f61df71e0/t/60edd33a6f118478735acbbc/1626198854176/THC+-+Accelerating+Change+-+July+2021.pdf.

Mori, Sayaka. (2021). https://x.com/sayakasofiamori/status/1422856371981094919.

Morin, Edgar and Anne Brigitte Kern. (1999). *Homeland Earth: A Manifesto for the New Millennium*. Trans. Sean M. Kelly and Roger LaPointe. Cresskill: Hampton Press.

Morris, Alex. (2012, August 28). "Formula One's Environmental Initiatives." *Triple Pundit* https://www.triplepundit.com/podium/formula-environmental-initiatives/.

Morteaux, Alexandre. (2024, June 28). "Quel bilan économique et environnemental des Jeux olympiques?" *Alternatives Economiques* https://www.alternatives-economiques.fr/bilan-economique-environnemental-jeux-olympiques/00111854.

Mosley, Max. (2015). *Formula One and Beyond: The Autobiography*. London: Simon and Schuster.

Motavalli, Jim. (2011, February 12). "A History of Greenwashing: How Dirty Towels Impacted the Green Movement." *AOL News* https://www.aol.com/article/2011/02/12/the-history-of-greenwashing-how-dirty-towels-impacted-the-green/19628686/?gen=1.

Mourão, Paulo. (2017). *The Economics of Motorsports: The Case of Formula One*. Cham: Palgrave Macmillan.

Moutiez, Julia. (2024, July 16). "Nager dans la Seine: Un loisir ancien qui refait surface à l'heure du réchauffement climatique." *The Conversation* https://the-conversation.com/nager-dans-la-seine-un-loisir-ancien-qui-refait-surface-a-lheure-du-rechauffement-climatique-234776.

M. S. (2019, November 14). "Youth Movements are the Fool's Gold of Politics." *Economist* https://www.economist.com/open-future/2019/11/14/youth-movements-are-the-fools-gold-of-politics.

Muhamad, Susana. (2023, December 13). Speech https://x.com/susanamuhamad/status/1735049874540744800?ref_src=twsrc%5Etfw%7Ctwcamp%5Etweetembed%7Ctwterm%5E1735049874540744800%7Ctwgr%5Ec0e274aa9c8c0603b25025d40ea0c902c7a15c7e%7Ctwcon%5Es1_&ref_url=https%3A%2F%2Fwww.climatechangenews.com%2F2023%2F12%2F13%2Fdubai-deal-ministers-and-observers-react-to-the-uae-consensus%2F.

Mullan, Donal, Iestyn Barr, Neil Brannigan, Nuala Flood, Oliver R. Gibson, Catherine Hambly, Alan T. Kennedy-Asser, Aimée C. Kielt, Tom Matthews, and Madeleine Orr. (2025). "Extreme Heat Risk and the Potential Implications for the Scheduling of Football Matches at the 2026 FIFA World Cup." *International Journal of Biometeorology* 69: 753–63.

Mullen, Tom. (2023, September 19). "Why Crowdsourcing Industrial Espionage is Economical for America's Cup." *Forbes* https://www.forbes.com/sites/tmullen/2023/09/19/why-crowdsourcing-industrial-espionage-is-economical-for-americas-cup/.

Mullenbach, Lauren E. and Birgitta L. Baker. (2020). "Environmental Justice, Gentrification, and Leisure: A Systematic Review and Opportunities for the Future." *Leisure Sciences: An Interdisciplinary Journal* 42, nos. 5–6: 430–47.

Müller, Martin, David Gogishvili, Sven Daniel Wolfe, Christopher Gaffney, Miriam Hug, and Annick Leick. (2023). "Peak Event: The Rise, Crisis and Potential Decline of the Olympic Games and the World Cup." *Tourism Management* 95 https://doi.org/10.1016/j.tourman.2022.104657.

Müller, Martin, Sven Daniel Wolfe, David Gogishvili, Christopher Gaffney, Miriam Hug, and Annick Leick. (2022a). "The Mega-Events Database: Systematising the Evidence on Mega-Event Outcomes." *Leisure Studies* 41, no. 3: 437–45.

Müller, Martin, David Gogishvili, and Sven Daniel Wolfe. (2022b). "The Structural Deficit of the Olympics and the World Cup: Comparing Costs Against Revenues Over Time." *Environment and Planning A: Economy and Space* 54, no. 6: 1200–218.

Müller, Martin. (2015). "(Im-)mobile Policies: Why Sustainability Went Wrong in the 2014 Olympics in Sochi." *European Urban and Regional Studies* 22, no. 2: 191–209.

Muñana, Gustavo. (2024, July 27). "Extinction Rebellion Activists Arrested in Paris for Alleged Sabotage." *Inside the Games* https://www.insidethegames.biz/articles/1147127/extinction-rebellion-activists-arrested.

Münter, Leilani. (n.d.). https://www.leilani.green/.

Murphy, Andrew and Valentin Simon. (2021, April). *Private Jets: Can the Super-Rich Supercharge Zero-Emission Aviation?* Transport & Environment https://www.transportenvironment.org/wp-content/uploads/2021/05/202209privatejetsFINALwithaddendum.pdf.

Musto, Julia. (2024, October 29). "Every Email, TikTok and Text We Send is Killing the Environment. Here's How." *Independent* https://www.independent.co.uk/climate-change/email-tiktok-environment-climate-change-b2637645.html.

Mytton, David. (2020). "Hiding Greenhouse Gas Emissions in the Cloud." *Nature Climate Change* 10: 700–01.

Nader, Laura. (1972). "Up the Anthropologist—Perspectives Gained from Studying Up." *Reinventing Anthropology*. Ed. Dell H. Hymes. New York: Pantheon. 284–311.

Nagao, Futa, Marika Oi, and Suranjana Tewari. (2021, August 6). "Tokyo 2020: Olympic Athletes Struggle with Heat at Games." *BBC* https://www.bbc.com/news/av/world-asia-58110846.

Nájar, Alberto. (2024, October 2). "Qué pasó en la matanza de Tlatelolco hace 56 años por la que Claudia Sheinbaum pidió perdón en su primer día como presidenta de México." *BBC Mundo* https://www.bbc.com/mundo/articles/cnvde3j6q6vo.

Nakagawa, Satoko. (2018, September 2). "Criticism Grows Against Gov't 'Urban Mines' Plan for Tokyo 2020 Medals." *The Mainichi* https://mainichi.jp/english/articles/20180902/p2a/00m/0na/012000c.

Nandy, Ashis. (1998). "The Twilight of Certitudes: Secularism, Hindu Nationalism, and Other Masks of Deculturation." *Postcolonial Studies* 1, no. 3: 283–98.

Nash, June and Maria Patricia Fernandez, eds. (1983). *Women, Men and the International Division of Labor*. Albany: SUNY Press.

National Audit Office. (2022). *Grassroots Participation in Sport and Physical Activity* https://www.nao.org.uk/wp-content/uploads/2022/07/Grassroots-participation-in-sport-and-physical-activity.pdf.

National Grid. (2010). *National Grid Powers Up for World Cup 2010* https://www.nationalgrid.com/uk/Media+Centre/WorldCup2010.

National Grid. (2014a, May 30). National Grid Up for the World Cup www.nationalgrid.com/mediacentral/uk-press-releases/2014/national-grid-up-for-the-world-cup/.

National Grid. (2014b, July 11). World Cup Fans Switched on to Brazil https://media.nationalgrid.com/press-releases/archive/world-cup-fans-switched-on-to-brazil/.

National Grid. (2020, April 6). *The 'Lockdown Effect' on TV Viewing Habits and the Electricity Grid* https://www.nationalgrideso.com/news/lockdown-effect-tv-viewing-habits-and-electricity-grid.

National Institute of Standards and Technology. (n.d.). "Earthquake Mexico 1985." https://www.nist.gov/el/earthquake-mexico-1985.

National Oceanic and Atmospheric Administration. (2017, April 20). "Deepwater Horizon Oil Spill Settlements: Where the Money Went." https://www.noaa.gov/explainers/deepwater-horizon-oil-spill-settlements-where-money-went.

National Oceanic and Atmospheric Administration. (2024, September 10). "U.S. Sweltered Through its 4th-Hottest Summer on Record." https://www.noaa.gov/news/us-sweltered-through-its-4th-hottest-summer-onrecord#:~:text=Meteorological%20summer%20(June%20through%20August,fourth%2Dhottest%20summer%20on%20record.

Neeson, J. M. (1996). *Commoners: Common Right, Enclosure and Social Change in England, 1700–1829.* Cambridge: Cambridge University Press.

Neils, Jennifer (1992). *Goddess and Polis: The Panathenaic Festival in Ancient Athens.* Princeton: Princeton University Press.

Nelson, Nigel. (2012, December 2). "Rob the Builder: Thousands of Construction Workers Were Denied Jobs as They Were Put on a Sinister Blacklist." *Daily Mirror* https://www.mirror.co.uk/news/uk-news/blacklist-thousands-of-construction-workers-denied-1469233.

Nestler, Stefan. (2022, August 17). "How Football is Responding to the Energy Crisis." *Deutsche Welle* https://www.dw.com/en/how-the-bundesliga-is-responding-to-the-energy-crisis/a-62837284.

Newman, Michael. (2005). *Socialism: A Very Short Introduction.* Oxford: Oxford University Press.

New Weather Institute with Scientists for Global Responsibility. (2025a, February 3). *Dirty Tackle: The Growing Carbon Footprint of Football* https://www.new-weather.org/wp-content/uploads/2025/01/Dirty_tackle_The_growing_carbon_footprint_of_football.pdf.

New Weather Institute with Scientists for Global Responsibility. (2025b, July). *FIFA's Climate Blind Spot: The Men's World Cup in a Warming World* https://www.newweather.org/wp-content/uploads/2025/07/FIFAs_climate_blind_spot.pdf.

New Weather Institute. (2024a, July). *Polluters on the Podium: Which Sponsors of the Paris 2024 Olympics Are the Biggest Polluters?* https://www.newweather.org/wp-content/uploads/2024/07/PollutersonthePodiumJuly2024.pdf.

New Weather Institute. (2024b, September 18). *Dirty Money: How Fossil Fuel Sponsors Are Polluting Sport* https://www.newweather.org/wp-content/

uploads/2024/09/Dirty-Money-How-Fossil-Fuel-Sponors-Pollute-Sport-18-09-2024.pdf.

Nichols, Georgia and Mike Savage. (2017). "A Social Analysis of an Elite Constellation: The Case of Formula 1." *Theory, Culture & Society* 34, nos. 5–6: 201–25.

Nicholson-Smith, Donald. (2008, October 1). "Black Glove/White Glove: Revisiting Mexico's 1968." *openDemocracy* https://www.opendemocracy.net/en/black_glove_white_glove_revisiting_mexicos_1968/.

Nielsen, Kristian S., Jan M. Bauer, Ramit Debnath, Charles A. Emogor, Sonja M. Geiger, Sakshi Ghai, Wencke Gwordz, and Ulf J. J. Hahnel. (2024). "Underestimation of Personal Carbon Footprint Inequality in Four Diverse Countries." *Nature Climate Change*. https://doi.org/10.1038/s41558-024-02130-y.

Nixon, Robert. (2011). *Slow Violence and the Environmentalism of the Poor.* Cambridge, Mass: Harvard University Press.

Noble, Josh with Paul Caruana Galizia, Michael Acton, and Chris Campbell. (2025, March 13). "Does the Olympic Business Model Still Work?" *Financial Times* https://www.ft.com/content/7e0e48eb-7c47-4028-9d87-589284efeacd?emailId=c94b36ed-2372-451c-bcd2-5fb0337e9107&segmentId=9264b0f7-e7ac-8f9b-044f-c10729049333.

Noëth, Bart. (2024, April 21). "Emirates Airline's Staff Speak Out (Anonymously) After UAE Records Heaviest Rainfall in 75 Years." *Aviation24* https://www.aviation24.be/airlines/emirates-airline/staff-speak-out-anonymously-after-uae-records-heaviest-rainfall-in-75-years/.

NOlympics LA. (n.d.). https://nolympicsla.com/.

NOlympics LA. (2025). *Nølympics LA Public Opinion Poll 2025* https://nolympicsla.com/polling/.

Notre Affaire à Tous. (2022, November 2). *Plainte devant le Jury de Déontologie Publicitaire* https://climatecasechart.com/wp-content/uploads/non-us-case-documents/2022/20221102_18433_complaint.pdf.

Nunn, Samuel and Mark S. Rosentraub. (1997). "Sports Wars: Suburbs and Center Cities in a Zero-Sum Game." *Journal of Sport & Social Issues* 21, no. 1: 65–82.

Nussbaum, Martha C. (2006). *Frontiers of Justice: Disability, Nationality, Species Membership*. Cambridge, Mass: The Belknap Press of Harvard University Press.

Nussbaum, Martha C. (2013). *Political Emotions: Why Love Matters for Justice.* Cambridge, Mass.: The Belknap Press of Harvard University Press.

Nussbaum, Martha C. (2016). *Anger and Forgiveness: Resentment, Generosity, Justice*. New York: Oxford University Press.

Nye, David E. (1994). *American Technological Sublime*. Cambridge, Mass.: MIT Press.

O'Brien, Isabel. (2024, September 15). "Data Center Emissions Likely 662% Higher Than Big Tech Claims. Can it Keep Up the Ruse?" *Guardian* https://www.theguardian.com/technology/2024/sep/15/data-center-gas-emissions-tech.

Oceana. (2021). *Exposed: Amazon's Enormous and Rapidly Growing Plastic Pollution Problem* https://oceana.org/reports/amazon-report-2021/.

O'Connor, Francis G. and David W. DeGroot. (2024). "Heat-Related Illness in Athletes." *Journal of the American Medical Association* 332, no. 8: 664–65.

O'Connor, James R. (1998). *Natural Causes: Essays in Ecological Marxism.* New York: Guilford.

Oceana. (2025, March). *Coca-Cola's World with Waste: The Soft Drink Company's Enormous and Worsening Plastic Problem and How Reusable Packaging Can Solve it* https://oceana.org/wp-content/uploads/sites/18/2025/03/Coca-Colas-World-With-Waste-Oceana-Report-2025_reduced-size-1.pdf.

Odigie, Akhator Joel. (2024, October 15). Letter to Patrice Motsepe https://www.ituc-africa.org/IMG/pdf/2024_oct_letter_caf_eng.pdf.

ODOXA. (2023, November 12). *Les Jeux Olympiques 2024: regards croisés des Français et des Franciliens* https://www.odoxa.fr/sondage/jo-2024-1-francilien-sur-6-souhaite-louer-son-logement/.

Office of the Assistant Secretary for Health. (2019). *National Youth Sports Strategy* https://health.gov/sites/default/files/2019-10/National_Youth_Sports_Strategy.pdf.

Oficina Municipal de Dades and Gerència Municipal. (2024). *Els salaris a Barcelona. 2022* https://bcnroc.ajuntament.barcelona.cat/jspui/handle/11703/134076.

Ogan, Christine L. Manaf Bashir, Lindita Camaj, Yunjuan Luo, Brian Gaddie, Rosemary Pennington, Sonia Rana, and Mohammed Salih. (2009). "Development Communication: The State of Research in an Era of ICTs and Globalization." *Gazette* 71, no. 8: 655–70.

Oltermann, Philip. (2014, June 16). "Greenpeace Loses £3m in Currency Speculation." *Guardian* https://www.theguardian.com/environment/2014/jun/16/greenpeace-loses-3m-pounds-currency-speculation.

O'Neill, Francis, Geoff Dickson, Tim Ströbel, and Ashleigh-Jane Thompson. (2025). "Elite Athlete Activism, Advocacy, and Protest: A Scoping Review." *European Sport Management Quarterly* 25, no. 1: 145–73.

Onwurah, Chi. (2021, July 16). "Lewis Hamilton is Right About Diversity. But the Issue Goes Way Beyond Motorsport." *Guardian* https://www.theguardian.com/commentisfree/2021/jul/16/lewis-hamilton-motorsport-diversity-f1-commission-black-female-minorities.

O'Reilly, Karen. (2000). *The British on the Costa Del Sol.* London: Routledge.

Organisationskomitee für die XI. Olympiade Berlin 1936 E.V. (1937). *The XIth Olympic Games, Berlin, 1936: Official Report.* Ed. Friedrich Richter. Berlin: Wilhelm Limpert.

Orlove, Raphael. (2024, April 22). "F1 in its Streamer Era." *Road & Track* https://www.roadandtrack.com/car-culture/a60013306/f1-is-for-the-streamers/.

Orwell, George. (1945, December). "The Sporting Spirit." *Tribune* https://orwell.ru/library/articles/spirit/english/e_spirit.

Orwell, George. (1949). *Nineteen Eighty-Four* https://files.libcom.org/files/1984.pdf.

O'Shea, Michelle, Neil Perry, and Sarah Duffy. (2020, January 15). "Formula 1 Says It's Going Carbon Neutral but Fans Must Demand Greater Detail on How." *The Conversation* https://theconversation.com/formula-1-says-its-going-carbon-neutral-but-fans-must-demand-greater-detail-on-how-127328.

OneTreePlanted. (n.d.). https://bellerintrees.raisely.com/.

Organisation for Economic Co-operation and Development. (2011). *Environmental Claims: Findings and Conclusions of the OECD Committee on Consumer Policy.* OECD Green Growth Papers 2011-01. https://www.oecd-ilibrary.org/docserver/5k9h3633prbq-en.pdf?expires=1733738709&id=id&accname=guest&checksum=B641F414D2BB4A02D244BFF937445B3F.

Organisation for Economic Co-operation and Development. (2016). *The Ocean Economy in 2030* https://www.oecd.org/en/publications/the-ocean-economy-in-2030978926425l724-en.html.

Organisation for Economic Co-operation and Development. (2018). *Working Together for Local Integration of Migrants and Refugees in Barcelona* https://doi.org/10.1787/9789264304062-en.

Organisation for Economic Co-operation and Development. (2022). *Global Plastics Outlook: Economic Drivers, Environmental Impacts and Policy Options* https://www.oecd-ilibrary.org/docserver/de747aef-en.pdf?expires=1732255776&id=id&accname=guest&checksum=69258D4A3FF9BCA837D7EBCED4EAF6F5.

Organization of American States. (2023, March 6). "OAS and ACES Europa Promote Sport as a Tool for Social Change." https://www.oas.org/en/media_center/photonews.asp?sCodigo=FNE-123173.

Orsini, Alessandro. (2024). *Sociological Theory: From Comte to Postcolonialism.* Cham: Palgrave Macmillan.

Osborne, Mike. (2024, April 18). "Sending Australian Team to Paris Olympics Will Cost $25M+." *Sports Hounds* https://sportshounds.com.au/2024/04/18/sending-australian-team-to-paris-olympics-will-cost-25m/#:~:text=AOC%20chief%20executive%20Matt%20Carroll's,medallist%20will%20earn%20in%20Paris.

Ostarek, Markus, Brent Simpson, Cathy Rogers, and James Ozden. (2024). "Radical Climate Protests Linked to Increases in Public Support for Moderate Organizations." *Nature Sustainability* 7: 1626–632.

Ostrom, Elinor. (2000). "Reformulating the Commons." *Swiss Political Science Review* 6, no. 1: 29–52.

Otani, Teinosuke. (2018). *A Guide to Marxian Political Economy: What Kind of a Social System is Capitalism?* Trans. Michael Schauerte. Cham: Springer.

Oxfam International. (2024). *Carbon Inequality Kills: Why Curbing the Excessive Emissions of an Elite Few Can Create a Sustainable Planet for All* https://oxfamilibrary.openrepository.com/bitstream/handle/10546/621656/bp-carbon-inequality-kills-281024-en.pdf?sequence=1.

Oxford English Dictionary. (2002). https://www.oed.com/dictionary/greenwashing_n?tl=true.

Öztopcu, Asli. (2024). "Environmental Sustainability in Sport: Formula 1's SDGs Responsiveness." *Environment, Development and Sustainability* 26: 31075–1097.

Paché, Gilles. (2020). "Sustainability Challenges in Professional Football: The Destructive Effects of the Society of the Spectacle." *Journal of Sustainable Development* 13, no. 1: 85–96.

Paché, Gilles. (2023). "Le ballet des FIFA World Cups: d'une aberration à l'autre." *Revue de Management et de Stratégie*: 1–10

Paché, Gilles. (2024). "FIFA World Cups: An Own Goal Against Sustainability." *IUP Journal of Management Research* 24, no. 3: 51–58.

Painter, Corinne. (2024). "Reconstructing British Identity: Formula One, Michael Schumacher and the British Press at the Turn of the Century." *Language and Intercultural Communication* 24, no. 2: 77–89.

Pallot, Richard. (2021, June 22). "Amazon Destroying Millions of Items of Unsold Stock in One of its UK Warehouses Every Year, ITV News Investigation Finds." *ITV* https://www.itv.com/news/2021-06-21/amazon-destroying-millions-of-items-of-unsold-stock-in-one-of-its-uk-warehouses-every-year-itv-news-investigation-finds.

Palmer, Jordi. (2024, June 19). "L'altra cara de la Fórmula 1: mig miler de barcelonins es manifesten contra l'exhibició." *El Nacional.cat* https://www.elnacional.cat/ca/barcelona/altra-cara-formula-1-mig-miler-barcelonins-es-manifesten-contra-exhibicio_1238529_102.html.

Paraguassu, Lisandra and Fabio Teixeira. (2023, October 6). "Heineken-Owned Brewery Added to Brazil 'Slave Labor' List." *Reuters* https://www.reuters.com/business/retail-consumer/heineken-owned-brewery-added-brazil-slave-labor-list-2023-10-06/.

Paris 2024 Social Charter. (2024). https://press.paris2024.org/assets/paris-2024-social-charter-pdf-deac-7578a.html.

Parker, Andrew, Alessandro Lomi, and Paola Zappa. (2024). "Effect of Time Pressure on Informal Advice Relations Across Organizational Units: Evidence from a Study of Collaboration Within a Formula One Racing Team." *Organization Studies* 45, no. 10: 1441–465.

Parodi, Emilio. (2019, February 13). "Exclusive: BT Executives Knew of Accounting Fraud in Italy Unit—Prosecutors." *Reuters* https://www.reuters.com/article/us-bt-italy-exclusive/exclusive-bt-executives-knew-of-accounting-fraud-in-italy-unit-prosecutors-idUSKCN1Q216M/.

Parodi, Emilio. (2024, June 11). "Exclusive: Italian Prosecutors Probing Supply Chain of Around a Dozen Fashion Brands." *Reuters* https://www.reuters.com/markets/europe/italian-prosecutors-probing-supply-chain-around-dozen-more-fashion-brands-source-2024-06-11/.

Parsons, Talcott and Robert F. Bales in collaboration with James Olds, Morris Zelditch, Jr., and Philip E. Slater. (1955). *Family, Socialization and Interaction Process.* Glencoe: Free Press.

Pato, Ignacio. (2021, September 1). "In Livorno, Italy's Most Famous Left-Wing Football Club is Fighting to Survive." *Jacobin* https://jacobin.com/2021/09/livorno-football-club-italy-communism.

Paul Nkemngu, Acha-Anyi. (2012). "The 2010 FIFA World Cup as Sustainable Tourism: A Community Perspective." *Journal of Tourism & Hospitality* 1, no. 5. https://doi.org/10.4172/2167-0269.1000106.

Payne, Catherine J., Tim S. Jessop, Patrick-Jean Guay, Michele Johnstone, Megan Feore, and Raoul A. Mulder. (2012). "Population, Behavioural and Physiological Responses of an Urban Population of Black Swans to an Intense Annual Noise Event." *PLOS One* https://doi.org/10.1371/journal.pone.0045014.

Paz, Octavio. (1994). *El laberinto de la soledad.* Mexico City: Fondo de Cultura Económica.

Pepper, David. (2000). "Environmentalism." *Understanding Contemporary Society: Theories of the Present.* Ed. Gary Browning, Abigail Halcli, and Frank Webster. London: Sage Publications. 445–62.

Perdiguero, Jordi and Alex Sanz. (2020). "Cruise Activity and Pollution: The Case of Barcelona." *Transportation Research Part D* 78 https://doi.org/10.1016/j.trd.2019.11.010.

Perkins, Sean. (2000). "Exploring Future Relationships Between Football Clubs and Local Government." *Soccer & Society* 1, no. 1: 102–13.

Peters, Justin. (2021, July 24). "Olympics Jerk Watch: The International Olympic Committee." *Slate* https://slate.com/culture/2021/07/international-olympic-committee-jerks-ioc-corruption-covid-tokyo.html.

Pickford, N. (2014, July 14). "Cyclists are Miles Behind Formula 1 in the Environmental Race." *Hull Daily Mail* https://www.hulldailymail.co.uk/

Cyclists-miles-Formula-1-environmental-race/story-21460343-detail/story.html.

Pieth, Mark, Stefan Wehrenberg, and Rodney Dixon. (2024, May 22). "Submission to FIFA Regarding Saudi Arabia's Bid to Host the Men's 2034 World Cup." https://wehrenberg-law.ch/media/documents/Legal-Submission-to-FIFA-22-May-2024.pdf.

Piketty, Thomas. (2022). *A Brief History of Equality*. Trans. Steve Rendall. Cambridge, Mass.: The Belknap Press of Harvard University Press.

Pinchot, Gifford. (1910). *The Fight for Conservation*. New York: Doubleday, Page & Company.

Pindar. (1915). *The Odes of Pindar Including the Principal Fragments*. Trans. John Sandys. London: William Heinemann; New York: Macmillan.

Pine, Matthew K., Emily Hague, Anna Kebke, Lauren McWhinnie, and Charlotte R. Findlay. (2024). "Not so Silent Spectators: How Spectator Vessels at International Sailing Regattas Alter Marine Soundscapes." *Marine Pollution Bulletin* 202 https://doi.org/10.1016/j.marpolbul.2024.116309.

Piñera, Sebastián. (1978). *Evolución de la pobreza en México: (Períodos 1950–1963; 1963–1968)*. Proyecto Interinstitucional de Pobreza Crítica en América Latina https://repositorio.cepal.org/server/api/core/bitstreams/dc713c40-d4ed-45c8-99f9-4c52e10f7008/content.

Pinkerton, Byrd. (2021, September 14). "One Good Thing: I Didn't Watch Any Sports. Then I Watched Netflix's Drive to Survive." *Vox* https://www.vox.com/22652554/drive-to-survive-f1-racing-netflix.

Pirks, Natalie and Tom Grundy. (2024, September 27). "29,000 Gambling Ads in Premier League Weekend, Says Research." *BBC* https://www.bbc.com/sport/football/articles/c4g07v7x70lo.

Pistolesi, Alice and Monica Pelliccia. (2024, July 2). "Across Mexico, World Cup Infrastructure Threatens Biodiversity and Communities." *Mongabay* https://news.mongabay.com/2024/07/across-mexico-world-cup-infrastructure-threatens-biodiversity-and-communities/.

Pitts, Adrian and Hanwen Liao. (2013). "An Assessment Technique for the Evaluation and Promotion of Sustainable Olympic Design and Urban Development." *Building Research and Information* 41, no. 6: 722–34.

Plaschke, Bill. (2024, August 11). "Bring it On! Los Angeles Begins Countdown to 2028 Olympics." *Los Angeles Times* https://www.latimes.com/sports/olympics/story/2024-08-11/plaschke-la-2028-olympics-bring-it.

Plato. (1997). *Plato: The Complete Works*. Ed. John M. Cooper. Assoc. Ed. D. S. Hutchinson. Indianapolis: Hackett Publishing.

Pocréaux, Cyril. (2024a, November 21). "Les barbouzeries de Bernard Arnault : en immersion au procès du Squale." *Fakir* https://fakirpresse.info/les-barbouzeries-1-les-sbires-de-lvmh-a-la-barre.

Pocréaux, Cyril. (2024b, November 28). "Les barbouzeries de Bernard Arnault : en immersion au procès du Squale Partie 4." *Fakir* https://fakirpresse.info/les-barbouzeries-4-bernard-arnault-a-la-barre.

Poindexter, Owen. (2021, June 21). "Amazon's Aggressive Expansion into Sports Continues." *Front Office Sports* https://frontofficesports.com/amazon-wants-all-the-live-sports/.

Polloni, Camille. (2024, August 5). "Visé par une enquête pour meurtre, un policier fait du BMX à la cérémonie d'ouverture des JO." *Mediapart* https://www.mediapart.fr/journal/france/050824/vise-par-une-enquete-pour-meurtre-un-policier-fait-du-bmx-la-ceremonie-d-ouverture-des-jo.

Poinssot, Amélie. (2024, August 2). "Sur le village Olympique, le gaspillage alimentaire n'esta pas évité." *Mediapart* https://www.mediapart.fr/journal/economie-et-social/020824/sur-le-village-olympique-le-gaspillage-alimentaire-n-est-pas-evite.

Polenberg, Richard, ed. (2002). *In the Matter of J. Robert Oppenheimer*. Ithaca: Cornell University Press.

Political Economy Research Institute. (2024). *Toxic 100 Air Polluters Index* https://peri.umass.edu/toxic-100-air-polluters-index-current.

Poniatowska, Elena. (1971). *La noche de Tlatelolco: Testimonios de historia oral*. Mexico City: Ediciones Era.

Poniatowska, Elena. (1995). *Nothing, Nobody: The Voices of the Mexico City Earthquake*. Trans. Aurora Camacho de Schmidt and Arthur Schmidt. Philadelphia: Temple University Press.

Poole, Gavin. (2024, September 13). "London's Olympics Have Given the City a Legacy to be Proud of." *Guardian* https://www.theguardian.com/sport/2024/sep/13/londons-olympics-have-given-the-city-a-legacy-to-be-proud-of?CMP=Share_iOSApp_Other.

Porter, Catherine. (2024a, August 11). "Paris, Uncharacteristically Giddy, Bids Au Revoir to the Olympics." *New York Times* https://www.nytimes.com/2024/08/11/paris-closing-ceremony-los-angeles.html.

Porter, Rick. (2024b, January 23). "Ted Sarandos: WWE Deal Doesn't Change Netflix's Live Sports Stance." *Hollywood Reporter* https://www.hollywoodreporter.com/business/business-news/netflix-wwe-deal-doesnt-change-live-sports-strategy-1235805365/.

Portillo, Miguel. (2022, March 4). "Alonso: 'Estoy al 100% y hay que ser rápidos en la pista y en la fábrica'." *Forbes España* https://forbes.es/lifestyle/141598/alonso-estoy-al-100-y-hay-que-ser-rapidos-en-la-pista-y-en-la-fabrica%ef%bf%bc/.

Poynter, Gavin, Valerie Viehoff, and Yang Li, eds. (2016). *The London Olympics and Urban Development: The Mega-Event City*. London: Routledge.

Prarthana, Prakasah. (2024, September 25). "Bernard Arnault Sent a Memo Banning LVMH Staff from Talking to Certain News Outlets, Sparking Backlash

Among French Journalists—Including from Publications He Owns." *Fortune* https://fortune.com/europe/2024/09/25/bernard-arnault-memo-lvmh-staff-banning-french-media-news-outlets/.

Premier League. (2025, February 13). Written Evidence to UK Culture, Media and Sport Committee https://committees.parliament.uk/writtenevidence/135100/pdf/.

Preston, Michelle. (2020, June). *Hell for Leather*. Vegetarian & Vegan Foundation https://cdn.viva.org.uk/wp-content/uploads/2020/06/Hell-for-Leather.pdf.

Preuß, Holger. (2015). "A Framework for Identifying the Legacies of a Mega Sport Event." *Leisure Studies* 34, no. 6: 643–64.

Preza, Ana. (2012, February 29). "America's Cup Stirs Up Controversy in San Francisco." *Golden Gate Xpress* https://goldengatexpress.org/19459/uncategorized/americas-cup/.

Price, Ben and Gareth Bryer. (2024, April 27). "Brewing Giant Heineken Fells Orchard to Sell Land." *BBC* https://www.bbc.com/news/articles/cz5d6mj5gp8o.

Prieto, Alberto. (2024, October 10). "No a la Copa Amèrica porta la inversió pública en la regata a Antifrau." *MónEconomia* https://elmon.cat/moneconomia/empreses/no-a-la-copa-america-porta-la-inversio-publica-en-la-regata-a-antifrau-79610/.

Pritchett, Henry S. (1929). "Preface: Athletics, An Element in the Evolution of the American University." *American College Athletics*. Bulletin 23. Howard J. Savage, Harold W. Bentley, John T. McGovern, and Dean F. Smiley. New York: The Carnegie Foundation for the Advancement of Teaching. v–xxii.

Protect Our Winters. (n.d.). "About POW." https://protectourwinters.org/our-work/about-pow/.

Pubs Code Adjudicator. (2020, October 15). *PCA Investigation into Star Pubs & Bars* https://www.gov.uk/government/publications/pca-investigation-into-star-pubs-bars.

Purves, Richard I., Amber Morgan, and Nathan Critchlow. (2022, November 13). *Alcohol Sponsorship of Football: Examining the Nature of Sponsorship Relations for Professional Football Teams Across Countries with Varied Restrictions on Alcohol Marketing*. Scottish Health Action on Alcohol Problems https://shaap.org.uk/downloads/reports/465-alcohol-sponsorship-of-football-examining-the-nature-of-sponsorship-relations-for-professional-football-teams-across-countries-with-varied-restrictions-on-alcohol-marketing.html.

PwC. (2024, July). *Sports Industry? On Track for Growth? PWC Global Sports Survey*, 8th ed. https://www.pwc.co.uk/hospitality-leisure/documents/global-sports-survey-2024.pdf.

PwC. (2025, March). *PwC's Second Annual State of Decarbonization Report: Essential Ingredients for Achieving Your Climate Ambition* https://www.pwc.com/us/en/services/esg/library/assets/pwc-sustainability-decarbonization-2025.pdf.

Queensland Government. (n.d.). *The Big Picture* https://q2032.au/big-picture.

Quiggin, John. (2022). *Economics in Two Lessons: Why Markets Work So Well, and Why They Can Fail So Badly.* Princeton: Princeton University Press.

Quiggin, John. (2023, November 30). "What is the Actual Economic Benefit of the Grand Prix?" *Crikey* https://www.crikey.com.au/2023/11/30/grand-prix-2023-victorian-economic-benefit-ernst-young/.

Racinais, Sebastien, Mohammed Ihsan, Marie-Elaine Grant, Valentin Dablainville, Yohan Rousse, Wolfgang Schobersberger, Richard Budgett, and Lars Engebretsen. (2024). "Ice Challenge in Recent Summer Olympic Games." *British Journal of Sports Medicine* 58, no. 17: 943–45.

Rafferty, Max. (1970). "The Indestructible Sport." *Newton Graphic* https://archive.org/stream/athletethe196970unse/athletethe196970unse_djvu.txt.

Rainforest Action Network. (2018). *Broken Promises: A Case Study on How the Tokyo 2020 Games and Japanese Financiers are Fueling Land-Grabbing and Rainforest Destruction in Indonesia* https://www.ran.org/wp-content/uploads/2018/11/BrokenPromises.pdf.

Rainforest Action Network. (n.d.). "Make Top Ramen Deforestation Free." https://www.ran.org/campaign/demand-deforestation-free-instant-noodles/.

Raji, Khalid. (2022, November 17). "Qatar 2022: The Environmental Cost of the FIFA World Cup." *Earth* https://earth.org/qatar-2022/.

Raji, Lilian. (2024, July 28). "How Bernard Arnault and LVMH are Dominating the 2024 Paris Olympics." *Forbes* https://www.forbes.com/sites/lilianraji/2024/07/27/how-bernard-arnault-and-lvmh-are-dominating-the-2024-paris-olympics/.

Ramaswamy, Chitra. (2015, November 30). "Hosting the Olympics: The Competition Noone Wants to Win." *Guardian* https://www.theguardian.com/sport/shortcuts/2015/nov/30/hosting-olympics-hamburg-drop-out-2024-games.

Ramírez, Sheila. (2025). "Encuentro triunfal." *Robb Report: Best of the Best*: 70–80.

Rayman, Josh. (2023). "The End of Oil? Formula One's Changing Face." *The Geopolitical Economy of Sport: Power, Politics, Money, and the State.* Ed. Simon Chadwick, Paul Widdop, and Michael M. Goldman. London: Routledge. 242–49.

Reda, Ahmed and Laurent Viviez. (2021, September 15). *How Sports Events Can Transform the Saudi Arabian Economy.* EY https://www.ey.com/enpk/services/consulting/how-sports-events-can-transform-the-saudi-arabian-economy#:~:text=Sports'%20contribution%20to%20GDP%20has,sports%20services%20such%20as%20gyms.

Rédaction Île de France. (2022, August 4). "Paris 2024: ces infrastructures abandonnées et retardées en Seine-Saint-Denis." *Actu Seine-Saint-Denis* https://actu.fr/societe/carte-paris-2024-ces-infrastructures-abandonnees-et-retardees-en-seine-saint-denis52658423.html.

Reed, Dave. (2024, October 17). "Legends and Louis Vuitton Get Their Due at America's Cup Hall of Fame Night in Barcelona." *Sailing World* https://www.sailingworld.com/racing/legends-and-louis-vuitton-get-their-due-at-americas-cup-hall-of-fame-night-in-barcelona/.

Reed, Sada. (2018). ""I'm not a Fan. I'm a Journalist": Measuring American Sports Journalists' Sports Enthusiasm." *Journal of Sports Media* 13, no. 1: 27–47.

Regué, Júlia. (2020, October 12). "12-O en BCN: Ofrenda a Colón, bendición a la virgen del Pilar y 'Cara al sol'." *El Periódico* https://www.elperiodico.com/es/politica/20201012/12-octubre-barcelona-colon-cara-sol-8152690.

Reid, Caroline. (2008, May 3). "For Formula One, Sex Sells; But Not the Way Max Likes It." *The Spectator* https://www.spectator.co.uk/article/for-formula-one-sex-sells-but-not-the-way-max-likes-it/.

Reid, Caroline. (2015, March 19). "20 Brands That Defined F1." *Raconteur* https://www.raconteur.net/20-brands-that-defined-f1/.

Reid, Helen. (2024, April 12). "Nike Bosses Plan 'Biggest' Olympics Spend as Marketing Ramps Up." *Reuters* https://www.reuters.com/business/retail-consumer/nike-bosses-plan-biggest-olympics-spend-marketing-ramps-up-2024-04-12/.

RepRisk. (2024, October). *A Turning Tide in Greenwashing? Exploring the First Decline in Six Years* https://www.reprisk.com/research-insights/reports/a-turning-tide-in-greenwashing-exploring-the-first-decline-in-six-years.

Republic of South Africa, Department of Environmental Affairs and Tourism, Norwegian Agency for Development Cooperation, and Norwegian Embassy in South Africa. (2009). *Feasibility Study for a Carbon Neutral 2010 FIFA World Cup in South Africa* https://www.norway.org.za/NR/rdonlyres/3E6BB1B1FD2743E58F5B0BEFBAE7D958/114457/FeasibilityStudyforaCarbonNeutral2010FIFAWorldCup.pdf.

Reygadas, Luis. (2021). *Otros capitalismos son posibles.* Mexico City: Universidad Autónoma Metropolitana, Iztapalapa.

Reynolds, John. (2012, February 17). "BP's Brand Image Benefits from London 2012 Sponsorship, Claims Research." *Campaign* http://www.campaignlive.co.uk/article/1117665/bps-brand-image-benefits-london-2012-sponsorship-claims-research?srcsite=marketingmagazine.

Reynolds, Tom. (2018, October 20). "Vera Caslavska and the Forgotten Story of her 1968 Olympics Protest." *BBC* https://www.bbc.com/sport/olympics/45900544.

Rich, Adrienne. (1997). "The Springboard." *A Whole Other Ball Game: Women's Literature on Women's Sport*. Ed. Joli Sandoz. New York: Noonday Press. 1.

Richards, Giles. (2021a, March 31). "Nico Rosberg: 'To Do Good You Need to Get Out There, You Can't Sit in a Cave'." *Guardian* https://www.theguardian.com/sport/2021/mar/31/nico-rosberg-to-do-good-you-need-to-get-out-there-you-cant-sit-in-a-cave.

Richards, Giles. (2021b, November 26). "Climate Emergency Accelerates F1's Efforts to Clean Up Its Image." *Guardian* https://www.theguardian.com/sport/2021/nov/26/climate-emergency-accelerates-f1-efforts-to-clean-up-image.

Richards, Giles. (2023a, May 21). "After the Flood, Storms Lie Ahead for Formula One in Race to Hit Carbon Zero." *Guardian* https://www.theguardian.com/sport/2023/may/21/after-the-flood-storms-lie-ahead-for-formula-one-in-race-to-hit-carbon-zero.

Richards, Giles. (2023b, July 5). "Next F1 Season Will be Longest Calendar of 24 Races in Sport's History." *Guardian* https://www.theguardian.com/sport/2023/jul/05/next-f1-season-will-be-longest-calendar-of-24-races-in-sports-history.

Richards, Giles. (2023c, December 30). "Ben Sulayem's Mishaps and Meddling Leave F1 Asking if FIA is Fit for Purpose." *Guardian* https://www.theguardian.com/sport/blog/2023/dec/30/ben-sulayems-mishaps-and-meddling-leave-f1-asking-if-fia-is-fit-for-purpose.

Richards, Giles. (2024a, January 23). "F1 Reveals Madrid to Replace Barcelona as Host of Spanish Grand Prix From 2026." *Guardian* https://www.theguardian.com/sport/2024/jan/23/f1-reveals-madrid-to-replace-barcelona-as-host-of-spanish-grand-prix-from-2026.

Richards, Giles. (2024b, March 22). "FIA Unable to Escape Scrutiny with Indictments Overshadowing Season." *Guardian* https://www.theguardian.com/sport/2024/mar/22/fia-formula-one-f1-christian-horner-susie-wolff-lewis-hamilton.

Richards, Giles. (2024c, July 16). "Ralf Schumacher Coming Out Will Have a Positive Impact on 'Evolving' F1." *Guardian* https://www.theguardian.com/sport/article/2024/jul/16/ralf-schumacher-formula-one-impact.

Richards, Giles. (2024d, September 19). "Lewis Hamilton Unhappy with 'Racial Element' of FIA President's Comments." *Guardian* https://www.theguardian.com/sport/2024/sep/19/lewis-hamilton-unhappy-with-racial-element-of-fia-presidents-comments.

Richards, Giles. (2024e, November 28). "F1 'Running Out of People' Says Sacked Race Steward in Rebuke to FIA President." *Guardian* https://www.theguardian.com/sport/2024/nov/28/f1-running-out-of-people-says-sacked-race-steward-in-rebuke-to-fia-president.

Richards, Giles. (2025, March 14). "Sun, Fun and a Favourite Son: Melbourne Makes a Full-Throttle Return to the Top of the F1 Calendar." *Guardian* https://www.theguardian.com/sport/2025/mar/14/melbourne-formula-one-f1-grand-prix-2025-guide-albert-park.

Richta, Radovan and a Research Team. (1969). *Civilization at the Crossroads: Social and Human Implications of the Scientific and Technological Revolution*, 3rd ed. Prague: International Arts and Sciences Press.

Richter, Felix. (2024, December 11). "The World Cup is FIFA's Quadrennial Cash Cow." *Statista* https://www.statista.com/chart/30976/fifa-annual-revenue/.

Rico, Anna, Julia Martínez-Blanco, Marc Montlleó, Gustavo Rodríguez, Nuno Tavares, Albert Arias, and Jordi Oliver-Sol. (2019). "Carbon Footprint of Tourism in Barcelona." *Tourism Management* 70: 491–504.

Riggs, Doug. (1986). *Keelhauled: Unsportsmanlike Conduct and the America's Cup*. Newport: Seven Seas Press.

Rigolon, Alessandro, Mariela Fernandez, Brandon Harris, and William Stewart. (2022). "An Ecological Model of Environmental Justice for Recreation." *Leisure Sciences: An Interdisciplinary Journal* 44, no. 6: 655–76.

Rio Declaration on Environment and Development. (1992). http://www.unesco.org/education/nfsunesco/pdf/RIOE.PDF.

Riisager-Simonsen, Christian, Gianna Fabi, Luc van Hoof, Noél Holmgren, Giovanna Marino, and Dennis Lisbjerg. (2022). "Marine Nature-Based Solutions: Where Societal Challenges and Ecosystem Requirements Meet the Potential of Our Oceans." *Marine Policy* 144 https://www.sciencedirect.com/science/article/pii/S0308597X22002457#bib1.

Rius-Ulldemolins, Joaquim, Francisco Torres Pérez, and Gil-Manuel Hernàndez. (2021). "Regional Entrepreneurialism, 'Creative City' and Strategic Projects in València." *The Hidden Side of the Creative City: Culture Instrumentalization, Political Control and Social Reproduction in Valencia*. Ed. Joaquim Rius-Ulldemolins, Juan Arturo Rubio-Arostegui, and Verònica Gisbert-Gracia. Cham: Palgrave Macmillan. 25–58.

Roan, Dan. (2022, July 26). "London 2012: Ten Years On, What Lessons Can We Learn from London's Legacy?" *BBC* https://www.bbc.co.uk/sport/olympics/62299772.

Roan, Dan. (2023, December 14). "Why is Saudi Arabia Spending so Much on Sport?" *BBC* https://www.bbc.com/sport/67713269.

Roan, Dan. (2024, September 14). "England 'Could Face Euros Ban' Over Regulator Plan." *BBC* https://www.bbc.com/sport/football/articles/c9wkjnvpy2ko.

Roba i Giner, Diana, Ana Miranda Paz, Cristina Guarda, Carolina Morace, David Cormand, Catarina Vieira, Pernando Barrena, Li Andersson, Rainer van Lanschot, Chloé Ridel, Estelle Ceulemans, erik Marquardt, Mélissa Camara, and Emma Fourreau. (2024, December 6). Letter to Gianni Infantino https://www.eunews.it/wp-content/uploads/2025/01/MEPs-on-FIFA-2034-World-Cup-in-Saudi-Arabia-and-2030-in-Morocco-Portugal-and-Spain.pdf.

Roberts, William O., Lawrence E. Armstrong, Michael N. Sawka, Susan W. Yeargin, Yuval Heled, and Francis G. O'Connor. (2023). "ACSM Expert Consensus Statement on Exertional Heat Illness: Recognition, Management, and Return to Activity." *Current Sports Medicine Reports* 22, no. 4: 134–49.

Robertson, Angeline. (2021). *Shady Routes: How Big Retail and Their Carriers Pollute Along Key Ocean Shipping Corridors.* STAND.earth/Pacific Environment https://shipitzero.org/wp-content/uploads/2021/11/FINAL-Ship-it-Zero-Campaign-Report.pdf.

Robinson, Joshua and Jonathan Clegg. (2024, November 12). "How 'Drive To Survive' Turned F1 Into The Real Housewives of Monte Carlo." *Forbes* https://www.forbes.com/sites/forbesdigitalcovers/2024/03/08/the-formula-book-excerpt-drive-to-survive-f1-real-housewives-of-monte-carlo-max-verstappen-christian-horner/.

Roces, Pablo R. (2024, February 2). "Ayuso resucita el Proyecto de los Juegos Olímpicos para Madrid en 2036: "Estamos en ello"." *GranMadrid* https://www.elmundo.es/madrid/2024/02/02/65bcdcb2fdddffa64d8b4587.html.

Rocha, Claudio M. (2023). "Rio 2016 Olympic Legacy for Residents of Favelas: Revisiting the Case of Vila Autódromo Five Years Later." *Social Sciences* 12, no. 3 https://doi.org/10.3390/socsci12030166.

Rodda, John. (2008). "'Prensa, Prensa': A Journalist's Reflections on Mexico '68." *Bulletin of Latin American Research* 29, no. 1: 11–22.

Rodgers, Dale. (2020, May 16). "How F1's Earth Dream Turned Nightmare." *The Race Torque* https://www.theracetorque.com/2020/05/how-f1s-earth-dream-turned-nightmare/.

Rodriguez, Cecilia. (2024, August 23). "Tensions and Hostility Against America's Cup Sailing Competition 'For The Rich' in Barcelona." *Forbes* https://www.forbes.com/sites/ceciliarodriguez/2024/08/23/tensions-and-hostility-against-americas-cup-sailing-competition-for-the-rich-in-barcelona/.

Rodríguez, Alejandra. (2022a, June 27). "Heineken responde a AMLO: Se dice abierta al diálogo por escasez de agua." *El Financiero* https://www.elfinanciero.com.mx/empresas/2022/06/27/heineken-responde-a-amlo-se-dice-abierta-al-dialogo-por-escasez-de-agua/.

Rodríguez, David. (2022b, April 4). "Escepticismo entre los expertos sobre el impacto económico de eventos como el Mobile o la Copa América para Catalunya." *Público* https://www.publico.es/sociedad/escepticismo-expertos-impacto-economico-eventos-mobile-copa-america-catalunya.html.

Rodríguez, Jesús. (2024a, September 9). "La Copa Amèrica falseja les dades d'audiència i de visitant." *Directa* https://directa.cat/la-copa-america-falseja-les-dades-daudiencia-i-de-visitants/.

Rodríguez, Jesús. (2024b, September 17). "La inversió pública oculta de la Copa Amèrica." *Directa* https://directa.cat/la-inversio-publica-oculta-de-la-copa-america/.

Rodríguez, Jesús. (2024c, October 10). "Un centenar d'entitats presenten una denúncia contra la Copa Amèrica a l'Oficina Antifrau." *Directa* https://directa.cat/un-centenar-dentitats-presenten-una-denuncia-contra-la-copa-america-a-loficina-antifrau/.

Rodríguez, Jesús. (2024d, October 11). "Un espectacle de drons per dopar les xifres d'assistència a la Copa Amèrica." *Directa* https://directa.cat/un-espectacle-de-drons-per-dopar-les-xifres-dassistencia-a-la-copa-america-de-vela/.

Rodríguez-Santos, Carmen, Stephane Ganassali, Francesco Casarin, Paoloa Cerchiello, Gunnar Mau, and Astrid Siebels. (2008). "How Perceived Image of a Sponsorship Event Influences on the Brand Perceptions: An International Research of the America's Cup and Louis Vuitton." *International Journal of Management Studies* 10, no. 3: 261–81.

Rojo, Juan J. (2016). *Revisiting the Mexican Student Movement of 1968: Shifting Perspectives in Literature and Culture Since Tlatelolco.* New York: Palgrave Macmillan.

Roland, Romain. (1920, March 19). "Un livre de Raymond Lefebvre: *Le Sacrifice d'Abraham.*" *l'Humanité* https://gallica.bnf.fr/ark:/12148/bpt6k299613c/f1.item.

Rolland, Rosalind M., Susan E. Parks, Kathleen E. Hunt, Manuel Castellote, Peter J. Corkeron, Douglas P. Nowacek, Samuel K. Wasser, and Scott D. Kraus. (2012). "Evidence That Ship Noise Increases Stress in Right Whales." *Proceedings of the Royal Society B: Biological Sciences* 279: 2363–368.

Ronay, Barney. (2024, November 15). "Saudi Arabia's World Cup: How Close Could FIFA Get to Corporate Manslaughter?" *Guardian* https://www.theguardian.com/football/2024/nov/15/saudi-arabias-world-cup-how-close-could-fifa-get-to-corporate-manslaughter?CMP=Share_iOSApp_Other.

Ropa, Anastasija, Ludmila Malahova, and Monta Jakovleva. (2024). "Animal Ethics or Sport Ethics? A Historical Approach to Professional Ethics in Equestrian Sports." *Cheiron: The International Journal of Equine and Equestrian History* 4, no. 1: 95–114.

Roper, Gemma. (2024, February 27). "Drive to Survive: Claire Williams Helps Make Season Six the Second-Best for Women's Speaking Time." *Females in Motorsport* https://www.femalesinmotorsport.com/post/drive-to-survive-

claire-williams-helps-make-season-six-the-second-best-for-women-s-speaking-time.
Rosa, Anton and Alberto Prieto. (2024, November 21). "El Parlament bloqueja l'auditoria dels comptes de la Copa Amèrica." *Tót Barcelona* https://www.totbarcelona.cat/economia/parlament-no-investiga-copa-america-567042/.
Rosa, Anton. (2025, February 18). "El Puerto reconoce que la Copa América dejó un agujero de 3,5 millones." *Tót Barcelona* https://www.totbarcelona.cat/es/economia-es/el-puerto-reconoce-que-la-copa-america-dejo-un-deficit-de-35-millones-592069/.
Rosenberg, Michael. (2024, August 11). "Paris Olympics Perfectly Captured the Spirit of its Host City." *Sports Illustrated* https://www.si.com/olympics/paris-olympics-perfectly-captured-spirit-of-host-city.
Ross, Andrew. (2009). *Nice Work If You Can Get It: Life and Labor in Precarious Times.* New York: New York University Press.
Ross, Hamish. (2023). "Throwing a Lifeline to an Old Sport: A Modern Review of the America's Cup Deed of Gift." *International Sports Law Journal* 23: 240–58.
Rossi, Raffaello, Jamie Wheaton, Maria Moxey, and Edoardo Tozzi. (2023). *New Season, More Self-Regulation, More Marketing: The Prevalence of Gambling Adverts During the Opening Weekend of the English Premier League 2023/24.* University of Bristol https://www.bristol.ac.uk/media-library/sites/business-school/documents/BRISTOL-UNI-GAMBLING-Report2023-2.pdf.
Rotillon, Gilles. (2024, July 12). "Paris 2024 : « Les événements sportifs comme les JO posent la question de leur pérennité, étant donné leur impact environnemental ».'' *Le Monde* https://www.lemonde.fr/idees/article/2024/07/12/paris-2024-les-evenements-sportifs-comme-les-jo-posent-la-question-de-leur-perennite-etant-donne-leur-impact-environnemental62492493232.html.
Rouget, Antton. (2024, August 17). "La Seine était complètement souillée quand la minister Oudéa-Castéra s'y est baignée." *Mediapart* https://www.mediapart.fr/journal/france/170824/la-seine-etait-completement-souillee-quand-la-ministre-oudea-castera-s-y-est-baignee.
Ruiz, José Luis. (2022, February 19). "El subidón de la Fórmula 1." *Marca* https://www.marca.com/motor/formula1/2022/02/19/6210af9bca4741ee4f8b45bd.html.
Ruta, Alessandro. (2021). "Livorno, la ciudad «más roja», en el centenario del PCI." *Naiz* https://www.naiz.eus/eu/hemeroteca/7k/editions/7k_2021-09-26-07-00/hemeroteca_articles/livorno-la-ciudad-mas-roja-en-el-centenario-del-pci.
Ruta, Dino and Beatrice Manzoni. (2015). "City Capacity Building—Preparing to Exploit the Legacy of a Large-Scale Sports Event: The Case of Valencia and the 2007 America's Cup." *Routledge Handbook of Sport and Legacy: Meeting the*

Challenge of Major Sports Events. Ed. Richard Holt and Dino Ruta. Asst. Ed. James Painter. Abingdon: Routledge. 231–44.

Sabaté, Arsen. (2024, October 11). "Una ceremonia deslumbrante para ocultar el negocio capitalista y depredador de la Copa América de vela." *Izquierda Diario* https://www.izquierdadiario.es/Una-ceremonia-deslumbrante-para-ocultar-el-negocio-capitalista-y-depredador-de-la-Copa-America-de.

Saccage. (2021, January). *Un recensement effectué par des Associations et des Collectifs Citoyens* https://saccage2024.noblogs.org/files/2021/03/saccage-2024-argumentaire.pdf.

Saccage. (2024a, July 29). "Les Jeux des arrestations arbitraires." https://saccage2024.noblogs.org/archives/3164.

Saccage. (2024b, September). *Pourquoi faire un bilan?* https://saccage2024.noblogs.org/files/2024/09/Pourquoi-faire-un-bilan-.pdf.

Saccage. (2024c, October). *Les jeux olympiques et paralympiques de Paris 2024: Bilan des saccages. Chapitre 2: Le saccage social* https://saccage2024.noblogs.org/files/2024/10/Bilan-de-Paris-2024-saccage-social.pdf.

Saccage. (2024d, October). *Les jeux olympiques et paralympiques de Paris 2024: Bilan des saccages. Chapitre 3: Le saccage écologique* https://saccage2024.noblogs.org/files/2024/10/Bilan-des-Saccages-2024-ECOLOGIQUE.pdf.

Sacristán, Claudia. (2024, July 14). ""No puedo ir ni a la playa de mi ciudad": los barceloneses, contra de la Copa América de Vela." *El Confidencial* https://www.elconfidencial.com/espana/cataluna/2024-07-14/no-puedo-ir-playa-ciudad-barceloneses-copa-america_3919767/.

Said, Edward. (1993). "An Interview with Edward Said." *boundary 2* 20, no. 1: 1–25.

Saleh, Tariq. (2024, August 13). "Significant Paris 2024 Audiences for WBD, BBC, NBC." *Sportcal* https://www.sportcal.com/media/significant-paris-2024-audiences-for-wbd-bbc-nbc/.

Salinas, Arango, Natalia Andrea. (2018). "Encrucijada de la violencia asociada al fútbol: Entre el desagrado y la complacencia." *Trabajo Social* 20, no. 1: 49–68.

Samuel, Shalini and Wendy Stubbs. (2013). "Green Olympics, Green Legacies? An Exploration of the Environmental Legacies of the Olympic Games." *International Review for the Sociology of Sport* 48, no. 4: 485–504.

Sandel, Michael J. (2005). *Public Philosophy: Essays on Morality in Politics*. Cambridge, Mass.: Harvard University Press.

Sandel, Michael J. (2007). *The Case Against Perfection: Ethics in the Age of Genetic Engineering*. Cambridge, Mass.: Harvard University Press.

Sanderson, Tom. (2024, November 9). "FC Barcelona Announces New Nike Record Sponsorship Deal." *Forbes* https://www.forbes.com/sites/tomsanderson/2024/11/09/fc-barcelona-announces-historic-new-nike-record-sponsorship-deal/.

Sanghera, Mandeep. (2024, October 14). "Legal Complaint Filed Against FIFA's 'Abuse of Dominance'." *BBC* https://www.bbc.com/sport/football/articles/c981203e61qo.

Santamarina Campos, Beatriz and Albert Moncusí Ferré. (2013). "De huertas y barracas a galaxias faraónicas. Percepciones sociales sobre la mutación de la ciudad de Valencia." *Papers: Revista de sociología* 98, no. 2: 365–91.

Sant'Anna School of Advanced Studies, Institute of Management. (2020). *The Environmental Awareness and Behaviour of Professional Football Supporters: An Empirical Survey* https://lifetackle.eu/assets/files/LIFETACKLER eportonsupporterssurvey.pdf.

Santini, Dominique and Holly Henderson. (2021). "The Winners and Losers in the Race to Environmental Sustainability: A Ranking of Summer Olympic International Federation Progress." *Emerald Open Research* 3, no. 12 https://doi.org/10.35241/emeraldopenres.14195.1.

Santos, Célia, Arnaldo Coelho, and Alzira Marques. (2024). "A Systematic Literature Review on Greenwashing and its Relationship to Stakeholders: State of Art and Future Research Agenda." *Management Review Quarterly* 74, no. 3: 1397–421.

Sarkar, Sahotra. (2021). "Origin of the Term *Biodiversity*." *BioScience* 71, no. 9: 893.

Sartre, Jean-Paul and Bertrand Russell. (1968, November 1). "Jean-Paul Sartre and Bertrand Russell Speak-out on Mexico and its Olympics." *Los Angeles Free Press*: 2.

Sartre, Jean-Paul. (2004). *Critique of Dialectical Reason. Volume 1: Theory of Practical Ensembles.* Trans. Alan Sheridan-Smith. Ed. Jonathan Ree. London: Verso.

Saudi Arabia. (2024). *Vision & Legacy* https://saudi2034bid.com/vision-legacy/#tournament-concept.

Saward, Joe. (2023, January). "Green Notebook in Front of a Christmas Tree." *JoeBlogsF1* https://joesaward.wordpress.com/2023/01/02/green-notebook-in-front-of-a-christmas-tree/.

Scanlon, Thomas F. (2018). "Class Tensions in the Games of Homer: Epeius, Euryalus, Odysseus, and Iros." *Bulletin of the Institute of Classical Studies* 61, no. 1: 5–20.

Scanlon, Thomas F. (2021). "Overview and Approaches." *The Oxford Handbook of Sport and Spectacle in the Ancient World.* Ed. Alison Futrell and Thomas F. Scanlon. Oxford: Oxford University Press. 3–13.

Schaerlaeckens, Leander. (2025, March 12). "Gianni Infantino and Donald Trump Have Taken the 2026 World Cup for Themselves." *Guardian* https://www.theguardian.com/football/2025/mar/12/gianni-infantino-donald-trump-2026-world-cup.

Schäfer, Mike S. (2012). “Online Communication on Climate Change and Climate Politics: A Literature Review.” *Wiley Interdisciplinary Reviews: Climate Change* 3, no. 6: 527–43.

Scharpf, Adam, Christian Gläßel, and Pearce Edwards. (2023). “International Sports Events and Repression in Autocracies: Evidence from the 1978 FIFA World Cup.” *American Political Science Review* 117, no. 3: 909–26.

Schenk, James. (n.d.). “What is the Most Popular Section in a Newspaper?” *AuthorsCast* https://authorscast.com/what-is-the-most-popular-section-in-a-newspaper.

Schimmel, Kimberly S., C. Lee Harrington, and Denise D. Bielby. (2007). “Keep Your Fans to Yourself: The Disjuncture between Sport Studies’ and Pop Culture Studies’ Perspectives on Fandom.” *Sport in Society* 10, no. 4: 580–600.

Schlosberg, David. (2007). *Defining Environmental Justice: Theories, Movements, and Nature.* Oxford: Oxford University Press.

Schor, Juliet B. (2010). *Plenitude: The New Economics of True Wealth.* New York: Penguin.

Schumpeter, Joseph Alois. (1966). *Imperialism & Social Classes: Two Essays.* Trans. Heinz Norden. Cleveland: Meridian Books.

Schwenkenbecher, Anne. (2014). “Is There an Obligation to Reduce One’s Individual Carbon Footprint?” *Critical Review of International Social and Political Philosophy* 17, no. 2: 168–88.

Scott, Mike. (2013, August 5). “Could Innovation in Formula One Drive Sustainable Technology?” *Guardian* https://www.theguardian.com/sustainable-business/innovation-formula-one-sustainable-technology.

Sedghi, Ami, George Arnett, and Chris Moran. (2014, April 17). “*The Guardian*’s Top 100: Which Articles Have Been Most Popular?” *Guardian* https://www.theguardian.com/news/datablog/2014/apr/17/the-guardians-top-100-which-articles-have-been-most-popular.

Selectra. (2020, January 22). *What Happens When Everyone in the UK Watches the World Cup at the Same Time?* https://selectra.co.uk/energy/news/world/world-cup-2018-electricity-surges.

Selleck, Emily. (2024, October 7). “What We Can Expect from LVMH’s Billion-Dollar F1 Deal.” *Motorsport* https://www.motorsport.com/Culture/news/formula-1-lvmh-sponsorship/10660515/.

Senate Finance Committee. (2024, May 20). *Insufficient Diligence: Car Makers Complicit with CCP Forced Labor* https://www.finance.senate.gov/imo/media/doc/insufficient_diligence_-_car_makers_complicit_with_ccp_forced_labor1.pdf.

Sexton, Sarah, Nicholas Hildyard, and Larry Lohmann. (2005, April 7). “We’re a Small Island: The Greening of Intolerance.” *The Corner House* https://www.thecornerhouse.org.uk/sites/thecornerhouse.org.uk/files/SmallIsland.pdf.

Seyfang, Gill. (2005). "Shopping for Sustainability: Can Sustainable Consumption Promote Ecological Citizenship?" *Environmental Politics* 14, no. 2: 290–306.

Shah, Dhwanil and Anthony Lamont Williams. (2024). ""Drive to Survive" Drives New Fans to Formula 1? Studying Viewer Experiences of a Sports Documentary and its Influence on the Sport's Fandom." *Communication & Sport* https://doi.org/10.1177/21674795241280209.

Shakespeare, William. (1994). *Complete Works of William Shakespeare*. N.p.: World Library.

Shanahan, Abigail. (2024, October 21). "Experiencing the Louis Vuitton 37th America's Cup as a Team Family Member and Volunteer." https://www.americascup.com/news/3772_EXPERIENCING-THE-LOUIS-VUITTON-37TH-AMERICA-S-CUP-AS-A-TEAM-FAMILY-MEMBER-AND-VOLUNTEER.

Sharma, Dinesh C. (2024, November 30). "40 Years After Bhopal: Lingering Health Effects and No Closure." *The Lancet* 404: 2145–46.

Shehabi, Arman, Sarah Josephine Smith, Alex Hubbard, Alexander Newkirk, Nuoa Lei, Md AbuBakar Siddik, Billie Holecek, Jonathan G. Koomey, Eric R. Masanet, and Dale A. Sartor. (2024, December). *2024 United States Data Center Energy Usage Report* https://eta.lbl.gov/publications/2024-lbnl-data-center-energy-usage-report.

Shenker, Jack. (2023, March 6). "The Existential Question for Climate Activists: Have Disruption Tactics Stopped Working?" *Guardian* https://www.theguardian.com/commentisfree/2023/mar/06/british-eco-activists-disruption-extinction-rebellion.

Shepard, Richard F. (1974, November 17). "Time to End Sports, Opiate of the Masses." *New York Times* https://www.nytimes.com/1974/11/17/archives/time-to-end-sports-opiate-of-the-masses-will-it-never-end.html#:~:text=Karl%20Marx%20once%20described%20religion,in%20the%20proliferation%20of%20tedium.

Shepherd, Lauren Lassabe. (2023). *Resistance from the Right: Conservatives and the Campus Wars in Modern America*. Chapel Hill: University of North Carolina Press.

Sher, Chloe and Caroline Fusco. (2024). "Sports and Sustainable Development: The Troubling Absence of Meat Sourcing Policies in the Sports Sector." *Frontiers in Sports and Active Living* 6 https://doi.org/10.3389/fspor.2024.1341810.

Sherter, Alain with Anne Marie Lee and Associated Press. (2024, September 24). "Visa Monopolizes Debit Cards, Hurting Consumers and Businesses, Justice Department Alleges." *CBS* https://www.cbsnews.com/news/visa-debit-cards-justice-department-lawsuit/.

Shilling, Chris. (1991). "Educating the Body: Physical Capital and the Production of Social Inequalities." *Sociology* 25, no. 4: 653–74.

Ship It Zero. (2024). *Ship It Zero Shipping Decarbonization Report Card* https://shipitzero.org/report-card/.

Shivji, Issa G. (2007). *Silence in NGO Discourse: The Role and Future of NGOs in Africa.* Nairobi: Fahamu.

Short, John Rennie. (2015, August 3). "The Green and the Gold: Can We Soften the Environmental Impact of the Olympics?" *The Conversation* https://theconversation.com/the-green-and-the-gold-can-we-soften-the-environmental-impact-of-the-olympics-45530.

Siemaszko, Corky. (2021, July 24). "Heat Wave Hits Tokyo as Olympic Organizers Battle to Keep Covid Rates Down." *NBC* https://www.nbcnews.com/news/olympics/heat-wave-hits-tokyo-olympic-organizers-battle-keep-covid-rates-n1274899.

Silic, Dario. (2022). "Is the FIFA World Cup in Qatar Profitable? *SSBM* https://www.ssbm.ch/is-the-fifa-world-cup-2022-in-qatar-profitable/.

Silk, Michael, David L. Andrews, and Cheryl L. Cole, eds. (2005). *Sport and Corporate Nationalisms.* Oxford: Berg.

Silverman, Alex and John Ourand. (2022, June 14). "MLS Goes with Apple in Landmark 10-Year Global Media Rights Deal." *Sports Business Journal* https://www.sportsbusinessjournal.com/Daily/Issues/2022/06/14/Media/MLS-TV-rights.aspx.

Sim, Walter. (2018, November 12). "Environment Group Accuses Tokyo Olympics Organisers of Using Illegally Sourced Wood." *Straits Times* https://www.straitstimes.com/asia/east-asia/environment-group-accuses-tokyo-olympics-organisers-of-using-illegally-sourced-wood.

Simmel, Georg. (1949). "The Sociology of Sociability." Trans. Everett C. Hughes. *American Journal of Sociology* 55, no. 3: 254–61.

Simpson, Brent, Robb Willer, and Matthew Feinberg. (2022). "Radical Flanks of Social Movements Can Increase Support for Moderate Factions." *PNAS Nexus* 1: 1–11.

Sin, Ben. (2014, February 12). "All the Modern Olympic Mascots in One Adorable Infographic." *Sports Illustrated* https://www.si.com/extra-mustard/2014/02/12/olympics-mascots-infographic.

Singh, Parminder, Ashok Kumar Balaraman, Rachana Mehta, and Sanjit Sah. (2025). "Urgent Need for Enhanced Food Safety Protocols in Fast-Food Supply Chains: Lessons from the Recent Multi-State *E. coli* Outbreak." *Infectious Diseases* 57, no. 1: 109–11.

Sklair, Leslie. (2016). "The Transnational Capitalist Class, Social Movements, and Alternatives to Capitalist Globalization." *International Critical Thought* 6, no. 3: 329–41.

Skóra, Paulina. (2024, July 3). "England: West Ham Close to New London Stadium Lease Agreement? Property Value Declines." *StadiumDB* https://stadiumdb.

com/news/2024/07/england_west_ham_close_to_new_london_stadium_lease_agreement_property_value_declines.

Smith, Anthony D. (2000). *The Nation in History: Historiographical Debates about Ethnicity and Nationalism.* Oxford: Polity Press.

Smith, Luke. (2023, February 11). "How Do F1's Sustainability Goals Fit with a 23-Race Calendar?" *Motorsport* https://us.motorsport.com/f1/news/how-do-f1s-sustainability-goals-fit-with-a-23-race-calendar/10430282/

Smith, Kirk R., Alistair Woodward, Bruno Lemke, Matthias Otto, Cindy J. Chang, Anna A. Mance, John Balmes, and Tord Kjellstrom. (2014). "The Last Summer Olympics? Climate Change, Health, and Work Outdoors." *The Lancet* 388, no. 10045: P642–44.

Smith, Luke. (2021, July 16). "'Drive to Survive' on Netflix has Ignited Formula 1." *New York Times* https://www.nytimes.com/2021/07/16/sports/autoracing/drive-to-survive-netflix-formula-one.html.

Smith Maguire, Jennifer. (2006). "Exercising Control: Empowerment and the Fitness Discourse." *Sport, Rhetoric, and Gender: Historical Perspectives and Media Representations.* Ed. Linda K. Fuller. New York: Palgrave Macmillan. 119–30.

Smythe, Dallas. (1954). "Reality as Presented by Television." *Public Opinion Quarterly* 18, no. 2: 143–56.

Soble, Caroline and Mark Lowes. (2024). "Narrative Storytelling as a Fan Conversion Tool in the Netflix Docuseries *Drive to Survive.*" *Communication & Sport* https://doi.org/10.1177/21674795241238158.

Solans, Maria. (2024, October 23). "La plataforma No a la Copa Amèrica que la prova marxi de Barcelona: "Bon vent i barca nova"." *Betevé* https://beteve.cat/societat/plataforma-no-a-la-copa-america-celebra-prova-marxi-barcelona/.

Soler, Jaume. (2025, March 10). "Grant Dalton llama "comunista" al alcalde de Barcelona." *Mundo Deportivo* https://www.mundodeportivo.com/vela/20250310/1002423002/grant-dalton-llama-comunista-alcalde-barcelona.html.

Solidarity. (2001). "Paris May 1968." *Beneath the Paving Stones: Situationists and the Beach, May 1968.* Comp. Dark Star. Edinburgh: AK Press/Dark Star. 64–96.

Solnit, Rebecca. (2021, August 23). "Big Oil Coined 'Carbon Footprints' to Blame Us for Their Greed. Keep Them on the Hook." *Guardian* https://www.theguardian.com/commentisfree/2021/aug/23/big-oil-coined-carbon-footprints-to-blame-us-for-their-greed-keep-them-on-the-hook.

Solon, Olivia. (2011, February 14). "F1 Needs to Have Sex with Sustainability." *Wired* https://www.wired.co.uk/news/archive/2011-02/14/f1-sustainability.

Source Material. (2025, April 9). "Big Tech's Data Centre Push Will Take Water from the World's Driest Areas." https://www.source-material.org/amazon-microsoft-google-trump-data-centres-water-use/.

Spanne, Autumn and *The Daily Climate*. (2014, June 19). "Brazil World Cup Fails to Score Environmental Goals." *Scientific American* https://www.scientificamerican.com/article/brazil-world-cup-fails-to-score-environmental-goals/?WT.mcid=send-to-friend.

Spencer, Herbert. (2005). *Essays on Education and Kindred Subjects*. London: Everyman's Library/New York: Dutton.

Spinoza, Baruch. (2002). *Complete Works*. Ed. Michael L. Morgan. Trans. Samuel Shirley. Indianapolis: Hackett.

SponsorPulse. (2023). *FIFA World Cup 2022 Sponsor Impact Report. Market: USA* https://insights.sponsorpulse.com/hubfs/FIFA%20World%20Cup%202022%20Sponsor%20Impact%20Report.pdf?_gl=1*u5y7jp*_gcl_au*NDQyMjg2ODY5LjE3MzM1OTY5OTg.*_ga*OTM4Mzc5MTMzLjE3MzM1OTY5OTg.*_ga_392WGBYBSG*MTczMzU5Njk5OC4xLjEuMTczMzU5NzAzOC4yMC4wLjA.

SponsorUnited. (2025, May). *Formula 1 Report: 2024 & 2025* https://cdn.prod.website-files.com/646f83ebeb2de03609e8a7bb/68373a6fcd198d8281b706e9_SponsorUnited%20Formula%201%20Turnkey%20Report%202024-25.pdf?utm_campaign=13471865-2025%20%E2%80%94%20F1%202024-25&utm_medium=email&_hsenc=p2ANqtz-_55IyPiB3upYBxOfQm91WD2PWyLzs32eUxdy5J44B3Tcim51raS18ha7x3tmggdPZrwLFLXTuBm-sExs-1n6DNIdPPhg&_hsmi=363846356&utm_content=363846356&utm_source=hs_automation.

Sport Positive Leagues. (2024, April). *Premier League Clubs Environmental Sustainability Report* https://www.sportpositive.org/wp-content/uploads/2024/05/Sport-Positive-Leagues_Premier-League-Report_Final2.pdf.

SportBusiness. (2023). *Global Media Report 2023* https://www.sportbusiness.com/2023/11/global-value-of-sports-media-rights-nears-56bn-sportbusiness-report-reveals/.

SportBusiness. (2024). *Global Media Report 2024* https://www.sportbusiness.com/global-media-report-2024/.

Spurgeon, Brad. (2010, June 30). "The Formula One Environment Initiative." *New York Times* https://archive.nytimes.com/rendezvous.blogs.nytimes.com/2010/06/30/the-formula-one-environment-initiative/.

Srinivas, Nidhi. (2022). *Against NGOs: A Critical Perspective on Civil Society, Management and Development*. Cambridge: Cambridge University Press.

STAND.earth. (2024, September). *Prime Polluter*. STAND.earth, Pacific Environment, Clean Mobility Collective, and Ship It Zero https://stand.earth/wp-content/uploads/2024/09/PRIME_POLLUTER-09.2024.pdf.

Stanton, John, David Lockwood, and Katie Gornall. (2021, November 12). "Premier League: Should Clubs Stop Flying to Domestic Matches for Environmental Reasons?" *BBC* https://www.bbc.co.uk/sport/football/59213173.

Starcevic, Seb. (2025, April 28). "EU Backs Olympic Bid for 2036, While London Wants the Games in 2040." *Politico* https://www.politico.eu/article/eu-olympics-bid-2036-london-2040-glenn-micallef/.

Steinberg, Brian. (2024, April 29). "Fox, Disney, Warner Will Tap Streaming Sports Venture in TV Upfront (EXCLUSIVE)." *Variety* https://variety.com/2024/tv/news/fox-disney-warner-tv-streaming-sports-venture-upfront-ad-sales-1235984530/.

Stevenson, Rachel and Agencies. (2008, October 12). "Blair Intervened Over F1 Tobacco Ban Exemption, Documents Show." *Guardian* https://www.theguardian.com/politics/2008/oct/12/tonyblair-labour.

Stockholm Environment Institute, Climate Analytics, E3G, International Institute for Sustainable Development, and United Nations Environment Programme. (2023). *The Production Gap: Phasing Down or Phasing Up? Top Fossil Fuel Producers Plan Even More Extraction Despite Climate Promises* https://doi.org/10.51414/sei2023.050.

Stoddart, Brian. (1988). "Sport, Cultural Imperialism, and Colonial Response in the British Empire." *Comparative Studies in Society and History* 30, no. 4: 649–73.

STOP. and FM. (2021). *Driving Addiction: Tobacco Sponsorship in Formula One, 2021* https://exposetobacco.org/wp-content/uploads/TobaccoSponsorshipFormula-One-2021.pdf.

STOP. (2022). *Driving Addiction: F1, Netflix and Cigarette Company Advertising* https://exposetobacco.org/wp-content/uploads/F1-Netflix-Driving-Addiction.pdf.

STOP. and Formula Money. (2024). *Driving Addiction: A Race for Future Generations. How F1 and Big Tobacco Marketing Align to Recruit Young Consumers* https://exposetobacco.org/wp-content/uploads/Race-For-Future-Generations.pdf.

Stratton, Jon. (1986). "Australia—This Sporting Life." *Power Play: The Commercialisation of Australian Sport. Essays in the Sociology of Australian Sport.* Ed. Geoffrey Lawrence and David Rowe. Sydney: Hale & Iremonger. 85–114.

Sturm, Damion, Tom Kavanagh, and Robert E. Rinehart. (2021). "They Are Not 'Team New Zealand' or the 'New Zealand' Warriors! An Exploration of Pseudonationalism in New Zealand Sporting Franchises." *Sport in Society* 24, no. 11: 2019–035.

Sturm, Damion. (2022). "The Formula One Paradox: Macho Male Racers and Ornamental Glamour 'Girls'." *Sports, Gender and Mega-Events*. Ed. Katherine Dashper. Leeds: Emerald Publishing. 113–30.

Sturm, Damion. (2023). "Processes of Greenwashing, Virtue Signaling and Sportwashing in Contemporary Formula One: Formula Façade?" *The Future of Motorsports: Business, Politics and Society*. Ed. Hans Erik Næss and Simon Chadwick. Abingdon: Routledge. 167–82.

Sturrock, Leanne. (2018, July 12). "The Environmental Impact of the World Cup—Are FIFA Scoring Sustainability Goals?" *thegreatProjects* https://www.thegreatprojects.com/blog/environmental-impact-of-the-world-cup.

Subcomandante Marcos. (2017). *Professionals of Hope: The Selected Writings of Subcomandante Marcos*. New York: The Song Cave.

Sugawara, Makoto, Yoshiaki Manabe, Fumihiro Yamasawa, and Yuri Hosokawa. (2022). "Athlete Medical Services at the Marathon and Race Walking Events During Tokyo 2020 Olympics." *Frontiers in Sports and Active Living* 4 https://doi.org/10.3389/fspor.2022.872475.

Sullivan, John. (2009, August 1). "'Greenwashing' Gets His Goat." *Times Herald-Record* https://www.recordonline.com/story/news/2009/08/01/greenwashing-gets-his-goat/51926672007/.

Sullivan, Sean P. (2024). "Market Definition." *Research Handbook on Abuse of Dominance and Monopolization*. Ed. Pinar Akman, Or Brook, and Konstantinos Stylianou. Cheltenham: Edward Elgar. 21–25.

Suñe, Ramón. (2024, October 24). "La patronal Foment del Treball luchará por retener la Copa del América en Barcelona." *La Vanguardia* https://www.lavanguardia.com/local/barcelona/20241024/10048731/patronal-foment-treball-luchara-retener-copa-america-barcelona.html.

Suriñach, Jordi and Esther Vayá, eds. (2025a, March). *Impacte economic de la 37a Copa América de vela, Barcelona 2024* https://www.barcelonacapitalnautica.org/wp-content/uploads/2025/03/Report-1-Impacte-Economic_.pdf.

Suriñach, Jordi and Esther Vayá, eds. (2025b, March). *Valoració social de la 37a Copa Amèrica de vela, Barcelona 2024* https://www.barcelonacapitalnautica.org/wp-content/uploads/2025/03/Report-2-Impacte-social_.pdf.

Suriñach, Jordi and Esther Vayá, eds. (2025c, March). *Anàlisi qualitativa de la 37a Copa Amèrica de vela, Barcelona 2024* https://www.barcelonacapitalnautica.org/wp-content/uploads/2025/03/Report-3-Llegat_.pdf.

Suzuki, Naofumi, Tetsuo Ogawa, and Nanako Inaba. (2018). "The Right to Adequate Housing: Evictions of the Homeless and the Elderly Caused by the 2020 Summer Olympics in Tokyo." *Leisure Studies* 37, no. 1: 89–96.

Swanton, Christine. (2010). "Heideggerian Environmental Virtue Ethics." *Journal of Agricultural & Environmental Ethics* 23, nos. 1–2: 145–66.

Swiss Federal Department of Foreign Affairs and United Nations. (2004). *The Global Compact. Who Cares Wins: Connecting Financial Markets to a Changing*

World https://web.archive.org/web/20220901042632/https://www.unepfi.org/fileadmin/events/2004/stocks/who_cares_wins_global_compact_2004.pdf.

Swithinbank, Robin. (2022a, January 28). "Pressure Builds on the Watch Industry Over Green Claims." *Financial Times* https://www.ft.com/content/cca70499-291a-4b95-813f-253584446c33.

Swithinbank, Robin. (2022b, May 13). Is Sustainability Message From Watchmakers Mere Green-Washing?" *Watchpro* https://www.watchpro.com/is-sustainability-message-from-watchmakers-mere-green-washing/.

Sylt, Christian. (2015, August 25). "Is Formula One the World's Most Sustainable Sport?" *Forbes* https://www.forbes.com/sites/csylt/2015/08/25/is-formula-one-the-worlds-most-sustainable-sport/?sh=5357d4d02952.

Sylt, Christian. (2017, March 13). "The $1 Billion Cost of Housing an F1 Race." *Forbes* https://www.forbes.com/sites/csylt/2017/03/13/the-1-billion-cost-of-hosting-an-f1-race/.

Sylt, Christian. (2019, April 23). "Formula One Gender Pay Gap Revealed to be Above UK Average at All but One British Team." *Independent* https://www.independent.co.uk/f1/formula-one-gender-pay-gap-statistics-standings-uk-motorsport-a8882946.html.

Szalai, Georg. (2024a, January 17). "Amazon to Buy Minority Stake in Diamond Sports, Sinclair to Pay $495M in Settlement." *Hollywood Reporter* https://www.hollywoodreporter.com/business/business-news/amazon-diamond-sports-investment-reorganization-sinclair-settlement-1235790585/.

Szalai, Georg. (2024b, October 31). "Peacock Loss Narrows to $436M as Revenue Rises, Subs Jump to 36M on Paris Olympics Boost." *Hollywood Reporter* https://www.hollywoodreporter.com/business/business-news/comcast-q3-earnings-peacock-loss-nbcuniversal-paris-olympics-1236047497/.

Tagsold, Christian. (2010). "Modernity, Space, and National Representation at the Tokyo Olympics 1964." *Urban History* 37, no. 2: 289–300.

Taibo II, Paco Ignacio. (2022). *68*. N.p.: Titivillus.

Takanobu, Saito. (2023, January 20). *How the Public Perceived the Tokyo Olympic and Paralympic Games*. NHK Broadcasting Culture Research Institute https://www.nhk.or.jp/bunken/english/research/yoron/20230130_3.html.

Talbot, Adam. (2025). "Olympics for Whom? Winners and Losers of Mega-Events." *Events and Society: Bridging Theory and Practice*. Ed. Mike Duignan. Abingdon: Routledge. 131–38.

Taylor, Charles. (2004). *Modern Social Imaginaries*. Durham: Duke University Press.

TePoel, Dain. (2017). "Digital Sport History, with Costs: An Ecocentric Critique." *Journal of Sport History* 44, no. 2: 350–66.

The Business Research Company. (2024). *Sports Global Market Report* https://www.thebusinessresearchcompany.com/report/sports-global-market-report.

The Climate Coalition. (2018). *Game Changer: How Climate Change is Impacting Sports in the UK* https://www.theclimatecoalition.org/gamechanger.

The League of Fans. (2011). *Sports Manifesto* http://leagueoffans.org/2011/06/29/the-sportsfans-manifesto.

The Shift Project. (2025, February). *Décarbonons le sport: Un premier applicatif au football et au rugby* https://theshiftproject.org/wp-content/uploads/2025/02/Rapport-final-Decarbonons-le-Sport-combine.pdf.

The Tokyo Organising Committee of the Olympic and Paralympic Games. (2021). *Update to the Sustainability Pre-Games Report* https://gtimg.tokyo2020.org/image/upload/production/cas22cdv09dc0h9yaf5t.pdf.

Thomas, Dylan. (2014). *The Poems of Dylan Thomas: Centenary Edition*. Ed. John Goodby. New York: New Editions.

Thomas, Ian. (2024, July 18). "Inside Evolving FIFA 2026 World Cup Plan to Cash in on the Rise of High-End Sports Tourism." *CNBC* https://www.cnbc.com/2024/07/18/fifa-2026-world-cup-plan-to-cash-in-on-high-end-sports-tourism.html.

Thompson, E. P. (1967). "Time, Work-Discipline, and Industrial Capitalism." *Past & Present* 38: 56–97.

Thompson, E. P. (1974). "Patrician Society, Plebeian Culture." *Journal of Social History* 7, no. 4: 382–405.

Thomson, Alana, Graham Cuskelly, Kristine Toohey, Millicent Kennelly, Paul Burton, and Liz Fredline. (2019). "Sport Event Legacy: A Systematic Quantitative Review of Literature." *Sport Management Review* 22, no. 3: 295–321.

Thomson, Ian and Robert G. Boutilier. (2011). "Social License to Operate." *SME Mining Engineering Handbook*. Ed. Peter Darling. Littleton: Society for Mining, Metallurgy and Exploration. 1779–796.

Thrall, A. Trevor, Jaime Lollio-Fahkreddine, Jon Berent, Lana Donnelly, Wes Herrin, Zachary Paquette, Rebecca Wenglinski, and Amy Wyatt. (2008). "Star Power: Celebrity Advocacy and the Evolution of the Public Sphere." *International Journal of Press/Politics* 13, no. 4: 362–85.

Ti, Jinsong, Yu Zheng, Weidong Duan, Haobin Zhao, Yanqing Qin, Guangting Yin, Liangwen Xie, Wenshuai Dong, Xiaochong Lu, and Zhaopeng Song. (2024). "Carbon Footprint of Tobacco Production in China Through Life-Cycle-Assessment: Regional Compositions, Spatiotemporal Changes and Driving Factors." *Ecological Indicators* 165 https://doi.org/10.1016/j.ecolind.2024.112216.

Till, Brian D., Sarah M. Stanley, and Randi Priluck. (2008). "Classical Conditioning and Celebrity Endorsers: An Examination of Belongingness and Resistance to Extinction." *Psychology & Marketing* 25, no. 2: 179–96.

Timms, Aaron. (2021, August 2). "NBC Paid $7.75bn for its Olympic Rights ... and We Got Televisual Vomit." *Guardian* https://www.theguardian.com/sport/2021/aug/02/nbc-olympic-coverage-peacock-replays-primetime?CMP=ShareiOSAppOther.

Timms, Aaron. (2025, July 13). "How the Rightwing Sports Bro Conquered America." *Guardian* https://www.theguardian.com/sport/ng-interactive/2025/jul/13/sports-bro-announcer-pat-mcafee-espn.

Tippett, Anna. (2019, June 24). "Formula 1: There Can Be No Equality in Sport while Women's Bodies are Used for Promotions." *The Conversation* https://theconversation.com/formula-1-there-can-be-no-equality-in-sport-while-womens-bodies-are-used-for-promotions-117443.

Todaro, Dina Lucia and Riccardo Torelli. (2024). "From Greenwashing to ESG-Washing: A Focus on the Circular Economy Field." *Corporate Social Responsibility and Environmental Management* 31, no. 5: 4034–046.

Tóffano Pereira, Rodrigo Pinheiro, Viachaslau Filimonau, and Glaydston Mattos Ribeiro. (2019). "Score a Goal for Climate: Assessing the Carbon Footprint of Travel Patterns of the English Premier League Clubs." *Journal of Cleaner Production* 227: 167–77.

Tolstoy, Leo. (1987). *A Confession and Other Religious Writings*. Trans. Jane Kentish. London: Penguin.

Tomizawa, Roy. (2016, February 3). "Olympia Cigarettes: How the 1964 Tokyo Olympics Led to Increased Cases of Lung Cancer in Japan Today." *The Olympians from 1964 to 2020* https://theolympians.co/2016/02/03/olympia-cigarettes-how-the-1964-tokyo-olympics-led-to-increased-cases-of-lung-cancer-in-japan-today/.

Tomizawa, Roy. (2019). *1964—The Greatest Year in the History of Japan: How the Tokyo Olympics Symbolized Japan's Miraculous Rise from the Ashes*. Carson City: Lioncrest.

Tomlinson, Alan. (1996). "Olympic Spectacle: Opening Ceremonies and Some Paradoxes of Globalization." *Media, Culture & Society* 18, no. 4: 583–602.

Tomlinson, Alan. (2014a). *FIFA (Fédération Internationale de Football Association): The Men, the Myths and the Money*. London: Routledge.

Tomlinson, Alan. (2014b). "Olympic Legacies: Recurrent Rhetoric and Harsh Realities." *Contemporary Social Science* 9, no. 2: 137–58.

Tomlinson, Alan, ed. (2017). *The Olympic Legacy: Social Scientific Explorations*. Abingdon: Routledge.

Totten, Mick. (2013). "Sport Activism and Political Praxis Within the FC Sankt Pauli Fan Subculture." *Play Left Wing* https://www.playleftwing.org/wp-content/uploads/2013/10/Fan-Culture-and-Sport-Activism1.pdf.

Touma, Rafqa. (2023, March 15). "Animal Activists and Commercial Industry at Odds After Nike Halts Use of Kangaroo Leather." *Guardian* https://www.

theguardian.com/business/2023/mar/15/animal-activists-and-commercial-industry-at-odds-after-nike-halts-use-of-kangaroo-leather.

Tovey, Alan. (2014, November 1). "Formula One's Vast Costs Are Driving Small Teams to Ruin." *Telegraph* http://www.telegraph.co.uk/sport/motorsport/formulaone/11203136/Formula-Ones-vast-costs-are-driving-small-teams-to-ruin.html.

Towler, Christopher C., Nyron N. Crawford, and Robert A. Bennett III. (2020). "Shut Up and Play: Black Athletes, Protest Politics, and Black Political Action." *Perspectives on Politics* 18, no. 1: 111–27.

Tramuta, Lindsey. (2024, June 26). "How Paris Hopes the Summer Olympics Will Transform the City—For Good." *Condé Nast Traveler* https://www.cntraveler.com/story/paris-olympics-sustainability.

Transparency International, ed. (2016). *Global Corruption Report: Sport.* Abingdon: Routledge.

Transport & Environment. (2024a, August). *"Cruisezillas": How Much Bigger Can Cruise Ships Get?* https://www.transportenvironment.org/uploads/files/2024-08CruisezillaFINAL.docx.pdf.

Transport & Environment. (2024b, November 28). *How Much Europe's Ships Emitted in 2023* https://www.transportenvironment.org/articles/shipping-2023-emissions.

Tranter, Paul J. and Mark Lowes. (2009). "Life in the Fast Lane: Environmental, Economic, and Public Health Outcomes of Motorsport Spectacles in Australia." *Journal of Sport and Social Issues* 33, no. 2: 150–68.

Treaty for the Prohibition of Nuclear Weapons in Latin America and the Caribbean (Tlatelolco). (n.d.) United Nations Office for Disarmament Affairs and Treaties Database https://treaties.unoda.org/t/tlatelolco.

Treepongkaruna, Sirimon, Hue Hwa Au Yong, Steen Thomsen, and Khine Kyaw. (2024). "Greenwashing, Carbon Emission, and ESG." *Business, Strategy and the Environment* 33, no. 8: 8526–539.

Trendafilova, Sylvia, Kathy Babiak, and Kathryn Heinze. (2013). "Corporate Social Responsibility and Environmental Sustainability: Why Professional Sport is Greening the Playing Field." *Sport Management Review* 16, no. 3: 298–313.

Trotsky, Leon. (1925). *Where is Britain Going?* https://www.marxists.org/archive/trotsky/britain/wibg/ch08.htm.

Trotsky, Leon. (1957). *Literature and Revolution.* Trans. Rose Strunsky. New York: Russell & Russell.

Trump, Donald J. (2025, July 31). *Fact Sheet: President's Council on Sports, Fitness, and Nutrition, and the Reestablishment of the Presidential Fitness Test* https://www.whitehouse.gov/fact-sheets/2025/07/fact-sheet-presidents-council-on-sports-fitness-and-nutrition-and-the-reestablishment-of-the-presidential-fitness-test/.

Turner, Andrew, Rob Arnold, and Tracey Williams. (2020). "Weathering and Persistence of Plastic in the Marine Environment: Lessons from LEGO." *Environmental Pollution* 262 https://doi.org/10.1016/j.envpol.2020.114299.

Tsujii, Koki, Tomonari Akamatsu, Ryosuke Okamoto, Kyoichi Mori, Yoko Mitani, and Naoya Umeda. (2018). "Change in Singing Behavior of Humpback Whales Caused by Shipping Noise." *PLoS ONE* 13, no. 10 https://doi.org/10.1371/journal.pone.0204112.

Tweedale, Alistair. (2018, November 21). "Fifa Talks a Good Game But its Drive for Sustainability Highlights Football's Hypocrisy." *Telegraph* https://www.telegraph.co.uk/football/2018/11/21/fifa-talks-good-game-drive-sustainability-highlights-footballs/.

Tyson, Alec and Brian Kennedy. (2023, October 25). *Climate Activism*. Pew Research Center https://www.pewresearch.org/science/2023/10/25/climate-activism/.

Tzanelli, Rodanthi. (2023). *Space, Mobility, and Crisis in Mega-Event Organisation: Tokyo Olympics 2020's Atmospheric Irradiation*. London: Routledge.

UEFA. (n.d.). https://www.uefa.com/sustainability/environment/climate-advocacy/.

UEFA. (2020). "Venues Confirmed for EURO 2020." https://www.uefa.com/uefaeuro-2020/news/025e-0fac6d3ee9e4-85b1a76389ea-1000/.

UEFA. (2024). *Environmental, Social and Governance Report: Post-Event Assessment* https://www.efdn.org/wp-content/uploads/2024/11/uefa_euro_2024_esg_report_compressed.pdf.

UEFA. (2025, March 7). *The European Club Finance and Investment Landscape* https://ecfil.uefa.com/2024.

Ukhova, Daria, Virve Marionneau, Janne Nikkinen, and Heather Wardle. (2024). "Public Health Approaches to Gambling: A Global Review of Legislative Trends." *The Lancet Public Health* 9, no. 1: e57–67.

Underdown, David. (1985). *Revel, Riot, and Rebellion: Popular Politics and Culture in England 1603–1660*. Oxford: Oxford University Press.

Underwood, John. (1968, September 30). "Games in Trouble." *Sports Illustrated* https://vault.si.com/vault/1968/09/30/games-in-trouble.

Une delegation inattendue. (n.d.). *Revendication du sabotage de lignes TGV quelques heures avant la cérémonie d'ouverture des Jeux Olympiques de Paris 2024* https://reporterre.net/IMG/pdf/lettrerevendicationsabotage.pdf.

Union Royale Belge des Sociétés de Football Association ASBL v Jean-Marc Bosman, Royal Club Liégeois SA v Jean-Marc Bosman and others, and Union des Associations Européennes de Football (UEFA) v Jean-Marc Bosman. (1995). C-415/93, ECLI:EU:C:1995:463.

United Nations. (n.d.). *Greenwashing—The Deceptive Tactics Behind Environmental Claims* https://www.un.org/en/climatechange/science/climate-issues/greenwashing.

United Nations. (2020). *Recovering Better: Sport for Development and Peace Reopening, Recovery and Resilience Post-COVID-19* https://social.desa.un.org/sites/default/files/migrated/22/2020/12/Final-SDP-recovering-better.pdf.

United Nations. (2024, May 7). "World Football Day." Resolution 78/281 https://docs.un.org/en/A/RES/78/281.

United Nations Climate Change. (n.d.). *The Paris Agreement* https://unfccc.int/process-and-meetings/the-paris-agreement.

United Nations Climate Change Conference. (2018). *Sports for Climate Action Framework* https://unfccc.int/sites/default/files/resource/SportsforClimateActionDeclarationandFramework.pdf.

United Nations Commission on Sustainable Development. (1998). *Chapeau for Business and Industry.* Background Paper No. 1 https://www.un.org/documents/ecosoc/cn17/1998/background/ecn171998-bp1.htm.

United Nations Environment Programme and Global Alliance for Buildings and Construction. (2025). *Global Status Report for Buildings and Construction 2024/25—Not Just Another Brick in the Wall* https://wedocs.unep.org/20.500.11822/47214.

United Nations Environment Programme. (n.d.). *Our Planet is Choking on Plastic* https://www.unep.org/interactives/beat-plastic-pollution/#:~:text=Because%20right%20now%2C%20a%20lot,dumps%20or%20the%20natural%20environment.

United Nations' High-Level Expert Group on the Net-Zero Emissions Commitments of Non-State Entities. (2022). *Integrity Matters: Net Zero Commitments by Business, Financial Institutions, Cities and Regions* https://www.un.org/sites/un2.un.org/files/high-levelexpertgroupn7b.pdf.

United States Department of Housing and Urban Development. (2023). *The 2023 Annual Homelessness Assessment Report (AHAR) to Congress* https://www.huduser.gov/portal/sites/default/files/pdf/2023-ahar-part-1.Pdf.

United States Environmental Protection Agency. (2023, June). *Tailpipe Greenhouse Gas Emissions from a Typical Passenger Vehicle* https://nepis.epa.gov/Exe/ZyPDF.cgi?Dockey=P1017FP5.pdf.

University of East London. (2015, December). *Olympic Games Impact Study—London 2012 Post-Games Report* https://www.uel.ac.uk/sites/default/files/post-games-ogi-report.pdf.

Vaccaro, Adam. (2015, April 2). "They Just Don't Want the Olympics." *Boston.com* http://www.boston.com/news/local-news/2015/04/02/they-just-dont-want-the-olympics.

Vadiati, Niloufar. (2020). *The Employment Legacy of the 2012 Olympic Games.* Cham: Palgrave Macmillan.

Valvur, Andy. (2018, November 14). "6 Cities That Rejected the Olympics." *DW* https://www.dw.com/en/6-cities-that-rejected-the-olympics/a-46289852.

Vamplew, Wray. (2021). *Games People Played: A Global History of Sport.* London: Reaktion Books.

van Beemen, Olivier. (2019). *Heineken in Africa: A Multinational Unleashed.* London: Hurst.

Vargas Llosa, Mario. (1997). *Making Waves.* Trans. and Ed. John King. New York: Farrar, Strauss and Giroux.

Varoufakis, Yanis. (2002). *Foundations of Economics: A Beginner's Companion.* New York: Routledge.

Vásquez, Alex. (2024, December 21). "Mexico City to Boost Mobility, Security Ahead of 2026 FIFA World Cup." *Business Standard* https://www.business-standard.com/world-news/mexico-city-to-boost-mobility-security-ahead-of-2026-fifa-world-cup-124122100164_1.html.

Vaughan, Roger. (2024, August 25). "Not Only Boring, it is Dehumanizing." *Scuttlebutt Sailing News* https://www.sailingscuttlebutt.com/2024/08/25/not-only-boring-it-is-dehumanizing/.

Veblen, Thorstein. (2007). *The Theory of the Leisure Class.* Oxford: Oxford University Press.

Veitch, Sarah. (2013). "Mythology as Hallmark Event Legacy: The Endurance of America's Cup Mythology in the City of Fremantle." *Journal of Sport & Tourism* 18, no. 2: 67–80.

Venzon, Andrea and Colombe Cahen-Salvador. (2019, November 1). "Party Politics is Dying, So Make Citizen Movements the New Unifier." *Economist* https://www.economist.com/open-future/2019/11/01/party-politics-is-dying-so-make-citizen-movements-the-new-unifier.

Vernon, Jezz, Catherine Johnson, Joanne Evans, Rob Collier, Richard Ellis, Abhaya Sumanasena, Graham Mills, Awnit Kumar, Tim Jacks, Bertrand Moullier, and Alex Taylor. (2024). *Future of TV Distribution: A Report Prepared by the University of Exeter with the University of Leeds, MTM, 3 Reasons and Real Wireless.* Department for Culture, Media and Sport https://assets.publishing.service.gov.uk/media/672cafe262831268b0b1a2f4/Future_of_TV_Distribution_FINAL__7_Nov_2024_-accessible.pdf.

Vettel Sebastian. (n.d.). https://sebastianvettel.de/en/start/.

Vicente, Sandra, Victòria Oliveres, and Pau Rodríguez. (2024, August 28). "El alquiler de temporada alcanza niveles inéditos en la Barceloneta coincidiendo con la Copa América." *El Diario* https://www.eldiario.es/catalunya/alquiler-temporada-alcanza-niveles-ineditos-barceloneta-coincidiendo-copa-america11591525.html.

Vicente, Sandra. (2023, October 10). "65 entidades se unen contra la Copa América de Barcelona: "Es la síntesis del expolio contra la ciudad"." *El Diario* https://www.eldiario.es/catalunya/65-entidades-unen-copa-america-barcelona-sintesis-expolio-ciudad_1_10585988.html.

Vidal, John. (2011, September 15). "Greenpeace at 40: A Global Brand in Good Health or an Out-of-Touch Bureaucracy?" *Guardian* https://www.theguardian.com/environment/2011/sep/15/greenpeace-40-year-anniversary.

Viñas, Carles and Natxo Parra. (2020). *St. Pauli: Another Football is Possible.* Trans. Luke Stobart. London: Pluto Press.

Virilio, Paul. (1991). *The Aesthetics of Disappearance.* Trans. Philip Beitchman. N.p.: Semiotext(e).

Vivar, Robero Fuentes. (2019, September 24). "Deporte y cambio climático." *Milenio* https://www.milenio.com/opinion/roberto-fuentes-vivar/las-otras-competencias/deporte-y-cambio-climatico.

Vlessing, Etan. (2024, July 18). "CBS Sports Nabs U.S. TV Rights to English Football League Games." *Hollywood Reporter* https://www.hollywoodreporter.com/business/business-news/cbs-sports-english-football-league-1235952038/.

Volkery, Carsten. (2012, August 13). "'Golden Games' Give Britain Much-Needed Boost." *Der Spiegel* https://www.spiegel.de/international/world/praise-for-closing-ceremony-of-the-london-olympic-games-a-849720.html.

Wågan, Frida Austmo. (2024). "Regenerating the Pitch: Environmental Discourses in Football." *Annals of Leisure Research* 28, no. 1: 73–94.

Wainwright, Oliver. (2022, June 30). "'A Massive Betrayal': How London's Olympic Legacy Was Sold Out." *Guardian* https://www.theguardian.com/uk-news/2022/jun/30/a-massive-betrayal-how-londons-olympic-legacy-was-sold-out?CMP=ShareiOSAppOther&fbclid=IwAR3vG%E2%80%A6.

Walker, Ali. (2024a, December 17). "FIFA Chief Infantino Warned by EU Lawmakers Over Saudi Arabia World Cup." *Politico* https://www.politico.eu/article/fifa-saudi-arabia-world-cup-eu-lawmakers-football-gianni-infantino-aramco-tournament/.

Walker, Alissa. (2018, June 7). "The Olympics Fixed LA's Traffic Problem—Can the 2028 Games Do it Permanently?" *Curbed* https://la.curbed.com/2018/6/7/17419270/olympics-2028-los-angeles-1984-traffic.

Walker, Josephine. (2024b, August 7). "Olympic Swimmers are Guzzling Coke to Fight Bacteria in Poop-Filled Seine River." *Daily Beast* https://www.thedailybeast.com/olympic-swimmers-are-guzzling-coke-to-fight-bacteria-in-poop-filled-seine-river.

Wallace, Julianne Poock, Eric Wiedenman, and Robert J. McDermott. (2019). "Physical Activity and Climate Change: Clear and Present Danger?" *Health Behavior and Policy Review* 6, no. 5: 534–45.

Walters, Geoff and Richard Tacon. (2011). *Corporate Social Responsibility in European Football: A Report Funded by the UEFA Research Grant Programme* http://www.sportbusinesscentre.com/wp-content/uploads/2012/08/CSR2.pdf.

Wang, Shixin, Tiexi Chen, Jing-Jia Luo, Meng Gao, Hongchao Zuo, Fenghua Ling, Jianlin Hu, Chaoxia Yuan, Yuanjian Yang, Lina Wang, Huaming Huang, Naiang Wang, Yaojun Li, and Toshio Yamagata. (2024). "Warming Climate is Helping Human Beings Run Faster, Jump Higher and Throw Farther Through Less Dense Air." *Climate and Atmospheric Science* 7. https://doi.org/10.1038/s41612-024-00635-z.

Wardle, Heather, Louisa Degenhardt, Virve Marionneau, Gerda Reith, Charles Livingstone, Malcolm Sparrow, Lucy T. Tran, Blair Biggar, Christopher Bunn, Michael Farrell, Viktorija Kesaite, Vladimir Poznyak, Jianchao Quan, Jürgen Rehm, Angela Rintoul, Manoj Sharma, Jeremy Shiffman, Kristiana Siste, Daria Ukhova, Rachel Volberg, Joana Salifu Yendork, and Shekhar Saxena. (2024). *The Lancet Public Health Commission on Gambling* https://doi.org/10.1016/S2468-2667(24)00167-1.

Ware, Alyn. (2021, July 29). "Tokyo's Games Are Harming the Nuclear Weapons Ban Movement." *The Nation* https://www.thenation.com/article/society/tokyo-olympics-nuclear-weapons/.

Warren, Gina. (2020). "Mega Sports Events Have Mega Environmental and Social Consequences." *Missouri Law Review* 85, no. 2 https://papers.ssrn.com/sol3/papers.cfm?abstractid=3463119.

Wartuny, Geffin. (2024). "Journal Review of Capitalism in European Soccer." *LATTE: A Journal of Language, Culture, and Technology* 1, no. 1: 17–23.

Watson, Bruce. (2016, August 20). "The Troubling Evolution of Corporate Greenwashing." *Guardian* https://www.theguardian.com/sustainable-business/2016/aug/20/greenwashing-environmentalism-lies-companies.

Watts, Jonathan. (2019, February 25). "Concrete: The Most Destructive Material on Earth." *Guardian* https://www.theguardian.com/cities/2019/feb/25/concrete-the-most-destructive-material-on-earth.

Weber, Max. (1981). "Some Categories of Interpretive Sociology." Trans. E. E. Graber. *Sociological Quarterly* 22, no. 2: 151–80.

Weber, Max. (1985). ""Churches" and "Sects" in North America: An Ecclesiastical Socio-Political Sketch." Trans. Colin Loader. *Sociological Theory* 3, no. 1: 7–13.

Weber, Max. (1992). *The Protestant Ethic and the Spirit of Capitalism.* Trans. Talcott Parsons. London: Routledge.

Weinstein, Roy, Peter Quies, Ryan Smart, Jeremy De Gracia, and Joe Hale. (2024, June). *Projected Economic Impact of FIFA World Cup 26™ County of Los Angeles June-July 2026.* Micronomics Economic Research and Consulting https://static1.squarespace.com/static/61b53e492ea58d13b806ccb3/

t/668c5701d897eb08bf8a2355/1720473346077/FWC_26_LA-Economic_Impact_Report-Micronomics%282024-06-25%29.pdf.

Wells, H. G. (1902). *Anticipations of the Reaction of Mechanical and Scientific Progress Upon Human Life and Thought*, 2nd ed.. London: Chapman & Hall.

WePlayGreen. (n.d.). https://weplaygreen.com/.

Weprin, Alex. (2021, August 11). "Sports Rights' Streaming Wave May Finally End Pay-TV Bundle." *Hollywood Reporter* https://www.hollywoodreporter.com/business/business-news/sports-rights-tv-amazon-disney-viacomcbs-1234995721/.

Weprin, Alex. (2024, January 12). "Amazon Wants to Turbocharge Its Sports Advertising Business, Too." *Hollywood Reporter* https://www.hollywoodreporter.com/business/business-news/amazon-sports-advertising-business-growth-1235786459/.

Wertheim, Jon. (2023, July 26). "Paris 2024 is One Year Away. Do the French Care?" *Sports Illustrated* https://www.si.com/olympics/2023/07/26/paris-2024-olympics-one-year-away-daily-cover.

West, Lindy. (2016, August 10). "How to Talk about Female Olympians Without Being a Regressive Creep—A Handy Guide." *Guardian* https://www.theguardian.com/commentisfree/2016/aug/09/female-olympians-guide-gaffes-athletes-sports-makeup-shorts-marital-status-lindy-west.

Westerveld, Jay. (2024, April 16). "Earth Day 2024: Greenwashing is the Order of the Day." *The Chronicle* https://www.chroniclenewspaper.com/opinion/columns/earth-day-2024-greenwashing-is-the-order-of-the-day-FF3304676.

Weston, Barney. (2022, May 6). "Football Must Do More to Tackle Climate Change: This is How Clubs and Fans Can Help." *Guardian* https://www.theguardian.com/football/blog/2022/may/06/football-must-do-more-to-tackle-climate-change-this-is-how-clubs-and-fans-can-help.

Whaly, Ghada. (2021). "Foreword." *Global Report on Corruption in Sport 9: Illegal Betting and Sport.* United Nations Office on Drugs and Crime. 2 https://www.unodc.org/documents/corruption/Publications/2022/Global_Report_on_Corruption_in_Sport_Chapter_9.pdf.

Wharton, Dave. (2024, November 14). "L.A. Olympics Boss Reynold Hoover Uses Wartime Problem-Solving Skills to Plan 2028 Games." *Los Angeles Times* https://www.latimes.com/sports/olympics/story/2024-11-14/olympics-boss-is-treating-the-games-like-wartime.

Wharton, Dave. (2025, January 15). "Can Fire-Torn L.A. Handle the World Cup, Super Bowl and Summer Olympics?" *Los Angeles Times* https://www.latimes.com/sports/olympics/story/2025-01-15/la-fires-world-cup-super-bowl-2028-summer-olympics.

Wheatley, Magnus. (2024). *There is No Second: The Definitive Account of the First Race in 1851 for What Would Become 'América's Cup'*. Newport: Seahorse Publishers.

Wheeler, David. (2024, December 8). "Female Footballers Have Shown Us How—Let's Build a Sport Free of Fossil Fuel Deals." *Guardian* https://www.theguardian.com/football/2024/dec/08/female-footballers-have-shown-us-how-lets-build-a-sport-free-of-fossil-fuel-deals.

Whiting, Robert. (2014, October 24). "Negative Impact of 1964 Olympics Profound." *Japan Times*https://www.japantimes.co.jp/sports/2014/10/24/olympics/negative-impact-1964-olympics-profound/.

Whiting, Robert. (2021). *Tokyo Junkie: 60 Years of Bright Lights and Back Alleys…And Baseball.* Berkeley: Stone Bridge Press.

Wilburn, Kathleen M. and Ralph Wilburn. (2011). "Achieving Social License to Operate Using Stakeholder Theory." *Journal of International Business Ethics 4, no. 2:* 3–16.

Wilby, Robert L., Madeleine Orr, Duncan Depledge, Richard Giulianotti, George Havenith, Jamie A. Kenyon, Stephen A. Mears, Donal J. Mullan, and Lee Taylor. (2023). "The Impacts of Sport Emissions on Climate: Measurement, Mitigation, and Making a Difference." *Annals of the New York Academy of Science* 1519, no. 1: 20–33.

Wilk, Richard R. and Lisa Cliggett. (2007). *Economies and Cultures: Foundations of Economic Anthropology*, 2nd ed. Cambridge, Mass.: Perseus.

Williams, Bernard. (2011a). *Ethics and the Limits of Philosophy.* London: Routledge.

Williams, Michael. (1998, January 19–25). "Ad Deals Strike a Cultural Match." *Variety*: A8.

Williams, Michael. (1999, September 20–26). "Soccer Kicky for Congloms." *Variety*: 1, 104.

Williams, Raymond. (2005). *Culture and Materialism.* London: Verso.

Williams, Raymond. (2011b). *Raymond Williams on Television: Selected Writings.* Ed. Alan O'Connor. London: Routledge.

Williams, Raymond. (2015a). *A Short Counter-Revolution—Towards 2000 Revisited*, 2nd ed. Ed. Jim McGuigan. London: Sage Publications.

Williams, Raymond. (2020a, September 1). "Ecology and the Labour Movement." *Raymond Williams Society* https://raymondwilliams.co.uk/2020/09/01/part-two-ecology-and-the-labour-movement-by-raymond-williams/.

Williams, Richard. (2015b, October 23). "F1 is in Deep Trouble and Bernie Ecclestone is Not the Man to Save it." *Guardian* https://www.theguardian.com/sport/blog/2015/oct/23/f1-bernie-ecclestone-max-mosley.

Williams, Richard. (2020b, November 14). "Lewis Hamilton: The Man from Stevenage Who Became the Moral Compass of F1." *Guardian* https://www.

theguardian.com/sport/2020/nov/14/lewis-hamilton-the-man-from-stevenage-who-became-the-moral-compass-of-f1.

Williams, Richard. (2021, August 29). "Louise King Obituary." *Guardian* https://www.theguardian.com/stage/2021/aug/29/louise-king-obituary.

Williams, Terri M., Susanna B. Blackwell, Outi Tervo, Eva Garde, Mikkel-Holger S. Sinding, Beau Richter, and Mads Peter Heide-Jørgensen. (2022). "Physiological Responses of Narwhals to Anthropogenic Noise: A Case Study with Seismic Airguns and Vessel Traffic in the Arctic." *Functional Ecology* 36: 2251–266.

Williamson, Jeannine. (2024, May 22). "Explora Journeys Showcases European Luxury at Sea in its Second 'Space Ship'." *Riviera* https://www.rivieramm.com/news-content-hub/news-content-hub/explora-journeys-showcases-european-luxury-at-sea-in-its-second-space-ship-80851.

Willsher, Kim. (2004, June 3). "Louis Vuitton's Links with Vichy Regime Exposed." *Guardian* https://www.theguardian.com/world/2004/jun/03/france.secondworldwar.

Willsher, Kim. (2024, August 10). "'We'll Just be Left with Memories': Can Olympic Success Help Transform the Poorest Suburbs in Paris?" *Guardian* https://www.theguardian.com/sport/article/2024/aug/10/well-just-be-left-with-memories-can-olympic-success-help-transform-the-poorest-suburbs-in-paris.

Wilson, Brian and Brad Millington. (2020). "Introducing a Sociological Approach to Sport, Environmental Politics, and Preferred Futures." *Sport and the Environment: Politics and Preferred Futures*. Ed. Brian Wilson and Brad Millington. Bingley: Emerald. 1–28.

Wilson, Elizabeth. (2022). *Unfolding the Past*. London: Bloomsbury Visual Arts.

Wilson, Jonathan. (2024, December 7). "Fifa's Infantino and Saudi Arabia 1, Football and Human Decency 0." *Guardian* https://www.theguardian.com/football/blog/2024/dec/07/fifa-infantino-saudi-arabia-world-cup-2034.

Wilson, Joseph. (2025, May 22). "An America's Cup Without an American Boat? American Magic Says it May Not Sail in Naples." *AP* https://apnews.com/article/americas-cup-sailing-american-magic-b16c447cef426b2a209385c0ea8af460.

Wilson, Sandra. (2012). "Exhibiting a New Japan: The Tokyo Olympics of 1964 and Expo '70 in Osaka. *Historical Research* 85, no. 227: 159–78.

Witt, Stephen. (2022, May 19). "Crypto.com Spent Its Way to the Top. Then the Market Crashed. Now What?" *GQ* https://www.gq.com/story/crypto-dot-com-explained.

Wittgenstein, Ludwig. (1965). "I: A Lecture on Ethics." *The Philosophical Review* 74, no. 1: 3–12.

Wolf, Eric R. (1982). *Europe and the People Without History*. Berkeley: University of California Press.

Wong, Gabriella. (2025, February 12). "Marshall Islands Football Jersey "Disappears" to Visualise Threat of Rising Seas." *Dezeen* https://www.dezeen.com/2025/02/12/marshall-islands-soccer-federation-playerlayer-kit/.

Woodcraft, Saffron, José Izcue-Gana, Georgios Melios, Nikolaos Tzivanakis, Elisabetta Pietristefani, Gillian Chan, Zaafir Hassan, and Jorge Pérez. (2024, July). *Prosperity in and Around London's Olympic Park: Findings from the Prosperity in East London 2021–2031 Longitudinal Study.* Institute for Global Prosperity, Procol UK, and London Prosperity Board https://discovery.ucl.ac.uk/id/eprint/10195859/1/Prosperity%20in%20Olympic%20Park%20and%20Fringes%20Bulletin,%20PieL%20Wave%201%20July%202024.pdf.

Woodford, Isabel. (2025, August 25). "'We Get Nothing': Residents Fight Back as Mexico Builds Big for 2026 World Cup." *Guardian* https://www.theguardian.com/global-development/2025/aug/25/we-get-nothing-residents-fight-back-as-mexico-builds-big-for-2026-world-cup?CMP=Share_iOSApp_Other.

Woods, Chelsea L. and Ashli Q. Stokes. (2019). "'For the Game, for the World': An Analysis of FIFA's CSR Initiatives." *Public Relations Inquiry* 8, no. 1: 49–85.

Worden, Minky, ed. (2008). *China's Great Leap: The Beijing Games and Olympian Human Rights Challenges.* New York: Seven Stories Press.

Worden, Minky. (2015). "Raising the Bar: Mega-Sporting Events and Human Rights." *Human Rights Watch* https://www.hrw.org/world-report/2015/country-chapters/global-1.

Worden, Minky. (2025, May 5). Letter to Gianni Infantino. Human Rights Watch https://www.hrw.org/news/2025/05/06/letter-fifa-re-human-rights-responsibilities-2026-world-cup.

World Anti-Doping Agency. (n.d.). https://www.wada-ama.org/.

World Athletics. (n.d.). https://www.worldathletics.org/home.

World Bank. (2017). *The Potential of the Blue Economy: Increasing Long-Term Benefits for Small Island Developing States and Coastal Least Developed Countries* https://openknowledge.worldbank.org/server/api/core/bitstreams/cee24b6c-2e2f-5579-b1a4-457011419425/content.

World Bank. (2024). *Poverty, Prosperity, and Planet Report 2024: Pathways Out of the Polycrisis.* Washington: World Bank.

World Commission on Environment and Development. (1987). http://www.un-documents.net/wced-ocf.htm.

World Economic Forum. (2024, March 26). "Paris Aims to Host the Most Sustainable Olympics Ever—Here's How the City is Preparing." https://www.weforum.org/agenda/2024/03/paris-most-sustainable-olympics-2024/.

World Health Organization. (2023, December 13). *Road Traffic Injuries* https://www.who.int/news-room/fact-sheets/detail/road-traffic-injuries.

World Health Organization. (2024). *Hooking the Next Generation: How the Tobacco Industry Captures Young Customers* https://escholarship.org/content/qt97d8k4xz/qt97d8k4xz.pdf.

World Intellectual Property Organization. (n.d.). https://www.wipo.int/portal/en/index.html.

World Sailing. (n.d.). *Sustainability Agenda 2030: A Bold Ambition for Sailing's Contribution to Global Sustainability* https://d7qh6ksdplczd.cloudfront.net/sailing/wp-content/uploads/2022/01/18094553/SustainabilityAgenda2030-23247.pdf.

World Sailing. (2024). *Ready for the Future: World Sailing Strategy 2025–2029* https://d7qh6ksdplczd.cloudfront.net/sailing/wp-content/uploads/2024/11/13112135/Ready-for-the-Future-2024-2029-World-Sailing-Strategy.pdf.

Wright, Richard Keith and Christopher Barron. (2021). ""It's Coming Home!" Leveraging Legacies in the City of Sails." *Place Event Marketing in the Asia Pacific Region*. Ed. Waldemar Cudny. London: Routledge. 143–61.

WWF. (n.d.). *WWF Guide to Greenwashing* https://www.wwf.org.uk/learn/guide-to-greenwashing.

WWF. (2004, July 16). *Environmental Assessment of the Athens 2004 Olympic Games*https://wwf.panda.org/wwf_news/?14215/Environmental-Assessment-of-the-Athens-2004-Olympic-Games.

WWF. (2023). *Time for Change: Demanding More Transparency and Responsibility in the Watch and Jewellery sector. Sustainability Rating and Industry Report 2023* https://www.wwf.ch/sites/default/files/doc-2023-11/_WWF_Watch%20and%20Jewellery%20Report%202023_Final.pdf.

Wyden, Ron and Richard Durbin. (2024, November 25). Letter to Gianni Infantino https://www.wyden.senate.gov/imo/media/doc/fifa_world_cup_2034_letter2.pdf.

Xu, Guoqi. (2008). *Olympic Dreams: China and Sports, 1895–2008*. Cambridge, Mass.: Harvard University Press.

Yanmar Power Technology Co., Ltd. Settlement Agreement Notices of Violation: EPES-2021-C00572 & EPES-2022-C00572. (2024, March). California Air Resources Board https://ww2.arb.ca.gov/sites/default/files/2024-03/yanmar_power_technology_co_ltd_sa.pdf.

Yates, Luke. (2015). "Rethinking Prefiguration: Alternatives, Micropolitics and Goals in Social Movements." *Social Movement Studies: Journal of Social, Cultural and Political Protest* 14, no. 1: 1–21.

Yeo, ShinJoung and Toby Miller. (2017). *Artists in Tech Cities*. SPACE https://spacestudios.org.uk/wp-content/uploads/2021/07/4501702tech-cities8web-2021.pdf.

YouGov. (2023). *The Global Sports Media Landscape: Sports Whitepaper 2023* https://commercial.yougov.com/rs/464-VHH-988/images/YouGov-Sports-Whitepaper-v4.pdf.

YouGov. (2024a, July 9). *Inclusivité, respect de l'environnement … Retour sur les engagements de Paris 2024* https://fr.yougov.com/sport/articles/49991-inclusivite-respect-de-lenvironnement-retour-sur-les-engagements-de-paris-2024.

YouGov. (2024b, August 7). *Impacto del turismo en la población residente* https://commercial.yougov.com/rs/464-VHH-988/images/YouGov-ES-Impacto-del-turismo-en-la-poblaci%C3%B3n-residente.pdf?version=0&utmmedium=email&utmsource=whitepaper&utmcampaign=WP-2024-08-ES-Turismofobia&mkttok=NDY0LVZISC05ODgAAAGV0KHp6pY9nLK2eypE1LjDUCM7U-KUYc8pGpfQswGs1FldEhCPFnn-XFQ3VvlOvZBVGRHPsWn3DSxSS9LtlwM3BGeICv50Mx4Mk5HTI2wimk.

Young, Iris Marion. (2000). *Inclusion and Democracy.* New York: Oxford University Press.

Young, Noel. (2012, August 9). "BP Olympic Commercials Cheer on Team USA … and Reputation Rises." *The Drum* https://www.thedrum.com/news/2012/08/09/bp-olympic-commercials-backing-team-usa-sees-its-reputation-rise.

Zacher, Hannes. (2024). "The Dark Side of Environmental Activism." *Personality and Individual Differences* 219 https://doi.org/10.1016/j.paid.2023.112506.

Zafaeiridou, Maria, Nicholas S. Hopkinson, and Nikolaos Voulvoulis. (2018). "Cigarette Smoking: An Assessment of Tobacco's Global Environmental Footprint Across its Entire Supply Chain." *Environmental Science & Technology* 52, no. 15: 8087–094.

Zafari, Zahra, Abooali Golzary, Kiana Rouhi, and Omid Mansourihanis. (2025). "From Conventional Approaches to Circular Systems: Evolution of Waste Management in Mega-Sporting Events." *Journal of the Air & Waste Management Association* https://doi.org/10.1080/10962247.2025.2462005.

Zerrenner, Kate. (2019, November 22). "Fasten Your Seat Belts: Formula 1 Racing to the Carbon Neutral Finish Line." *Triple Pundit* https://www.triplepundit.com/story/2019/fasten-your-seat-belts-formula-1-racing-carbon-neutral-finish-line/85706.

Zimmerman, Max. (2021, June 3). "How the 1964 Olympic Games Changed Tokyo Forever." *Bloomberg* https://www.bloomberg.com/news/features/2021-06-03/how-the-tokyo-olympics-in-1964-changed-japan-s-capital-forever.

Zipes, Jack. (2023). *Buried Treasures: The Power of Political Fairy Tales.* Princeton: Princeton University Press.

Zirin, Dave and Jules Boykoff. (2019, July 23). "These Women Have Lost Their Homes to the Olympics in Tokyo—Twice." *The Nation* https://www.thenation.com/article/archive/tokyo-olympics-displacement/.

Zirin, Dave and Jules Boykoff. (2022, July 19). "Organizers Don't Care That the 2028 Olympics Could be on Fire." *The Nation* https://www.thenation.com/article/society/la-olympics-climate-change/.

Zirin, Dave and Jules Boykoff. (2023, June 27). "The Paris Olympics Are Developing a Familiar Stench." *The Nation* https://www.thenation.com/article/world/paris-olympics-corruption/.

Zirin, Dave. (2011, June 8). "Let Them Play: Behind FIFA's Decision to Ban Iran's National Women's Team." *The Nation* www.thenation.com/blog/161228/let-them-play-behind-fifas-decision-ban-womens-team-iran.

Zirin, Dave. (2020, January 17). "The Australian Open is the Tip of a Melting Iceberg." *The Nation* https://www.thenation.com/article/environment/australian-open-fires-climate-change/.

Zolov, Eric. (2004). "Showcasing the 'Land of Tomorrow': Mexico and the 1968 Olympics." *The Americas* 61, no. 2: 159–88.

Zyvatkauskas, Caz. (2007, February 3). "Theatre Critic." *Economist*: 18.

Index

T. Miller, J. P. Carañana, *Global Sports Go Green—Or Do They?*, Global Culture and Sport Series,
https://doi.org/10.1007/978-3-032-14207-8

Kolibri 360 GmbH
Ferdinand-Jühlke-Straße 7
99095 Erfurt, Deutschland
produktsicherheit@kolibri360.de